PHYSICS

MED EASY

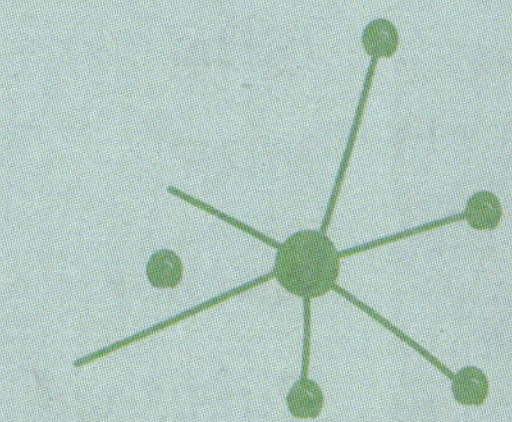

By Dr. Manish Raj (MR Sir)

Class Notes in Handwritten Format

Hello... Hello... Hello...
Aap Sabhi Bacho ko MR ka
Bahut Bahut Namaskar

You are Not Here to Prove Yourself,
You are Here to Improve Yourself

EDITION : Second

ISBN : 978-81-19352-94-4

MRP : 525/-

Author : Dr. Manish Raj

Published By : Physicswallah Private Limited

For Feedback : **publication@pw.live**

(This Book shall only be Used for Educational Purpose.)

Printed at Krown Printers

Preface

Physics Wallah (PW) strongly believes in conceptual and fun-based learning. PW provides highly exam-oriented content to bring quality and clarity to the students.

A plethora of Competitive Exams Books is available in the market but PW professionals are continuously working to provide the supreme book for Competitive Exams to our students.

Books adopt a multi-faceted approach to master and understanding the concepts by having a rich diversity of questions asked in the examination and equip the students with the knowledge for the competitive exam.

The main objective of the book is to provide a large number of quality problems with varying cognitive levels to facilitate teaching-learning of concepts that are presented through the book.

It has become popular among aspirants because of its easy to understand language and clear illustrations with diagrams, mnemonics, flow charts and tables.

Factor in the content that should be checked thoroughly including the quality of language or ease of understanding or comprehension, coverage of topics and chapters from the Competitive Exams latest syllabus by NTA.

Students can benefit themselves by attempting the exercise given in the book for the self-assessment and also mastering the basic techniques of problem-solving.

Mastering the Physics Wallah (PW) study material curated by the PW team, the students can easily qualify for the exam with a Top Rank in the Competitive Exams.

In each chapter, for better understanding, theory and examples are according to the latest syllabus for Competitive Exams.

- To strengthen the concept of the topic at the zenith level

This Book Consists of

- Formulas
- Mnemonics
- Important Examples
- Special Tips and Tricks

About the Author

Dr. Manish Raj, Ph.D., hailing from Bihar, is a renowned physics educator, popularly known as "M.R. sir" among his students. With over 12 years of teaching experience, he is celebrated for his unwavering commitment to simplifying the intricacies of physics and making it accessible to learners of all levels.

Dr. Manish Raj completed his education from Hindu College, University of Delhi. His academic journey has equipped him with a profound knowledge of the subject, and he continues to inspire students with his expertise.

With a remarkable teaching career spanning more than a decade, Dr. Raj has guided countless students at institutions like PW and others. His innovative teaching methodologies have not only garnered recognition but also consistently yielded outstanding results.

MR sir's dedication to teaching has touched the lives of an astounding 5 million students and continues to grow. His impact on the world of physics education is immeasurable.

MR sir is recognized nationally for his teaching style and is regarded as a prominent professor of physics. He has served as a mentor to several successful NEET aspirants, including those securing AIR 24, 28,39, 42, 59, 67, and many more in NEET 2023 and AIR 28, 34,35, 54, 61, 94, and many more in NEET 2024. His own academic achievements include qualifications in CSIR NET (All India rank 135), GATE, JAM, and a remarkable All India Rank 04 in the Delhi University Exam.

MR sir's commitment to education earned him the prestigious 'Star Teacher of the Year' award from INKZOID FOUNDATION in 2022, 2023 and 2024 also. He achieved this recognition for teaching complete NEET Physics on alternate days for the longest 24 hours in 2022 and 18 + 12 hours continuously in 2023 on YouTube. He is also an esteemed Josh Talk and TEDx speaker, sharing his insights and passion for physics education with a broader audience.

He wishes to extend his heartfelt gratitude to his dedicated students, including Mayur Pawar, Aryan Patel and others, whose contributions were invaluable in bringing this resource to fruition.

With an unrelenting dedication to physics education, MR Sir is excited to share his knowledge and expertise with students through this book. He aims to help students master physics concepts, empowering them to excel in their academic pursuits.

Contents

1 Basic Maths

1 Binomial theorem

✦ $(1 + x)^2 = 1 + 2 \times 1x + x^2$

if $x <<< 1$ then

$(1 + x)^2 = 1 + 2x$

MR* feel

(Career + love)^2 = Carrier + 2 love

Because carrier >>> love

✦ $[x+\Delta x]^n = x^n\left[1 + \frac{\Delta x}{x}\right]^n = x^n\left[1 + n\frac{\Delta x}{x}\right]$

$\Delta x <<<< x$.

✦ $(1 - x)^n = 1 - nx$

✦ $(1 - x)^{-n} = 1 + nx$

✦ $(1 + x)^{-n} = 1 - nx$

Q.1. The expression of gravitational potential energy is $U = -\frac{GMm}{R + h}$ where, R is the radius of the planet and h is the height above the surface. Approximate the expression by using binomial expansion when $h \ll R$.

Sol. $U = -\frac{GMm}{R + h} = -\frac{GMm}{R(1 + h/R)}$

For $h \ll R$

$(1 + h/R)^{-1} \approx 1 - \frac{h}{R}$

Therefore,

$U \approx -\frac{GMm}{R}\left(1 - \frac{h}{R}\right)$

2 Important formulae

✦ $(a + b)^2 = a^2 + b^2 + 2ab$

✦ $(a - b)^2 = a^2 + b^2 - 2ab$

✦ $a^2 - b^2 = (a + b)(a - b)$

✦ $(r - a)^2(r + a)^2 \approx (r^2 - a^2)^2$

Multiplication

✦ $\left(\frac{a}{b}\right)\left(\frac{c}{d}\right) = \frac{ac}{bd}$, Ex.: $\left(\frac{2}{3}\right)\left(\frac{5}{7}\right) = \frac{10}{21}$

Division

✦ $\frac{a/b}{c/d} = \frac{ad}{bc}$, Ex.: $\frac{2/3}{3/4} = \frac{(2)(4)}{(3)(3)} = \frac{8}{9}$

Addition

✦ $\frac{a}{b} \pm \frac{c}{d} = \frac{ad \pm bc}{bd}$

Ex.: $\frac{1}{2} + \frac{5}{6} = \frac{(1)(6) + (2)(5)}{(2)(6)} = \frac{16}{12} = \frac{4}{3}$

3 Componendo and Dividendo Theorem

In Ratio and proportion problems. It states that for any four numbers a, b, c, and d. if;

$$\frac{a}{b} = \frac{c}{d}$$

Then,

$$\frac{a + b}{a - b} = \frac{c + d}{c - d}$$

4 AP series

Next term = Previous term + Common difference

$a,\ a + d,\ a + 2d,\ a + 3d,\ a + 4d.....$

Common difference

$d = n^{th}$ term $- (n-1)^{th}$ term

Ex: 2, 5, 8, 11, 14, 17, ...

$d = 5 - 2 = 3$

n^{th} term;

$$a_n = a + (n-1)\,d$$

(a_n → n^{th} term; a → 1st term; $(n-1)$ → no. of term; d → Common diff.)

+ For last term, an = l

Sum of n term;

+ $S_n = \frac{n}{2}\left[2a + (n-1)\,d\right]$ (n → no. of terms.)
+ $S_n = \frac{n}{2}$ (1st term + n^{th} term)

NOTE:-

+ n = no. of terms not last term.
+ Sum of 1st n-natural numbers

$$1 + 2 + 3 + 4 + ... + n = \frac{n(n+1)}{2}$$

+ Sum of Squares of 1st n-natural numbers

$$1^2 + 2^2 + 3^2 + 4^2 + ... + n^2 = \frac{n\,(n+1)\,(2n+1)}{6}$$

+ Sum of Cubes of 1st n-natural numbers

$$= \left[\frac{n(n+1)}{2}\right]^2$$

5 GP series

Next term = Previous term × Common ratio

$a\ ,\ ar\ ,\ ar^2\ ,\ ar^3\ ,\ ar^4$

Ex: 16, 8, 4, 2, 1, 1/2, 1/4, so on

Common ratio;

$$r = \frac{n^{th} \text{ term}}{(n-1)^{th} \text{ term}}$$

n^{th} term;

$T_n = ar^{n-1}$,

Sum of infinites terms;

$S_\infty = \frac{a}{1-r}$, valid when $r < 1$.

Q.2. Find the sum of given infinite series

(i) $1, \frac{1}{2}, \frac{1}{4}, \frac{1}{8}, \frac{1}{16}, ...$

(ii) $1, \frac{-1}{2}, \frac{1}{4}, \frac{-1}{8}, \frac{1}{16}, \frac{-1}{32}, ...$

Sol. (i) $r = \frac{1/4}{1/2} = \frac{1}{2}$, Sum $= \frac{1}{1-\frac{1}{2}} = \frac{1}{1/2} = 2$

(ii) $r = -\frac{1}{2}$, Sum$= \frac{1}{1-(-\frac{1}{2})} = \frac{1}{3/2} = \frac{2}{3}$

6 Quadratic equation

$ax^2 + bx + c = 0$

a, b, & c are constant in which a can not be zero

$$X = \frac{-b \pm \sqrt{b^2-4ac}}{2a.}$$

Sum of roots, $x_1 + x_2 = \frac{-b}{a}$

Products of roots, $x_1 \cdot x_2 = \frac{c}{a}$

Q.3. Find roots of equation x2 − 5x + 6 = 0; find value of a, b & c by comparing with ax2 + bx + c = 0

Sol. $a = 1,\ b = -5\ \&\ c = 6$

$$X = \frac{-(-5) \pm \sqrt{(-5)^2 - 4\times1\times6}}{2\times1}$$

$X_1 = \frac{5+\sqrt{1}}{2} = 3$ (Taking + sign)

$X_2 = 2$ (Taking – sign)

Q.4. $x^2 - 4x = 0$; then find roots of equation.

Sol.

$x^2 = 4x$
$x = 4$ wrong

$x(x - 4) = 0$
$x = 0$; $x = 4$ correct

Q.5. $x^2 - 4x + 3 = 0$; then find roots.

Sol. $x^2 - 3x - x + 3 = 0$

$x(x-3) - 1(x - 3) = 0$

$(x - 3)(x - 1) = 0$

$x = 3, x = 1$

7 Logarithms

$\log_y x = \log x$ on the base y

$\log_e x = 2.303 \log_{10} x$

Properties of Logarithms

- $\log_a (xy) = \log_a x + \log_a y$
- $\log\left(\frac{x}{y}\right) = \log x - \log y$
- $\log_y x = \frac{1}{\log_x y}$
- $\log_e x^{1/n} = \frac{1}{n} \log_e x$
- $\log_e x^n = n \log_e x$
- $\log_b a \times \log_a b = 1$
- $\log_a a = 1$

Some numerical value of Logarithms

- $\log_e 1 = 0$
- $\log_{10} 2 = 0.30$
- $\log_{10} 1 = 0$
- $\log_{10} 3 = 0.48 \approx 0.5$
- $\log_e(\sin 90°) = 0$
- $\log_{10} 5 + \log_{10} 20 = 2$
- $\log_2 3 = \frac{\log_{10} 3}{\log_{10} 2} = \frac{48}{30}$
- Concept of Anti-log

$\log e^x = Y$

By taking Anti-log
(convert into concept of power)

$x = e^y$

MR* ka tadka

log → Concept of Power

$2^3 = 8 \Rightarrow \log 2^8 = 3$ (Power: 3, Base: 2, Result: 8)

Base wahi rahega (Power ⇌ Result interchange hoga)

8 Exponents

A quantity multiplied by itself one or more times is indicated by an exponent. The exponent, shown as a superscript, denotes how many times the multiplication occurs, for example:

$$x = x^1$$
$$x \cdot x = x^2$$
$$x \cdot x \cdot x = x^3$$
$$x \cdot x \cdot x \cdot x = x^4 \text{ ...... so on}$$

The concept of exponents with a few examples are shown in table:

Base x	x^2	x^3	x^4
1	$1^2 = 1$	$1^3 = 1$	$1^4 = 1$
2	$2^2 = 4$	$2^3 = 8$	$2^4 = 16$
3	$3^2 = 9$	$3^3 = 27$	$3^4 = 81$
4	$4^2 = 16$	$4^3 = 64$	$4^4 = 256$
5	$5^2 = 25$	$5^3 = 125$	$5^4 = 625$

6	$6^2 = 36$	$6^3 = 216$	$6^4 = 1296$
7	$7^2 = 49$	$7^3 = 343$	$7^4 = 2401$
8	$8^2 = 64$	$8^3 = 512$	$8^4 = 4096$
9	$9^2 = 81$	$9^3 = 729$	$9^4 = 6561$
10	$10^2 = 100$	$10^3 = 1000$	$10^4 = 10000$

Rule of Exponents

- If Power of any non-zero number is zero then result will be one.

Ex: $8^0 = 1$.

- Negative Property of exponent (x is non zero number)

$$x^n = \frac{1}{x^{-n}} \Rightarrow \frac{1}{x^n} = x^{-n}$$

$$\frac{1}{10^3} = 10^{-3}$$

Product Property of Exponent

$$x^n x^m = x^{n+m}$$

Ex: $10^3 \times 10^4 = 10^7$

Division Property

$$\frac{x^n}{x^m} = x^{n-m}$$ Ex: $\frac{10^3}{10^2} = 10^{3-2}$

Power of a Power:

$$(x^n)^m = x^{nm}$$

Ex: $(10^2)^3 = 10^6$

Ex: $10^2 + 10^3 = 100 + 1000 = 1100$

Fractional exponent

$$(x)^{3/2} = (x^3)^{1/2}$$

- The powers of 10 from 10^{-6} to 10^6:

$10^0 = 1$

$10^1 = 10$	$10^{-1} = 0.1$
$10^2 = 100$	$10^{-2} = 0.01$
$10^3 = 1000$	$10^{-3} = 0.001$
$10^4 = 10,000$	$10^{-4} = 0.0001$
$10^5 = 100,000$	$10^{-5} = 0.00001$
$10^6 = 1,000,000$	$10^{-6} = 0.000001$

- Multiplication with fraction.

$0.5 = \frac{1}{2}$	$1.33 \times 12 = \frac{4}{3} \times 12 = 16$
$0.6 = \frac{6}{10}$	$0.25 \times 16 = \frac{1}{4} \times 16 = 4$
$0.4 = \frac{4}{10}$	$0.75 \times 16 = \frac{3}{4} \times 16 = 12$
$0.66 = \frac{2}{3}$	$0.33 \times 15 = \frac{1}{3} \times 15 = 5$

$1.33 = \frac{4}{3} \Rightarrow 0.75 = \frac{3}{4} \Rightarrow 0.33 = \frac{1}{3}$

Important property

$2^\infty = \infty$	$e^\infty = \infty$
$1^\infty = 1$	$e^{-\infty} = 0$
$4^{-\infty} = 0$	$e^0 = 1$

Ex: $(8)^{2/3} = (8)^{(1/3)\times 2} = (2)^{3\times(1/3)\times 2} = 2^2 = 4$

$(32)^{3/5} = (2^5)^{3/5} = 2^3 = 8$

9 Square root

A square root represents a quantity that, when multiplied by itself, gives the original number. It is indicated by the radical symbol ($\sqrt{}$) or by an exponent of $\frac{1}{2}$. For example: $\sqrt{x} = x^{\frac{1}{2}}$.

Square root	Square root of Decimal
$\sqrt{1} = 1$	$\sqrt{0.01} = 0.1$
$\sqrt{4} = 2$	$\sqrt{0.04} = 0.2$
$\sqrt{9} = 3$	$\sqrt{0.09} = 0.3$
$\sqrt{16} = 4$	$\sqrt{0.16} = 0.4$
$\sqrt{25} = 5$	$\sqrt{0.25} = 0.5$
$\sqrt{36} = 6$	$\sqrt{0.36} = 0.6$
$\sqrt{49} = 7$	$\sqrt{0.49} = 0.7$
$\sqrt{64} = 8$	$\sqrt{0.64} = 0.8$

$\sqrt{81} = 9$	$\sqrt{0.81} = 0.9$
$\sqrt{100} = 10$	$\sqrt{1.00} = 1.0$
$\sqrt{121} = 11$	$\sqrt{1.21} = 1.1$
$\sqrt{144} = 12$	$\sqrt{1.44} = 1.2$
$\sqrt{169} = 13$	$\sqrt{1.69} = 1.3$
$\sqrt{196} = 14$	$\sqrt{1.96} = 1.4$
$\sqrt{225} = 15$	$\sqrt{2.25} = 1.5$
$\sqrt{256} = 16$	$\sqrt{2.56} = 1.6$
$\sqrt{289} = 17$	$\sqrt{2.89} = 1.7$
$\sqrt{324} = 18$	$\sqrt{3.24} = 1.8$
$\sqrt{361} = 19$	$\sqrt{3.61} = 1.9$
$\sqrt{400} = 20$	$\sqrt{4.00} = 2.0$

Square Roots involving Even exponents

Expression	Simplified form
$\sqrt{10^6}$	$10^{6/2} = 10^3$
$\sqrt{10^{-4}}$	$10^{-4/2} = 10^{-2}$
$\sqrt{5 \times 10^4}$	$\sqrt{5} \times \sqrt{10^4} = 2.24 \times 10^2$
$\sqrt{1.3 \times 10^{-8}}$	$\sqrt{1.3} \times \sqrt{10^{-8}} = 1.14 \times 10^{-4}$
$\sqrt{9 \times 10^{-2}}$	$\sqrt{9} \times \sqrt{10^{-2}} = 3 \times 10^{-1} = 0.3$

Square Roots involving Odd exponents

Expression	Simplified form
$\sqrt{10^5}$	$\sqrt{10 \times 10^4} = 3.16 \times 10^2$
$\sqrt{10^{-3}}$	$\sqrt{10 \times 10^{-4}} = 3.16 \times 10^{-2}$
$\sqrt{3 \times 10^5}$	$\sqrt{30 \times 10^4} = 5.48 \times 10^2$
$\sqrt{6 \times 10^{-7}}$	$\sqrt{60 \times 10^{-8}} = 7.75 \times 10^{-4}$
$\sqrt{0.000025}$	$\sqrt{25 \times 10^{-6}} = 5 \times 10^{-3}$

10 Cube root

The cube root of a number x is a value that, when multiplied by itself three times (raised to the power of three), gives the number x. Mathematically, if y is the cube root of x, then:

$$y^3 = x$$

$$y = \sqrt[3]{x}.$$

Example Description	Cube root	Because
Cube root of 8	$\sqrt[3]{8} = 2$	$2^3 = 2 \times 2 \times 2 = 8$
Cube root of 27	$\sqrt[3]{27} = 3$	$3^3 = 3 \times 3 \times 3 = 27$
Cube root of 64	$\sqrt[3]{64} = 4$	$4^3 = 4 \times 4 \times 4 = 64$
Cube root of 125	$\sqrt[3]{125} = 5$	$5^3 = 5 \times 5 \times 5 = 125$
Cube root of −8	$\sqrt[3]{-8} = -2$	$(-2)^3 = -2 \times -2 \times -2 = -8$

11 Trigonometry

Angle
- → Arc = $R\theta$; Algebraic function
- → $\sin\theta/\cos\theta/\tan\theta$; Trigo. function

$R\theta$, θ, R

→ Angle is dimensionless.

→ For algebraic function, we always use S.I. unit radian but for trigonometric function we may use radian or degree.

→ $180° = \pi$ rad

$$1° = \frac{\pi}{180} \text{ rad; or } 1\text{rad} = \frac{180°}{\pi}$$

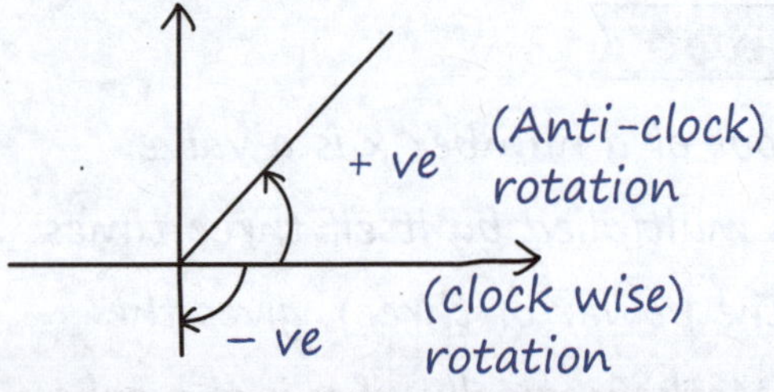

Q.6. Total Angle moved by object in π-rotation?

Sol. – $\theta = \pi(2\pi) = 2\pi^2$ rad.

Trigonometric function:

A right triangle has two perpendicular sides. The hypotenuse, which is opposite the right angle, is always the longest side. The three main trigonometric functions—sine, cosine, and tangent—are defined based on the angles in a right triangle as follow:

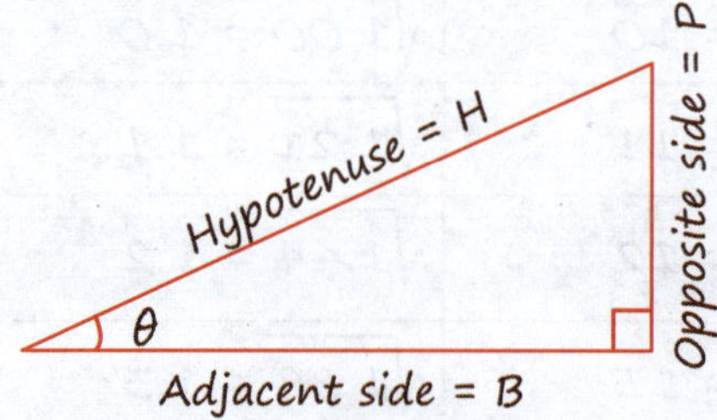

- $\text{Sin } \theta = \frac{P}{H}$, $\text{Cos } \theta = \frac{B}{H}$, $\text{Tan } \theta = \frac{P}{B}$
- $\text{Sin}\theta = \frac{1}{\text{Cosec } \theta}$, $\text{Sec}\theta = \frac{1}{\text{Cos } \theta}$, $\text{Tan}\theta = \frac{1}{\text{Cot } \theta}$
- From Pythagoras theorem

$$P^2 + B^2 = H^2$$

Angle θ	0°	30°	45°	60°	90°	120°	135°	150°	180°
Sin θ	0	$1/2$	$1/\sqrt{2}$	$\sqrt{3}/2$	1	$\sqrt{3}/2$	$1/\sqrt{2}$	$1/2$	0
Cos θ	1	$\sqrt{3}/2$	$1/\sqrt{2}$	$1/2$	0	$-1/2$	$-1/\sqrt{2}$	$-\sqrt{3}/2$	–1
Tan θ	0	$1/\sqrt{3}$	1	$\sqrt{3}$	Not define (a)	$-\sqrt{3}$	–1	$-1/\sqrt{3}$	0

Trigonometric quadrant graph:

+y axis

II Quadrant
+sin θ
+cosec θ

I Quadrant
All

–x axis — +x axis

I Quadrant
+tan θ
+cot θ

IV Quadrant
+cos θ
+sec θ

–y axis

- $\text{Sin } (90 + \theta) = \cos \theta$
- $\sin (180 - \theta) = \sin \theta$
- $\sin (90 - \theta) = \cos \theta$
- $\cos (180 - \theta) = -\cos \theta$
- $\cos (90 - \theta) = \sin \theta$
- $\cos (90 + \theta) = -\sin \theta$
- $\sin (-\theta) = -\sin \theta$
- $\cos (-\theta) = \cos \theta$
- $\tan (-\theta) = -\tan \theta$

Unique Relation

- $\text{Sin}^2\theta + \text{Cos}^2\theta = 1 \Rightarrow \quad 1 + \cot^2\theta = \text{Cosec}^2\theta$
- $\text{Tan}^2\theta + 1 = \text{Sec}^2\theta$

Some Important Triangles

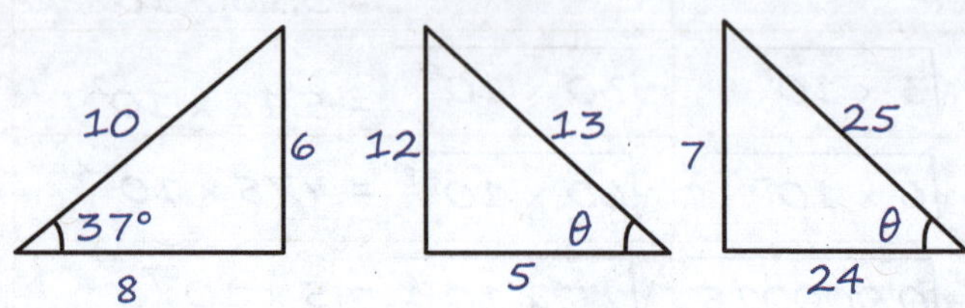

✦ 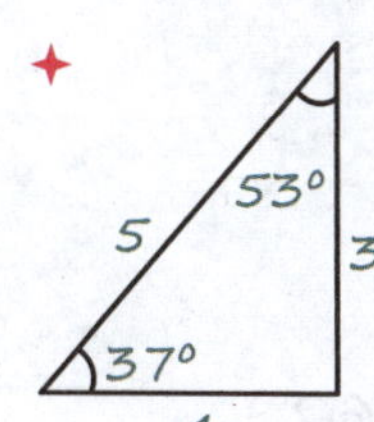

$\text{Sin } 37° = \frac{3}{5}$ $\text{Cos } 37° = \frac{4}{5}$

$\text{Sin } 53° = \frac{4}{5}$ $\text{Cos } 53° = \frac{3}{5}$

$\text{Cos }(-60°) = \frac{1}{2} \Rightarrow \text{Sin }(-30°) = -\frac{1}{2}$

$\tan(-135°) = 1$

Trigonometric formulae

✦ Sin (A + B) = Sin A cos B + cos A sin B

✦ Sin (A − B) = Sin A cos B − cos A sin B

✦ Cos (A + B) = Cos A cos B − Sin A sin B

✦ Cos (A − B) = Cos A cos B + Sin A sin B

✦ $\tan(A + B) = \frac{\tan A + \tan B}{1 - \tan A \tan B}$

✦ $\tan(A - B) = \frac{\tan A - \tan B}{1 + \tan A \tan B}$

(a) A = B = θ

$\text{Sin }(A + B) = \text{Sin } 2\theta = 2\sin\theta \text{ Cos } \theta$

$\text{Cos }(A + B) = \text{Cos } 2\theta = \text{Cos}^2\theta - \text{Sin}^2\theta$

(b) $2 \text{ Cos}^2\theta = 1 + \text{Cos }(2\theta)$.

$2 \text{ Sin}^2\theta = 1 - \text{Cos }(2\theta)$

✦ If Angle is Small:-

$\text{Sin } \theta \approx \theta$ $\tan \theta \approx \theta$ $\cos \theta = 1$

Note:

Sin (2°) = 2° (wrong)

$\text{Sin }(2°) = 2 \times \frac{\pi \text{rad}}{180°} = \frac{\pi}{90°} \text{ rad}$

$\text{Cos }(4°) = 1$

$\tan 3° = \frac{\pi \text{rad}}{60}$

Sin Law

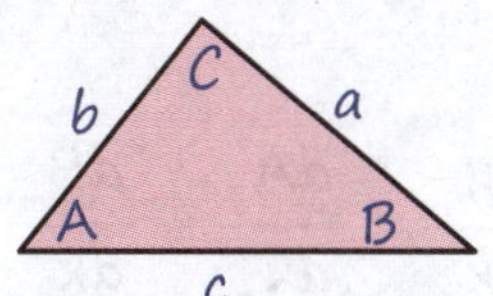

$$\frac{a}{\sin A} = \frac{b}{\sin B} = \frac{c}{\sin C}$$

Some trignometric function and their maximum & minimum value

Trigonometric function	Maximum Value	Minimum Value
$Y = 3 \sin \theta$	$Y_{max} = 3$	$Y_{min} = -3$
$Y = 4 \sin (5\theta)$	$Y_{max} = 4$	$Y_{min} = -4$
$Y = 3 \sin \theta + 4 \cos \theta$	$Y_{max} = 5$	$Y_{min} = -5$
$Y = 3 \sin \theta + 4 \sin \theta$	$Y_{max} = 7$	$Y_{min} = -7$
$Y = 5 - 2 \sin \theta$	$Y_{max} = 7$	$Y_{min} = 3$

Q.7. Force acting on object $F = \frac{4}{3\sin\theta + \cos\theta}$

Then find minimum magnitude of force.

Sol. $y = A\sin\theta + B\cos\theta$

$$Y_{max} = +\sqrt{A^2 + B^2}$$

$$Y_{min} = -\sqrt{A^2 + B^2}$$

$$F_{min} = \frac{4}{(3\sin\theta + \cos\theta)_{max}}$$

$$F_{min} = \frac{4}{\sqrt{9 + 1}} = \frac{4}{\sqrt{10}}$$

12 Phasor diagram

A phasor is a rotating vector that represents a sinusoidal function in terms of its magnitude (amplitude) and phase angle. Phasors are typically represented in the complex plane with the real axis corresponding to the cosine component and the imaginary axis corresponding to the sine component.

The phase difference

$$\Delta\phi = \theta_2 - \theta_1$$

✦ Vector representation of trigonometric function

Cos θ
-30° sin (θ + 60°)
or Cos (θ − 30°)
60°
-Sin θ
sin (θ) Let
-Cos θ

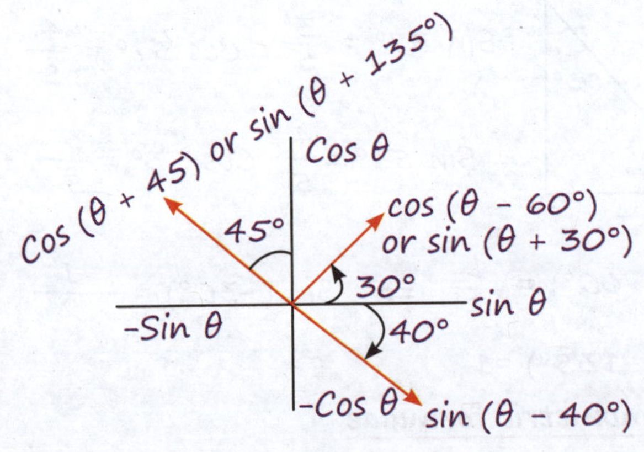

Equation-1	Equation-2	Phase difference
$I = I_0 \sin(\theta + \pi/3)$	$I = I_0 \sin(\theta - \pi/6)$	$\phi = \left\|\left(\theta - \frac{\pi}{6}\right) - \left(\theta + \frac{\pi}{3}\right)\right\| = 90°$
$I = I_0 \sin(\theta + \pi/3)$	$I = I_0 \cos(\theta - \pi/6)$	$\phi = 0°$
$I_1 = I_0 \sin(\theta)$	$I = I_0 \cos(\theta + \pi/6)$	$\phi = 2\pi/3$
$I_1 = \sin(\theta - \pi/3)$	$I = I_0 \cos(\theta + \pi/3)$	$\phi = \frac{7\pi}{6} = 210°$
$I_1 = \sin(\theta - 60°)$	$I = I_0 \cos(\theta - 30°)$	$\phi = \frac{2\pi}{3} = 120°$

13 Differentiation

DC = – ve

If x and y are variables, then the rate of change in y w.r.t. x $= \frac{dy}{dx}$

= Slope of y–x graph.

- $\frac{d^2y}{dx^2}$ = Double diff^n of y w.r.t. 'x'

  = The rate of change in $\left(\frac{dy}{dx}\right)$ w.r.t. 'x'

  = Slope of Slope

  = Change in slope w.r.t. 'x'

Differentiation of some function

- If $y = x^n$, then $\frac{dy}{dx} = \frac{d\,x^n}{dx} = nx^{n-1}$
- If y = Constant, $\frac{dy}{dx} = 0$
- $\frac{d}{dx}(\sin x) = \cos x$
- $\frac{d}{dx}(\cos x) = -\sin x$
- $\frac{d}{dx}(\tan x) = \sec^2 x$
- $\frac{d}{dx}(\text{cosec } x) = -\text{cosec } x \cot x$
- $\frac{d}{dx}(\sec x) = \sec x \tan x$
- $\frac{d}{dx}(\cot x) = -\text{cosec}^2 x$
- $\frac{d}{dx}(\log_e x) = \frac{d}{dx}(\ln x) = \frac{1}{x}$
- $\frac{d}{dx}[\sin(90°)] = 0$
- $\frac{d}{dx}(e^x) = e^x$
- $\frac{d}{dx}(e^2) = 0$ $\quad (\because e^2$ is const)

Rules

(i) Addition Rule:-

$Y = A + B, \quad \frac{dy}{dx} = \frac{dA}{dx} + \frac{dB}{dx}$

(ii) Substraction Rule:-

$Y = A - B$, $\frac{dy}{dx} = \frac{dA}{dx} - \frac{dB}{dx}$

(iii) Multiplication Rule:-

$Y = AB$, $\frac{dy}{dx} = \frac{A\,dB}{dx} + \frac{B\,dA}{dx}$

(iv) Division Rule:-

$Y = \frac{A}{B}$, $\frac{dy}{dx} = \frac{B\left(\frac{dA}{dx}\right) - A\left(\frac{dB}{dx}\right)}{B^2}$

Q.8. If $Y = t^2$, find $\frac{dy}{dx}$

Sol: $\frac{dy}{dx} = \frac{dt^2}{dx}$, $\frac{dy}{dx} = 2t\frac{dt}{dx}$

The MR*

Outside Inside Rule

$Y = f(z(x)) = y$ is function of z and z is a function of x.

$\frac{dy}{dx} = \left(\begin{array}{l}\text{differentiation}\\ \text{of outer function}\\ \text{keep inside as it is}\end{array}\right) \times \left(\begin{array}{l}\text{diff}^n \text{ of Inner}\\ \text{fu}^n \text{ w.r.t x}\end{array}\right)$

Q.9. Find $\frac{dy}{dx}$ for given function.

(i) $y = e^{5x}$

(ii) $y = (x^2 + 4)^3$

(iii) $y = e^{-4x}$

(iv) $y = \sin(4x^2)$

(v) $y = A\sin(wt - kx)$

(i)

$Y = e^{(5x)}$

$\frac{dy}{dx} = 5e^{5x}$

(iii)

$Y = e^{-4x}$

$\frac{dy}{dx} = -4e^{-4x}$

(ii)

$Y = (x^2+4)^3$

$\frac{dy}{dx} = 3(x^2+4)^2 \frac{d(x^2+4)}{dx}$

$= 3(x^2+4)^2 \times 2x$

(iv)

$Y = \sin(4x^2)$

$\frac{dy}{dx} = \cos(4x^2) \times 8x$

(v) $Y = A\sin(\omega t - kx)$

$\frac{dy}{dx} = A\cos(\omega t - kx) \times (-k)$

Q.10. $y = \sin(3x)$; then find $\frac{dy}{dx}$

Sol. $\frac{dy}{dx} = \cos(3x)\frac{d(3x)}{dx}$

$= 3\cos(3x)$

Q.11. If radius of sphere is increasing $1/\pi$ m/s then find rate of change in volume w.r.t. time when radius is 3m.

Sol. $V = \frac{4}{3}\pi R^3$

$\frac{dv}{dt} = \frac{4}{3}\pi 3R^2 \frac{dR}{dt}$

$= 4\pi R^2\left(\frac{1}{\pi}\right)$

$\left.\frac{dv}{dt}\right| = 4R^2 = 4(3)^2 = 4 \times 9 = 36$

14 Maxima and minima

MR* for maxima/minima

- For location of maxima/minima put $\frac{dy}{dx}$ (slope) = 0 and find value where x will be max^m/min^m.
- For exact maxima and minima dont check double differentiation. Just put value of 'x' and find 'y'.
- Double differentiation check nahi karna just 'x' ki value put kark 'y' nikala jo 'y' jayda wo maximum 'y' ko kam wo minimum 'y'.

Maxima

$\frac{dy}{dx} = 0$; $\frac{d^2y}{dx^2} = -ve$

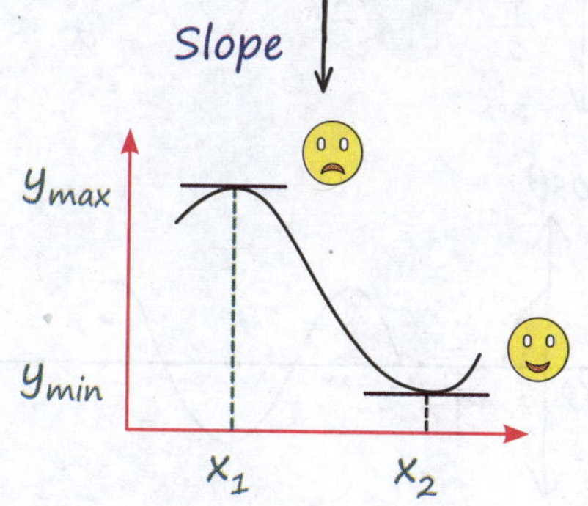

Minima

$$\frac{dy}{dx} = 0; \quad \frac{d^2y}{dx^2} = +ve$$

Slope ↑

15 Integration

→ Area under the curve → Inverse of differentiation

$$\int x^n \, dx = \frac{x^{n+1}}{n+1} + C \quad \text{Not valid for } n = -1$$

Integration of some function:

- $\int (u + v).\, dx = \int u.dx \int + v.dx$
- $\int \sin x \, dx = -\cos x + c.$
- $\int \cos x \, dx = \sin x + c.$
- $\int e^x dx = e^x + c$
- $\int 1/x \, dx = \ln x + c.$
- $\int \sec^2 dx = \tan x + c$
- $\int e^{3x} dx = \frac{e^{3x}}{3} + c.$

Graphical representation of sinθ & cosθ

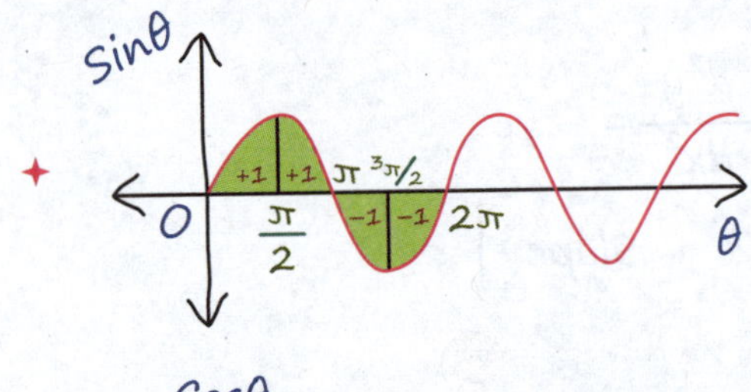

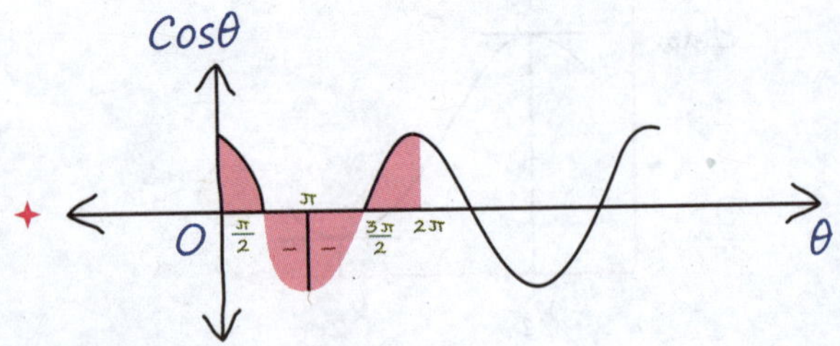

Chain Rule → MR*

Applicable when power of x is one

$$\int y \, dx = \frac{\text{Integration of outer function keep inside as it is.}}{\text{Coefficient of } (x)}$$

$$\int (2x+3)^4 \, dx = \frac{(2x+3)^5}{5[2]} + C$$

$$\int \sin(3x-4) \, dx = \frac{-\cos(3x-4)}{3} + C$$

16 Co-ordinate geometry and graph

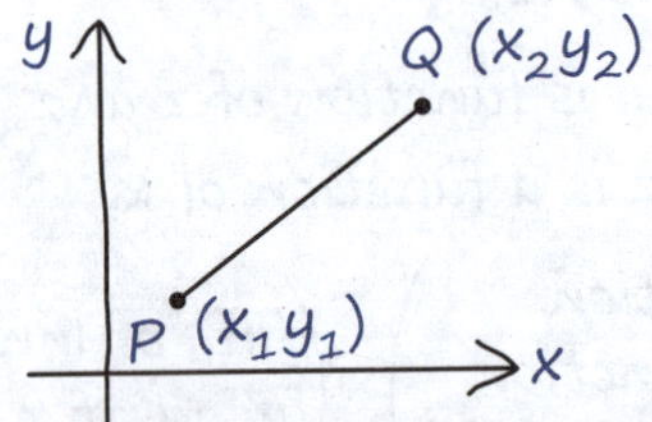

Distance (PQ) between two points in x-y plane

$$PQ = \sqrt{(x_2 - x_1)^2 + (y_2 - y_1)^2}$$

Slope (m) of straight line

$$m = \tan\theta = \frac{y_2 - y_1}{x_2 - x_2}$$

NOTE:-

- Slope of straight line remains same at all the point
- If $0° \leq \theta < 90°$ then slope is positive
- If $90° < \theta \leq 180°$ then slope is negative
- If $\theta = 90°$ then slope is infinite
- If $\theta = 0°$ then slope is zero
- If straight line parallel to x-axis then slope zero

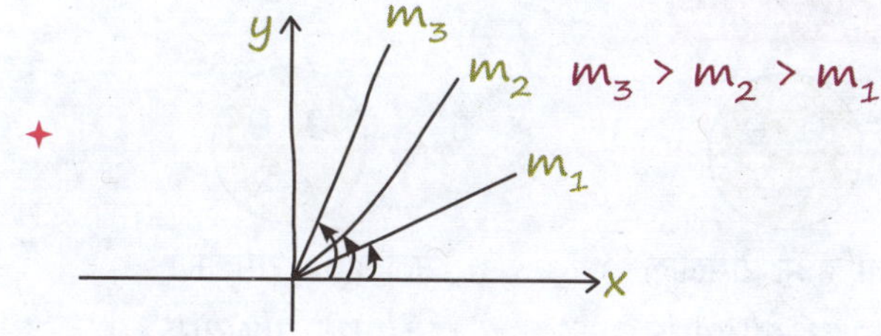

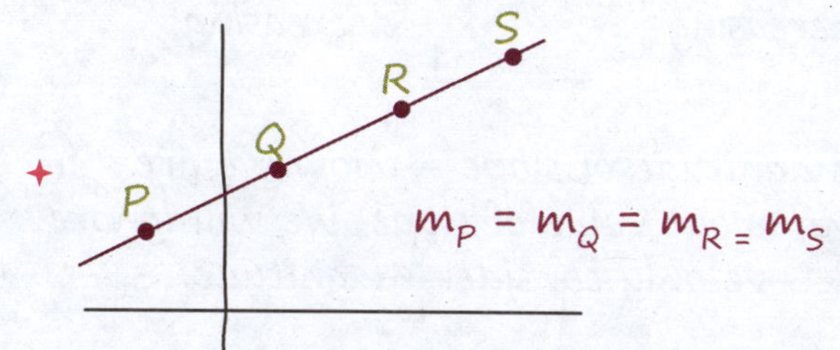

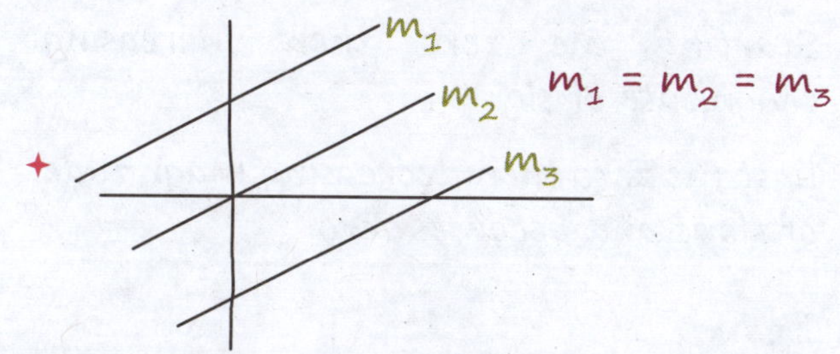

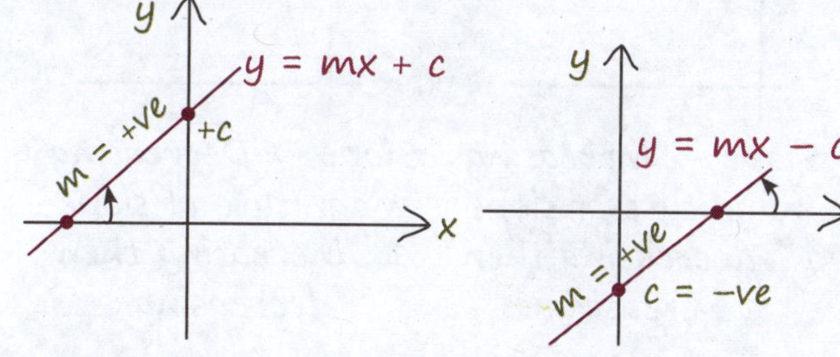

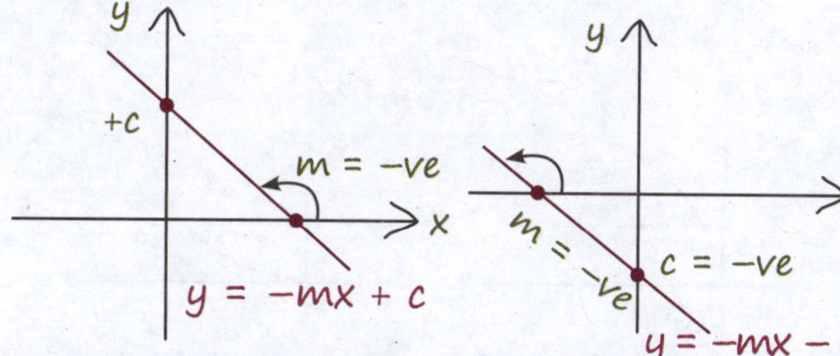

- If two straight line perpendicular to each other then product of their slope is −1.

17 Rectangular Hyperbola

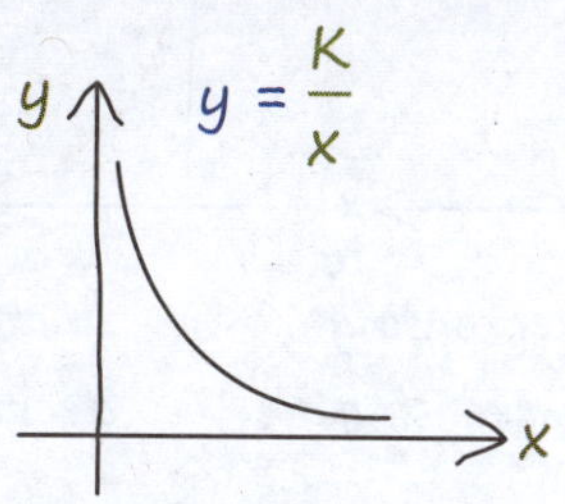

K is value Jitna Jayda graph utna upar shift hoga.

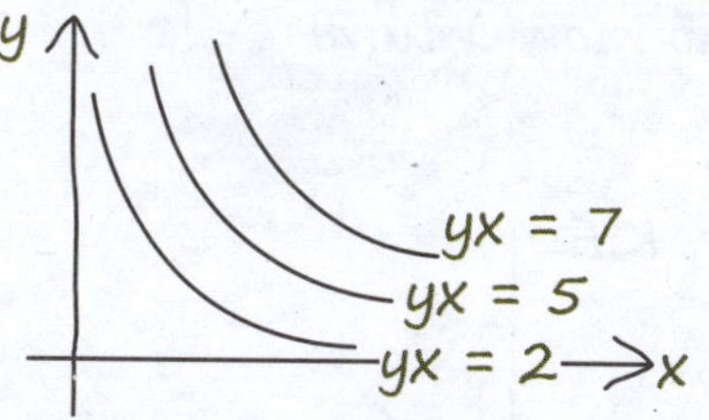

x ka pawer jitna jayda graph utna niche jayga.

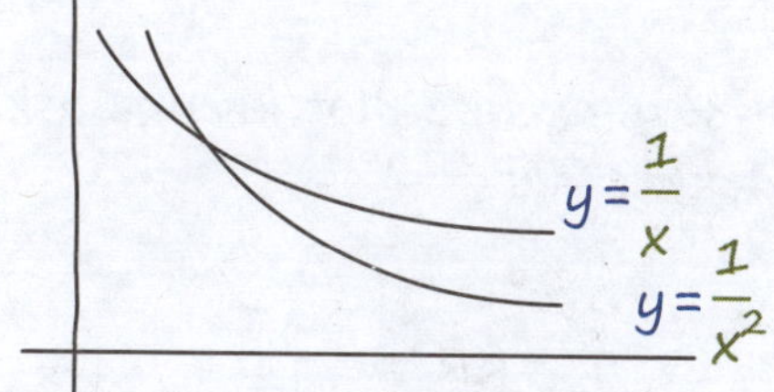

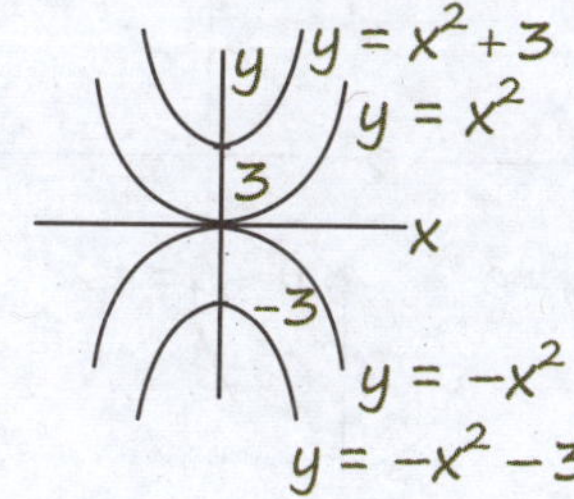

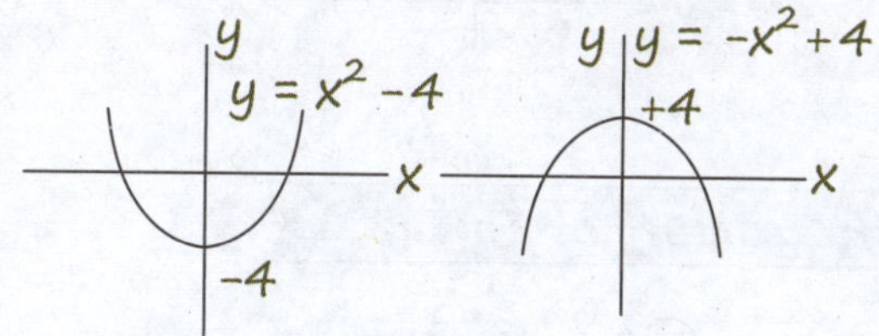

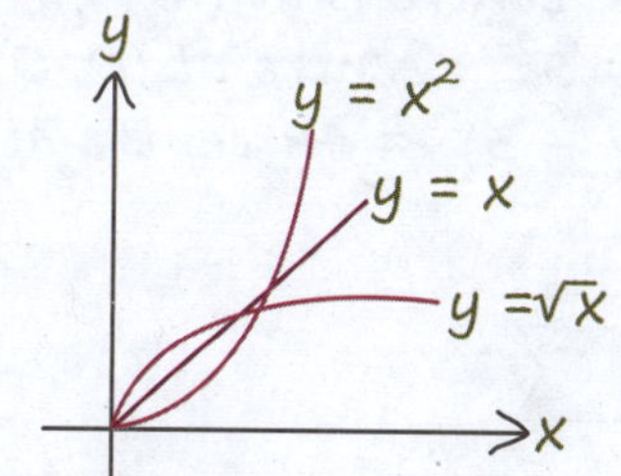

- graph for PV = nRT

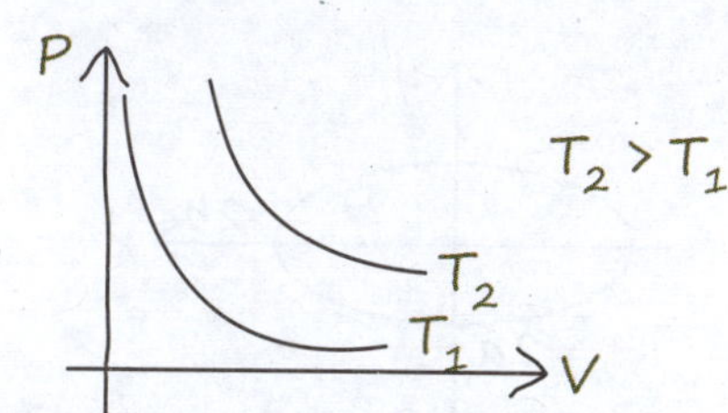

✦ K.E. $= \frac{P^2}{2m}$ graph b/w K.E. and $\frac{1}{m}$ for constant momentum.

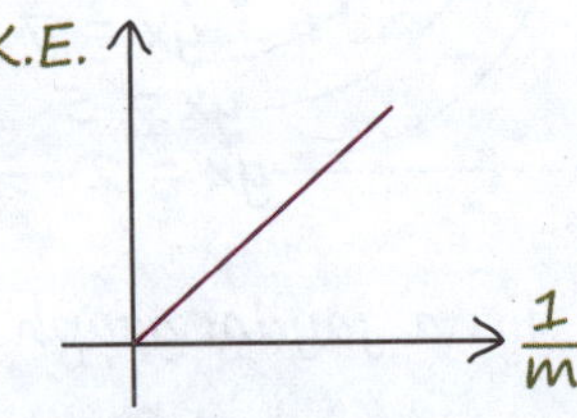

MR*

Jisko x- & y-axis pe plot krenge uska power dekhte hai.

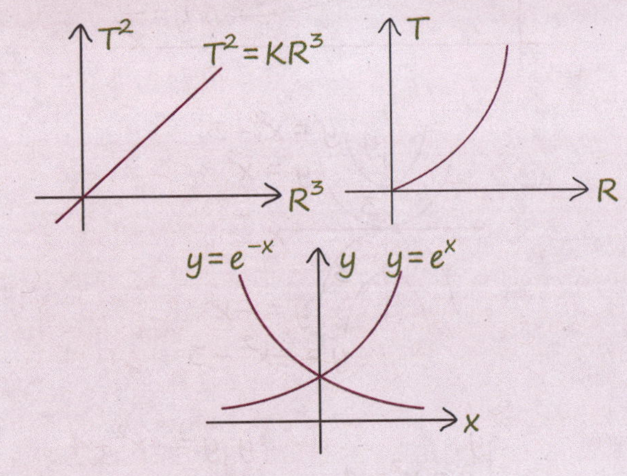

18 Equation of Circle

$(x - x_0)^2 + (y - y_0)^2 = R^2$

R is radius & centre is at (x_0, y_0)

Ex: $x^2 + y^2 = 5^2$ centre at (0, 0) R = 5

$(x + 4)^2 + (y - 3)^2 = 49$ centre at (−4, 3)

R = 7

19 Ellipse

$$\frac{x^2}{a^2} + \frac{y^2}{b^2} = 1$$

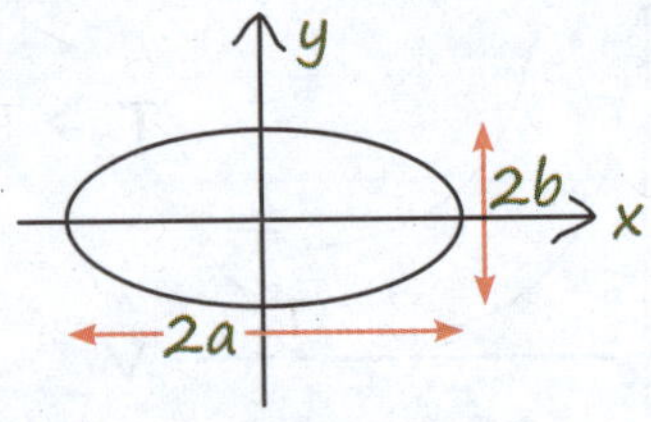

MR* For Slope

हँसता हुआ रामलाल

Slope always increasing

रोता हुआ रामलाल

Slope always decreasing

For magnitude of slope → Now we are talking about value of slope, we will ignore +ve & −ve only consider magnitude.

MR* → Locate where slope is zero

✦ Starting me zero then increasing magnitute of slope.

✦ Last me zero then decreasing magnitude of slope and becomes zero.

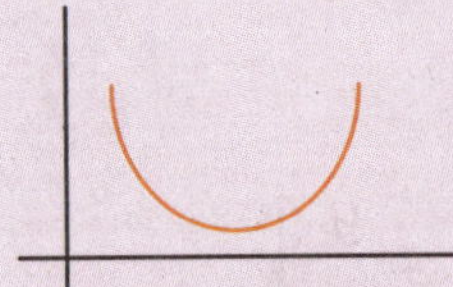

Slope → Increasing magnitute of slope 1st decreasing then increasing

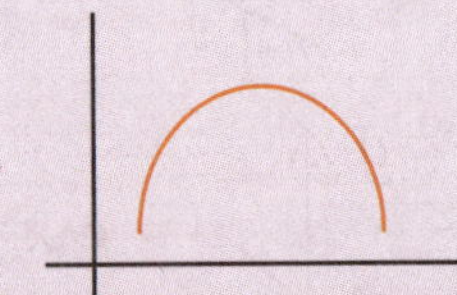

Slope → Decreasing magnitute of slope 1st increasing then decreasing

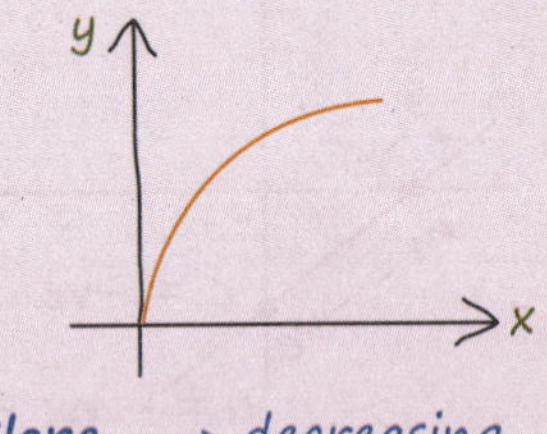

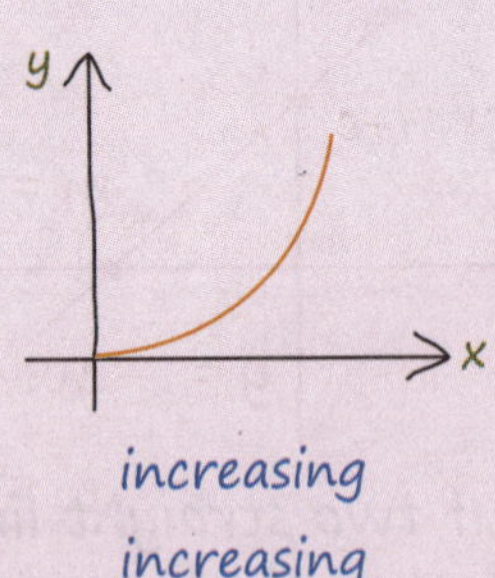

Slope	→ decreasing	increasing
Magnitute	→ decreasing	increasing

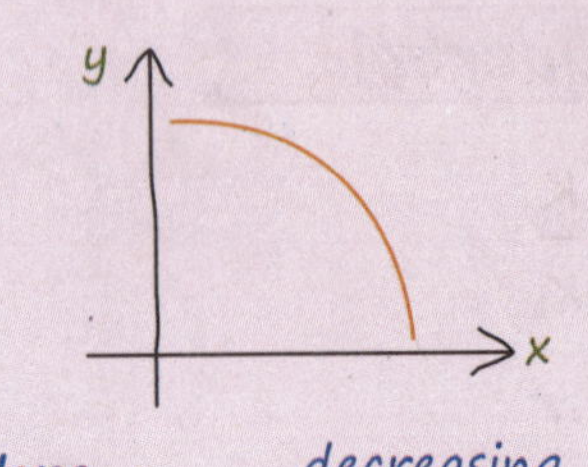

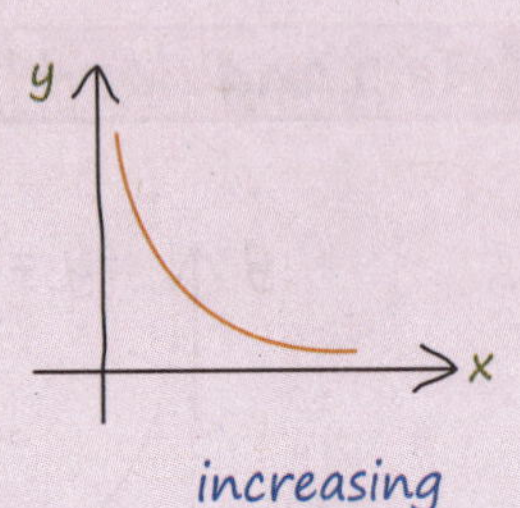

Slope →	decreasing	increasing
Magnitute →	increasing	decreasing

20 Some Basic Geometry Shapes

Equilateral Triangle of side (a)

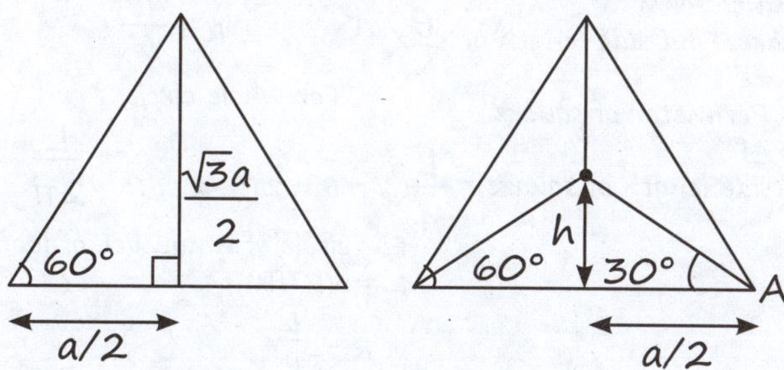

✦ Distance from base to centre

$$\tan 30^\circ = \frac{h}{a/2}$$

$$h = \frac{a}{2\sqrt{3}}$$

✦ Distance from centre to corner

$$= \frac{\sqrt{3}a}{2} \times \frac{2}{3} = \frac{a}{\sqrt{3}}$$

$$\text{Area} = \frac{\sqrt{3}a^2}{4}$$

Hexagonal of side 'a'

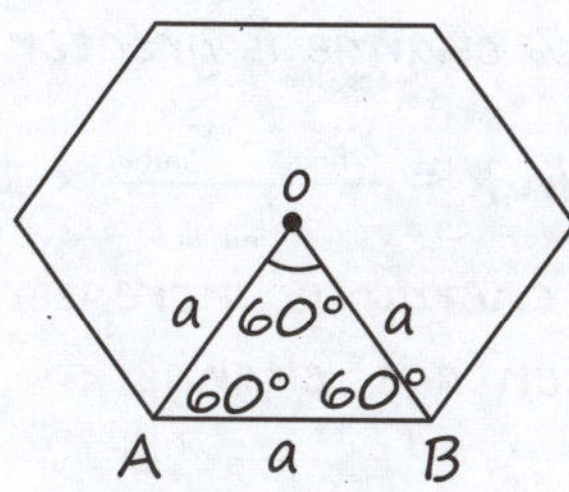

Distance from centre to corner = a

Square

Rectangle

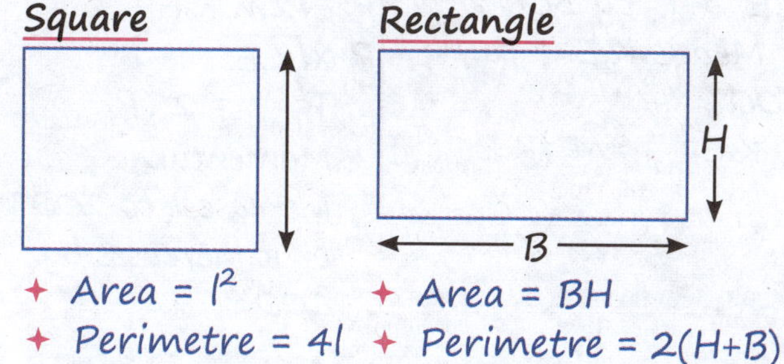

✦ Area = l^2

✦ Perimetre = $4l$

✦ Area = BH

✦ Perimetre = 2(H+B)

Trapezoid

Triangle

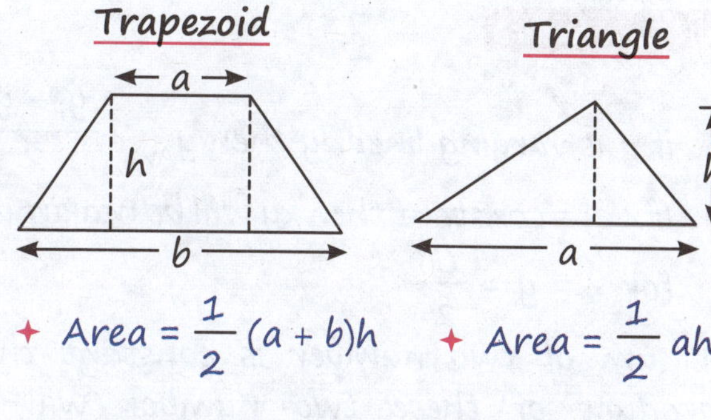

✦ Area = $\frac{1}{2}(a+b)h$

✦ Area = $\frac{1}{2}ah$

Circle

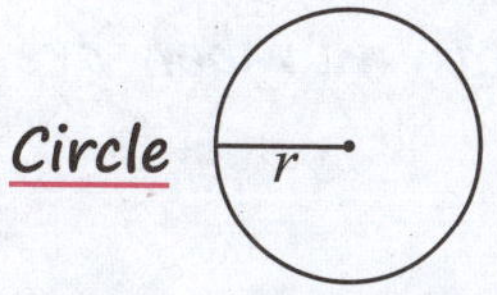

Elipse

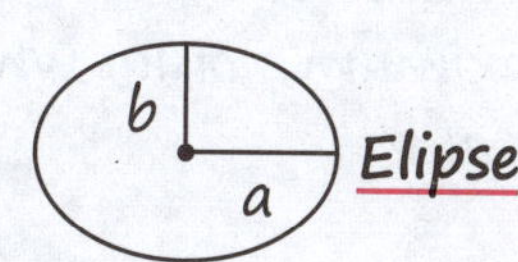

✦ Circumference $= 2\pi r$

✦ Area $= \pi r^2$

✦ Area $= \pi ab$

Cube

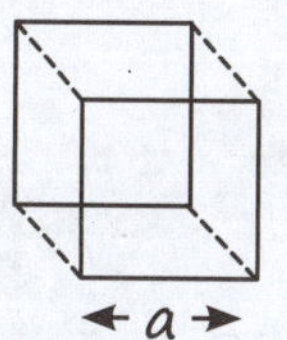

Cylinder

✦ Area = $6a^2$

✦ Volume = a^3

✦ Area = $2\pi r^2 + 2\pi rh$

✦ Volume = $\pi r^2 h$

Cone

Area = $\pi r^2 + \pi rl$

Volume = $\frac{1}{3}\pi r^2 h$

where, $l = \sqrt{r^2 + h^2}$

Sphere

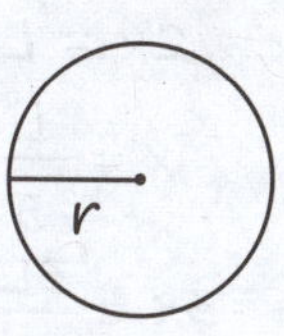

Area = $4\pi r^2$

Volume = $\frac{4}{3}\pi r^3$

21 Average of a varying quantity

If $y = f(t)$ then

$$\langle y \rangle = \frac{\int_{t_1}^{t_2} y\,dt}{\int_{t_1}^{t_2} dt} = \frac{\int_{t_1}^{t_2} y\,dt}{t_2 - t_1}$$

Y may be any physical quantity.

MR* ka tadka

- if y is varying linearly then $y_{Avg} = \frac{y_i + y_f}{2}$
- If x+y = constant then xy will be maximum for $x = y = \frac{C}{2}$

If sum of two number is constant then product of these two number will be maximum, only when both number are equal.

22 Important concept and Question

- Length of Rod 20m is divided in the ratio 1 : 3 : 5 then find length of each Part.

$$1x + 3x + 5x = 20$$

$$9x = 20$$

$$x = \frac{20}{9}$$

$x = \frac{20}{9}$ m $3x = \frac{20}{9}$ m $5x = \frac{100}{9}$ m

- Spring of lengh L and spring constant k divided into length ratio 3 : 2, then the ew spring constant of bigger part

$$3x + 2x = L \qquad k \propto \frac{1}{l}$$

$$x = \frac{L}{5} \qquad \therefore k_1 l_1 = k_2 l_2$$

$$3x = \frac{3L}{5} \qquad KL = K' \frac{3L}{5}$$

$$\boxed{K' = \frac{5}{3}K}$$

- If n identical liquid sphere of radius r cobine to form single large sphere the find redius of bigger sphere

Total volume will be consant

$$\therefore 3x \frac{4}{3}\pi r^3 = \frac{4}{3}\pi R^3$$

$$\therefore nr^3 = R^3$$

$$\boxed{R = (n)^{\frac{1}{3}} r}$$

- If wire of Length L is Bended as shown find Radius R.

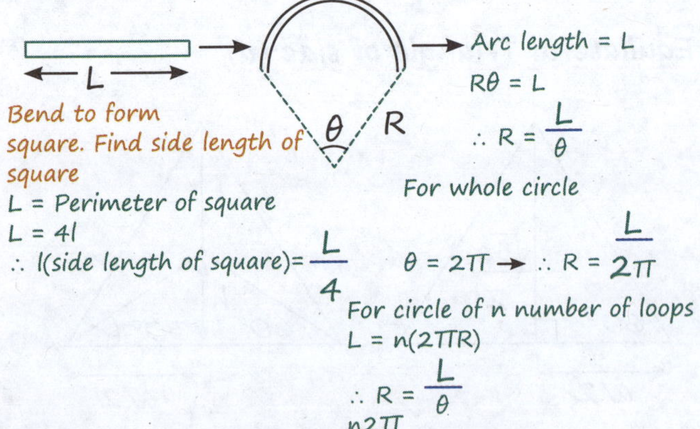

23 Percentage changes

- Case 1: If % changes is less than 5%

$y = kx^n$ [K = constant]

then $\boxed{\frac{\Delta y}{y} = n\frac{\Delta x}{x}}$

If Momentum of object is increased by 3%. Find % change in kinetic energy.

$$\Rightarrow K.E = \frac{P^2}{2m} \quad \therefore \frac{\Delta K.E.}{K.E.} = 2\frac{\Delta P}{P}$$

$$\therefore \left(\frac{\Delta K.E.}{K.E.} \times 100\%\right) = 2\left(\frac{\Delta P}{P} \times 100\%\right)$$

$$= 2\,(3)$$

percetage change in Kinetic Energy = 6%

- Case 2: If % change is greater than 5%

$$\% \text{ change in } x = \frac{X_{final} - X_{initial}}{X_{initial}} \times 100\%$$

If kinetic energy is increased by 300%, then percentage change in momentum will be

Sol.

MR*

K.E → ↑ by 300% It Means K.E. ↑ to 400% $\therefore K.E_{final} = 4\,(K.E_{initial})$	$P = \sqrt{2m\,K.E}$ $P \propto \sqrt{K.E.}$ $\therefore P_{final} = 2\,(P_{initial})$ ∴ Momentum increasest to 200% or it increases by 100 %

If kinetic energy is decreased by 19% then % change in Momentum?

Sol.

MR*

K.E → ↓es by 19% ∴ K.E is reduced to 81% ∴ $K.E_f = \frac{81}{100} K.E_i$ $P \propto \sqrt{K.E.}$ ∴ $P_{final} = \frac{9}{10} P_{final}$	∴ Momentum is reduced to $\left(\frac{9}{10} \times 100\%\right)$ ∴ Momentum is reduced by 10% (or reduced to 90%)

Linear mass density:

- Mass per unit length

$\lambda = \frac{dm}{dL} \Rightarrow dm = \lambda dL$

Scalar Unit kg/m

If λ = constant, mass = $\lambda \times$ (length)

Q. 12 Find mass of rod if λ = 3 kg/m and length 12m

Ans. $dm = \lambda dL$

$\int dm = \int 3dL$

mass = $3(L)_0^{12} = 3(12-0) = 36$ kg.

Surface mass density or Areal mass density

$\sigma = \frac{dm}{dA} \Rightarrow$ unit kg/m^2

Q. 13 Find mass of a disc if $\sigma = \pi$ kg/m^2 of Radius R = 2m.

Ans. mass = σ Area = $\pi (\pi R)^2 = \pi^2 (2)^2 = 10 \times 4 = 40$ kg

Areal mass density (s)

- Mass per unit Area

$\sigma = \frac{dM}{dA}$ $dM = \sigma dV$

↳ kg/m^2

if σ = constant, mass = $\sigma \times$ Area

Q. 14 A Disc of Radius R/2 cut from a disc of mass (M,R) as shown in Fig. then find mas of remaining portion.

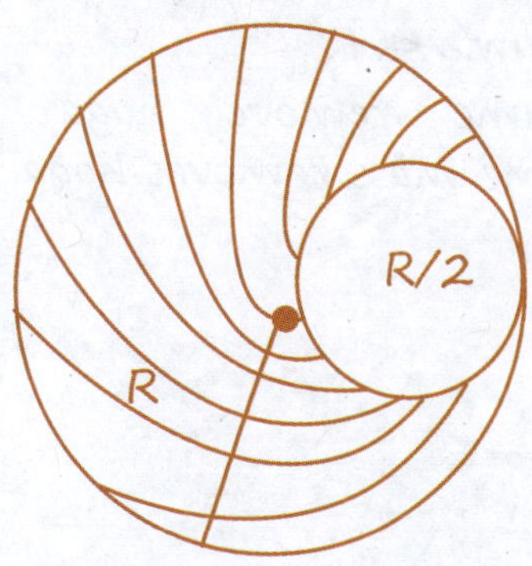

Ans. $\sigma = \frac{M}{\pi R^2}$ Areal mass density

Mass of removed part m^1 = σ Area of removed

mass of remaining = $M - \frac{M}{4} = \frac{3M}{4}$

$\frac{M}{\pi R^2} \times \pi \left(\frac{R}{2}\right)^2 = \frac{M}{4}$

Volume mass density (ρ)

- Mass per unit volume

$\rho = \frac{dM}{dV}$ $dM = \rho dV$

↳ kg/m^3

if ρ = constant, mass = $\rho \times$ volumes

Q. 15 A sphere of Radius R/2 is cut from a sphere of mass m and Radius R as shown in figure.

Find mass of remaining part.

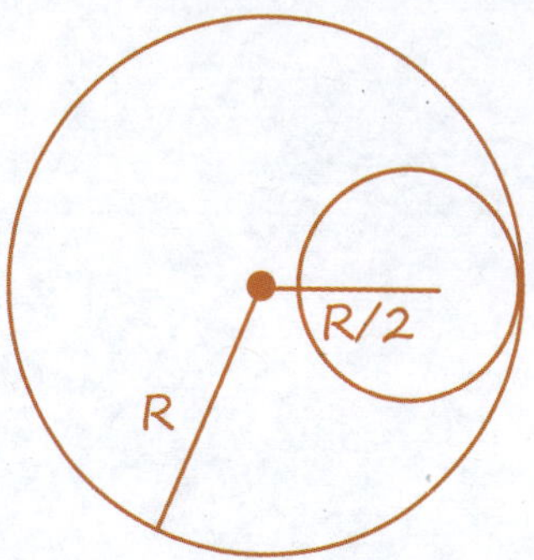

Sol. Mass = ρ Volume

Same

Mass $\propto$ volume $\propto R^3$

Jitna volume remove hoga same propotion me mass removes hoga.

Volume of removed part $= \frac{4}{3}\pi\left(\frac{R}{2}\right)^3$

$= v/8$

Mass of removed part $= \frac{m}{8}$

Remaining mass $= m - \frac{m}{8} = \frac{7m}{8}$

MR*

'जब टूट के बिखर जाओगे समझ जाओ अब निखर जाओगे'

2 Vector

1 Vectors

A vector is a quantity that has both magnitude and direction. Vectors are used to represent physical quantities that cannot be fully described by a single number alone. Examples of vector quantities include displacement, velocity, acceleration, and force.

Scalar Quantity	Vector Quantity
✦ Having Magnitude only	✦ Having Magnitude and direction
✦ Follow simple algebric addition	✦ follow triangle law of vector addition.
✦ Can be changed only by changing its value	✦ Can be changed by changing magnitude only, or changing dir^n only or changing both.
Ex-Speed, time, Mass, Volume, density current, etc.	Ex-Force, Velocity, current density, torque etc.

Properties of Vectors

- ✦ In a vector +ve sign and sign −ve indicate direction only.

  Ex: +5N and −5N, same magnitude of force in opposite direction.
- ✦ Angle between vector – When two vectors are placed head to head or tail to tail then smaller angle between vector is called angle between vector.

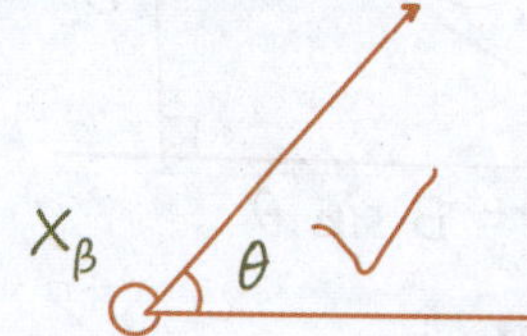

- ✦ Vector can be shifted parallel to itself by keeping magnitude and direction fixed.
- ✦ Rotation of vector not allowed it will change meaning of vector.
- ✦ If angle between $\vec{A}$ and $\vec{B}$ vector is θ then angle between $\vec{A}$ and $-\vec{B}$ is $(180°-\theta)$.
- ✦ All equal vectors are parallel but all parallels are not equal.
- ✦ All opposite (Negative) Vectors are Antiparallel but all antiparallel are not Opposite Vector

2 Type of Vectors

Type	Magnitude	Direction\Angle
Equal Vector	Same	Same (θ = 0)
Parallel Vector	May or May not same	Same (θ = 0)
Opposite Vector or Negative Vectors	Same	Opposite θ = 180°
Anti-parallel Vector	May or May not same	θ = 180° opposite
Orthogonal	May same	θ = 90°
Zero/Null Vector	Zero	any direction
Unit Vectors	One	$\hat{A} = \frac{\vec{A}}{A}$

Components of Vector in 2-D (effect of Vector)

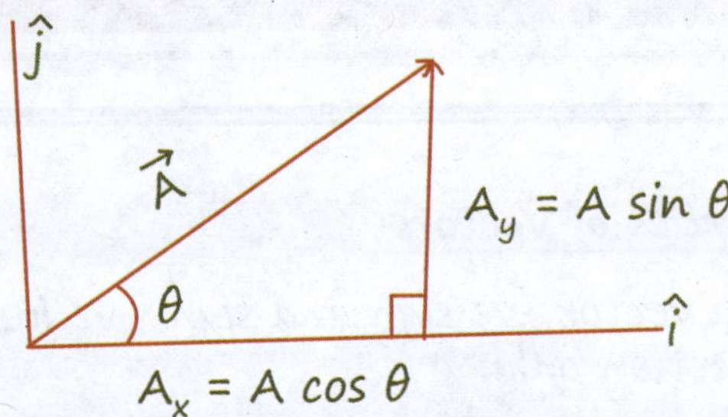

- $\vec{A} = A_x\hat{i} + A_y\hat{j}$

  $\vec{A} = A\cos\theta\ \hat{i} + A\sin\theta\ \hat{j}$

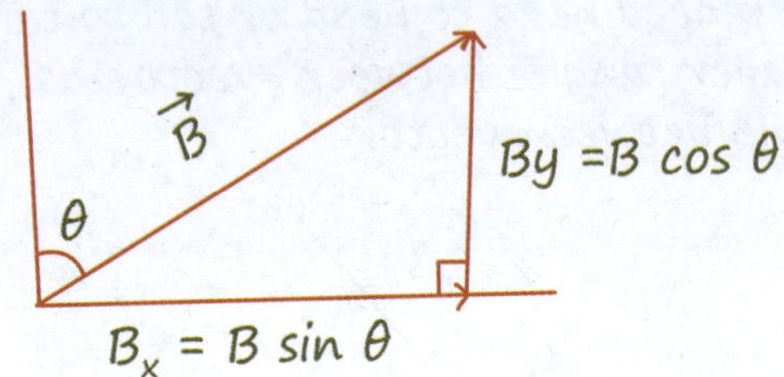

- $\vec{B} = B_x\hat{i} + B_y\hat{j}$

  $= B\sin\theta\ \hat{i} + B\cos\theta\ \hat{j}$

Magnitude of resultant Vector

$A = \sqrt{A_x^2 + A_y^2}$ or $B = \sqrt{B_x^2 + B_y^2}$

Direction:

$\theta = \tan^{-1}\left(\frac{A_y}{A_x}\right)$ or $\theta = \tan^{-1}\left(\frac{B_x}{B_y}\right)$

Rectangular component of a vector in 3D

$\vec{A} = A_x\hat{i} + A_y\hat{j} + A_z\hat{k}$

Magnitude

$|\vec{A}| = \sqrt{A_x^2 + A_y^2 + A_z^2}$

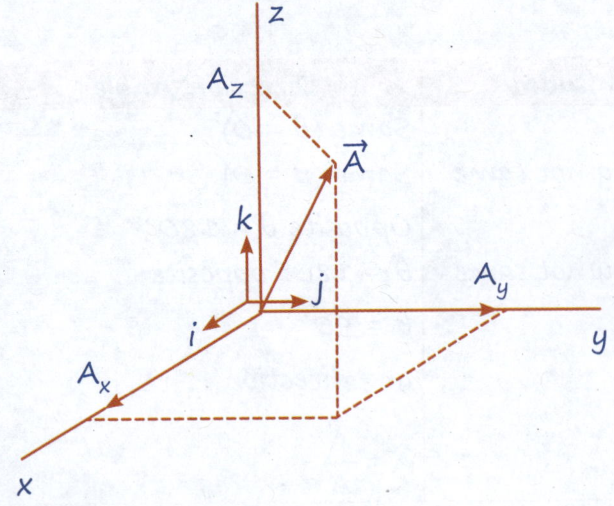

- If vector is making an angle α, β and γ from x, y and z-axis respectively then $\cos^2\alpha + \cos^2\beta + \cos^2\gamma = 1$; $\sin^2\alpha + \sin^2\beta + \sin^2\gamma = 2$

Direction Cosine

$\cos\alpha = \frac{A_x}{A}$ $\quad \cos\beta = \frac{A_y}{A}$ $\quad \cos\gamma = \frac{A_z}{A}$

3 Vectors addition

- It is the process of combining two vectors by placing the tail of one vector at the head of the other. i.e.,

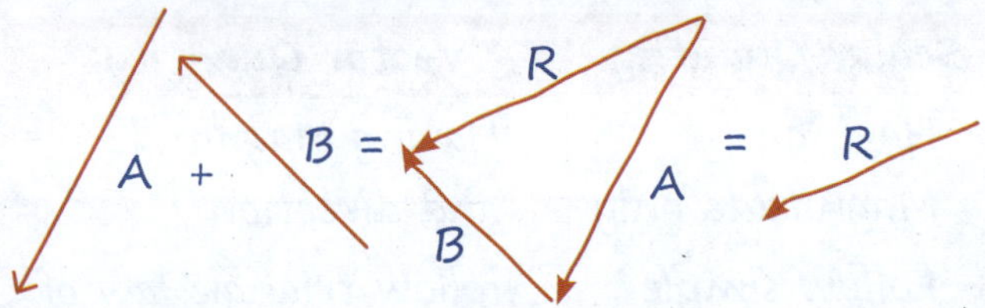

Polygon Law of vector addition

- Start tail of next vector from head of previous vector and so on.

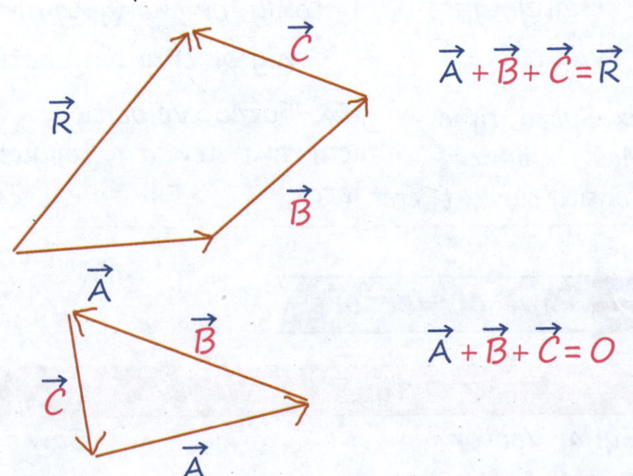

Triangle Law of Vector addition

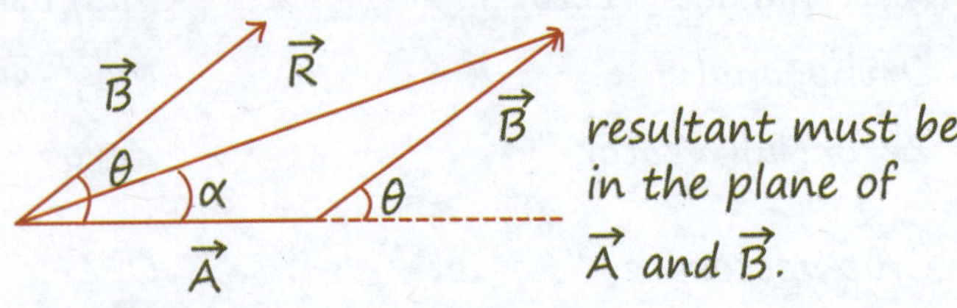

resultant must be in the plane of $\vec{A}$ and $\vec{B}$.

- $|\vec{R}|=\sqrt{A^2+B^2+2AB\cos\theta}$
- $\tan\alpha = \dfrac{B\sin\theta}{A+B\cos\theta}$
- If $\theta = 0°$, $R_{max} = A + B$ | $\theta = 90°$, $R = \sqrt{A^2+B^2}$ | $\theta = 180°$, $R_{min} = A - B$
- $A - B \le R \le A + B$
- If $|\vec{A}| = |\vec{B}| = A$ and Angle b/w them θ

  $|\vec{R}| = 2A\cos(\theta/2)$ $|\vec{D}| = 2A\sin(\theta/2)$

Vector Subtraction

Angle B/w $\vec{A}$ & $\vec{B}$ is θ then $\vec{D} = \vec{A} - \vec{B}$

Magnitude

$|\vec{D}|=\sqrt{A^2+B^2-2AB\cos\theta}$

- If $\theta = 0°$, $D_{min} = A - B$ | $\theta = 90°$, $D = \sqrt{A^2+B^2}$ | $\theta = 180°$, $D = A + B$
- $A - B \le D \le A + B$

$\theta = 0°$	$\theta = 60°$	$\theta = 90°$	$\theta = 120°$	$\theta = 180°$
$R = 2A$	$R = \sqrt{3}A$	$R = \sqrt{2}A$	$R = A$	$R = 0$
$D = 0$	$D = A$	$D = \sqrt{2}A$	$D = \sqrt{3}A$	$D = 2A$

Properties of vector addtion and substraction:

- Magnitude of Vector addition and subtraction are same at 90°.
- $\vec{A} + \vec{B} = \vec{B} + \vec{A}$ Commutative
- $n(\vec{A} + \vec{B}) = n\vec{A} + n\vec{B}$ distributive
- $\vec{A} - \vec{B} \ne \vec{B} - \vec{A}$
- $\vec{A} = A_x\hat{i} + A_y\hat{j} + A_z\hat{k}$ and $\vec{B} = B_x\hat{i} + B_y\hat{j} + B_z\hat{k}$ then $\vec{A} + \vec{B} = (A_x + B_x)\hat{i} + (A_y + B_y)\hat{j} + (A_z + B_z)\hat{k}$
- If $|\vec{A} + \vec{B}| = |\vec{A}| = |\vec{B}|$ then angle between $\vec{A}$ and $\vec{B}$ is 120°
- If $|\vec{A}| + |\vec{B}| = |\vec{A} + \vec{B}|$ then angle between $\vec{A}$ and $\vec{B}$ is zero.
- If $\vec{A} + \vec{B} = \sqrt{A^2 + B^2}$ then angle between $\vec{A}$ and $\vec{B}$ is 90°.
- If $|\vec{A} + \vec{B}| = |\vec{B} - \vec{A}|$ then angle between $\vec{A}$ and $\vec{B}$ is 90°.

4 Displacement vector

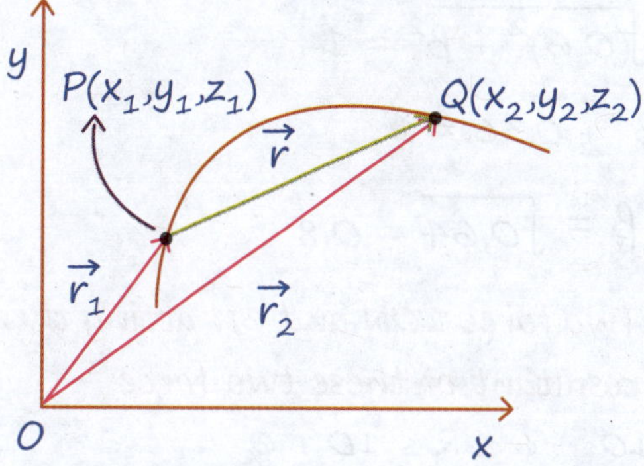

- If a particle moves from the initial position (x_1, y_1, z_1) to the final position (x_2, y_2, z_2), the displacement vector r is given by:

$r = (x_2 - x_1)i + (y_2 - y_1)j + (z_2 - z_1)k$

or $r = \Delta xi + \Delta yj + \Delta zk$

Magnitude of displacement vector

- $r = \sqrt{(x_2 - x_1)^2 + (y_2 - y_1)^2 + (z_2 - z_1)^2}$

Q.1. Ramlal is moving with velocity 6m/s along east and pinky with 6 m/s at 30° east of north then relative velocity of pinky w.r.t Ramlal.

Sol. $\vec{V}_{PR} = \vec{V}_P - \vec{V}_R$ same vector ka subtraction

at 60° $|\vec{V}_{PR}| = 6$ m/s

Q.2. Find change in Speed and velocity

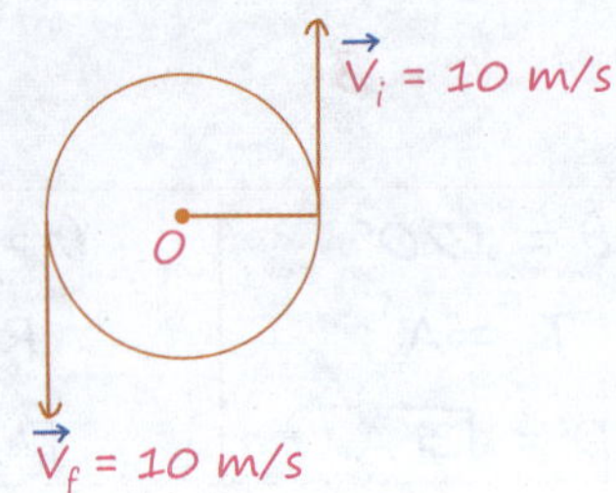

Sol. Change in speed = 0 magnitude of change in velocity = 20 m/s

Q.3. If $\vec{A} = 0.6\hat{i} + \beta\hat{j}$ is a unit vector then find value of β.

Sol. $|\vec{A}| = 1$ if A is unit vector

$$\sqrt{(0.6)^2 + \beta^2} = 1$$

$$\beta^2 + 0.36 = 1$$

$$\beta = \sqrt{0.64} = 0.8$$

Q.4. Two force 10N and 6N acting then find resultant of these two force.

Sol. $10 - 6 \le R \le 10 + 6$

R will between 4N to 16N

Q.5. Find the angle which a vector $\hat{i} + \hat{j} + \sqrt{2}\hat{k}$ makes with x, y and z-axis

$$A = \sqrt{1^2 + 1^2 + (\sqrt{2})^2} = 2$$

$A_x = 1$

$A_y = 1$

$A_z = \sqrt{2}$

$$\cos\alpha = \frac{A_x}{A} \quad \cos\beta = \frac{A_y}{A} \quad \cos\gamma = \frac{A_z}{A}$$

$\alpha = 60°$ $\quad \beta = 60°$ $\quad \gamma = 45°$

Q.6. In which of the following combination of three force resultant will be zero.

(a) 3N, 7N, 8N

(b) 2N, 5N, 1N

(c) 3N, 12N, 7N

(d) 4N, 5N, 10N

Sol. (a) Sum of two smaller must be greater or equal to (3rd).

5 Scalar Product (Dot Product)

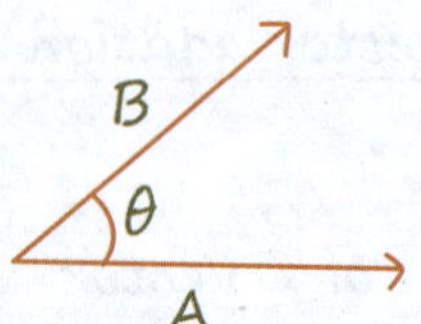

- $\vec{A}\cdot\vec{B} = A(B\cos\theta) = A$(Component of B along A)

$= (A\cos\theta)B = B$(Component of A along B)

Component of B along A $= \dfrac{\vec{A}\cdot\vec{B}}{A}$

Component of A along B $= \dfrac{\vec{A}\cdot\vec{B}}{B}$

- Result of dot product is always scalar.

$\hat{i}\cdot\hat{i} = 1 \quad \hat{j}\cdot\hat{j} = 1 \quad \hat{k}\cdot\hat{k} = 1$

$\hat{i}\cdot\hat{k} = 0 \quad \hat{j}\cdot\hat{i} = 0 \quad \hat{k}\cdot\hat{j} = 0$

$\vec{A}\cdot\vec{B} = (A_x\hat{i} + A_y\hat{j} + A_z\hat{k})\cdot(B_x\hat{i} + B_y\hat{j} + B_z\hat{k})$

$= A_xB_x + A_yB_y + A_zB_z$

Application of dot Product

- To Find Angle B/W vectors

$\vec{A}\cdot\vec{B} = AB\cos\theta$

$$\cos\theta = \frac{\vec{A}\cdot\vec{B}}{AB}$$

+ To check unit vector

If $\vec{A}$ is a unit vector then $\vec{A}\cdot\vec{A} = 1$

+ To check perpendicular vector (orthogonal)

If $\vec{A}\cdot\vec{B} = AB\cos 90° = 0$

$\vec{A}\cdot\vec{B} = 0 \quad (\vec{A} \perp r\ \vec{B})$

+ To find component of one vector along other.

$\vec{A}\cdot\vec{B} = A(B\cos\theta)$

$B\cos\theta = \dfrac{\vec{A}\cdot\vec{B}}{A}$ = Comp^n of B along A

Q.7. If $\vec{A} = 4\hat{i} + 2\hat{j} - 3\hat{k}$ and $\vec{B} = \hat{i} + 3\hat{j} - 2\hat{k}$ then find $\vec{A}.\vec{B}$

(1) 10

(2) 16

(3) 3

(4) 14

Sol. (2) $\vec{A}.\vec{B} = (4\hat{i} + 2\hat{j} - 3\hat{k})\cdot(\hat{i} + 3\hat{j} - 2\hat{k})$

$= 4 + 6 + 6$

$= 16$

Q.8. A force $\vec{F} = 4\hat{i} + 5\hat{j}$ newton displaces a particle through $\vec{S} = 3\hat{i} + 6\hat{j}$ meter. Then find work done.

(1) 24 Joule

(2) 12 Joule

(3) 28 Joule

(4) 42 Joule

Sol. (1) $W = \vec{F}.\vec{S} = (4\hat{i} + 5\hat{j})\cdot(3\hat{i} + 6\hat{j})$

$= 12 + 30$

$= 42\ J$

6 Cross-Product : [Vector Product]

+ $\vec{A}\times\vec{B} = AB\sin\theta\ \hat{n}$

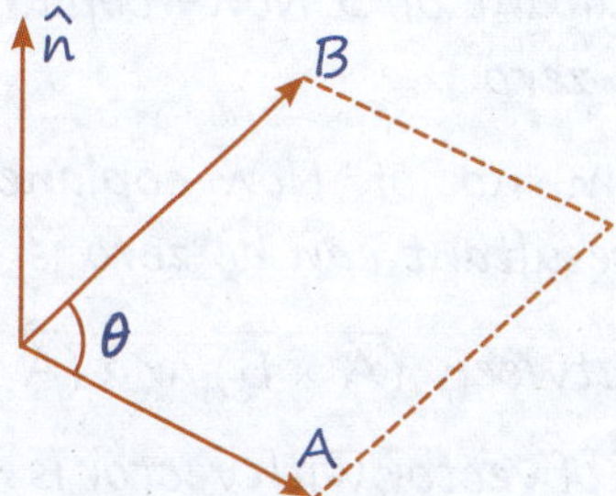

+ $\hat{n}$ is direction of $\vec{A}\times\vec{B}$ which is perpendicular to $\vec{A}$ & $\vec{B}$.

+ $(\vec{A}\times\vec{B})\cdot\vec{A} = 0 \quad (\vec{A}\times\vec{B})\cdot\vec{B} = 0$

+ $B\sin\theta = \dfrac{\vec{A}\times\vec{B}}{A}$ = component of B perpendicular of A

+ $\vec{R} = \vec{A}\times\vec{B}$

Place your finger of right hand along $\vec{A}$ and slap $\vec{B}$ then thumb will represent $\vec{R}$.

$\hat{i}\times\hat{i} = 0 = \hat{j}\times\hat{j} = \hat{k}\times\hat{k}$

$\hat{i}\times\hat{j} = \hat{k} \qquad \hat{j}\times\hat{i} = -\hat{k}$

$\hat{j}\times\hat{k} = \hat{i} \qquad \hat{k}\times\hat{j} = -\hat{i}$

$\hat{k}\times\hat{i} = \hat{j} \qquad \hat{i}\times\hat{k} = -\hat{j}$

+ $\vec{A}\times\vec{B} = \begin{vmatrix} i & j & k \\ A_x & A_y & A_z \\ B_x & B_y & B_z \end{vmatrix}$

$= i(A_yB_z - A_zB_y)$

$- J(A_xB_z - A_zB_x)$

$+ K(A_xB_y - A_yB_x)$

+ Unit vector does not have any unit only have direction and magnitude one.

- Minimum no. of vectors whose resultant can be zero is '2'.
- Minimum no of unequal vectors whose resultant can be zero is 3.
- The resultant of 3 Non- coplaner vectors can't be zero.
- Minimum no of Non-coplaner, vectors whose resultant can be zero is 4.
- Angle between $(\vec{A} \times \vec{B})$ and $(\vec{A} + \vec{B})$ is 90°
- Division of vector with vector is not possible
- Division of magnitude of vector is possible
- Vector can be divided by scalar.
- If vector multiplied by positive scalar then magnitude change and direction remains same.
- If vector multiplied by negative scalar then magnitude change and direction becomes opposite.

Scalar triple Product

$R = (\vec{A} \times \vec{B}) \cdot \vec{C}$ Result R will be scalar and R will be zero if any of these two vector becomes parallel.

Area of parallelogram

$$\text{Area} = |A \times B|$$

Area of triangle:

$$\text{Area} = \frac{1}{2}|A \times B|$$

The condition for coplanarity is

$$A \cdot (B \times C) = 0$$

Q.9. Force acting on object $\vec{F} = 5\hat{i} + 3\hat{j} - 7\hat{k}$ position vector $\vec{r} = 2i + 2j - k$ then find torque ?? (NEET 2022)

Sol. $\vec{\tau} = \vec{r} \times \vec{F} = \begin{vmatrix} \hat{i} & \hat{j} & \hat{k} \\ 2 & 2 & -1 \\ 5 & 3 & -7 \end{vmatrix}$

$\hat{i}(-14 - (-3) - \hat{j}(-14 - (-5)$ $+ \hat{k}(6 - 10)$

$-11\hat{i} + 9\hat{j} - 4\hat{k}$

Q.10. If $|\vec{A} \times \vec{B}| = \sqrt{3}\, \vec{A} \cdot \vec{B}$ then angle between $\vec{A}$ and $\vec{B}$ is?

Sol. $AB \sin\theta = \sqrt{3}\, AB \cos\theta$

$\tan\theta = \sqrt{3} \Rightarrow \theta = 60^\circ$

7 Lami's Theorem (Laws of motion) for 3 Vectors

When,

$$\frac{A}{\sin\alpha} = \frac{B}{\sin\beta} = \frac{C}{\sin\gamma}$$

Then, $\vec{A} + \vec{B} + \vec{C} = 0$

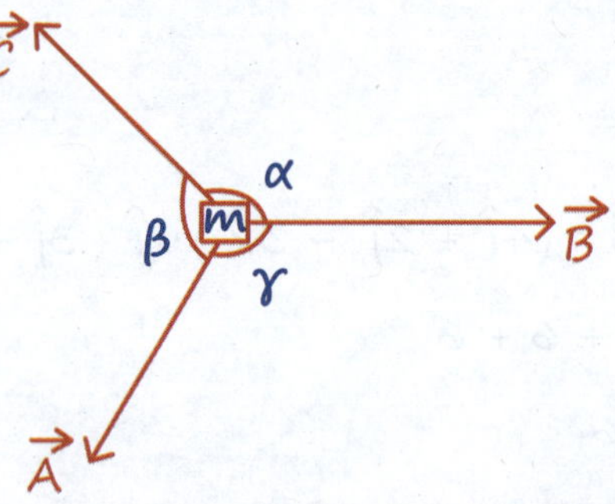

Q. 11. As shown in figure tension in the horizontal cord is 30 N. The weight W and tension in the string OA in Newton are

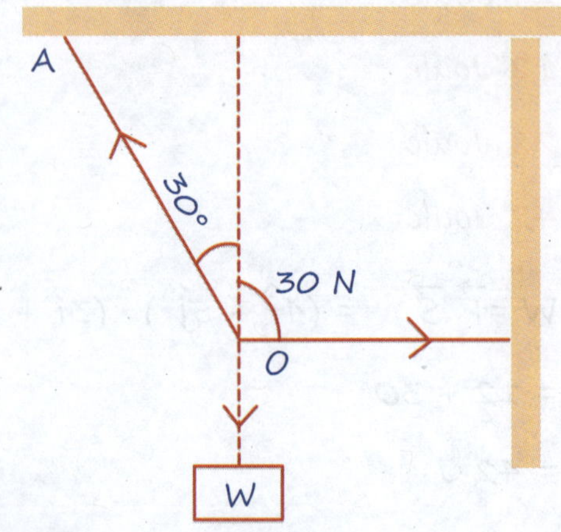

(a) $30\sqrt{3}, 30$ (b) $30\sqrt{3}, 60$

(c) $60\sqrt{3}, 30$ (d) None of these

Ans. $\dfrac{30}{\sin 150°} = \dfrac{T_1}{\sin 90°}$

$\dfrac{T_1}{1} = \dfrac{30}{1/2}, T_1 = 60\text{ N}$

$\dfrac{W}{\sin 120°} = \dfrac{30}{\sin 150°}$

$\dfrac{W}{\sqrt{3}/2} = \dfrac{30}{1/2} = 30\sqrt{3}$

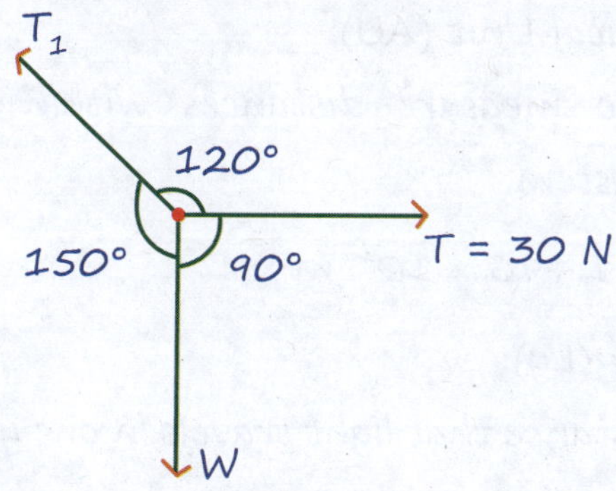

Q. 12. If three force acting on the object as shown in figure. Then find net force on object.

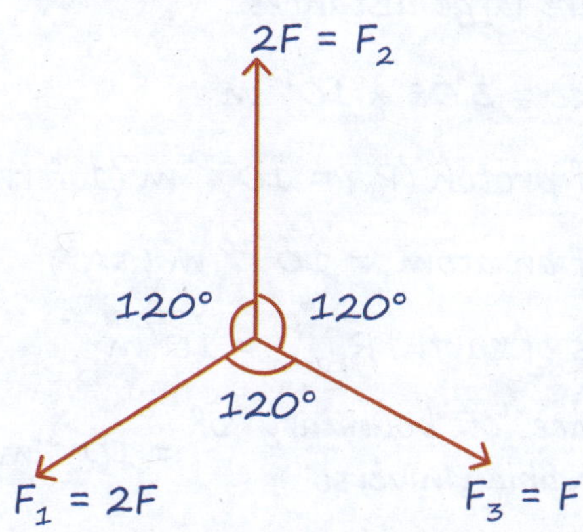

Sol. Net force on this object (point)

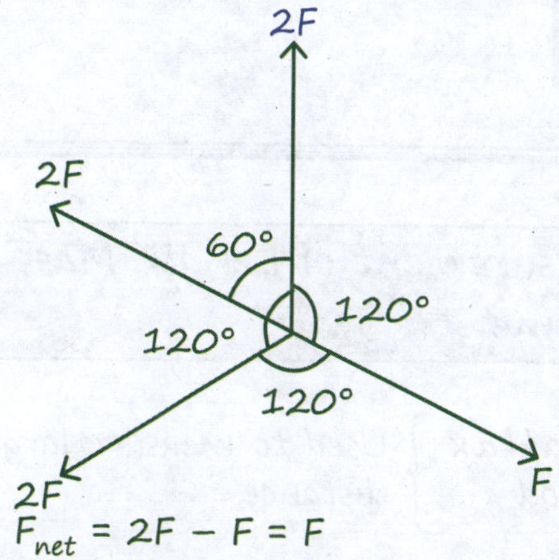

$F_{net} = 2F - F = F$

Q. 13. For two vector $\vec{A}$ and $\vec{B}$ vector addition $\vec{R} = \vec{A} + \vec{B}$ and vector subtraction $\vec{D} = \vec{A} - \vec{B}$ then angle between $\vec{A}$ and $\vec{B}$.

Condition	Angle
(i) $\lvert\vec{A} + \vec{B}\rvert = \lvert\vec{R}\rvert = \lvert\vec{A}\rvert = \lvert\vec{B}\rvert$	$\rightarrow \theta = 120°$
(ii) $\lvert\vec{A}\rvert = \lvert\vec{B}\rvert = \lvert\vec{A} - \vec{B}\rvert$	$\rightarrow \theta = 60°$
(iii) $A^2 + B^2 = \lvert\vec{A} - \vec{B}\rvert^2 = D^2$	$\rightarrow \theta = 90°$
(iv) $A^2 + B^2 = R^2$	$\rightarrow \theta = 90°$
(v) $\lvert\vec{A} + \vec{B}\rvert = \lvert\vec{A} - \vec{B}\rvert$	$\rightarrow \theta = 90°$
(vi) $\lvert\vec{A} + \vec{B}\rvert = \lvert\vec{A}\rvert + \lvert\vec{B}\rvert$ $\lvert\vec{R}\rvert = A + B$	$\rightarrow \theta = 0°$
(vii) $\lvert\vec{A} - \vec{B}\rvert = \lvert\vec{A}\rvert + \lvert\vec{B}\rvert$ $\lvert\vec{D}\rvert = A + B$	$\rightarrow \theta = 180°$
(viii) $\lvert\vec{A} + \vec{B}\rvert = \lvert\vec{A}\rvert - \lvert\vec{B}\rvert$ $\lvert R\rvert = A - B$	$\rightarrow \theta = 180°$
(ix) $\lvert\vec{A} - \vec{B}\rvert = \lvert\vec{A}\rvert - \lvert\vec{B}\rvert$	$\rightarrow \theta = 0°$

MR*

'अपनी पढाई छोड़ जो तेरे पीछे चला आयेगा।
वो खुद का ना हो सका, तेरा क्या हो जायेगा।। '

3 Units & Dimensions

1 Measurement of Length Mass & time

Parallax Method } Used to measure large distance

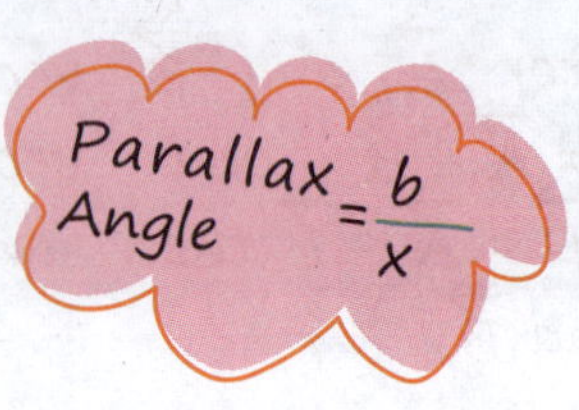

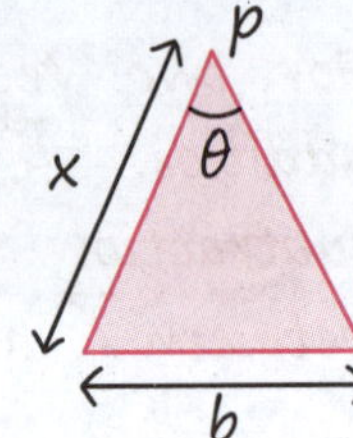

Where,

- θ = The angle formed between two lines of sight to an object when observed from two distinct positions.
- b = The distance between the two observation points
- x = The actual distance to the object being measured

Note:

The conversion between degrees, minutes, and seconds to radians helps in precise calculation and understanding of the small angles involved in parallax measurements.

- $1^\circ = 60' = 1.745 \times 10^{-2}$ rad
- $1' = 60' = 2.91 \times 10^{-4}$ rad
- $1'' = 4.85 \times 10^{-6}$ rad

Measuring Very Small Sizes:

Optical microscope:

For small objects visible with light.

Tunneling microscope:

Uses quantum tunneling for atomic-level objects.

Electron microscope:

Uses electrons for high-resolution imaging.

Measurements:

Astronomical Unit (AU):

- Used to measure distances within our solar system.

  $1AU = 1.496 \times 10^{11}$m

Light year (Ly):

- The distance that light travels in one year.

  $1Ly = 9.46 \times 10^{15}$m

Parsec:

- Another unit used in astronomy to measure large distances.

  $1parsec = 3.08 \times 10^{16}$m

- Size of proton (P^+) = 10^{-15}m (1 fermi)
- Size of an atom, = 10^{-10} m (1A°)
- Radius of earth, $R_{Earth} = 10^7$m
- Distance of boundary of Observable Universe = 10^{26} m

2 Conversion of Units

- nu = constant, n = measure value of Physical quantity, u = unit of that Physical quantity.

  $n \propto \frac{1}{u}$

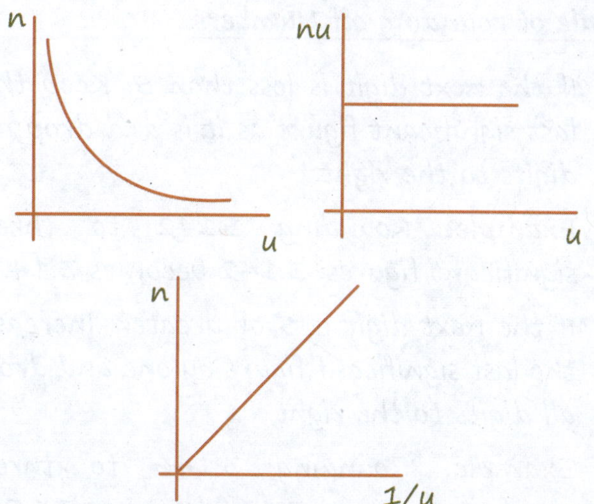

→ Only use to find value of physical quantity in new system of unit, if value is know in one unit.

Measurement of Mass & Time

- 1 amu = $\frac{1}{12}$ Mass of C^{12} atom
- 1 amu = 1.66×10^{-27} kg
- e^- mass = 9.1×10^{-31} kg
- Earth mass:- 10^{25} kg
- Observable Universe $=10^{55}$ kg.

Q.1. Convert 18 km/hr in m/s.

Sol. $n_1u_1 = n_2u_2$

$18 \text{ km/hr} = n_2 \text{ m/s}$

$$\frac{18 \times 10^3 \text{m}}{60 \times 60 \text{s}} = n_2 \text{ m/s}$$

$$n_2 = 18 \times \frac{5}{18} = 5$$

Q.2. If unit of length is y m in new system of unit then find value of x m^2 area in new system of unit.

Sol. un = cost

$n_1u_1 = n_2u_2$

$xm^2 = n_2y^2m^2$

$$n_2 = \frac{x}{y^2}$$

3 SI System (International System of Units)

The SI (International System of Units) is a modern version of the metric system, which is used worldwide for measuring different things. It is based on seven fundamental units, called base units. Each base unit has a specific symbol and represents a fundamental quantity in science and everyday life. Here is a simple explanation of these base units, their symbols, and what they measure:

7 Base/Fundamental Units:-			
No.	Quantity	Unit	Symbol
1.	Length	Meter	m
2.	Mass	Kilogram	kg
3.	Time	Second	S
4.	Temperature	Kelvin	K
5.	Elec. Current	Ampere	A
6.	Luminous Intensity	Candela	cd
7.	Amt. of Sub^s	Mole	Mol
2 supplementary Units:-			
No.	Quantity	Unit	Symbol
1.	Plane Angle	radian	rad
2.	Solid Angle	Steradian	sr

Plane Angle

It is the ratio of the arc length (ds) to the radius (r) of the circle. It is measured in radians, which is a dimensionless unit.

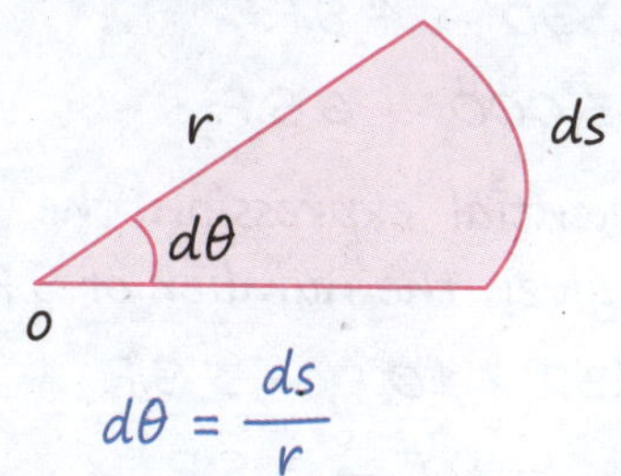

$$d\theta = \frac{ds}{r}$$

Solid Angle

It is the ratio of the area (dA) on the surface of a sphere to the square of the radius (r) of the sphere. It is measured in steradians (sr), which is also a dimensionless unit.

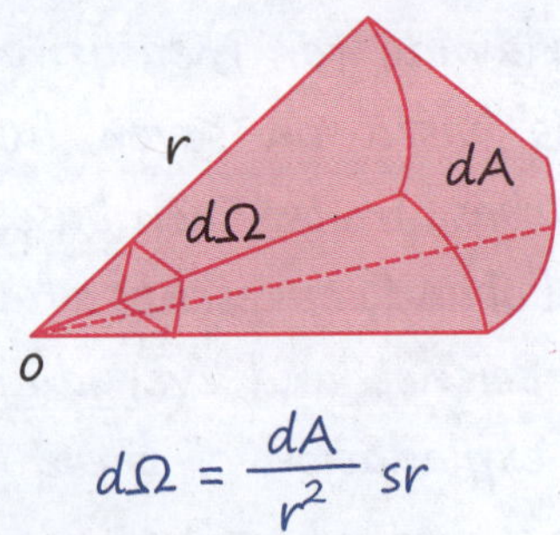

$$d\Omega = \frac{dA}{r^2}\text{ sr}$$

4 Significant Figures

The digits in a number that contribute to its precision and accuracy are known as significant figure.

Properties of significant figure:

- All non-zero digits are significant.

  eg:- 42.3 → 3 S.F.

  243.4 → 4 S.F.

- Zero between two non-zero digits is significant.

  eg:- 4.03 → 3 S.F.

  243.4 → 4 S.F.

- Leading Zero or zeros placed to left are never significant.

  eg:- 0.543 → 3 S.F.

  0.006 → 1 S.F.

- Trailing zeros or zero placed to the right of the number after decimal are significant.

  eg:- 4.330 → 4 S.F.

  343.000 → 6 S.F.

- In exponential expression the numerical portion given the number of S.F.

  eg:- 1.32×10^{-2} → 3 S.F.

  1.32×10^{4} → 3 S.F.

Rule of rounding off Numbers

- If the next digit is less than 5: Keep the last significant figure as it is and drop all digits to the right.

  Example: Rounding 3.142 to three significant figures: 3.142 becomes 3.14.

- If the next digit is 5 or greater: Increase the last significant figure by one and drop all digits to the right.

  Example: Rounding 3.146 to three significant figures: 3.146 becomes 3.15.

Addition & Substraction:-

Final result should have same no. of decimal placed as that of original no. with minimum no. of decimal places.

3.1421
0.241
0.09
3.4731
Ans:- 3.47.

9.99
− 0.00 99
9 · 9801
Ans: 9.98
Ans:- 9.98

Multiplication & Division:-

The no. of S.F. equals the smallest no. of S.F. in any of the original no.

51.028
× 1.31
66.84668
Ans:- 66.8

$\frac{25.5}{5}$ Ans:- 66.8

5 Dimensional Analysis

Dimension of physical quantity are power to which units of base quantity are raised.

eg:- $[M]^a\ [L]^b\ [T]^c\ [A]^a\ [K]^C$

Dimension of some physical quantity

- Mass → M
- Length → L
- Time → T

- Velocity → LT^{-1}
- Acceleration → LT^{-2}
- Force → MLT^{-2}
- Energy → ML^2T^{-2}
- Power → ML^2T^{-3}
- Force gradiant → MT^{-2}

-:Applications:-

1. **Checking the Correctness of various formulae:-**

   eg:- Z = A +B

   [Z] = [A] = [B]

2. **Conversion of one system of unit to another.**

   $n_1 U_1 = n_2 U_2$

   eg:- $n_1 [M_1^A L_1^B T_1^C] = n_2[M_2^A L_2^B T_2^C]$

   $$n_1 = n_2 \left[\frac{M_2}{M_1}\right]^A \left[\frac{L_2}{L_1}\right]^B \left[\frac{T_2}{T_1}\right]^C$$

MR*

Different physical quantity ka dimension nikalne ke liye force and energy ka dimension yad rakna hai. Avi tension nahi lena aage ke chapter ke sath yad hota jayga.

3. **Creation of new formule**

Using the principle of homogeneity of dimension, we can drive the formula of physical quantity:

Assume the physical quantity Y depends on a, b, and c, which have dimensions [a], [b], and (c) respectively. Write the equation in the form:

$$Y = k \cdot a^x \cdot b^y \cdot c^z$$

Equate the Dimensions:

Substitute the dimensional formulas of each quantity into the equation and equate the dimensions on both sides.

Solve for the Exponents: Solve the resulting system of equations to find the values of x, y, and z.

Q.3. If momentum (p), area (A) and time (t) are to be fundamental quantities, energy has the dimensioanl formula

(a) $[p^1A^{-1}t^{-1}]$ (b) $[p^2A^1t^1]$

(c) $[p^1A^{-1/2}t^1]$ (d) $[p^1A^{-1/2}t^{-1}]$

Sol. Fundamental quantites are momentum (p), area (A) and time (t). therefore

$E = kp^xA^yt^z$

Substituting the dimensions

$[ML^2T^{-2}] = [ML\,T^{-1}]^a\,[M^0L^2T^0]^b\,[M^0L^0T]^c$

Comparing power

$a + 2b = 2$

$a = 1$

$-a + c = -2$

Solving these equations

$a = 1,\ b = \frac{1}{2},\ c = -1$

Hence, $E = pA^{1/2}T^{-1}$

Note:

- Formula of force to find dimension of different physical quantity

$F = G\frac{m_1m_2}{r^2}$, $F = \frac{q_1q_2}{4\pi\varepsilon_0 r^2}$

$F = Kx$, $F = \frac{\mu_0 I_1 I_2 d}{4\pi r}$

$F = qE$

$F = qvB$

$F = 6\pi\eta rv$

$F = Sl$

+ Formula of energy to find dimn of different physical quantity

$E = hf$

$\frac{H}{t} = \frac{kA\Delta T}{l}$

$E = \frac{3}{2}k_BT$

Stress = γ × Strain

$E = \frac{Q^2}{2C}$

$Q = ms\Delta T$

$H = I^2Rt$

$Q = mL$

$E = \frac{1}{2}LI^2$

$PV = nRT$

6 Dimensional Formula

+ Pressure = stress = Young's modules $= ML^{-1}T^{-2}$
+ Work = Energy = Torque = $M L^2 T^{-2}$
+ Power P = $M L^2 T^{-3}$
+ Gravitational constant G = $M^{-1} L^3 T^{-2}$
+ Force constant = Spring constant = $M T^{-2}$
+ Coefficient of viscosity = $ML^{-1} T^{-1}$
+ Latent heat L = $L^2 T^{-2}$
+ Electric potential = $\frac{P}{I} = M L^2 T^{-3} A^{-1}$
+ Resistance = $\sqrt{\frac{\mu_0}{\varepsilon_0}} = M L^2 T^{-3} A^{-2}$
+ Capacitance = $M^{-1} L^{-2} T^4 A^2$
+ Permittivity $\varepsilon_0 = M^{-1} L^{-3} T^4 A^2$
+ Angular momentum = planck's constant $= M^1 L^2 T^{-1}$

Time Period:-

$$T \alpha \sqrt{\frac{L}{g}} \alpha \sqrt{\frac{M}{k}} \alpha \sqrt{\frac{R}{g}}$$

$$\frac{L}{R} = RC = \sqrt{LC}$$

MR*

Resistance = $R = \omega L = \frac{1}{\omega C}$

$\omega = \frac{2\pi}{T}$

Time = $\frac{L}{R} = \sqrt{LC} = RC$

Dimensionless Quantities:-

+ Strain
+ Refractive index
+ Relative density
+ Plane Angle
+ Solid Angle
+ Poissons ratio
+ Exponential function
+ Trigonometry function
+ Relative permittivity
+ Pure number
+ Efficiency
+ Length gratiant
+ Coef. of friction

MR*

Pressure = Stress = Young modulus

= Bulk modulus

$= \frac{1}{2}$ strain × stress = modulus of regitity

$= \frac{B^2}{2\mu_0} = \frac{1}{2}\varepsilon_0 E^2$ = energy density $= \frac{nRT}{V}$

dimensionally addition, substraction ko equal le ke solve karte hai.

Kisi be dimensionless function ya quantity ko one likh sakte hai.

Q.4. If velocity V = Ax + Bt + C find dimension of A, B and C.

MR*

Sol. V = Ax = Bt = C

$$A = \frac{V}{x} = T^{-1}$$

$$B = \frac{V}{t} = LT^{-2}$$

$$C = V = LT^{-1}$$

Q.5. Force $F = \alpha e^{-\beta t}$ then find dimension of α and β.

Sol. $F = \alpha$ ⟶ $\alpha = MLT^{-2}$ | $\beta t = 1$ ⟶ $\beta = T^{-1}$

Q.6. Acceleration $a = \alpha t + \frac{\beta}{t - \delta}$ find dimension of α, β and δ.

Sol. $a = \alpha t = \frac{\beta}{t = \delta}$ ← MR* Ka feel

$$\Rightarrow \alpha = \frac{a}{t} = LT^{-3}$$

$$\Rightarrow \delta = t$$

$$\Rightarrow a = \frac{\beta}{t}$$

$$\beta = \alpha t = LT^{-1}$$

Q.7. Fill in the blanks with correct statement, according to given statement

Dimension	(1)	(2)	(c) A physical quantity have dimension	(d) A physical quantity does not have dimension
Unit	(a) A physical quantity have unit	(b) A physical quantity does not have unit	(3)	(4) ....

MR*

Ans. (1) May have dimension/may be dimensionless

(2) Must be dimensionless/does not have dimension

(3) Must have unit

(4) May or may not have unit.

Q.8. Fill in the blanks with correct statement, according to given statement

Physically correctness	(1)	(2)	(c) Equation is physically wrong	(d) Equation is physically correct
Dimensional correctness	(a) Equation is dimensional wrong	(b) Equation is dimensional correct	(3)	(4) ....

MR*

Ans. (1) Must be physically wrong

(2) May or may not physically correct

(3) May or may be dimensionally correct

(4) Must be dimensionally correct.

$$S_{nth} = u + \frac{a}{2}(2n - 1)$$

(S_{nth} → dimensionally correct because it is displacement in one sec.)

Q.9. If force, acceleration and time taken as fundamental physical quantity then find dimension of energy?

(a) $F^2A^{-1}T$ (b) FAT^2

(c) $F^{-1}AT^{-2}$ (d) $FA^{-1}T$

MR*

$E(ML^2T^{-2}) \rightarrow$ Mass ka dimension force hi dega ek mass energy me hai to F^1 hona chahiye.

Now L ka square hai ek length force dega ek acceleration hence A^1 hona chahiye.

Q.10. Planks constant (h), speed of light (c), gravitational constant (G) taken as fundamental quantity then dimension of length in terms of them.

(a) $\frac{\sqrt{hG}}{c^{3/2}}$ (b) $\sqrt{\frac{hc}{G}}$

(c) $\frac{\sqrt{hG}}{c^{5/2}}$ (d) $\frac{\sqrt{Gc}}{h^{3/2}}$

MR*

$M^0T^0l = h^xc^YG^Z$

We need dimension of length, then mass should be cancell out by arranging h, c and G. c me to mass hai nahi; $h \rightarrow ML^2T^{-1}$ and $G = M^{-1}L^3T^{-2}$ to h and G ko multiply karne se mass kat jayga. Hence option (b) and (d) wrong ho gaya. Now option (a) and (c) dono me root hai to root laga ke sirf length ka dimension likho phir c se divide kar ke ek length (L^1) sirf rakho.

Q.11. Dimension of critical velocity V of liquid flowing through the tube are expressed as $\eta^x\delta^yr^z$, where η is coefficient of viscosity, δ is density of liquid and r is radius of the tube then the value of x, y and z are given by.

(a) 1, 1, 1 (b) 1, −1, −1

(c) −1, −1, 1 (d) −1, −1, −1

MR*

Velocity me mass hai nahi to η, δ and r ko arrange kar velocity lena hai hence mass cancell, radius me bhi mass nahi hai, $\delta = ML^{-3}$ and $\eta = ML^{-1}T^{-3}$

δ and η divide karne se mass kat jayga to ek ka power positive ek ka negative hona chahiya.

Q.12. If energy (E), velocity (V) and time (T) are chosen as the fundamental quantities the dimensional formula of surface tension will be

(a) EV^2T^2 (b) EV^2T^{-2}

(c) $EV^{-2}T^{-2}$ (d) $E^{-2}V^2T^2$

MR*

MR* → Surface tension (MT^{-2}) Ramlal yaha length nahi to length katne ka socho. Sirf (c) me length kat ho raha hai.

Limitation of dimensional analysis:

(1) It is not use to derive dimensionless physicall quantity and constant.

(2) This can not decide weather the give quantity is vector or a scalar.

(3) It can not be use to derive an equation involving more than three physicall quantity.

(4) It can not derive dimensionless function having $\sin\theta$, $\cos\theta$, e^x etc.

(5) Can not use if one quantity depends on two other quantity having same dimension.

(6) It can not derive equation which cantain +ve and −ve terms.

7 Instruments

- Least Count (L.C) :-
- Smallest value that

mm Scale	Vernier Scale	Screw Gauge
↓	↓	↓
1mm	0.1mm	0.01mm

Vernier calipers:-

- L. C. = 1MSD - 1VSD
- If nVsD Coincides with (n-1) MSD then:-

  (n-1) MSD = nVSD

  $1 \text{ VSD} = \frac{n-1}{n} \text{ MSD}$

- $LC = 1 \text{ MSD} - \frac{n-1}{n} \text{ MSD} = \frac{1\text{MSD}}{n}$
- Total Reading = 1 MSR + coinciding VSR × LC

{MSD → Main Scale division
V.S.D → Vernier scale devision}

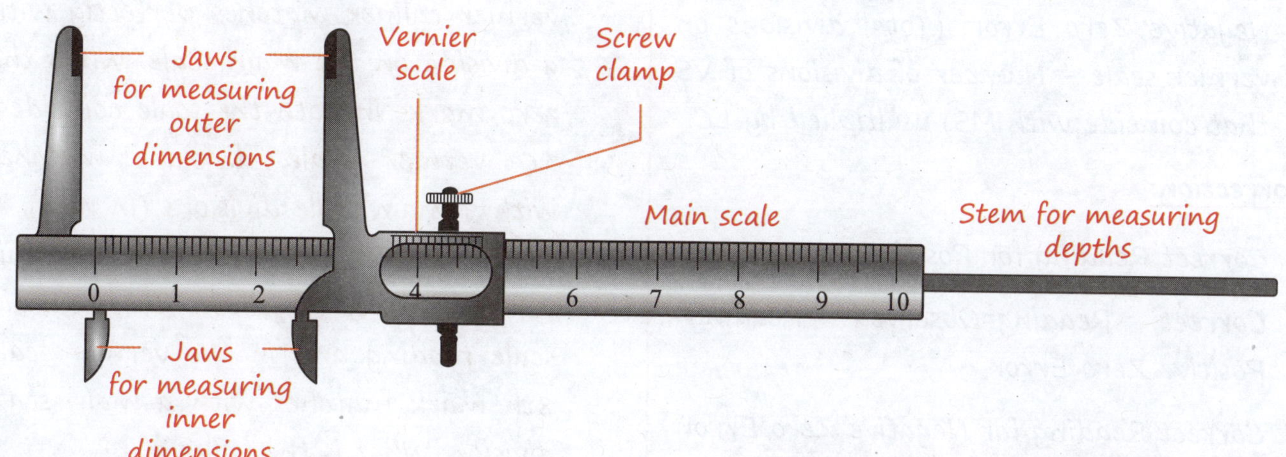

Screw gauge:-

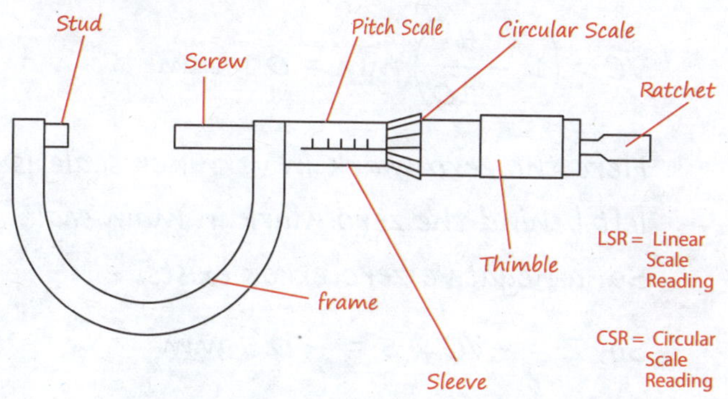

- $\text{Pitch} = \frac{\text{MSR}}{\text{no. of rotation}}$
- $\text{L.C.} = \frac{\text{Pitch}}{\text{Total no. of division on Circular Scale}}$
- Total Reading = 1 LSR + CSR × LC

{MSR → Main Scale division
CSR → Circular scale reading}

Measured length	Used instrument
1.56 cm	vernier calliper
6.8 cm	metre scale
8.96 mm	screw gauge
9.812 cm	screw gaye
8.3 mm	vernier calliper

Accuracy: It is the measure of how close the measured value is to the true value. Closness of measured and true value.

Precision: It tells us to what resolution or limit the quantity is measured.

Zero error:

Types:

- Positive Zero Error: When the zero of the vernier scale is to the right of the zero of the main scale.

- Negative Zero Error: When the zero of the vernier scale is to the left of the zero of the main scale.

Calculation:

- Positive Zero Error: Number of divisions of the vernier scale (VS) that coincide with the main scale (MS) multiplied by the least count (LC).
- Negative Zero Error: (Total divisions on vernier scale - Number of divisions of VS that coincide with MS) multiplied by LC.

Correction:

- Correct Reading for Positive Zero Error:
- Correct Reading=Observed Reading-Positive Zero Error
- Correct Reading for Negative Zero Error: Correct Reading = Observed Reading - Negative Zero Error

Q.13. In a vernier caliper, when the jaws are touching each other the zero marks in both the scale coincides, 10 vernier scale divisions matches with 9 main scale divisions (in mm). A rod of length L is measured with this vernier caliper. if It shows 55 in main scale reading and in the vernier scale 8th mark matches with a main scale division. What is the value of L?

Sol. The verniner constat,

$$VC = \left(1 - \frac{9}{10}\right) \text{mm} = 0.1 \text{ mm}$$

Here the zero mark in vernier scale matches with the zero mark in main scale. So, no zero error exists.

So, $Z = -VC \times 0 = -0$ mm

$\therefore L = MSR + xVC - Z$

$= 55 + (8 \times 0.1) - (-0) = 55.8$ mm

Q.14 In a vernier caliper, when the jaws are touching each other the zero mark on the vernier scale is left behind slightly of the zero mark on the main scale. At that moment the 5th division mark on vernier caliper matches perfectly with a division on the main scale. When the zero marks in both the scale coincides, 10 vernier scale divisions matches with 9 main scale divisions (in mm). A rod of length l is measured with this vernier caliper. It shows 55 in main scale reading and in the vernier scale 8th mark matches with a main scale division. What is the value of L?

Sol. The verniner constat,

$$VC = \left(1 - \frac{9}{10}\right) \text{mm} = 0.1 \text{ mm}$$

Here the zero mark in verniner scale is left behind the zero mark in main scale. So, a negative zero error exists.

So, $Z = -VC \times 5 = -0.5$ mm

$\therefore L = MSR + xVC - Z$

$= 55 + (8 \times 0.1) - (-0.5) = 56.3$ mm

Q.15. If true value of length is 6.57 m then which of the following reading is most accurate and most precise.

(a) 6.52 m (b) 6.61 m

(c) 6.513 m (d) 6.68 m

Sol. Most accurate (b), most precise (c)

8 Error in Measurement

Difference between true value & measured value of a quantity

Systematic Errors
Errors which tend to occur only in one direction, either positive or negative

Random Errors
Irregular and at random in magnitude & direction

Instrumental
Due to inbuilt defect of measuring instrument

Experimental
Limitation in experimental technique

Personal
Due to individual bias, Lack of proper setting of apparatus

Absolute Error:-

$$\Delta a = |a_i - a_{mean}|$$

$$a_{mean} = \frac{a_1 + a_2 \ldots + an}{n}$$

- Always positive
- Unit and dimension same as physical quantity
- Least count error can be taken as absolute error
- It can not tell about accuracy of measurment

Mean Absolute error

$$\Delta a_{mean} = \frac{\Delta a_1 + \Delta a_2 \ldots + \Delta a_n}{n}$$

Relative Error:-

$$\frac{\Delta a_{mean}}{a_{mean}}$$

$$\text{Percentage Error} = \frac{\Delta a_{mean}}{a_{mean}} \times 100$$

- Unit and dimension less
- It tell about accuracy of measurment
- Random error can be decreased by increasing no of obserbation.

$nx = cost^n$, n = no. of obserbation,

x = Random error

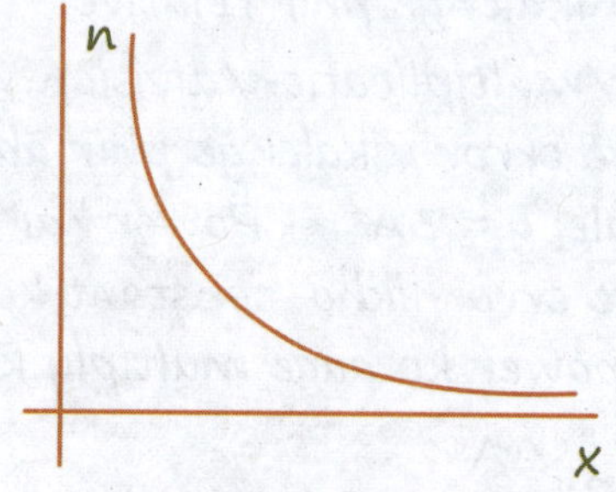

- In 5 reading random error is 3% and systematic error is 4%. If we increased no. of observation to 30 then random error 1/2% and systematic error remains 4%.

Combination of errors

Operations	Formula Z	Absolute error ΔZ	Relative error ΔZ/Z	Percentage error 100 × ΔZ/Z
Sum	$A + B$	$\Delta A + \Delta B$	$\frac{\Delta A + \Delta B}{A + B}$	$\frac{\Delta A + \Delta B}{A + B} \times 100$
Difference	$A - B$	$\Delta A + \Delta B$	$\frac{\Delta A + \Delta B}{A - B}$	$\frac{\Delta A + \Delta B}{A - B} \times 100$
Multiplication	$A \times B$	$A\Delta B + B\Delta A$	$\frac{\Delta A}{A} + \frac{\Delta B}{B}$	$\left(\frac{\Delta A}{A} + \frac{\Delta B}{B}\right) \times 100$
Division	$\frac{A}{B}$	$\frac{B\Delta A + A\Delta B}{B^2}$	$\frac{\Delta A}{A} + \frac{\Delta B}{B}$	$\left(\frac{\Delta A}{A} + \frac{\Delta B}{B}\right) \times 100$
Power	A^n	$nA^{n-1}\,\Delta A$	$n\frac{\Delta A}{A}$	$n\frac{\Delta A}{A} \times 100$
Root	$A^{1/n}$	$\frac{1}{n}A^{((1/n)-1)}\Delta A$	$\frac{1}{n}\frac{\Delta A}{A}$	$\frac{1}{n}\frac{\Delta A}{A} \times 100$

General Rule:-

$$Z = \frac{A^p B^q}{C^r},$$

Then max. frac^n relative error in Z will be:-

$$\frac{\Delta Z}{Z} = p\frac{\Delta A}{A} + q\frac{\Delta B}{B} + r\frac{\Delta C}{C}$$

MR*

- Addition/Substration me pahle absolute error nikalenge phir relative.
- Power/multiplication/division me pahle relative error nikalenge phir absolute.

Example: $y = 3A^2$ ← Power hai to direct relative error likho, constant ko remove karo, power ko aage multiply kar do.

$$\frac{\Delta y}{y} = 2\times\frac{\Delta A}{A}$$

Example:

$$y = \frac{2A^4\sqrt{B}}{C^3}$$

$$\frac{\Delta y}{y} = 4\times\frac{\Delta A}{A} + \frac{1}{2}\times\frac{\Delta B}{B} + 3\times\frac{\Delta C}{C}$$

Q.16. In an experiment the angles are required to be measured using an instrument. 29 divisions of the main scale exactly coincide with the 30 divisions of the vernier scale. If the smallest division of the main scale is half-a-degree (= 0.5°) then the least count of the instrument is:

(a) one minute (b) half minute

(c) one degree (d) half degree

Sol. MSD = 0.5°

30 VSD = 29 MSD

Least Count

= Length of 1 main scale division / No. of divisions of Vernier scale

= 0.5° / 30

= 0.5 × 60min / 30

= 1 min

Q.17. A vernier callipers has 1 mm marks on the main scale. It has 20 equal divisions on the Vernier scale which match with 16 main scale divisions. For this Vernier callipers, the least count

(a) 0.02 mm (b) 0.05 mm

(c) 0.1 mm (d) 0.2 mm

Sol. MSD = 1 mm

20 VSD = 16 MSD

⇒ VSD = 16/20 MSD

= 0.8 MSD = 0.8 mm

Least Count = MSD−VSD

= 1−0.8 = 0.2 mm

Q.18. If the error in the measurement of area of sphere is 3% then find percentage error in measurement of volume of sphere

Sol. $A = 4\pi r^2 \qquad V = \frac{4}{3}\pi r^3$

$$\frac{\Delta A}{A} = 2\frac{\Delta r}{r} \qquad \text{...(i)}$$

$$\frac{\Delta V}{V} = 3\frac{\Delta r}{r} \qquad \text{...(ii)}$$

(ii)/(i) $\frac{\Delta V}{V} \Big/ \frac{\Delta A}{A} = \frac{3}{2}$

$$\frac{\Delta V}{V} \times 100 = \frac{3}{2}\frac{\Delta A}{A} \times 100 = \frac{3}{2} \times 3$$

= 4.5%

Q.19. If $T = 2\pi\sqrt{\frac{\ell}{g}}$ then find percentage error in measurement of acceleration due to gravity.

Sol. Ignore constant

$$T^2 = \frac{\ell}{g},\ g = \frac{\ell}{T^2}$$

$$\frac{\Delta g}{g} = \frac{\Delta \ell}{\ell} + 2\frac{\Delta T}{T}$$

MR*

'खुल जायेंगे सभी रास्ते,
तू रुकावटों से लड़ तो सही।
सब होगा हासिल,
तू अपनी जिद्द पर अड़ तो सही।।'

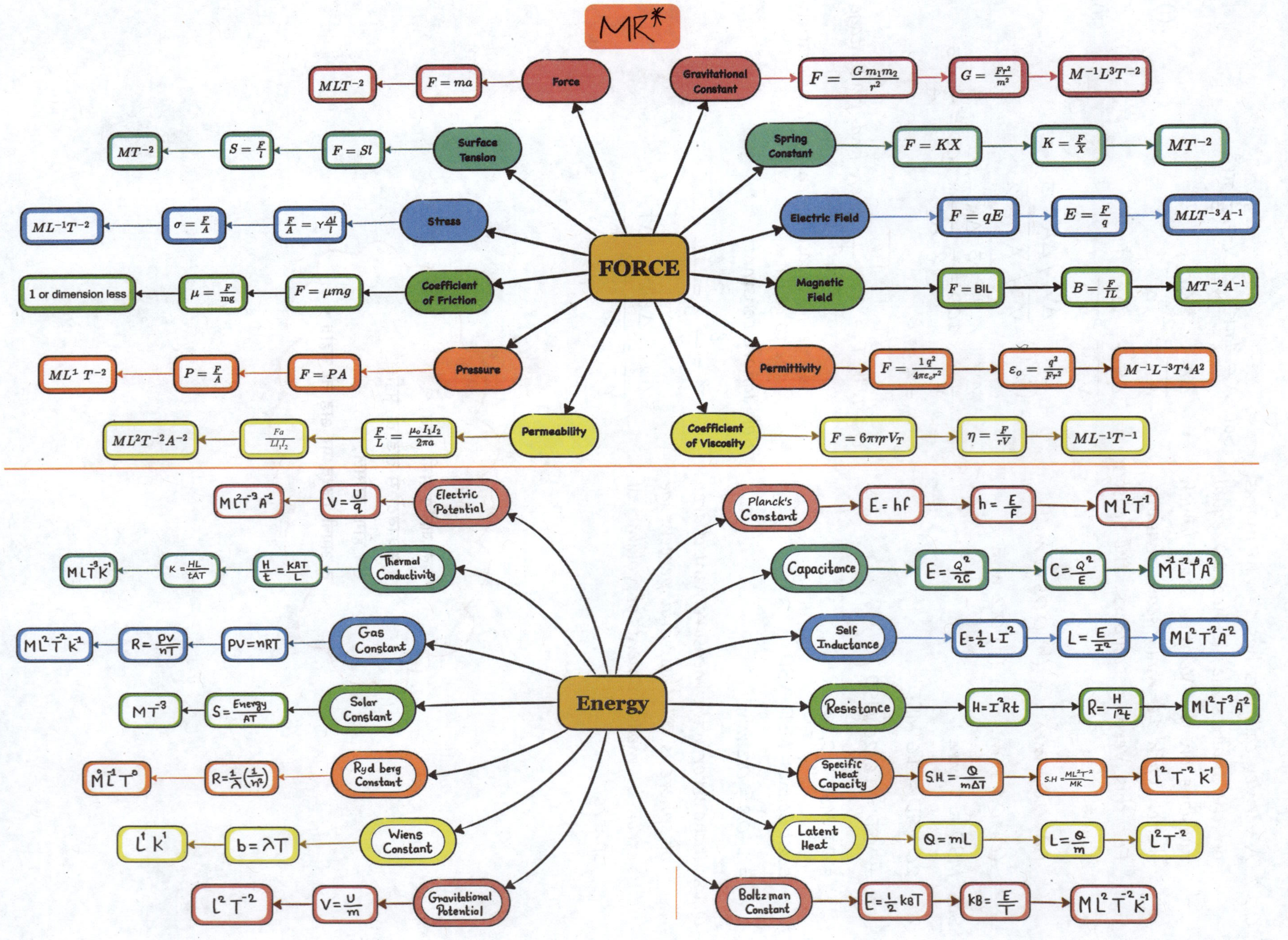
MR*
FORCE
Force
$F = ma$
MLT^{-2}
Gravitational Constant
$F = \frac{G m_1 m_2}{r^2}$
$G = \frac{Fr^2}{m^2}$
$M^{-1}L^3T^{-2}$
Surface Tension
$F = Sl$
$S = \frac{F}{l}$
MT^{-2}
Spring Constant
$F = KX$
$K = \frac{F}{X}$
MT^{-2}
Stress
$\frac{F}{A} = Y\frac{\Delta l}{l}$
$\sigma = \frac{F}{A}$
$ML^{-1}T^{-2}$
Electric Field
$F = qE$
$E = \frac{F}{q}$
$MLT^{-3}A^{-1}$
Coefficient of Friction
$F = \mu mg$
$\mu = \frac{F}{mg}$
1 or dimension less
Magnetic Field
$F = BIL$
$B = \frac{F}{IL}$
$MT^{-2}A^{-1}$
Pressure
$F = PA$
$P = \frac{F}{A}$
$ML^{1} T^{-2}$
Permittivity
$F = \frac{1}{4\pi\epsilon_o}\frac{q^2}{r^2}$
$\epsilon_o = \frac{q^2}{Fr^2}$
$M^{-1}L^{-3}T^4A^2$
Permeability
$\frac{F}{L} = \frac{\mu_o I_1 I_2}{2\pi a}$
$\frac{Fa}{LI_1I_2}$
$ML^2T^{-2}A^{-2}$
Coefficient of Viscosity
$F = 6\pi\eta r V_T$
$\eta = \frac{F}{rV}$
$ML^{-1}T^{-1}$
Energy
Electric Potential
$V = \frac{U}{q}$
$ML^2T^{-3}A^{-1}$
Planck's Constant
$E = hf$
$h = \frac{E}{f}$
ML^2T^{-1}
Thermal Conductivity
$\frac{H}{t} = \frac{KAT}{L}$
$K = \frac{HL}{tAT}$
$MLT^{-3}K^{-1}$
Capacitance
$E = \frac{Q^2}{2C}$
$C = \frac{Q^2}{E}$
$M^{-1}L^{-2}T^4A^2$
Gas Constant
$PV = nRT$
$R = \frac{PV}{nT}$
$ML^2T^{-2}K^{-1}$
Self Inductance
$E = \frac{1}{2}LI^2$
$L = \frac{E}{I^2}$
$ML^2T^{-2}A^{-2}$
Solar Constant
$S = \frac{Energy}{AT}$
MT^{-3}
Resistance
$H = I^2Rt$
$R = \frac{H}{I^2t}$
$ML^2T^{-3}A^{-2}$
Ryd berg Constant
$R = \frac{1}{\lambda}\left(\frac{1}{n^2}\right)$
$M^0L^{-1}T^0$
Specific Heat Capacity
$S.H = \frac{Q}{m\Delta T}$
$S.H = \frac{ML^2T^{-2}}{MK}$
$L^2T^{-2}K^{-1}$
Wiens Constant
$b = \lambda T$
L^1K^1
Latent Heat
$Q = mL$
$L = \frac{Q}{m}$
L^2T^{-2}
Gravitational Potential
$V = \frac{U}{m}$
L^2T^{-2}
Boltz man Constant
$E = \frac{1}{2}k_BT$
$k_B = \frac{E}{T}$
$ML^2T^{-2}K^{-1}$

4 Motion in 1-D

1 Motion

Motion in one dimension, also known as linear motion, is the movement of an object along a straight line. In this type of motion, only one coordinate out of three is needed to describe the position of the object.

Observer & Frame of Reference

- Observer → who takes observation and from where. it takes is called frame of reference.
- Observer always assume to be at rest.
- Nothing is at absolute rest or in absolute motion.
- Agar koi Gadhe pr baitha hai toh

  Gadha:- Frame of Ref. Uske upar joh baitha hoga woh observer!

Distance	Displacement
✦ Total Path length	✦ Shortest Path b/w initial and final position
✦ Scalar, Struggle	✦ Always straight line
✦ Can't decrease with time	✦ Vector, success
✦ Always positive	✦ Direction - From initial to final position
✦ Depends on path taken	✦ Can decrease with time
✦ Both have same unit and dimension	✦ May be +ve or −ve
	✦ Does not depends on path

- If we know only initial and final position then we can't calculate distance but can find displacement.
- If initial position (x_1, y_1, z_1) and final position (x_2, y_2, z_2) then displacement $= \sqrt{(x_2-x_1)^2 + (y_2-y_1)^2 + (z_2-z_1)^2}$

$\Rightarrow$ distance $\geq$ |displacement|

2 Distance & Displacement on Circular Path

- $\text{Disp}^m_{(x)} = 2R\sin\left(\frac{\theta}{2}\right)$ (O, θ, x, R)
- Arc = $\text{dist}^n = R\theta$ (Arc, R, θ)

Displacement (2nd floor)	Disp^m must be zero	Disp^m may or may not be zero	If disp^m is zero	If disp^m is not equal to zero then:-
	⇑	⇑	⇓	⇓
Distance (1st floor)	If dist^n is zero	If dist^n is not equal to zero then	Dist^n may or may not be zero	Dist^n must not equal to zero

Challenger Question

There is a cubical room. One insect is moving from one corner to other body. Diagonal, then find minimum distance

The Insect Can Fly:

- When the insect can fly, it can take the shortest possible path, which is the direct diagonal across the cube from one corner to the opposite corner.

Body Diagonal = $\sqrt{L^2 + L^2 + L^2} = \sqrt{3}L$

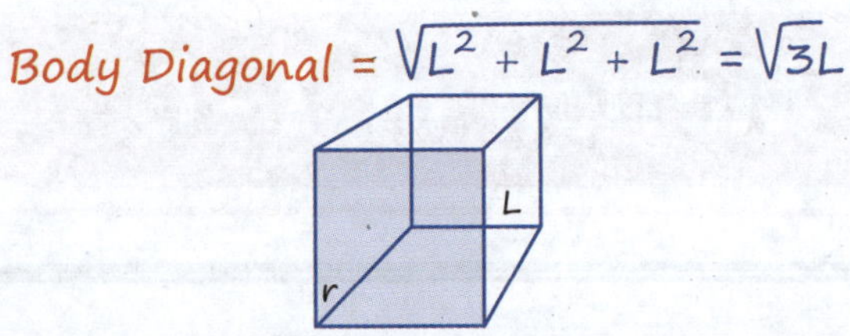

The Insect Cannot Fly:

- When the insect cannot fly, it has to travel along the surfaces of the cube.

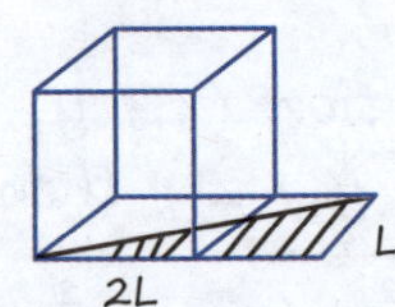

The insect will move along one face and then another. The shortest path on the surface of the cube

$$= \sqrt{L^2 + (2L)^2} = \sqrt{5}\,L$$

3 Average Speed (How fast in an interval not at instant)

$$V_{avg} = \frac{\text{Total distance}}{\text{Total time}}$$

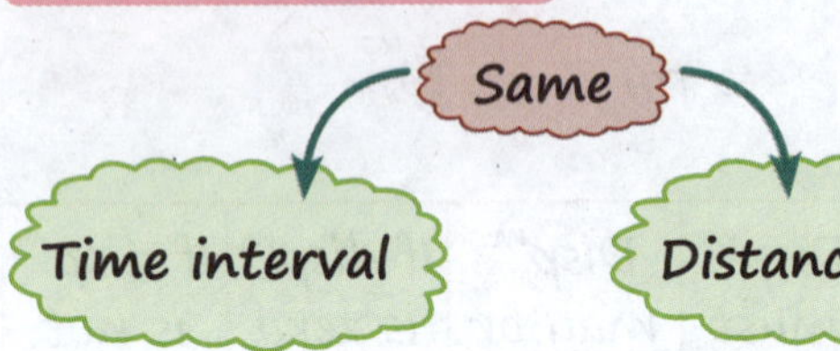

$$V_{avg} = \frac{V_1 t_1 + V_2 t_2}{t_1 + t_2}$$

When $t_1 = t_2 = t$

$$V_{avg} = \frac{V_1 + V_2}{2}$$

$$V_{avg} = \frac{X_1 + X_2}{\frac{X_1}{V_1} + \frac{X_2}{V_2}}$$

When $x_1 = x_2 = x$

$$V_{avg} = \frac{2V_1 V_2}{V_1 + V_2}$$

4 SPEED ((How Fast) Scalar, unit: m/s, only magnitude.)

- Instantaneous

$$v_{inst} = \frac{dx}{dt}$$

- Average

$$V_{avg} = \frac{\int s.\,dt}{\int dt}$$

$$\int \square = \frac{\int \square.\,dt}{\int dt}$$

For Uniform motion :- $s_{avg} = s_{inst.}$

Note:

By substituting the box (□) with a specific function, you can calculate the integral and apply it to find a quantity like average velocity.

5 VELOCITY (How fast and where)

Hum kitna Tez bhag rahe hai and kis direction me bhag rhe hai !

Inst.

$$\vec{V}_{inst.} = \frac{d\vec{x}}{dt}$$

Average

$$\vec{V}_{avg} = \frac{\int v.dt}{\int dt}$$

= Rate of change in position

= Slope of position time graph

= Inst. speed × direction

= How fast × where

On circular path, $V_{avg} = \frac{V\sin(\theta/2)}{(\theta/2)}$

$|\text{Avg speed}| \geq |\text{Avg Velocity}|$

$\text{Inst speed} = |\text{Inst Velocity}|$

Uniform motion

- Body moving with constant speed in fixed direction
- Uniform velocity, i.e., body travels equal displacement in equal interval of time.
- Acceleration zero
- Avg. velocity = Inst. velocity
- Must be straight line

Non-Uniform motion

- Velocity non-uniform: i.e., Body travels different displacements in equal intervals of time, or the same displacement in different intervals of time.
- Acceleration non-zero

- Velocity can be change by changing speed only or direction only or both
- In non-uniform speed may constant

Dimag me set feel ke sath.

- If velocity is uniform then → speed must be uniform.
  - Velocity = Speed + Direction = Constant
- If velocity is variable → Speed may or may not be variable
  - Velocity ko sirf direction change kar ke vary kar sakte hai
- If speed is uniform → then velocity may uniform
  - Direction ka nahi pata.
- If speed is variable → then velocity must be variable
- If avg. velocity is zero then avg. speed may or may not be zero.
- If avg. speed is zero, then avg. velocity must be zero.

6 Acceleration

Ye Motion Ka Feel Nai Hai ! Ye velocity me change ka feel hai.

- Accl^n opposite to motion is retardation.
- Negative acceleration does not mean retardation, retardation may be positive or negative.
- Per-sec velocity inject to body or per-sec velocity extract from body ka feel hai.
- Vector → direction of acceleration along change in velocity.

$$\vec{a}_{inst} = \frac{dv}{dt} = \frac{v.dv}{dx} = \frac{d^2x}{dt^2}$$

$$\vec{a}_{avg} = \frac{\vec{V_2} - \vec{V_1}}{\Delta t} = \frac{\int_{t_1}^{t_2} a_{ins}\, dt}{\int_{t_1}^{t_2} dt}$$

$\frac{d\vec{v}}{dt} = \vec{a}$ = The rate of change in velocity

$\left|\frac{d\vec{v}}{dt}\right| = |\vec{a}|$ = Magnitude of acc^n, $\frac{d|\vec{v}|}{dt}$ = Rate of change in speed.

u → , ← a	u → , → a	u → , a ↓ (90°)
$\theta = 180°$ $\vec{a}.\vec{u} = -ve$ speed↓ retardation Tangential acc^n	$\theta = 0°$ $\vec{a}.\vec{v} = +ve$ speed↑ Tangential acc^n	$\theta = 90°$ $\vec{a}.\vec{v} = 0$ at this instant only direction will change normal or centripetal acc^n

MR*

Bade aaram se

Uniform or constant non-zero acceleration.
Position (x) ∝ t^2
Velocity (v) ∝ t
Velocity v ∝ $\sqrt{x}$
If acceleration zero then velocity must be non zero constant

- approach to solve question

$\vec{a} = 0$	$\vec{a}$ = non zero constant	$\vec{a}$ = variable
$\vec{v}$ = constant uniform motion S = vt	Equation of motion applicable	Eq^n of motion is not applicable do integration or differentiation

MR*

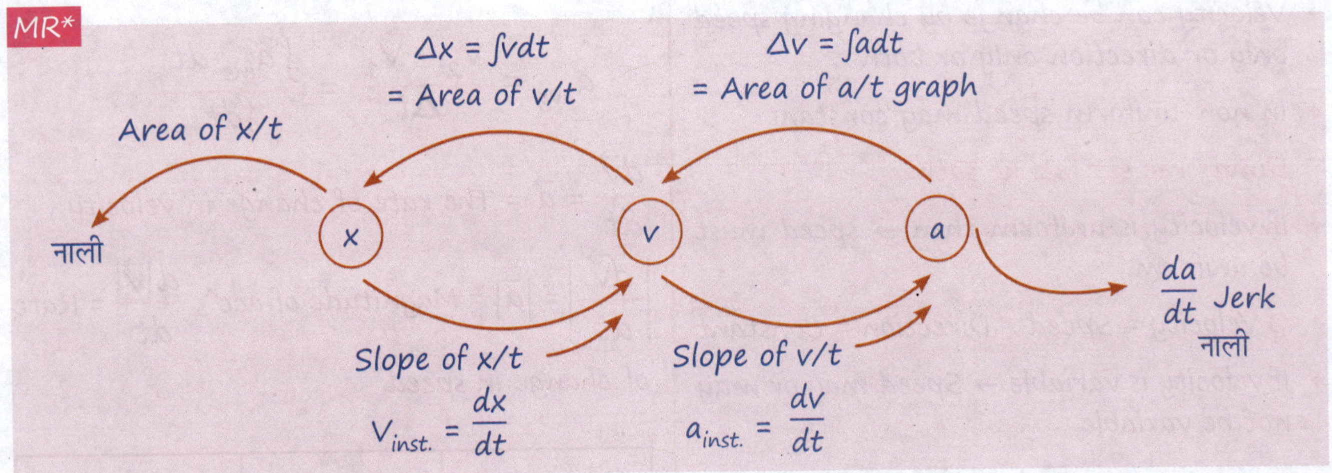

Q.1. Which of the following is correct for velocity and acceleration?

(a) Velocity increasing, acceleration decreasing

(b) Velocity decreasing, acceleration increasing

(c) Both increasing

(d) Both decreasing

(e) All of these

Sol. (e)

Q.2. If position $x = at^2 - bt^3$. Find the time when acceleration is zero ?

Sol. $x = at^2 - bt^3$

$$v = \frac{dx}{dt} = 2at - 3bt^2$$

$$a = \frac{dv}{dt} = 2a - 6bt = 0$$

$$2a = 6bt$$

$$t = \frac{a}{3b}$$

Q.3. If velocity $v \propto \sqrt{x}$ then which of the following function is correct for position time relation.

(a) $x \propto t$ (b) $x \propto t^2$

(c) $x \propto \sqrt{t}$ (d) $x \propto t^{3/2}$

Sol. MR* question me acc^n constant then option me acceleration constant option (b) i.e.,

$$\frac{dx}{dt} \propto t \Rightarrow \frac{d^2x}{dt} = \text{constand}$$

Q.4. If acceleration $a = \beta t^{3/2}$ then find velocity after time t if intial velocity is u.

Sol. Equation of motion is not valid

$$a = \frac{dv}{dt} = \beta t^{3/2}$$

$$\int_u^v dv = \beta \int t^{3/2}\, dt$$

$$v - u = \frac{\beta t^{5/2}}{5/2}$$

Q.5. If acceleration of object $a = \beta x^2$ then find velocity after x displacement, if initial velocity was zero.

Sol. $a = v\dfrac{dv}{dx} = \beta x^2$

$$\int_0^v v\,dv = \int_0^x \beta x^2 dx$$

$$\frac{v^2}{2} = \frac{\beta x^3}{3}$$

$$V = \sqrt{\frac{2\beta x^3}{3}}$$

Mr special*

Majduri se duri MR* hai jaruri

Position Ke Formula Mein Time ke dono term ko dekho agar dono term

+ve/+ve ya –ve/–ve sign rakhta hai toh woh U-turn Nai lenge ya $dist^n = |disp^m|$ agar sign +ve/-ve Rahi toh U-turn lenge aur distance $\neq |disp^m|$

yaad rahe 1-D mein U-turn keliye rukhna hoga (v = o) $\therefore dist^n \neq disp^n$

Note: To calculate $dist^n$, $disp^m$ from x-t eq^n:

- Moving Frame se body ko drop/release karne pr frame ka velocity share hojata hai but $accl^n$ nai!

Ex. $x = t^2 - 4t + 8$ then take v-t graph, plot it using "v" eq^n which we"ll get by differentiating "x" eq^n & then put time given from t_1 & t_2 & see graph calculate $dist^n$/ $disp^m$.

7 Motion with constant acceleration

$$\vec{v} = \vec{u} + \vec{a}t$$

$$v^2 - u^2 = 2\vec{a}\vec{s}$$

$$\vec{s} = \vec{x}_f - \vec{x}_c = \vec{u}t + \frac{1}{2}\vec{a}t^2$$

$$\vec{V}_{Avg} = \frac{\vec{v} + \vec{u}}{2} \qquad S = \frac{\vec{u} + \vec{v}}{2} \times t$$

$$S_{nth} = u + \frac{a}{2}(2n - 1)$$

$$\frac{S_{nth}}{S_n} = \frac{2n-1}{n^2} = \frac{2}{n} - \frac{1}{n^2}$$

Q.6. If velocity of object $V = \sqrt{25 - 8x}$ then find velocity and acceleration.

Sol. Acceleration is constant then compare velocity with 3rd equation of motion

$v^2 = 25 - 8x$ or $v^2 = u^2 + 2ax$

$u^2 = 25$ $\qquad -8x = 2ax$

$u = 5$ m/s $\qquad 2a = -8$

$a = -4$ m/s^2

Note:-

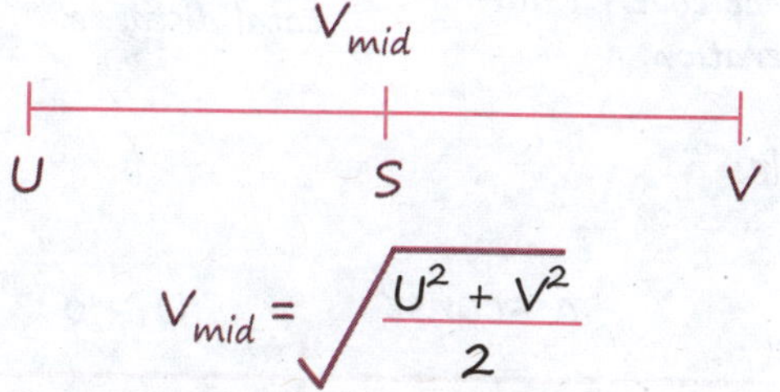

$$V_{mid} = \sqrt{\frac{U^2 + V^2}{2}}$$

Stopping Distance:-

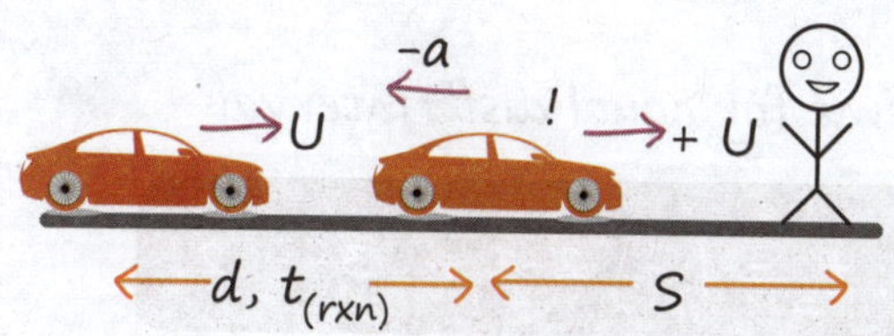

$d = Ut_{rx^n}$ — Reaction time

Stopping disance(S) $= \dfrac{U^2}{2a}$

Rest To Rest Motion:-

This concept explains the motion of an object that starts from rest, accelerates (α) to a maximum speed, and then decelerates (β) back to rest.

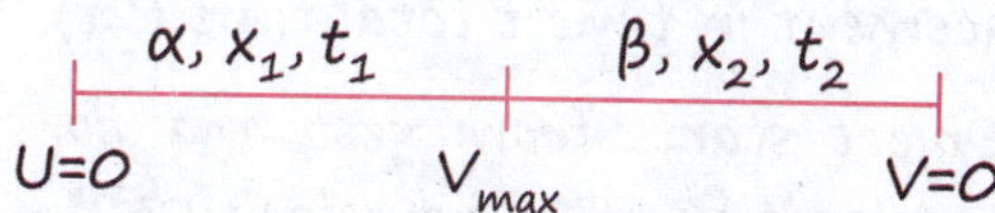

Time relationship:
$\alpha t_1 = \beta t_2$

- This means that the product of acceleration and the time taken to accelerate equals the product of deceleration and the time taken to decelerate.

Maximum velocity:
$$V_{max} = \left[\frac{\alpha\beta}{\alpha + \beta}\right]T$$

- This equation shows that the maximum velocity depends on the acceleration, deceleration, and total time of the motion.

Distance relationship

$\alpha x_1 = \beta x_2$

- This indicates that the product of acceleration and the distance covered during acceleration equals the product of deceleration and the distance covered during deceleration.

Total Distance (S):

$$s = \frac{1}{2}\left[\frac{\alpha\beta}{\alpha+\beta}\right]T^2$$

- This equation combines the effects of acceleration, deceleration, and total time to determine the total distance.

Example:

U = O   a = const^n U = o

$\leftarrow \frac{S}{3}, t_1 \rightarrow \leftarrow \frac{S}{3}, t_2 \rightarrow \leftarrow \frac{S}{3}, t_3 \rightarrow$

Ratio of time for equal dist^n interval:-

$$t_1 : t_2 : t_3 = 1 : \sqrt{2}-1 : \sqrt{3}-\sqrt{2}$$

Ratio of disp^m for equal time interval:-

$$S_{1st} : S_{2nd} : S_{3rd} = 1 : 3 : 5$$

$$S_{1s} : S_{2s} : S_{3s} = 1 : 4 : 9$$

$S_t : S_{next\ t} = 1 : 3$ or $x : 3x$

Ratio of displacement in time t and next same time intraval t, where motion starts from rest and constant acceleration

$S_t : S_{2t} = 1 : 4$ or $x : 4x$

displacement in time t total time (2t)

Q.7. Object starts from rest and constant acceleration attained velocity 32 m/s in 10 sec. then find displacement in next 10 sec.

Sol. $S = \frac{u+v}{2} \times t \Rightarrow S = \frac{0+32}{2} \times 10$

$S = 160$ m in 1st 10 sec.

Hence in next 10 sec.

it is $3 \times 160 = 480$ m

Q.8. Object starts from rest and constant acceleration moves 80 m in 7 sec. then find displacement in next 7 sec.

Ans. Displacement in next 7 sec = 3x

$= 3 \times 80 = 240$ m

8 Motion Under Gravity

(a = const^n)

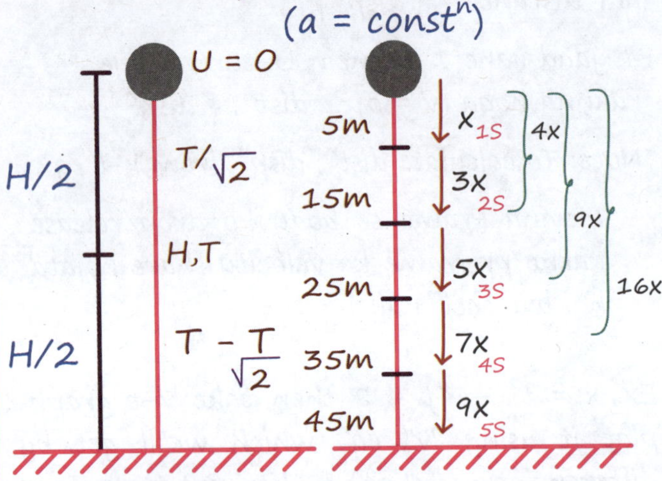

$S_{1st} : S_{2nd} : S_{3rd} = 1 : 3 : 5$

$S_{1s} : S_{2s} : S_{3s} = x : 4x : 9x$

Note:-

1. Time of Fight (T_F) $= \sqrt{\frac{2H}{g}}$
2. Velocity at ground :- $v = \sqrt{2gH}$

Graphs:-

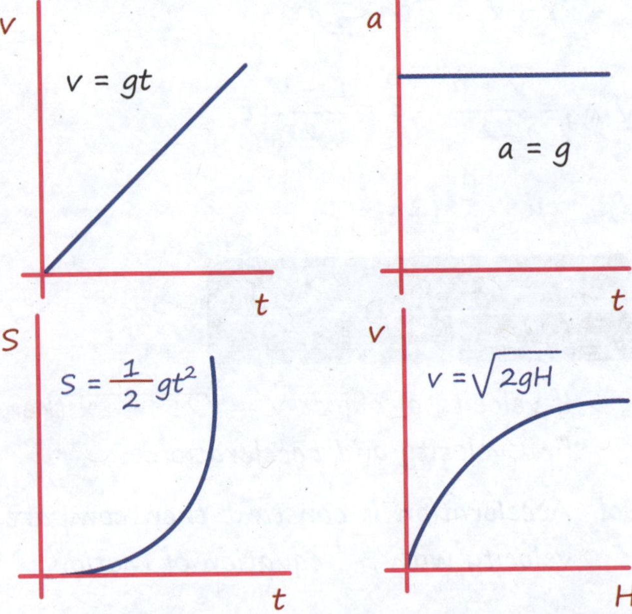

Motion Under Gravity from ground to ground

- Non-uniform motion (velocity = variable) with constant acceleration (g).
- At maximum height velocity zero and a = g.

$H_{max} = \frac{U^2}{2g}$ $T_f = \frac{2U}{g}$

$T_{up} = \frac{U}{g}$ $T_{down} = \frac{U}{g}$

find

1. Time of flight
2. Maximum height
3. Velocity at t = 7 sec
4. Displacement in 8 sec
5. Distance in 8 sec
6. Distance in 9th sec. downward journey
7. Distance in last sec of upward journey

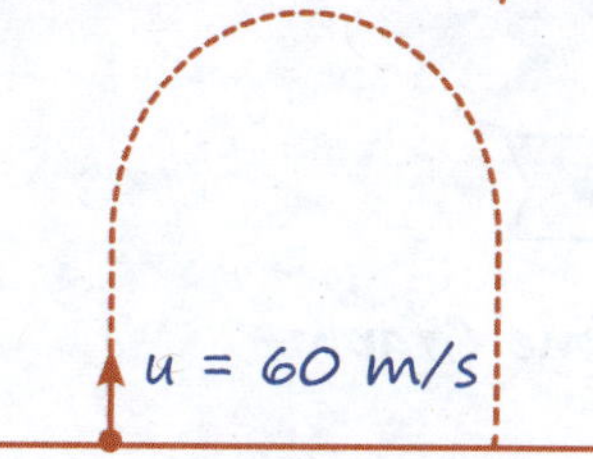

Sol.

1. Total time of flight $T = \frac{2u}{g} = \frac{2\times60}{10}$
   = 12 sec.
2. Maximum Height $H = \frac{u^2}{2g} = 180$ m
3. Velocity at t = 7 sec.
   $v = u + gt$
   $= 60 - 10\times7 = -10$ m/s
4. Displacement in 8 sec.
   $S = ut + \frac{1}{2}at^2 = 160$ m
5. Distance in 8 sec.
   at t = 6 sec. body comes to at rest and takes u-turn hence calculate distance 0 to 6 sec. then 6 to 8 sec
   S = 180 + 20 = 200 m
6. Distance in 9th sec. downward journey ka 3rd sec = 25 m. Use ratio.
7. Distance in last sec of upward journey = distance in 1st sec of downward journey = 5 m (always)

Q.9. A stone with weight W is thrown vertically upward into the air with initial velocity v_o. If a constant force, due to air drag acts on the stone throughout the flight & if the maximum height attain by stone is h and velocity when it strikes to the ground is u. Which one is correct?

(a) $h = v_o^2\left(1+\frac{f}{W}\right)/2g$, $v = v_o$

(b) $h = v_o^2/2g\left(1+\frac{f}{W}\right)$, v = zero

(c) $h = v_o^2/2g\left(1+\frac{f}{W}\right)$, $v = v_o\sqrt{\frac{W-f}{W+f}}$

(d) $h = v_o^2/2g\left(1+\frac{f}{W}\right)$, $v = v_o\sqrt{\frac{W+f}{W-f}}$

Sol.

MR*

If f = 0 then,

$H = \frac{v^2}{2g}$ and $v = v_o$

The MR*

If $f \neq 0$ $H < \frac{v^2}{2g}$ and $v < v_o$

Kam karne ke liye niche +ve hoga.

MR TABLE

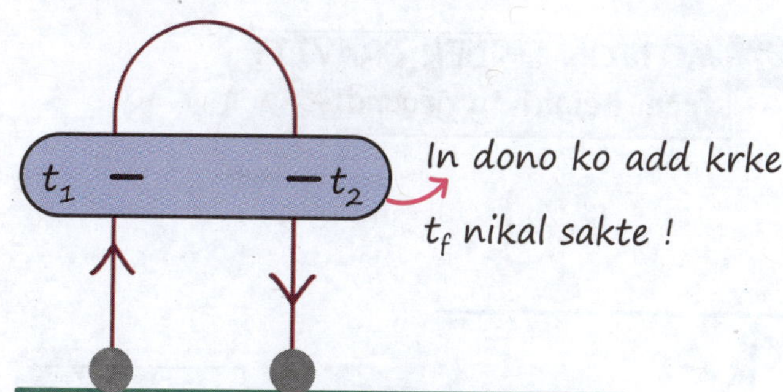

Object is at same height at t_1 and t_2.

That height $h = \frac{1}{2}gt_1t_2$

Ball is projected up with speed "U" graphs:-

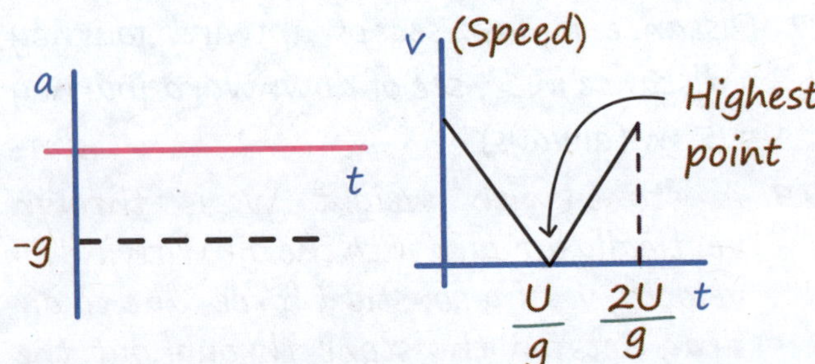

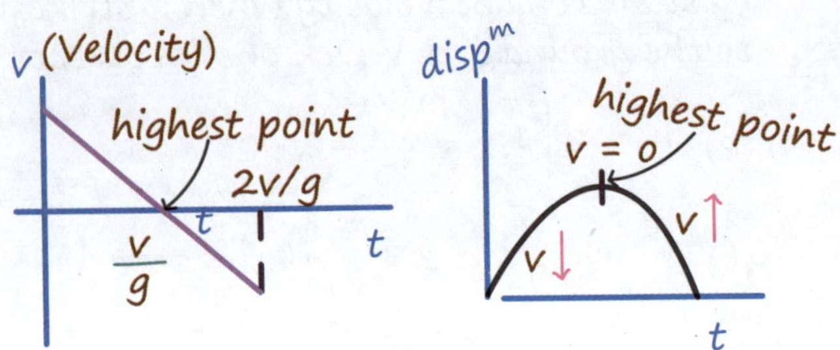

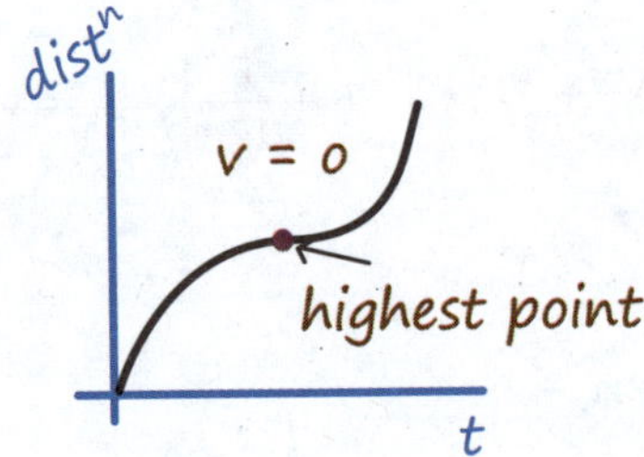

- If air friction in not ignored then:-

$$\frac{t_{up}}{t_{down}} = \sqrt{\frac{g - a}{g + a}}$$

- $t_{up} < t_{down}$
- $V_{projection} > V_{collision}$

9 MOTION UNDER GRAVITY from height to ground:-

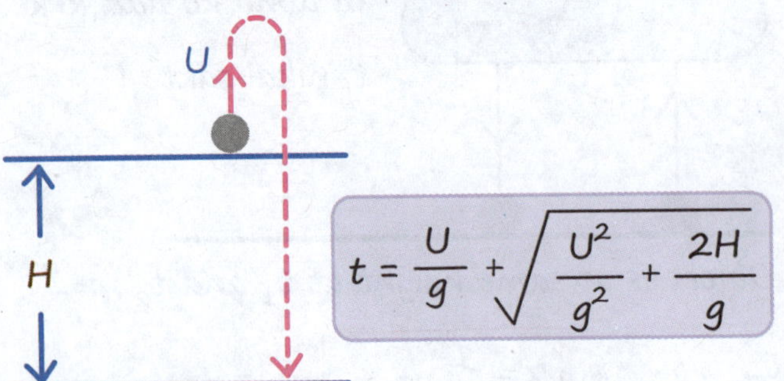

$$t = \frac{U}{g} + \sqrt{\frac{U^2}{g^2} + \frac{2H}{g}}$$

MR*

If $u = 0$ then it is like drop from height H then

$$t = \sqrt{\frac{2H}{g}}$$

If $H = 0$ then it is like ground to ground motion

$$t = \frac{2u}{g}$$

Q.10. Ball is projected up with speed "u" from height H. Then time of fight T_1. With same speed "u" it is projected downward then time of fight is t_2. find time of fight "t" when object is dropped from same height.

$$T = \sqrt{T_1 T_2}$$

10 Graph

Position - Time Graph:-

- For a bdoy at rest

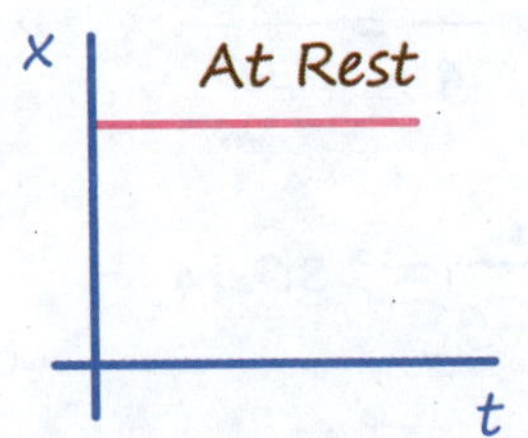

- Body moving with fonst velocity

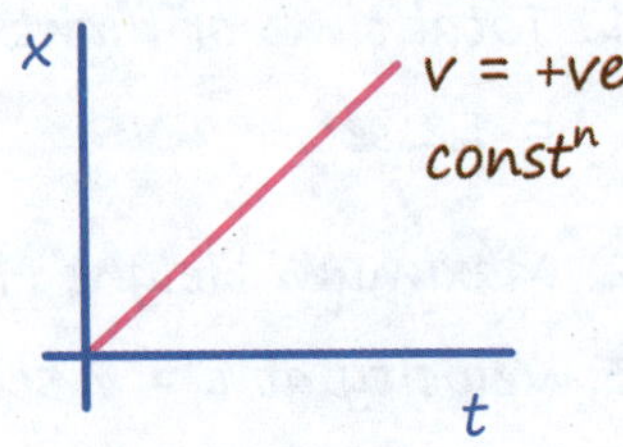

- motion with constant negative velocity

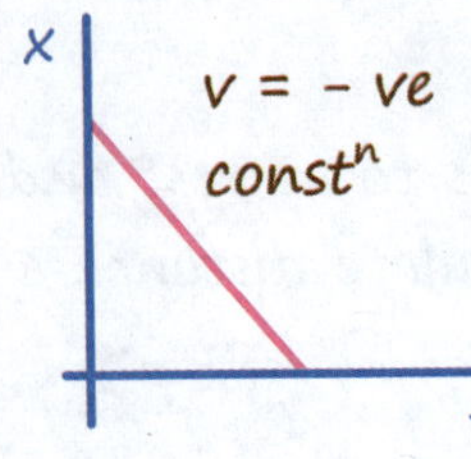

- Motion with constant acceleration

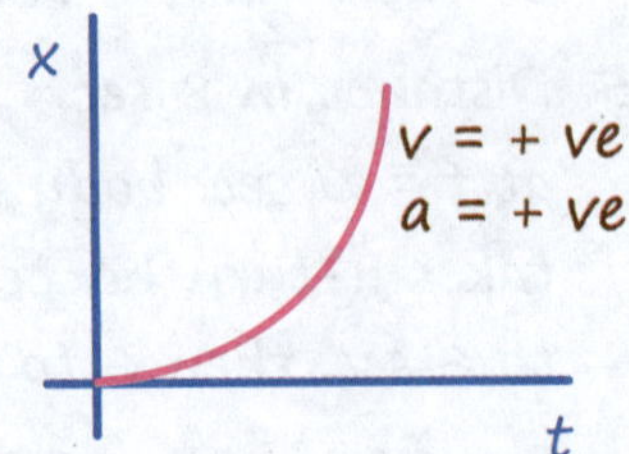

✦ Motion with constant retation

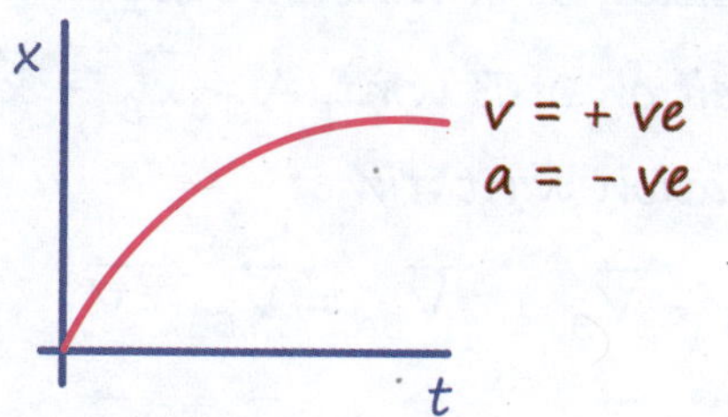

Velocity – Time Graph:-

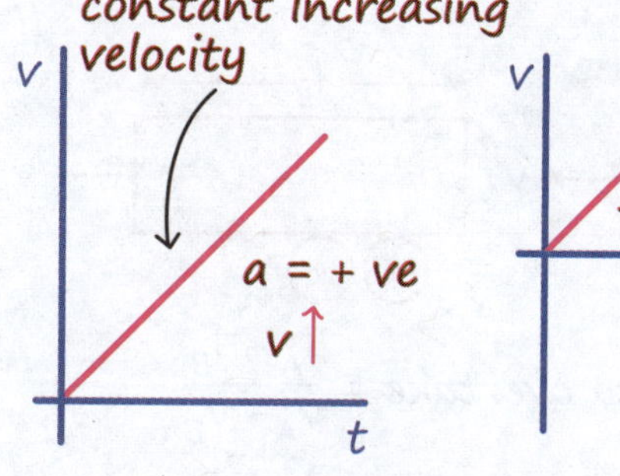

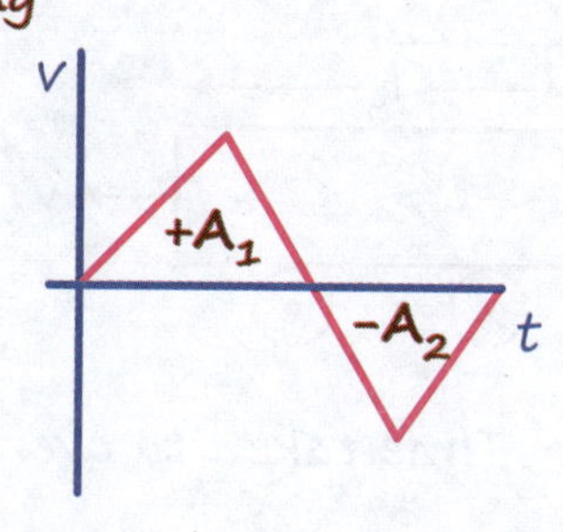

- Area = displacement = $A_1 - A_2$
- slope = acceleration
- but distance = $A_1 + A_2$

✦ Motion with constant velocity

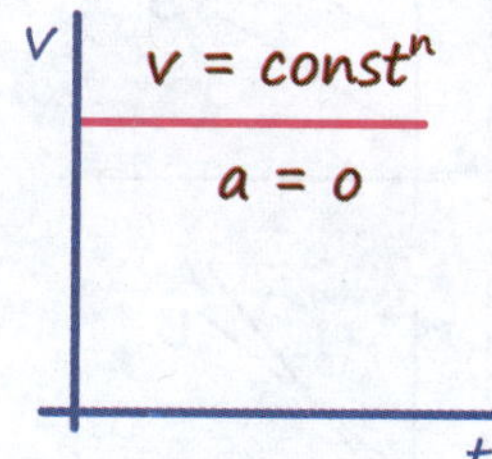

✦ Motion with constant retrodation with non-zero initial velocity

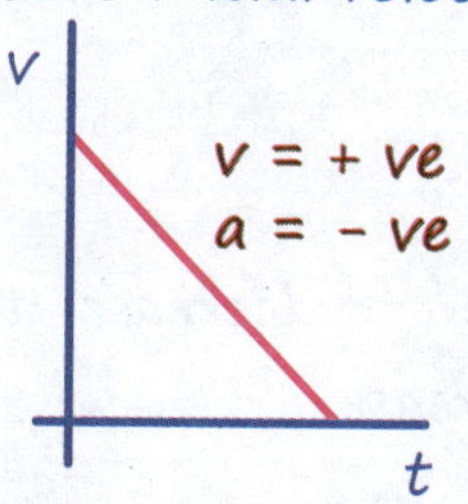

✦ Motion with increasing (a)

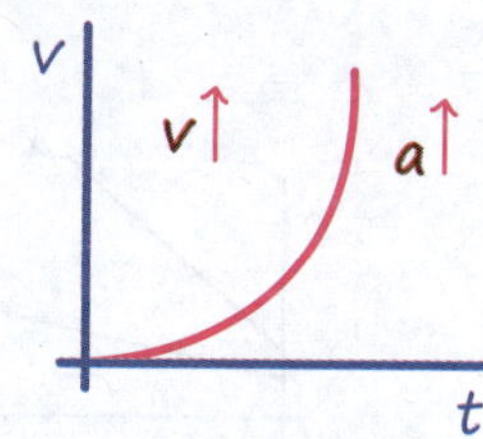

✦ Motion with increasing retordation

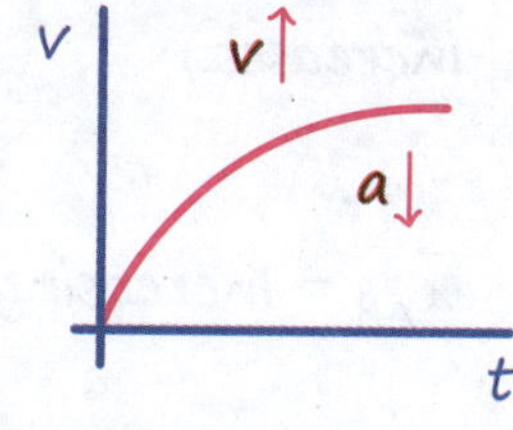

✦ Motion with decreasing (a)

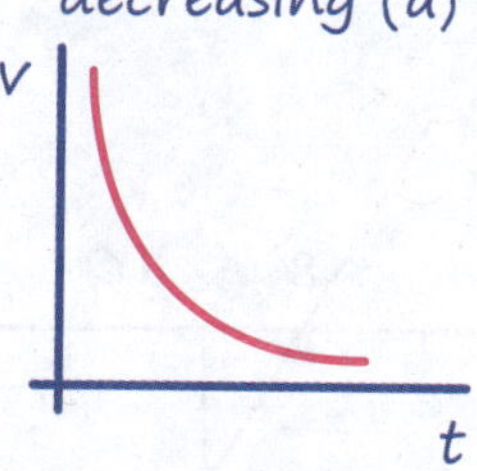

Velocity – Position Graph:-

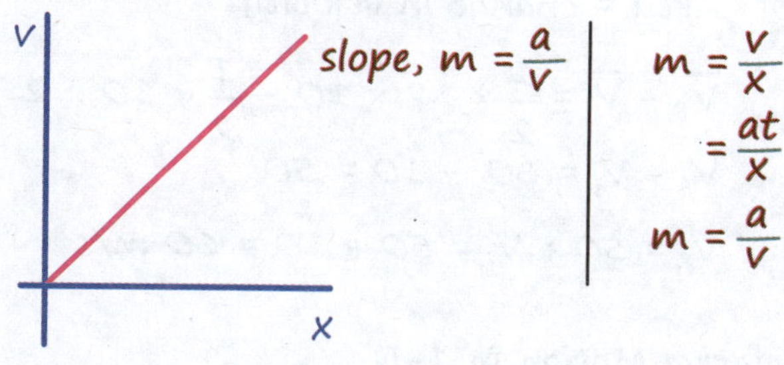

$$m = \frac{v}{x} = \frac{at}{x}$$

$$m = \frac{a}{v}$$

Acceleration time graph:-

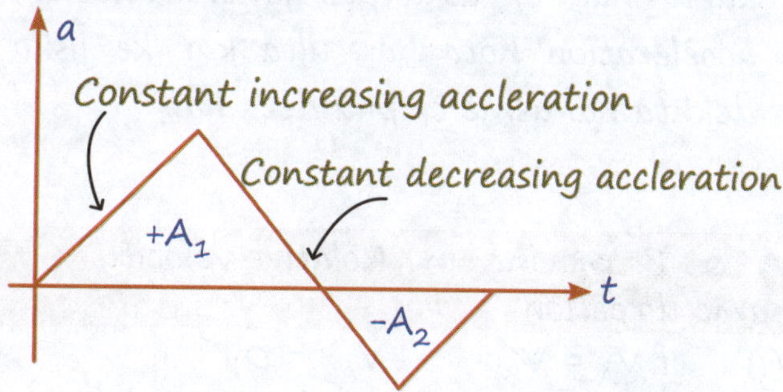

✦ Slope = नाली (Jerk)

✦ Area = Change in velocity = $A_1 - A_2$

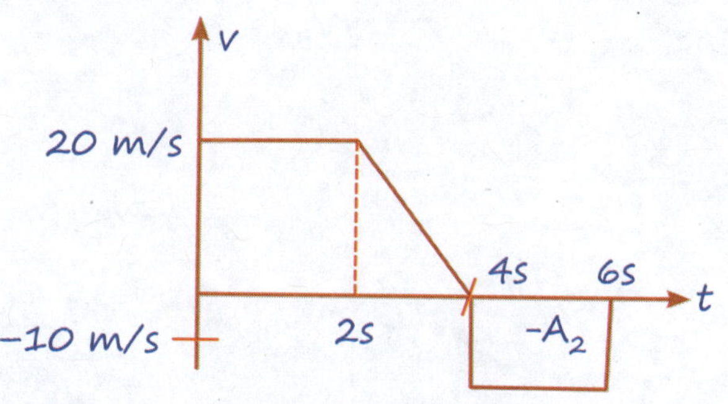

Distance = total area = 40 + 20 + 20 = 80 m

Displacement = 40 + 20 – 20 = 40 m

Q.11. If initial velocity of object is 10 m/s then find velocity at 10 sec

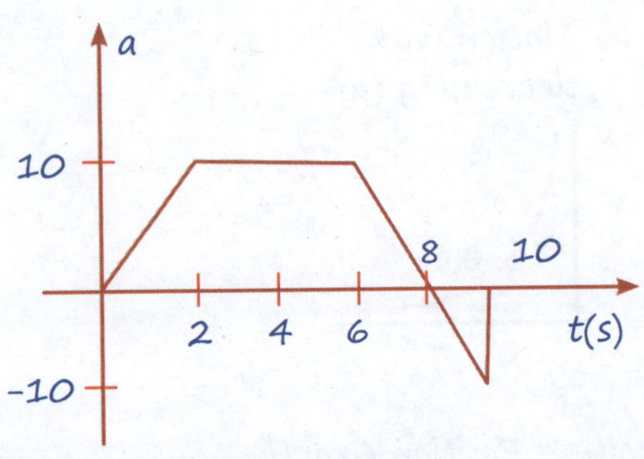

Sol. Area = change in velocity

$$\vec{V}_f - \vec{V}_i = \frac{1}{2} \times 12 \times 10 - \frac{1}{2} \times 10 \times 2$$

$$V_f - V_i = 60 - 10 = 50$$

$$V_f = 50 + V_i = 50 + 10 = 60 \text{ m/s}$$

Relative Motion in 1-D

- Observer khud ko hamesa rest me assume karta hai, or uska pas jo bhi velocity, acceleration hota hai, ulta kar ke jisko dekhta hai usme chipka deta hai.

$\vec{x}_{AB}$ = Position of A w.r.t. B = $\vec{x}_A - \vec{x}_B$

$\vec{x}_{BA}$ = Position of B w.r.t. A = $\vec{x}_B - \vec{x}_A$

differentiation w.r.t. time

$\vec{V}_{AB} = \vec{V}_A - \vec{V}_B$ $\quad$ $\vec{V}_{BA} = \vec{V}_B - \vec{V}_A$

$\vec{a}_{AB} = \vec{a}_A - \vec{a}_B$ $\quad$ $\vec{a}_{BA} = \vec{a}_B - \vec{a}_A$

$\vec{V}_{AB} = -\vec{V}_{BA}$ $\quad$ $\vec{a}_{AB} = -\vec{a}_{BA}$

A V_A $\quad$ B V_B

l_A $\quad$ l_B

A V_A $\quad$ V_B

- Time taken to overtake $= \dfrac{l_A + l_B}{V_A - V_B}$
- If they are moving opposite to each other $= \dfrac{l_A + l_B}{V_A + V_B}$

A & B moving in same direction	Relative velocity	Relative position	Graphical representation
(i) If $V_A = V_B$	$V_{AB} = 0$ (A & B apper at rest with one another)	$x_{AB} = \text{cost}^n$	x, A, B, t
(ii) If $V_A > V_B$	$V_{AB} = +ve$ $V_{BA} = -ve$	$\vec{x}_{AB}$ = decrease then increase (Initially, the relative position decreases as A catches up to B. After A overtakes B, it increases.)	x, A, B, t
(iii) If $V_A < V_B$	$V_{AB} = -ve$ $V_{BA} = +ve$	$\vec{x}_{AB}$ = Increasing	x, A, B, t

A & B moving in opposite direction with $V_A \neq V_B$	✦ If A is moving in the positive direction and B in the negative direction: $V_{AB} = V_A - (-V_B) = V_A + V_B$ ✦ If A is moving in the negative direction and B in the positive direction: $V_{AB} = -V_A - V_B$	Depending on the initial positions and directions, the relative position AB can increase or decrease rapidly.	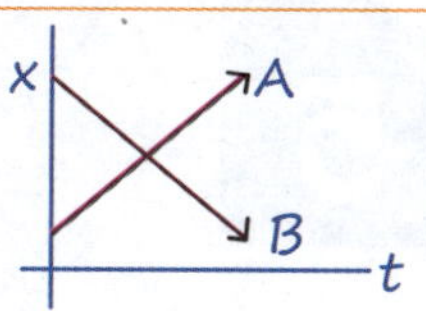

Motion of Object on the Moving Surface

1. Man is running on the surface of train with V_M in the direction of train (V_T) $V_{MG} = V_T + V_M$.

   If man is running in opposite direction then, $V_{MG} = V_T - V_M$

2. River is flowing with V_R and man is swimming with V_M in downstream then V_{MG} = Velocity of man w.r.t ground or effective velocity of man $= V_M + V_R$

   In upstream $V_{MG} = V_R - V_M$

3. Same as above in stair case.

- Motion under gravity of one object w.r.t other which is also in motion under gravity is uniform relative motion.

  $a_{AB} = 0 \qquad V_{AB} = Cost^n$

  S_{AB} = Increasing or decreasing linear

  Time of collision $= \dfrac{S_{AB}}{V_{AB}}$

Q.12. A ball is drop from 80 m height and another ball is projected with speed 40 m/s then they will collide.

Sol. $V_{relative} = 40$ m/s

$a_{relative} = 0 \qquad t = \dfrac{80}{40} = 2$ sec

$S_{relative} = 80$

Q.13. A ball thrown downward with speed 20 m/s and 30 m/s simultaneously, then find relative velocity and separation b/w them after 4 sec

Sol. $a_{AB} = 0$ $V_{BA} = 10$ m/s (const w.r.t bus)

$S_{BA} = V_{BA}t = 10 \times 4 = 40$ m

Q.14. A bus starts from rest moving with an acceleration of 2 m/s². A cyclist, 96 m behind the bus starts simultaneously towards the bus at 20 m/s. After what time will he be able to overtake the bus:-

Sol.

$u_c = 20$ m/s

$a_{CB} = 2\ m/s^2$ ←96 m→

$S = ut + \frac{1}{2}at^2$ (cyclist w.r.t bus)

$96 = 20t - \frac{1}{2}2t^2$

$t^2 - 20t + 96 = 0$ t = 12s and t = 8 sec

at 8 sec cyclist overtake bus and at 12 sec bus will again cross cyclist.

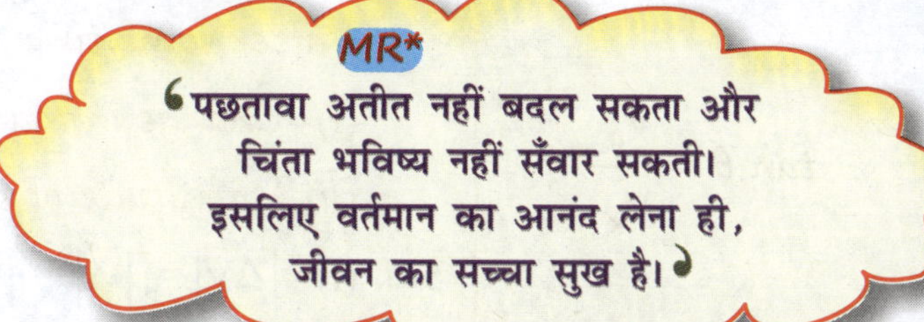

5 Motion in a Plane

1 2-D Motion

In this type of motion, two coordinates out of three needed to describe the position of the object.

$\Rightarrow [1\text{-}D]_{x\text{-}axis} + [1\text{-}D]_{y\text{-}axis}$

MR* feel

- Displacement, velocity and acceleration along x-axis Independent upon displacement, velocity and acceleration of y-axis
- Vector → Component of vector (तोड़ना) / Magnitude of vector (जोड़ना)

Velocity

$$\vec{V} = V_x\hat{i} + V_y\hat{j} \qquad |\vec{V}| = \sqrt{V_x^2 + V_y^2}$$

Acceleration

$$\vec{a} = a_x\hat{i} + a_y\hat{j} \qquad |\vec{a}| = \sqrt{a_x^2 + a_y^2}$$

Displacement

$$\vec{r} = \vec{r_2} - \vec{r_1} \qquad \vec{r} = [x_2 - x_1]\hat{i} + [y_2 - y_1]\hat{j},$$

here, $\vec{r} = \Delta x\hat{i} + \Delta y\hat{j}$ is position vector.

Direction of motion

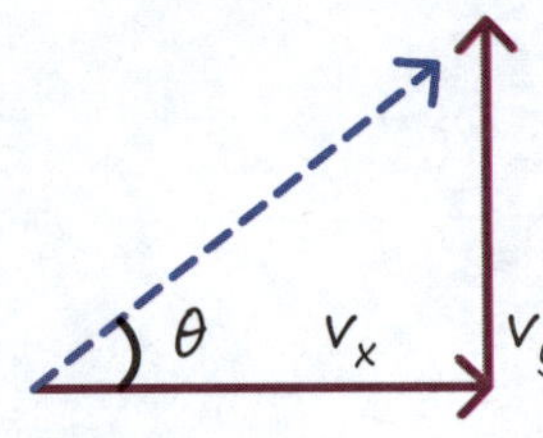

$$\tan\theta = \frac{V_y}{V_x}$$

Equation of Motion in a plane

$a = const^n$

- $\vec{V_x} = \vec{u_x} + \vec{a_x}\,t$
- $\vec{X} = \vec{u_x}t + \frac{1}{2}\vec{a_x}t^2$
- $\vec{V_x}^2 = \vec{u_x}^2 + 2\,\vec{a_x}\,x$
- $\vec{V_y} = \vec{u_y} + \vec{a_y}t$
- $\vec{y} = u_y t + \frac{1}{2}\vec{a_y}t^2$
- $\vec{V_y}^2 = \vec{u_y}^2 + 2\,\vec{a_y}\,y$
- x-axis and y-axis ka motion independent hota hai. Acceleration along x only change velocity of x-axis

$$\vec{V_x} = \frac{d\vec{x}}{dt} \qquad \vec{V_y} = \frac{d\vec{y}}{dt}$$

$$\vec{a_x} = \frac{d\vec{V_x}}{dt} \qquad \vec{a_y} = \frac{d\vec{V_y}}{dt}$$

○ Only time is same in both co-ordinate

Q. 1. Initial velocity of object $\vec{u} = 3\hat{i} + 4\hat{j}$ and acceleration $\vec{a} = 0.4\hat{i} + 0.3\hat{j}$ then find velocity after t = 10 sec.

Sol. $\vec{V_x} = \vec{U_x} + \vec{a_x}t = 3\hat{i} + 0.4 \times 10\hat{i} = 7\hat{i}$

$\vec{V_y} = \vec{U_y} + \vec{a_y}t = 4\hat{j} + 0.3 \times 10\hat{j} = 7\hat{j}$

$\vec{V} = 7\hat{i} + 7\hat{j}$

$|V| = 7\sqrt{2}$

Q. 2. If initial velocity of object $\vec{u} = 3\hat{i} + 4\hat{j}$ after some time $\vec{V} = 4\hat{i} + 3\hat{j}$ then find.

(i) Change in Magnitude of velocity.

(ii) Magnitude of change in velocity.

Sol. (i) $|\Delta\vec{V}| = |\vec{V_f}| - |\vec{V_i}| = 5 - 5 = 0$

(ii) $\overrightarrow{\Delta V} = \overrightarrow{V_f} - \overrightarrow{V_i} = 4i + 3j - 3i - 4j$

$\overrightarrow{\Delta V} = \hat{i} - \hat{j}$

$|\overrightarrow{\Delta V}| = \sqrt{2}$

Q. 3. x and y of the particle are $x = 5t - 2t^2$ and $y = 10t$, acceleration of particle at t=2s.

Sol. $V_x = \frac{dx}{dt} = 5 - 4t$

$V_y = \frac{dy}{dt} = 10$

$a_x = \frac{dV_x}{dt} = -4,\ a_y = 0$

$|\vec{a}| = 4 m/s^2$

Q. 4. Initial velocity of bus is 5m/s east after 2sec its velocity becomes 5m/s north then find acceleration.

Sol. $\vec{a} = \frac{\vec{V_f} - \vec{V_i}}{\Delta t} = \frac{5\hat{j} - 5\hat{i}}{2}$

N, W, E, S

$|\vec{a}| = \frac{5\sqrt{2}}{2} = \frac{5}{\sqrt{2}}$ (North-west)

2 Equation of Trajectory

Relation between x^{th} and y^{th} co-ordinate which are derived with the help of time.

Position Vector	Equation of Trajectory
(1) $\vec{r} = A \sin(\omega t)\hat{i} + A\cos(\omega t)\hat{j}$	$x^2 + y^2 = A^2$ (Circle)
(2) $\vec{r} = A \sin(\omega t)\hat{i} + B\cos(\omega t)\hat{j}$	$x^2/A^2 + y^2/B^2 = 1$ (Ellipse)
(3) $\vec{r} = A \sin(\omega t)\hat{i} + B\sin(\omega t)\hat{j}$	(Straight line) $Y = (B/A)\,x$
(4) $\vec{r} = 2t\,\hat{i} + 4t^2\,\hat{j}$	$Y = x^2$ (Parabola)

- Motion starts from rest and constant acceleration then path → straight line
- If Angle between velocity and acc^n is always 90° then path → circle
- If Angle between initial velocity and acceleration is between 0° or 180° then path → Parabolic.

3 Projectile motion

Non uniform motion with uniform acceleration

θ = with Horizontal

θ = with Vertical

sab change except range

sinθ ⌒ cosθ

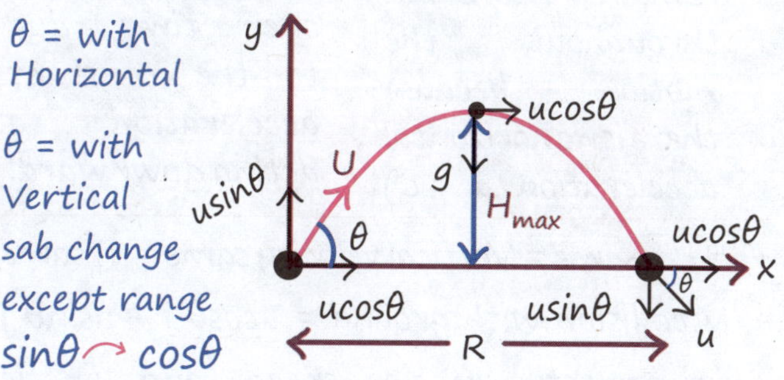

Time of Flight (Tf)

$$T_f = \frac{2\,u\sin\theta}{g} = \frac{2u_y}{g}$$

Range (R):

$$R = u_x T_f = u\cos\theta \,\frac{2\,u\sin\theta}{g}$$

$$R = \frac{u^2 \sin 2\theta}{g} = \frac{2u_x u_y}{g}$$

Maximum Height (H_max):

$$H_{max} = \frac{u^2 \sin^2\theta}{2g} = \frac{u_y^{\,2}}{2g}$$

x-axis (Horizontal components)	y-axis (Vertical components)
✦ At $t = 0$, $\vec{u}_x = U\cos\theta\,\hat{i}$	✦ $\vec{u}_y = U\sin\theta\,\hat{j}$
✦ At 't', $\vec{V}_x = u\cos\theta\,\hat{i}$	✦ $V_y = (u\sin\theta - gt)\,\hat{j}$, $a_y = -g\,\hat{j}$
✦ $\vec{a}_x = 0$ (througout motion)	✦ $y = u\sin\theta\, t - \frac{1}{2} gt^2$
✦ $x = u_x t = u\cos\theta\, t$	✦ $\lvert\Delta\vec{P}\rvert = (2\,mu\sin\theta)$
$(V_{avg}) = u\cos\theta$ ✦ The horizontal component of velocity $(u\cos\theta)$ remains constant throughout the motion because there is no horizontal acceleration $(a_x = 0)$	✦ The vertical component of the initial velocity is $u\sin\theta$. The vertical velocity changes over time due to the constant acceleration g acting downward.

✦ "H" = same = Vertical velocity same = T_f same

✦ V collision with ground = $u\cos\theta\,\hat{i} - u\sin\theta\,\hat{j}$

✦ in projectile motion speed (and kinetic energy) is minimum at the highest point.

✦ speed = $u\cos\theta$

✦ in projectile motion H=R When $u_y = 4u_x$ or $\tan\theta = 4$

Elevation angle of Max. Point from point of projection

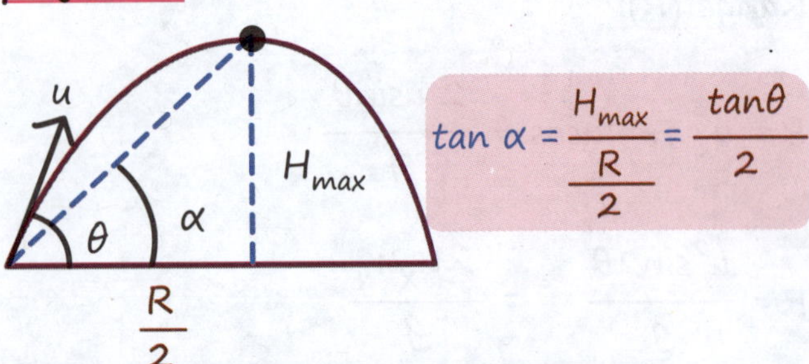

$$\tan\alpha = \frac{H_{max}}{\frac{R}{2}} = \frac{\tan\theta}{2}$$

4 Relation between Range & H_{max} ?

$$H = \frac{R\tan\theta}{4}$$

Speed at any point

Magnitude,

$$|\vec{V}| = \sqrt{(U\cos\theta)^2 + (U\sin\theta - gt)^2}$$

$$\vec{V} = U\cos\theta\,\hat{i} + U\sin\theta - gt\,\hat{j}$$

$$\text{Direction, } \tan\alpha = \frac{U\sin\theta - gt}{U\cos\theta}$$

Speed at point $\frac{H_{max}}{2}$:

$$V = \sqrt{(U\cos\theta)^2 + \left[\frac{U\sin\theta}{\sqrt{2}}\right]^2}$$

Condition of Max. horizontal Range

$$R_{max} = \frac{u^2}{g}$$

$\theta = 45°$ iske upar R↓

$$H = \frac{u^2}{4g}$$

5 Complementary Angle

✦ Two angles are complementary if their sum is 90°. In the context of projectile motion, if α and β are complementary angles, then:

$\alpha + \beta = 90°$;

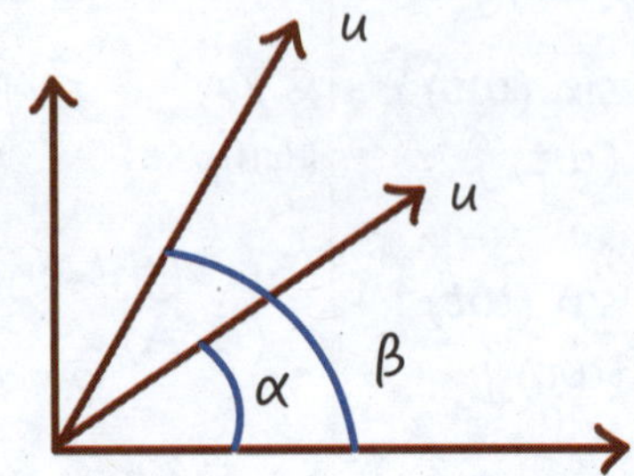

Range:

- When a projectile is launched at two complementary angles with the same initial speed, the horizontal ranges (R_1 and R_2) will be equal. i.e.,

$$R_1 = R_2$$

Heights Ratio:

IIT

$$\frac{H_1}{H_2} = \tan^2 \alpha$$

Times of Flight Ratio:

$$\Rightarrow \frac{T_1}{T_2} = \tan \alpha$$

Product of Heights:

$$H_1 H_2 = \frac{R^2}{16}$$

for same range, $T_1 T_2 = \frac{2R}{g}$

- 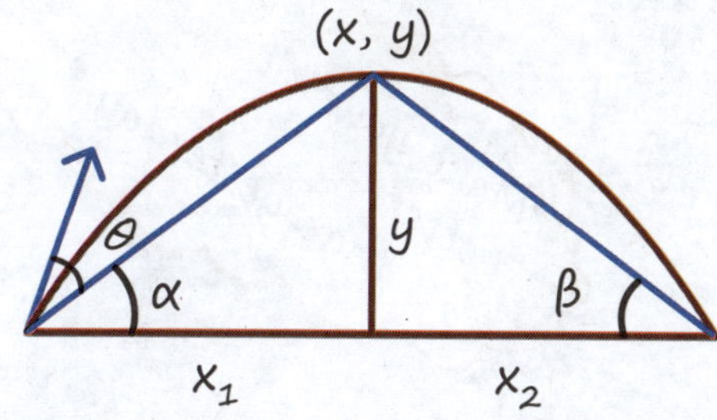

$\tan \theta = \tan \alpha + \tan\beta$, $\tan \theta = y \left[\frac{1}{x_1} + \frac{1}{x_2}\right]$

- Ball is projected with same speed at 42° and 47° then Range R_1 and R_2 respectively then $R_1 < R_2$

hint :- Angle 45° ke jitna pas range utna jyada.

6 Eq^n of Trajectory in Projectile Motion

$$y = x \tan\theta \left[1 - \frac{x}{R}\right]$$

$$R = \frac{2u^2 \cos\theta . \sin\theta}{g}$$

$$y = x \tan \theta - \frac{1}{2} \frac{gx^2}{u^2 \cos^2\theta}$$

Time at which particle moving ⊥^{er} to initial velocity

$$t = \frac{u}{g \sin\theta}$$

Note:-

($v \perp u$)

Only valid when $\theta \geq 45°$. For $\theta < 45°$ kabhi ⊥^{er} nai hoga!

Q. 5. Equation of trajectory $y = \sqrt{3}\, x - \frac{gx^2}{\sqrt{3}}$ then find range and angle of projectile.

Sol. $Y = x \tan\theta \left[1 - \frac{x}{R}\right]$ $\tan\theta = \sqrt{3}$, $\theta = 60°$

When $y = 0$ then $x=R$

$$0 = \sqrt{3}\, x - \frac{gx^2}{\sqrt{3}} \Rightarrow \sqrt{3}\, x = \frac{gx^2}{\sqrt{3}}$$

$$\Rightarrow x = R = \frac{3}{10} \text{ m.}$$

7 Horizontal Projectile motion from some height

$u_y = 0$, u_x

$X = U_x t$, $y = \frac{1}{2} g t^2$

$|\vec{V}|_{ground} = \sqrt{U^2 + (2gh)}$

$T_f = \sqrt{\frac{2H}{g}}$

$\vec{V} = U\hat{i} + gt\hat{j}$

g, y, +ve, H, x

MR "Raste me jitne sangharsh ho, manzil utna hi khubsoorat hota hai."*

Eq^n of Trajectory:-

$$y = \frac{1}{2} g \left[\frac{x}{U}\right]^2$$

Speed at any time "t" :-

$$|\vec{V}| = \sqrt{U^2 + (gt)^2}$$

✦

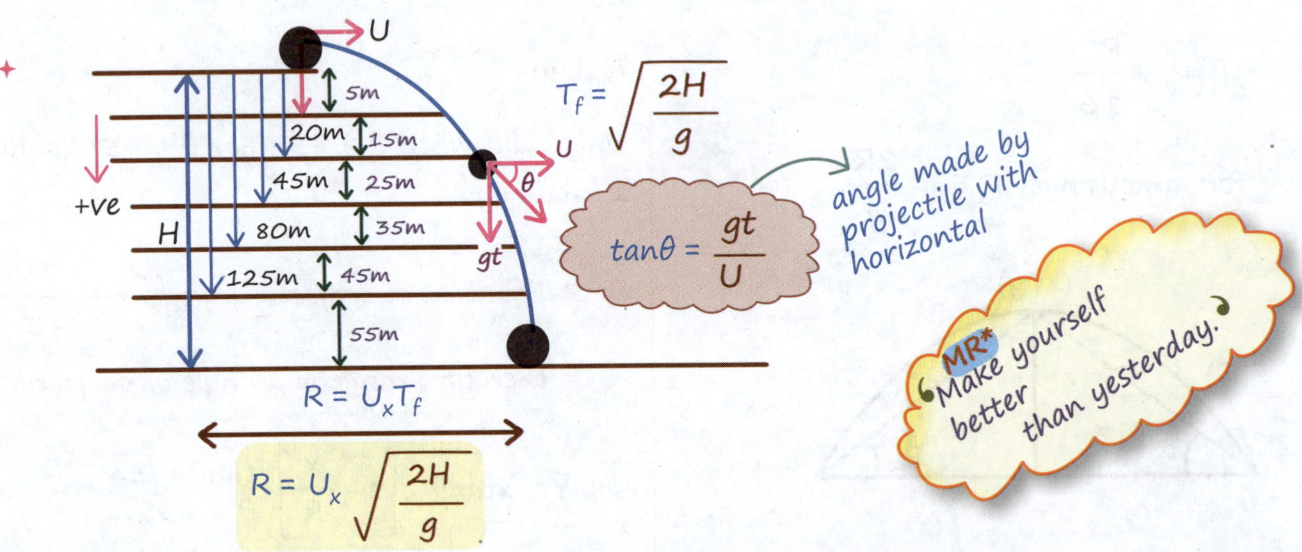

✦ distances like 5m, 15m, 25m, 35m, 45m, and 55m represent how far the object has fallen at each corrosponding equal time interval due to gravity.

Steps and Distances:

✦ The ball is projected horizontally from a height, and the steps are at regular intervals.

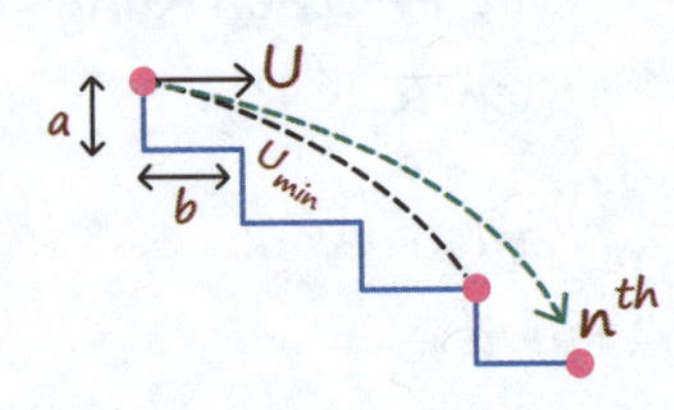

- a= The vertical distance between each step.
- b =The horizontal distance between each step.

✦ The minimum velocity of the ball lands at least on the nth step;

$$U_{min} = \sqrt{\frac{(n-1)\, b^2 g}{2a}}$$

✦ The exact velocity for the ball to land precisely on the nth step:

$$U = \sqrt{\frac{nb^2 g}{2a}}$$

8 Projectile Motion from some height at angle θ,

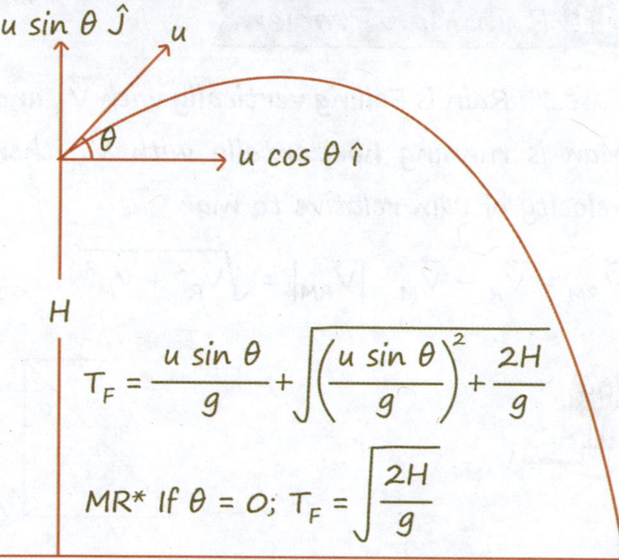

$$T_F = \frac{u \sin\theta}{g} + \sqrt{\left(\frac{u \sin\theta}{g}\right)^2 + \frac{2H}{g}}$$

MR* If $\theta = 0$; $T_F = \sqrt{\frac{2H}{g}}$

MR* If $H = 0$ $\quad T_F = \frac{2u \sin\theta}{g}$

Q. 6. Ball is projected in Horizontal direction with speed u then find time when distance moved in horizontal and vertical direction is same.

Sol. $x = ut$ & $y = \frac{1}{2} gt^2$

since, $x = y$ therefore,

$$ut = \frac{1}{2} gt^2 \quad t = \frac{2u}{g}$$

9 Relative Motion in a Plane

$$\vec{V}_{AB} = \vec{V}_A - \vec{V}_B$$

If Angle between V_A and V_B is θ then.

$$|\vec{V}_{AB}| = \sqrt{V_A^2 + V_B^2 - 2V_A V_B \cos\theta}$$

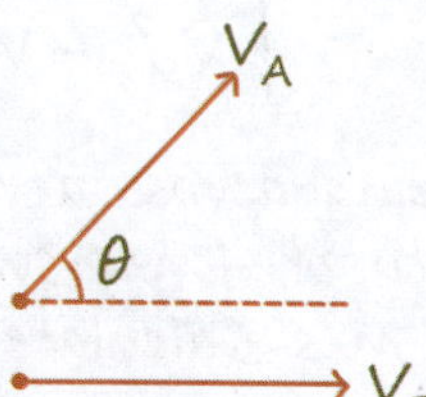

✦ Relation between V_A and V_B so that A always moving along y-axis.

$V_A \cos\theta = V_B$

Q. 7. Bus is moving in east with 30m/s and car in north with speed 40m/s then velocity of Car w.r.t bus.

Sol. $\vec{V}_{CB} = \vec{V}_C - \vec{V}_B = 40\hat{j} - 30\hat{i}$

$|\vec{V}_{CB}| = 50$m/s

= 37° west of north

$$\theta = \tan^{-1}\left(\frac{V_y}{V_x}\right) = \tan^{-1}\left(\frac{40}{30}\right)$$

10 River man Problem

V_{MR} = Velocity of Man w.r.t River

= Velocity of Man w.r.t still water

= Velocity by which man can swim

MR** → River apne dam pe cross kiya jata hai

→ Ramlal puri jan apne jan ki taraf lagayga river minimum time me cross karne ke liye. (Man will swim perpendicular to flow of river.)

$t_{min} = \frac{D}{V_{MR}}$, Drift along river $= V_R \times \frac{D}{V_{MR}}$

Shortest Path (Just want to reach opposite end)

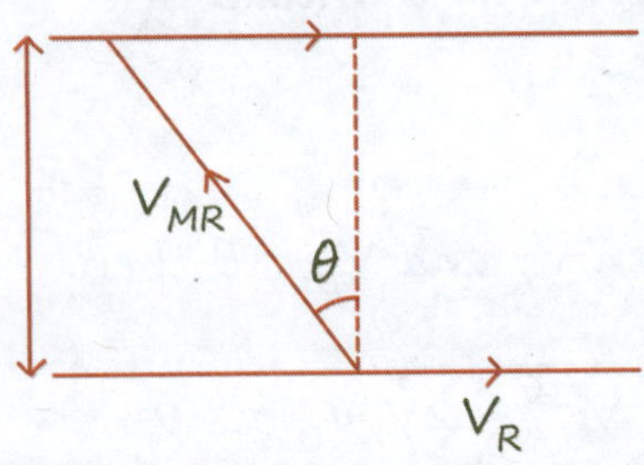

$[V_{MR} \sin\theta = V_R]$ for shortest path

$$\text{time} = \frac{D}{V_{MR}\cos\theta}$$

$$= \frac{D}{\sqrt{V_M^2 - V_R^2}}$$

Q. 8. A swimmer swimming across a river flowing at a velocity of 4m/s swims at the velocity of 2 m/s. Calculate the actual velocity of the swimmer and the angle.

Sol. The actual velocity of the swimmer can be found out as follows:

$$V_{actual} = \sqrt{2^2 + 4^2} = 4.47 \text{m/s}$$

The angle is calculated as follows:

$$\tan\theta = \frac{2}{4}$$

$$\theta = \tan^{-1}\frac{2}{4} = 26.57°$$

Q. 9. A boat takes 2 hours to travel 8 km and back in still water lake. With water velocity of 4 km/h, the time taken for going upstream of 8 km and coming back is

Sol. Velocity of boat $= \frac{8+8}{2} = 8$ km/h

Velocity of water = 4 km/h

$$\Rightarrow t = \frac{8}{8-4} + \frac{8}{8+4}$$

$$= \frac{8}{3}\text{h} = 160 \text{ minutes}$$

Q. 10. The speed of a swimmer in still water is 20 m/s. The speed of river water is 10 m/s and is flowing due east. If he is standing on the south bank and wishes to cross the river along the shortest path the angle at which he should make his strokes w.r.t north is given by:

Sol. $|\vec{V}_{SR}| = 20\text{m/s}$

$$|\vec{V}_{RG}| = 10\text{m/s}$$

$$\vec{V}_{SG} = \vec{V}_{SR} + \vec{V}_{RG}$$

So,

$$\sin\theta = \frac{|\vec{V}_{RG}|}{|\vec{V}_{SR}|}$$

$$\sin\theta = \frac{10}{20}$$

$$\Rightarrow \sin\theta = \frac{1}{2}$$

$$\Rightarrow \theta = 30° \text{ west}$$

11 Rain Man Problem

Case 1 : Rain is Falling vertically with $\vec{V}_R$ and Man is running horizontally with $\vec{V}_M$ then velocity of rain relative to man

$$\vec{V}_{RM} = \vec{V}_R - \vec{V}_M,\ |\vec{V}_{RM}| = \sqrt{V_R^2 + V_M^2}$$

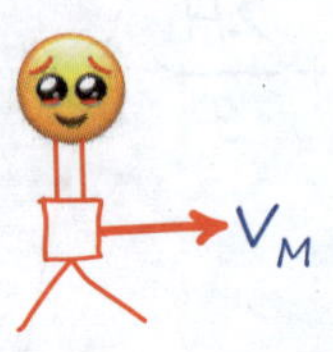

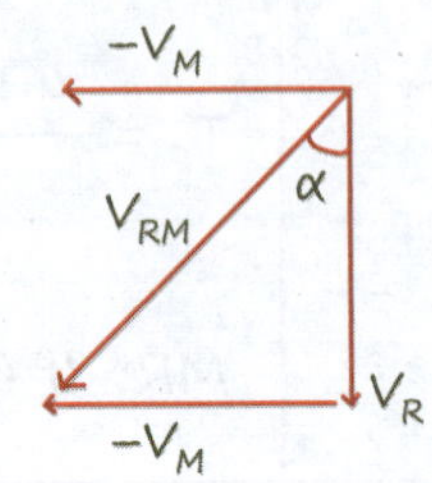

direction of umbrella from vertical

$$\tan\alpha = \frac{V_M}{V_R}$$

Case 2 : Rain is falling vertically at an angle θ then find velocity of man so that rain appears to fall vertically downward.

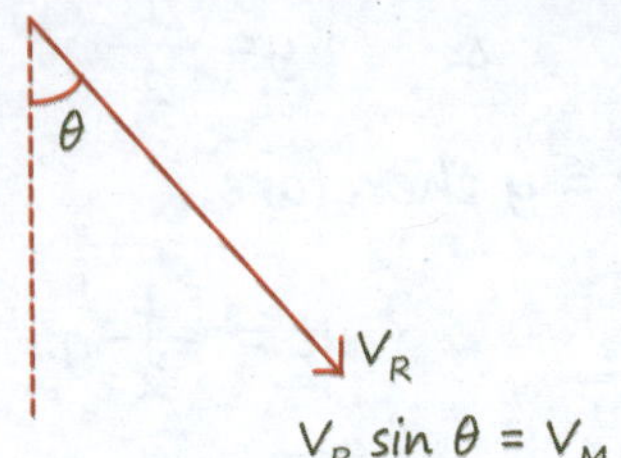

$V_R \sin\theta = V_M$

Condition of Collision : Position of A and B are $\vec{r}_A$ and $\vec{r}_B$ moving with velocity $\vec{V}_A$ and $\vec{V}_B$ then find condition of collision

$$\frac{\vec{r}_A - \vec{r}_B}{|\vec{r}_A - \vec{r}_B|} = -\frac{\vec{V}_A - \vec{V}_B}{|\vec{V}_A - \vec{V}_B|} \Leftrightarrow \hat{r}_{AB} = -\hat{V}_{AB}$$

Direction of relative velocity opposite to relative position. Hence relative velocity perpendicular to line joining of particle is zero.

Q. 11. Rain is falling vertically with a speed of $30 ms^{-1}$. A woman rides a bicycle with a speed of $10 ms^{-1}$ in the north to south direction. What is the direction in which she should hold her umbrella?

Sol. Here, v_C = Velocity of the cyclist

v_r = Velocity of falling rain

In order to protect herself from the rain, the woman must hold her umbrella in the direction of the relative velocity (v) of the rain with respect to the woman.

$\tan\theta = v_c/v_r = 10/30$

$\theta = 18°$

MR**

Component of their velocity perpendicular to line joining will be same.

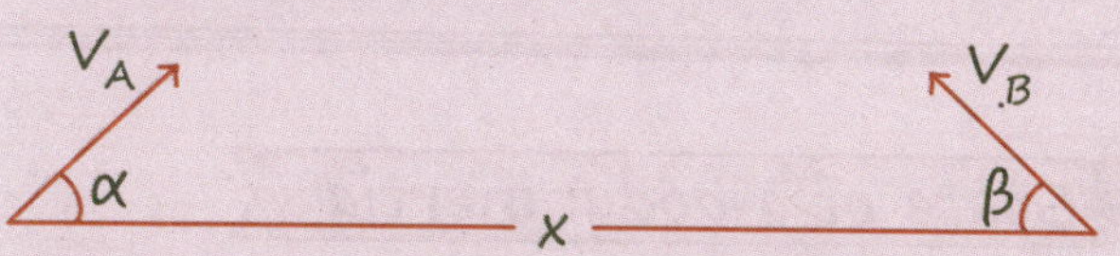

Condition of collision = $V_A \sin\alpha = V_B \sin\beta$

$$t = \frac{x}{V_A \cos\alpha + V_B \cos\beta}$$

\# Two object moving perpendicular to each other with same speed V; having initial separation d then

$$d_{min} = \frac{d}{\sqrt{2}} \quad \text{time} = \frac{d}{2V}$$

n-person is standing on the cornor of n-side polygon starts moving towards each other with same speed, then time when they will meet.

$$t = \frac{d}{V - V\cos\left(\frac{2\pi}{n}\right)}$$

n = 2	$t = \frac{d}{2V}$
n = 3 (Triangle)	$n = \frac{2d}{3V}$
n = 4 (Square)	$n = d/V$
n = 6 (Hexagon)	$t = \frac{2d}{V}$

MR*

"Main kisi se behtar banu kya fark padta hai...., Main kisi ka behtar karu usse se bahut fark padta hai."

6 Newton's Laws of Motion

1 State of a body: inertia

We define physical state of a body with the help of velocity.

1> State of rest ($\vec{v} = 0$)

2> State of uniform motion ($\vec{v}$ = const)

Inertia → Property of object, not a physical quantity, does not have unit and dimension.

Inertia ∝ Mass

2 Force

Force is anything that causes an object to start moving, stop moving, change direction, or change its shape.

Types of force

- **Contact Forces:** Require physical touch (friction, tension, spring, normal force).
- **Distant Forces:** Act over a distance without touching (electrostatic force, gravitational force, magnetic force).

Weight

Weight is how strongly gravity pulls on an object. It's why things fall to the ground when dropped formula of weight:

$$w = mg$$

Where, where m is the mass of the object, and g is the acceleration due to gravity.

Normal Reaction

When you place a book on a table, the table pushes up on the book with a force equal and opposite to the weight of the book. This push is the normal reaction.

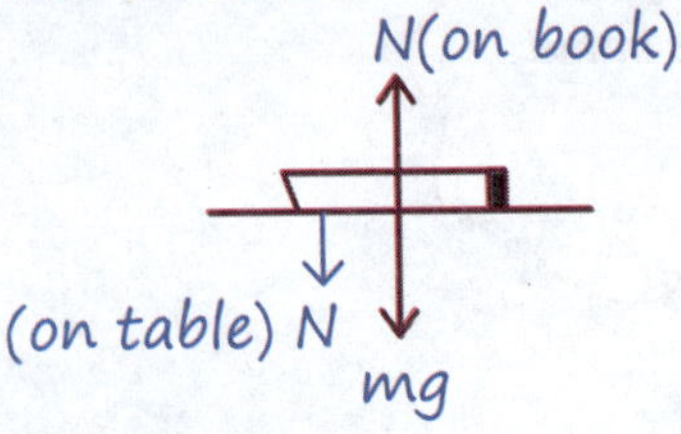

Tension:-

Tension is a force that is always present in a rope, string, or cable when it is pulled tight. It is a reactive force, meaning it responds to other forces acting on the rope.

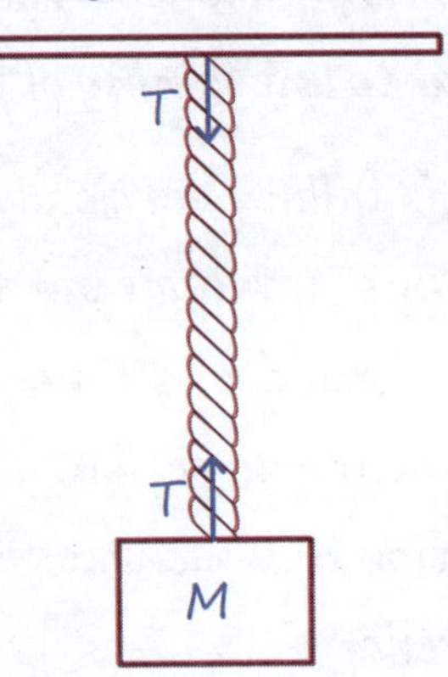

Key Points about tension

- Tension always pulls along the length of the rope or string.
- Tension is not an active force; it exists because other forces are acting.
- When a rope passes over a massless (ideal) or frictionless pulley, the tension in the rope remains the same throughout its length.
- If the tension force drops to zero, the rope becomes slack and no longer pulls.

3 Newton's Ist Law

Law of inertia: No external net force required to keep the body in same physical state, Net force required to change physical state.

For Constant velocity, $a = 0$,

$F_{net} = 0$ → Rest

Inertia of a body depends on its mass. The larger the mass, the larger is the inertia of the body.

4 Reading of spring

For ideal spring replace it by string & find tension that will be equivalent to spring force.

Ideal Spring:- Force is same at all Points!

5 For string

if a string is massless, the tension in it is the same every where. however, if a sting has a mass, tension at different points will be different.

6 Equilibrium of a body

- A body is in equilibrium when the combined effect of all the forces acting on it is zero. This means that the body will stay at rest or move with constant velocity.
- Mathematically, if you add up all the forces acting on the body and their sum is zero, the body is in equilibrium.

Example:

- Consider two forces, F_1 and F_2, acting on a particle.

$F_2 \longleftarrow O \longrightarrow F_1$

- For the particle to be in equilibrium, these two forces must cancel each other out. This means that the sum of these forces must be zero:

$$F_1 + F_2 = 0$$

- If F_1 is pointing in one direction, F_2 must be equal in magnitude but pointing in the opposite direction.

7 Lamis Theorem

If three forces act on a particle and keep it in equilibrium, each force is proportional to the sine of the angle between the other two forces.

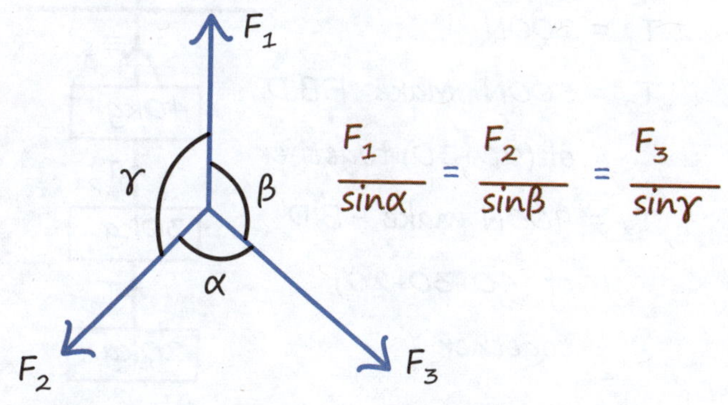

$$\frac{F_1}{\sin\alpha} = \frac{F_2}{\sin\beta} = \frac{F_3}{\sin\gamma}$$

Note:-

When,

$$\frac{F_1}{\sin\alpha} = \frac{F_2}{\sin\beta} = \frac{F_3}{\sin\gamma}$$

Then, $\vec{F_1} + \vec{F_2} + \vec{F_3} = 0$

Q.1. Find Tension in wire?

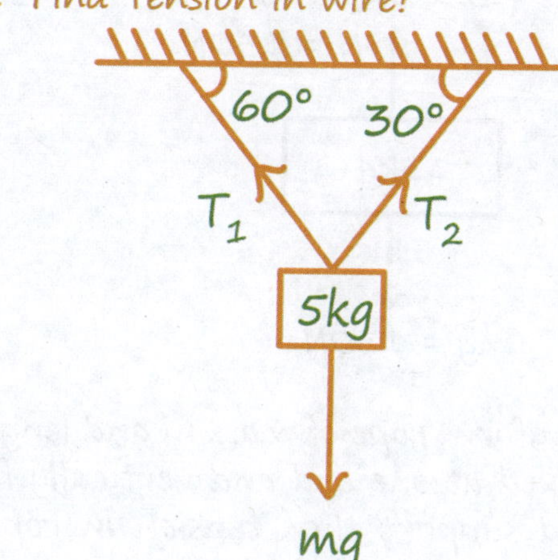

Sol.

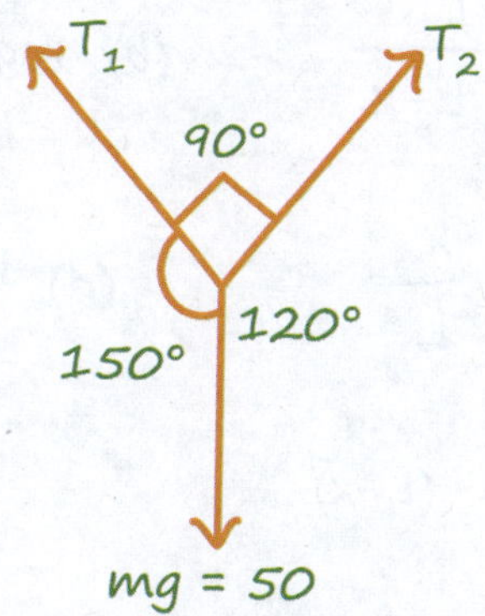

$$\frac{T_1}{\sin 120°} = \frac{T_2}{\sin 150°} = \frac{50}{\sin 90°}$$

$$T_1 = \frac{50\sqrt{3}}{2} \quad T_2 = 25\text{ N}$$

Q.2. T_1 = 200N

T_2 = 500N Make F.B.D. of (20+30) together

T_3 = 900N make F.B.D. of (40+30+20) together

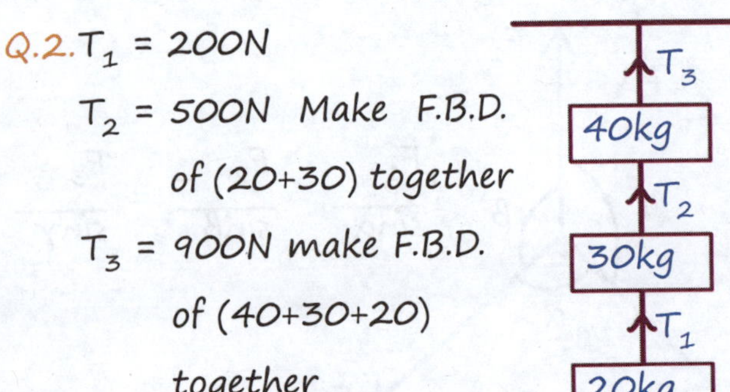

Q.3.

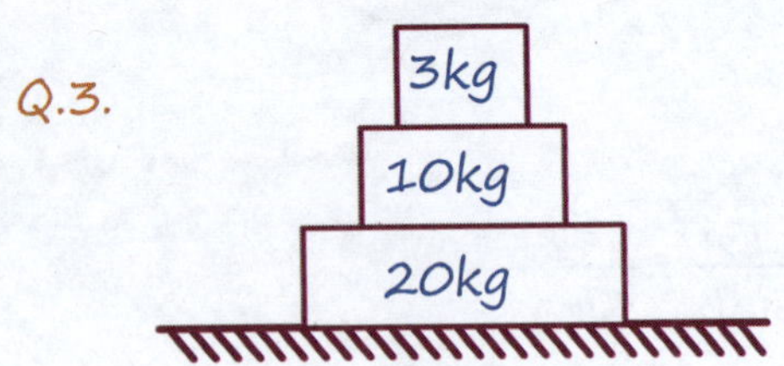

Contact force b/w 10kg. and 20kg. Make F.B.D. of (3+10) kg.

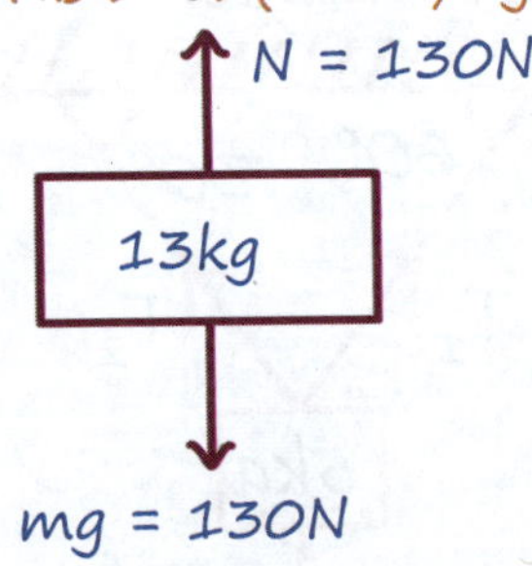

Q.4. A uniform rope of mass m and length L is fixed at one end and vertically from rigid support then tension in rope at distance x from rigid support.

(a) $mg\dfrac{L}{L+x}$ (b) $mg\dfrac{L+x}{L}$

(c) $mg\dfrac{x}{L}$ (d) $mg\left(\dfrac{L}{x}\right)$

(e) $\dfrac{mg}{L}(L-x)$

MR*

If x=0 T=mg and If x=L then T=0

Q.5. A man of mass m is standing on a plank of mass M. A light string Passing over a fixed smooth pulley connect man and Plank. find tension force exterted by man on string to keep block at rest.

F.B.D. of Man + Plank

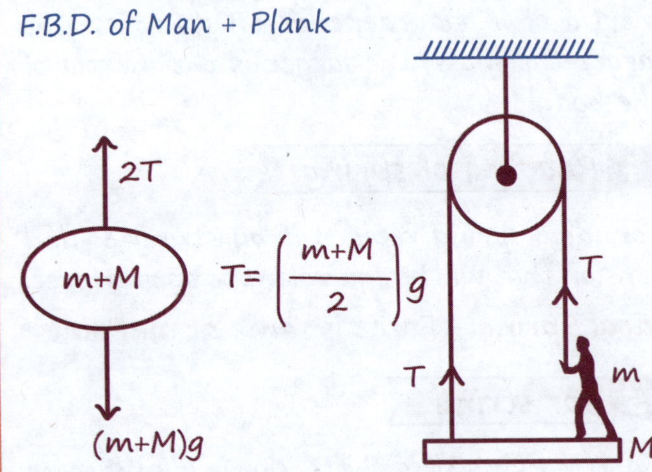

$$T = \left(\frac{m+M}{2}\right)g$$

Q.6. Release from rest then falling down find Normal reation b/w them

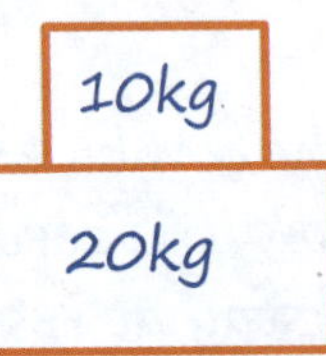

Ans. N=0

8 Newton's 2nd laW

Momentum → Quantity of motion contained in a body.

$$\vec{P} = M\vec{v}$$

→ Vector, Kg m/sec, parallel to velocity

→ frame dependent.

$\Delta P = \int F.dt$ Area

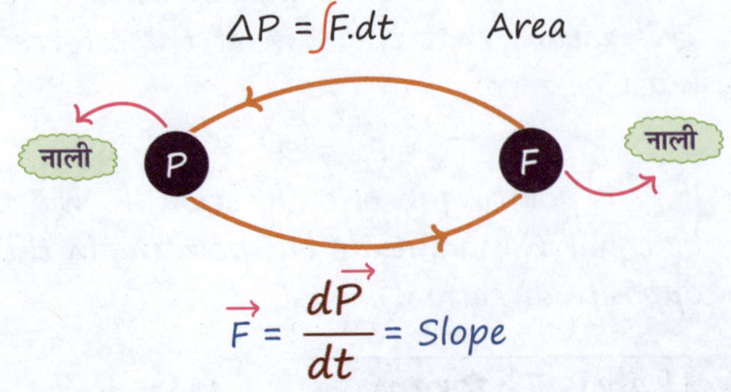

$$\vec{F} = \frac{d\vec{P}}{dt} = \text{Slope}$$

✦ Newton's Second Law of Motion states that the rate of change of linear momentum of an object is equal to the applied force and occurs in the direction of the applied force.

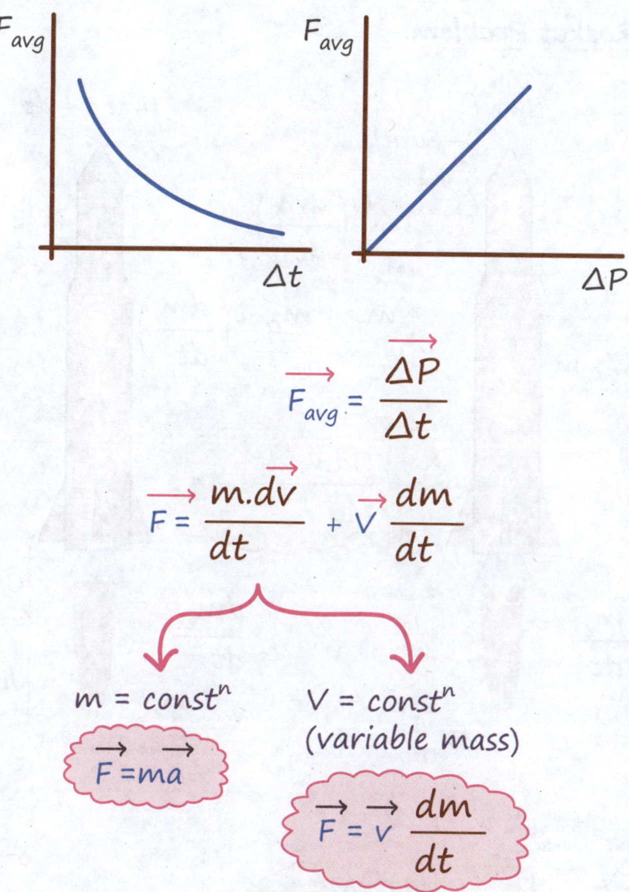

Ex:- Rocket Propulsion

Note:-

- Ball Rebounds With Same Speed Find Change In Momentum:-

$\vec{\Delta P} = 2mV_\perp$ $V_\perp$ = Velocity $\perp^{er}$ to Surface

- Ball change direction perpendicular with same speed then, $|\vec{\Delta P}| = \sqrt{2}\, mv$

- BuLLet plate question

$\vec{F} = \frac{2nmu}{\Delta t}$, $\vec{\Delta P} = 2nmu.$

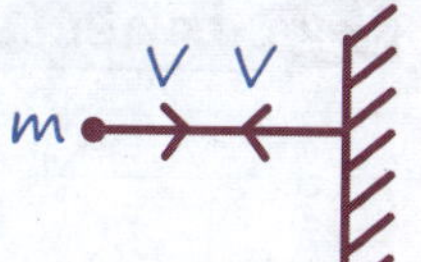

9 Impulse

[Change in momentum]

$\vec{I} = \int \vec{F}.\, dt = \vec{\Delta P} = \vec{F}_{avg}.\, \Delta t$

- Impulse is the result of a force acting over a certain period of time. It represents the change in momentum of an object.
- The SI unit of impulse is the newton-second (N-s) or kilogram meter per second (kg.m/s)
- The total impulse for a force applied over a time interval from t_1 to t_2 can be found by calculating the area under the force-time (F-t) graph for that interval.

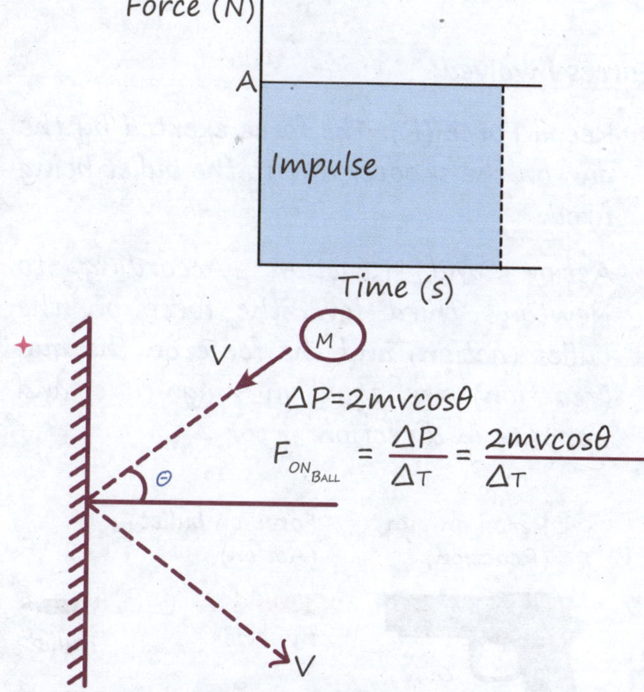

10 newton's 3rd law

[Action-Reaction]

- Action-Reaction pair acts on two diff. body but they should be of same nature.

1 F_{12} F_{21} 2

Mathematically, $\vec{F}_{12} = -\vec{F}_{21}$

11 conservation of momentum

If no external forces act on a system, the total linear momentum of the system remains constant.

$$F = \frac{dp}{dt} \quad (2^{nd} \text{ law})$$

$F_{ext} = 0 \qquad P = const^n$

Gun-Bullet System:-

In a gun-bullet system:

- The gun has mass M and recoils with velocity V_G.
- The bullet has mass m and is fired with velocity U_B.

Forces Involved:

- Recoil Force (R): The force exerted by the gun on the shooter due to the bullet being fired.
- Action and Reaction: According to Newton's third law, the force on the bullet (action) and the force on the gun (reaction) are equal in magnitude and opposite in direction.

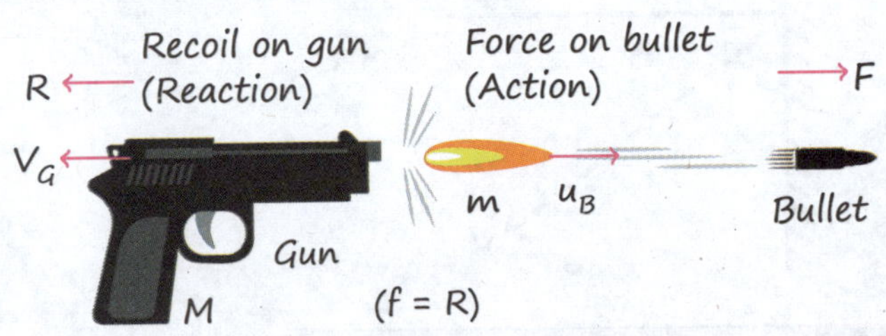

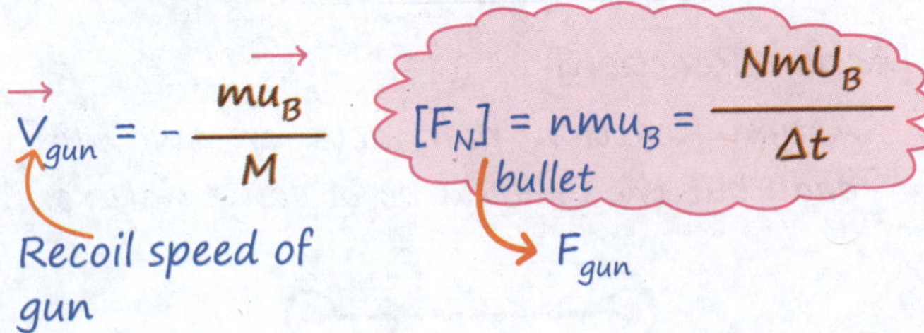

$n = \frac{N}{\Delta t}$ = No. of bullet fired per sec.

$|\vec{P}_{gun}| = |\vec{P}_{bullet}| \quad \frac{(KE)_{gun}}{(KE)_{bullet}} = \frac{m}{M} \quad K_B > K_G$

Rocket Problem:-

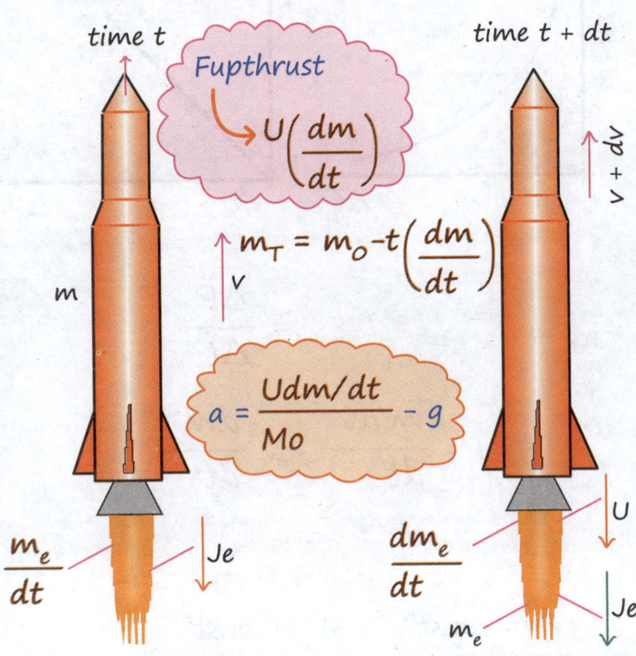

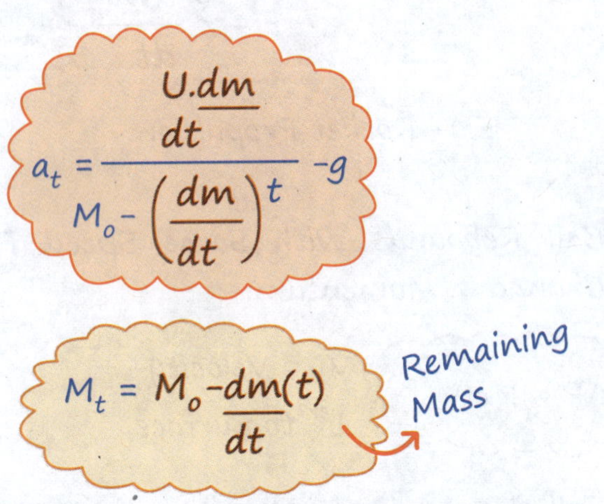

12 connected body dynamics

(i)

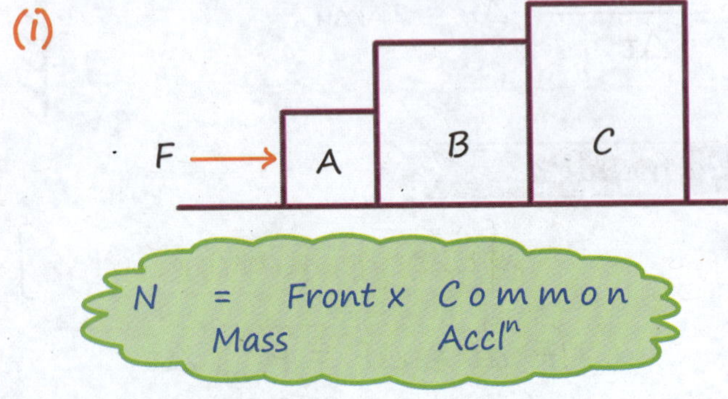

(ii)

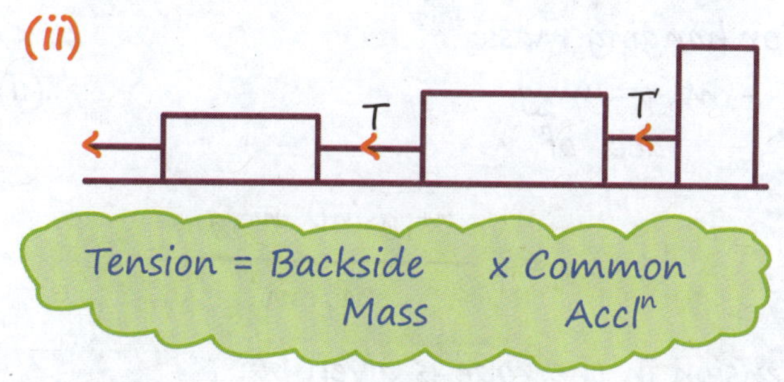

Q.7. Find acceleration and normal reaction.

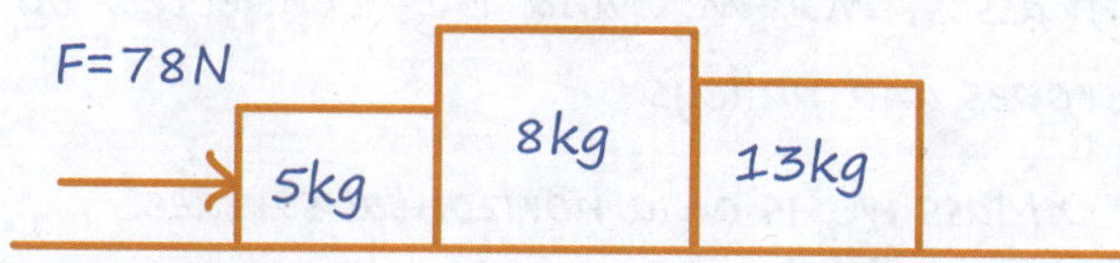

common acceleration $a = \frac{F_{net}}{\text{Total mass}} = \frac{78}{26} = 3\ m/s^2$

MR*

Normal on 13 kg

$N_1 = 13kg \times a = 13 \times 3 = 39N$

Normal on 8 kg by 5 kg

$N_2 = (8+13)a = 21 \times 3 = 63N$

(8+13) → front mass

Q. find T_1 and T_2

$F_1 = 90N$ ← 3kg —T_1— 7kg —T_2— 5kg → $F_2 = 60N$

Sol.

$$a = \frac{\text{Net Pulling force}}{\text{Net Mass}} = \frac{90-60}{15} = 2m/s^2$$

$90 - T_1 = 3 \times 2$ $\qquad T_1 = 90 - 6 = 84N$

For 7kg. $T_1 - T_2 = 7 \times 2;\ 84 - 14 = T_2$

$T_2 = 70N$

LIFT SAWAAL:-

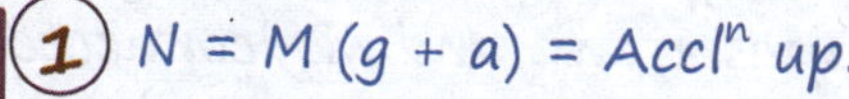

N

Ramlal

Mg

When lift accelerates upward

(1) $N = M(g + a) = Accl^n$ up.

When lift accelerates downward

(2) $N = M(g - a) = Accl^n$ down

Lift at rest or moving with a constant speed

(3) $N = Mg = Accl^n = 0$

$V = const^n$

13 Pulley Block System-I

- In a pulley block system, we have two masses,
- M_1 and M_2, connected by a rope over a pulley. Here's a breakdown of how the system works and the key concepts:
- *The acceleration of the system can be determined by the difference in the gravitational forces acting on the two masses divided by the total mass of the system:

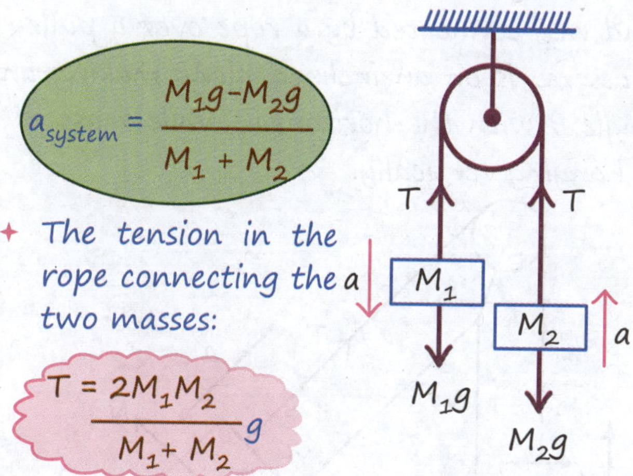

$$a_{system} = \frac{M_1g - M_2g}{M_1 + M_2}$$

- The tension in the rope connecting the two masses:

$$T = \frac{2M_1M_2}{M_1 + M_2}g$$

MR*

If $M_1 = M_2 \rightarrow a = 0$

If $M_1 = 0$ or $M_2 = 0 \rightarrow a = g$

⇒ Think MR* for tension ☺

If $M_1 = 0$	$M_2 = 0$	check also dimension of
$T = 0$	$T = 0$	acc^n and tension

14 Pulley Block System-II

In this type of pulley block system, we have two masses, m_1 and m_2, connected by a rope over a pulley, with one mass hanging and the other on a horizontal surface.

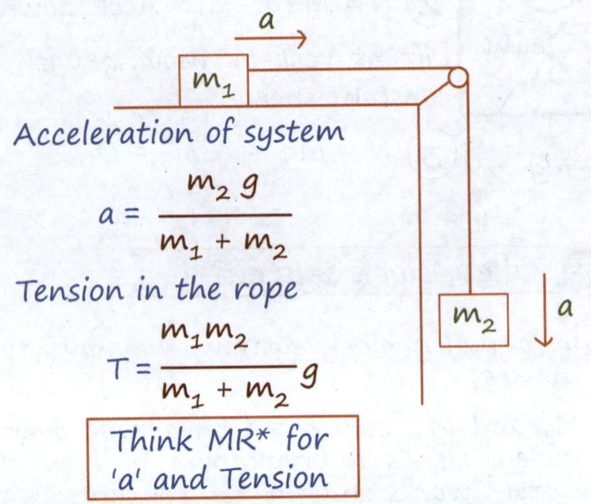

Acceleration of system

$$a = \frac{m_2 g}{m_1 + m_2}$$

Tension in the rope

$$T = \frac{m_1 m_2}{m_1 + m_2} g$$

Think MR* for 'a' and Tension

15 Inclined Plane Pulley Block System

In this system, we have two masses, m_1 and m_2, connected by a rope over a pulley. Mass m_1 is on an inclined plane making an angle θ with the horizontal, while mass m_2 is hanging vertically.

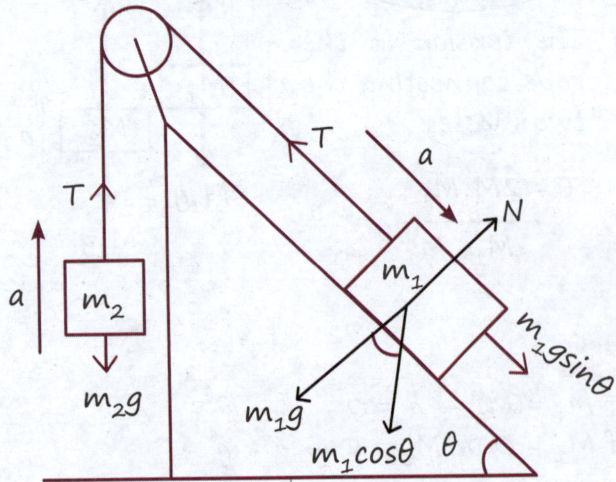

Newton's second law along the incline for M_1:

$$m_1 g \sin\theta - T = m_1 a \quad \text{...(i)}$$

For hanging mass.

$$T - m_2 g = m_2 g \quad \text{...(ii)}$$

acc^n of m_1

$$a = \frac{m_1 g \sin\theta - m_2 g}{m_1 + m_2}$$

tension in the rope is given by:

$$T = \frac{m_1 m_2 g(1 + \sin\theta)}{m_1 + m_2}$$

- In following arrangment, we have three masses, m_1, m_2, and m_3, connected by ropes and pulleys.
  - Mass m_1 is on a horizontal surface.
  - Mass m_2 hangs vertically on the right side.
  - Mass m_3 hangs vertically on the left side.
  - The system will move in such a way that the heavier of the hanging masses (m_2 or m_3) will determine the direction of the acceleration.

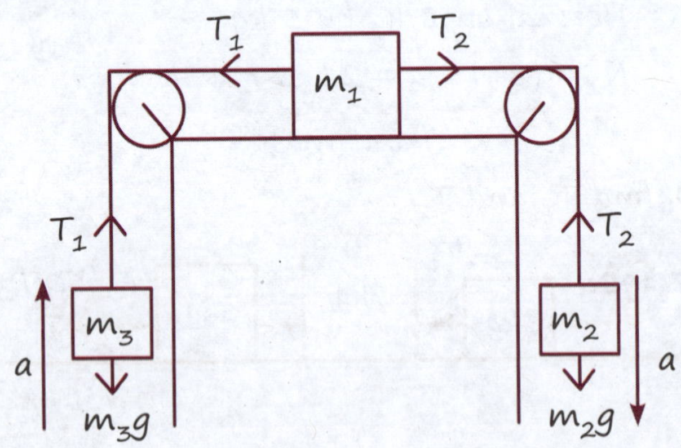

For m_2:

$$m_2 a = m_2 g - T_2$$

For m_3:

$$m_3 a = T_1 - m_3 g$$

For m_1:

$$T_2 - T_1 = m_1 a$$

After solving all equation:

$$\text{Acceleration: } a = \frac{(m_2 - m_3)g}{m_1 + m_2 + m_3}$$

- Tnesion in the Rope connecting m_1 and m_2:

$$T_2 = m_2 g\left(\frac{m_1 + 2m_3}{m_1 + m_2 + m_3}\right)$$

- Tension in the Rope Connecting m_1 and m_3

$$T_1 = m_3 g\left(\frac{m_1 + 2m_2}{m_1 + m_2 + m_3}\right)$$

16 Pseudo force concept

Pseudo force, also known as fictitious force, is an apparent force that arises when observing motion from a non-inertial (accelerating) frame of reference. It is used to apply Newton's laws of motion in non-inertial frames by introducing an additional force.

- Used to valid Laws of Motion in non-inertial frame (Accelerated frame).

$$\vec{F}_{pseudo} = -m\vec{a}_{frame}$$

- The pseudo force acts in the direction opposite to the acceleration of the reference frame.
- Apparent weight (Normal) and weight change nai hoga inertial/non inertial frame ho

Pendulum in Car

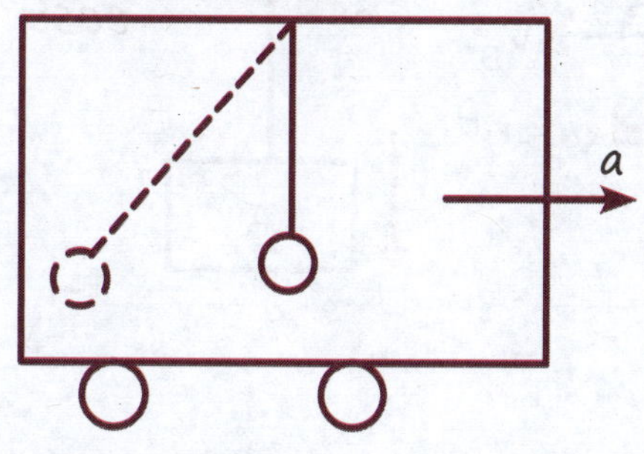

$$\sin\theta = \frac{a}{\sqrt{a^2+g^2}}$$

- Angle made by pendulum with the verticle $\theta = \tan^{-1}(a_o/g)$
- Tension in the pendulum string, $T = m\sqrt{a^2+g^2}$

Accl^n of incline plane so that block over it does not slip:-

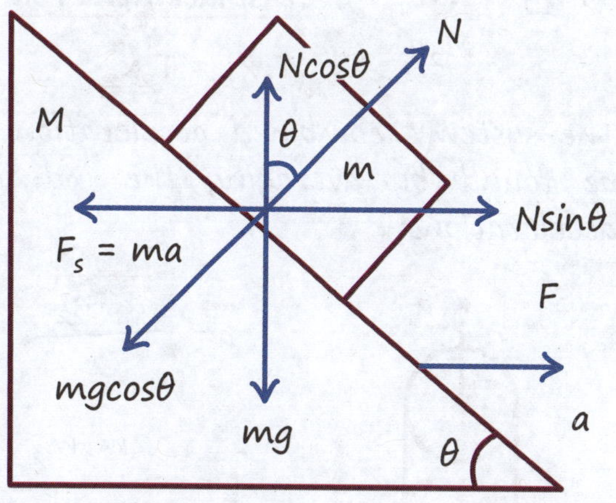

$$N = \frac{Mg}{\cos\theta}$$

$$(a = g\tan\theta) \Rightarrow f = (M + m)g\tan\theta$$

Accl^n of incline plane so that block over it can free fall:-

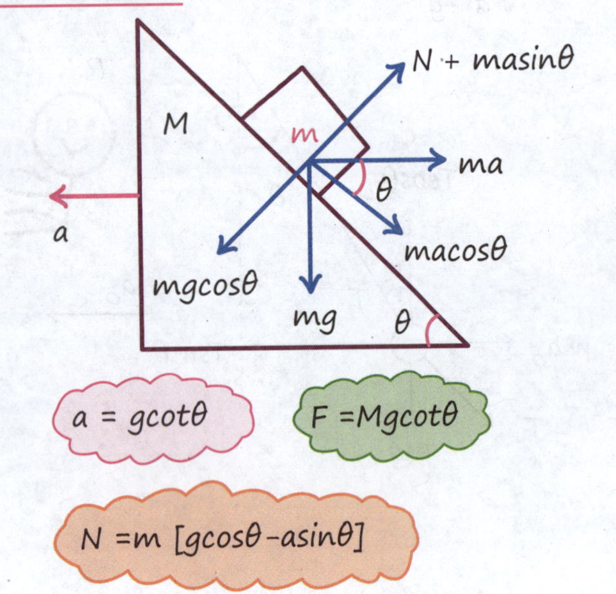

$a = g\cot\theta$

$F = Mg\cot\theta$

$N = m[g\cos\theta - a\sin\theta]$

17 Constraint Relation

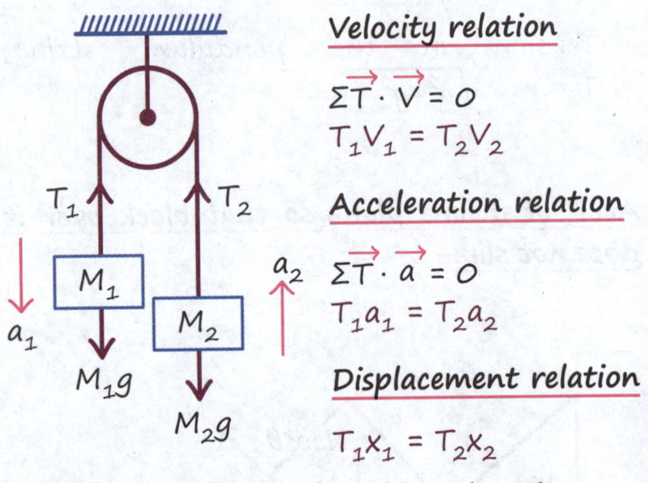

Velocity relation

$\Sigma \vec{T} \cdot \vec{V} = 0$

$T_1V_1 = T_2V_2$

Acceleration relation

$\Sigma \vec{T} \cdot \vec{a} = 0$

$T_1a_1 = T_2a_2$

Displacement relation

$T_1x_1 = T_2x_2$

- The system's combined acceleration can be found by averaging the individual accelerations:

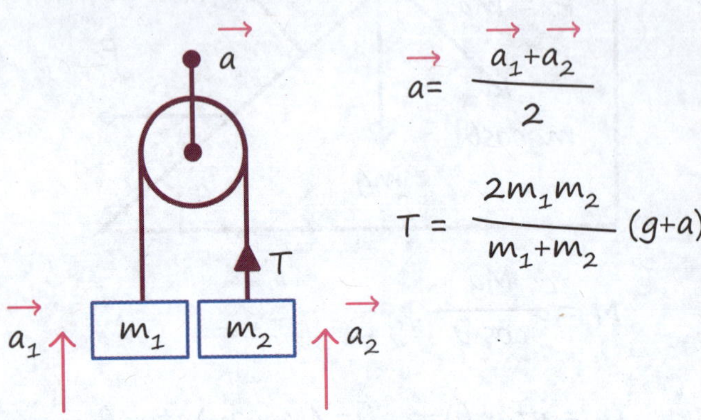

$$\vec{a} = \frac{\vec{a_1} + \vec{a_2}}{2}$$

$$T = \frac{2m_1m_2}{m_1 + m_2}(g+a)$$

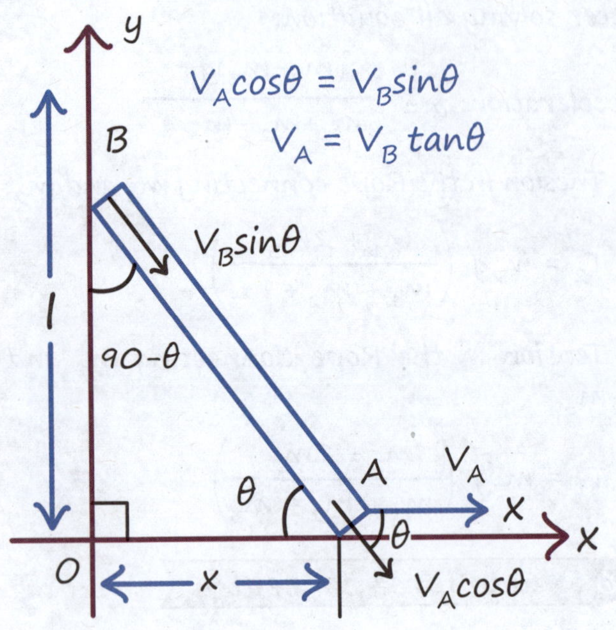

$V_A\cos\theta = V_B\sin\theta$

$V_A = V_B\tan\theta$

- Component of velocity along the length of Rod will be same.

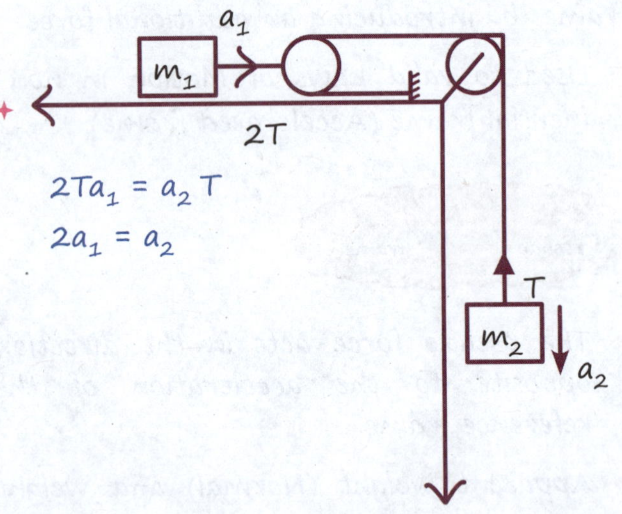

$2Ta_1 = a_2T$

$2a_1 = a_2$

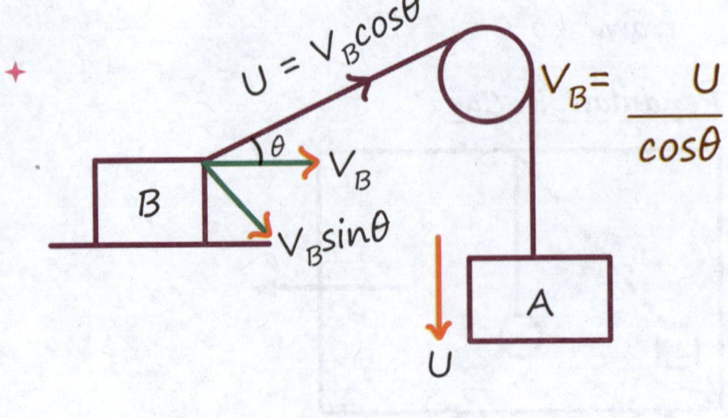

$$V_B = \frac{U}{\cos\theta}$$

Q. 8. Block is sliding on smooth inclined plane, then component of acceleration in vertical direction?

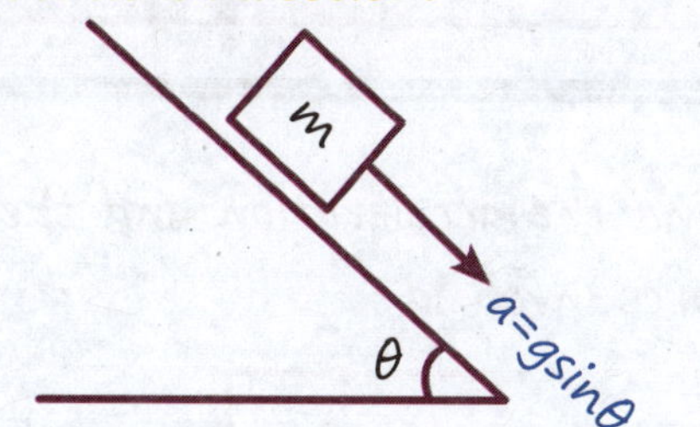

Sol.

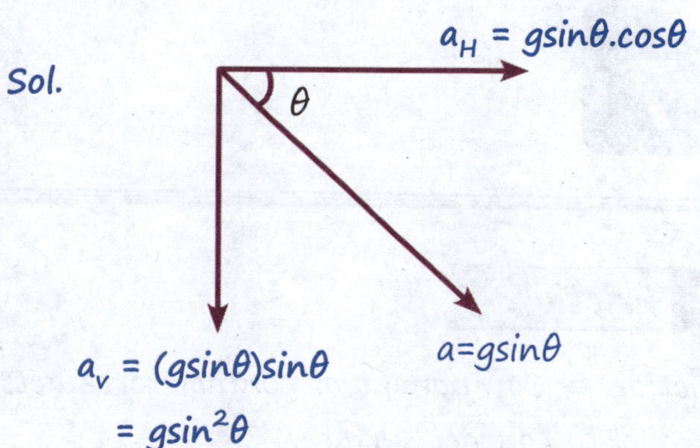

$a_v = (g\sin\theta)\sin\theta$

$= g\sin^2\theta$

"तुम किताबों के सामने झुक जाओ।
ये तुम्हारे सामने दुनियां झुका देंगी॥"

7 Friction

1 Friction

Friction → Component of contact force acts parallel to contact surface.

- Oppose relative motion or tendency of relative motion.

Type of Friction

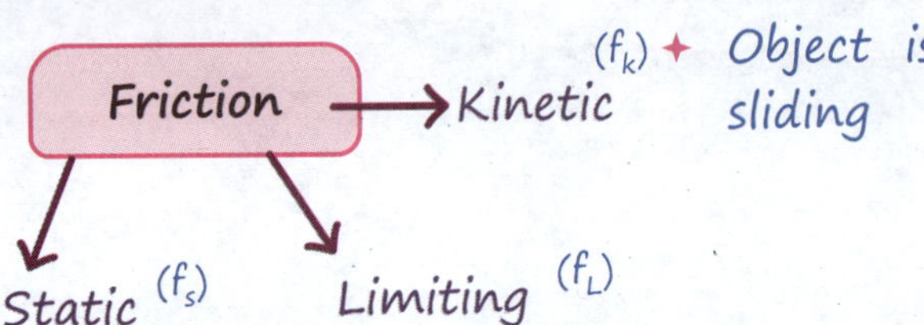

- Object is sliding (kinetic)
- Object is at rest, not sliding (static)
- Object about to move (limiting)

Self adjusting

Friction की औकाद

$f_s = F_{app}$ $\quad f_k = \mu_k N$ $\quad f_l = \mu_l N$

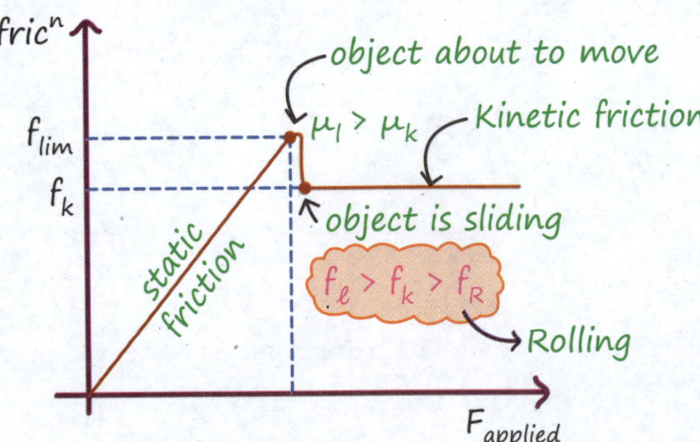

- Ramlal is walking in east then friction on Ramlal is static and direction along east.

MR*

For question solving

i> Find limiting friction Force.

ii> Compare it with Fapp

$F_{lim} > F_{app}$. [Rest] $[f_{app} = f_{static}]$

$F_{lim} < F_{app}$. [move]

Q. 1. Find the acceleration and the friction force involved

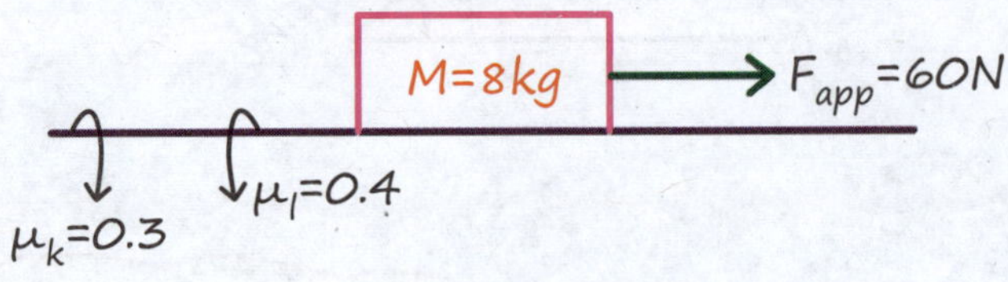

$$F_{limiting} = \mu N = \frac{4}{10} \times 80 = 32N$$

$F_{app} > F_{lim}$ object will move

$$a = \frac{60-\mu_k N}{M} = \left(\frac{60-24}{8}\right) m/s^2$$

$$f_{kinetic} = 24N$$

Q. 2. Find the acceleration and the friction force.

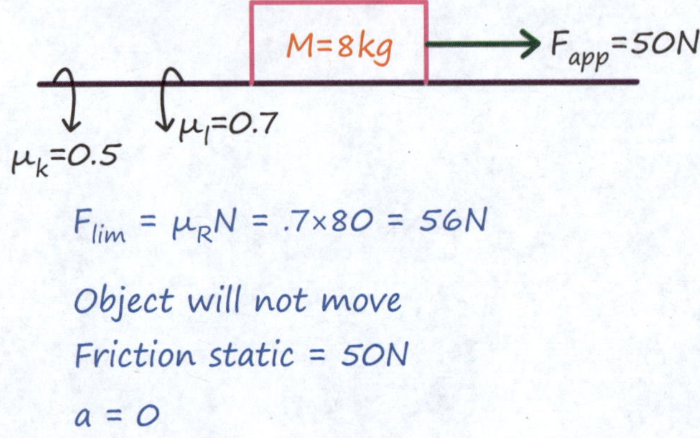

$$F_{lim} = \mu_R N = .7 \times 80 = 56N$$

Object will not move

Friction static = 50N

$a = 0$

- Object is thrown with velocity V_o on rough surface of coefficient of friction "μ" then stopping distance and time is:-

$$a = \mu g \qquad s = \frac{V_o^2}{2\mu g} \quad t = \frac{V_o}{\mu g}$$

- Object is moving with constant velocity as shown in figure then find a contact force between ground and block

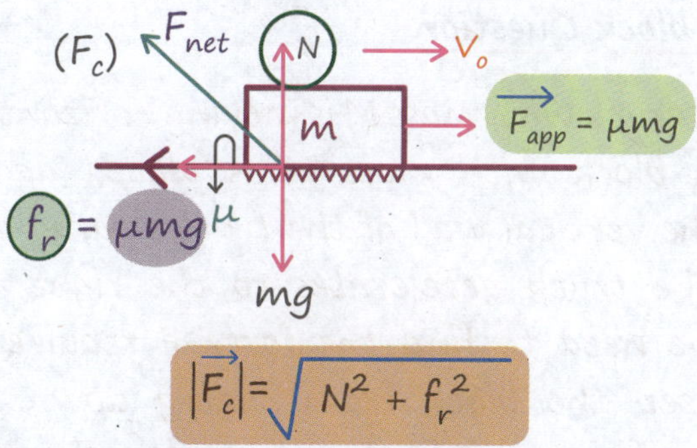

$$|\vec{F_c}| = \sqrt{N^2 + f_r^2}$$

- Since the object is moving at a constant velocity, the net force acting on it in the horizontal direction must be zero (Newton's first law of motion). This means the applied force F_{app} is balanced by the frictional force f_r.
- The contact force between the ground and the block is a combination of the normal force N and the frictional force f_r.

Example: Max. & Min value of F_{MR} such that block wont slide:-

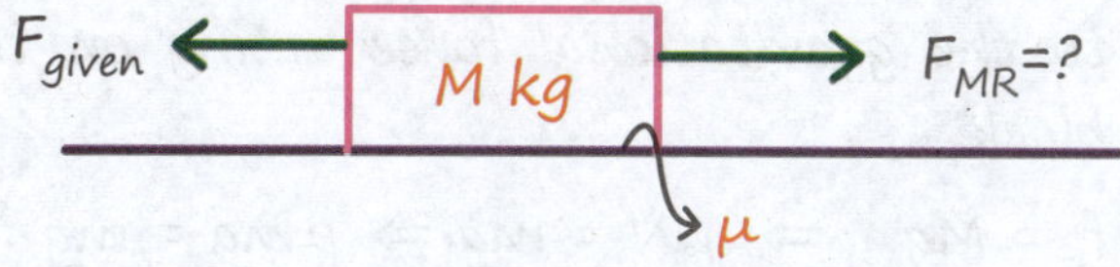

To prevent the block from sliding, the net force on the block must be within the limits of static friction. This gives us the following two conditions:

$$\left.\begin{aligned}(F_{MR})_{min} &= F_{given} - F_{lim}\\ (F_{MR})_{max} &= F_{given} + F_{lim}\end{aligned}\right\} F_{lim} = \mu N$$

2 Angle of friction

- The angle of friction is the angle between the normal force N and the resultant contact force F_C.

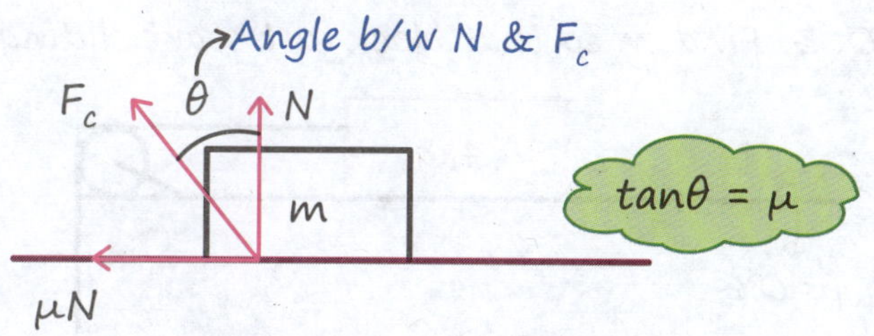

F_{min} require to move an object:-

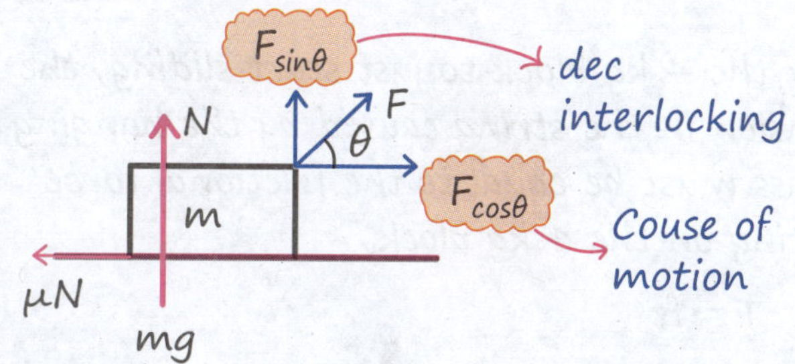

Net force along horizontal

$F\cos\theta = \mu N$

Net force along vertical

$N = mg - F\sin\theta$

Sdving for F

Table chain Question

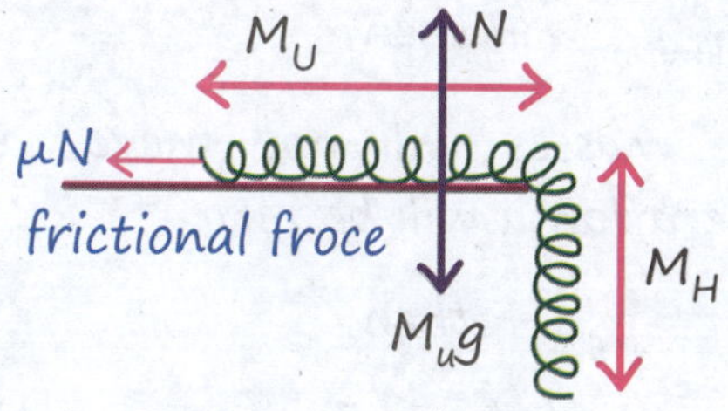

Tension,

$T = M_H g = \mu N$

$N = M_u g$

M_H = Hanging Mass. M_U = Upper Mass

Q. 3. Find m so that block just start sliding.

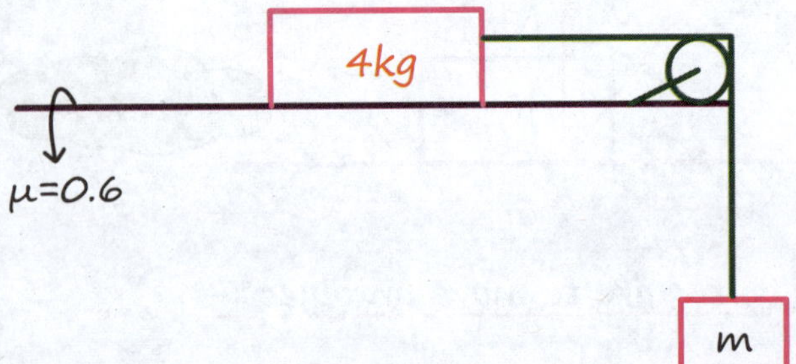

For the 4 kg block to just start sliding, the tension in the string caused by the hanging mass must be equal to the frictional force acting on the 4 kg block.

$T = fr$

$mg = f_r = \mu N$

$mg = 0.6 \times 4g$

$m = 2.4$ kg

Q. 4. Find acceleration and tension in wire?

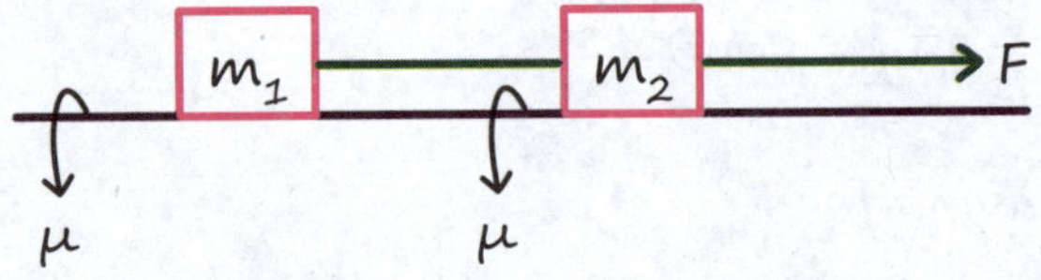

Net trictional force on two masses.

$F_{limiting} = f_l 1 + f_l 2$

$= \mu m_1 g + \mu m_2 g$

Condition for motion

If $F_{limiting} > F$ then,

- The masses will not move, and the acceleration a will be zero.

If $F_{limiting} < F$ then,

- The applied force F is greater than the limiting frictional force, the system will start to move, and we can calculate the acceleration using Newton's second law:

$$a = \frac{F - f_{lim}}{m_1 + m_2}$$

Now find tension by making f.b.d. of m_1

Car block Question

- A truck with mass M is moving horizontally. A block with mass m is placed against the vertical wall of the truck's cargo area. The truck accelerates to the right, and we need to find the force F required to keep the block from sliding down. The coefficient of friction between the block and the wall is μ.

1. Determine the acceleration a that the truck must have to prevent the block from sliding down.
2. Using this acceleration, find the total force F that must be applied to the truck.

- For the block to stay in place and not slide down, the frictional force must be equal to the gravitational force acting on the block.

$f_r = Mg \Rightarrow \mu N = mg \Rightarrow \mu ma = mg$

$$\mu = \frac{g}{a}$$

$$F = (M + m)\, a = \frac{(M + m)g}{\mu}$$

3 Friction Force on inclined plane

$f_{lim} = \mu N = \mu mg\cos\theta$

Angle of repose

Maximum angle (θ) on a **rough** inclined plane with **horizontal** such that the block kept on it remains at **rest**

$\mu = \tan\theta$ **Just about to slide**

MR*

When object is placed on rough inclined plane.

- Object chalega ki nai??

$\mu = \tan\theta$ Just slide

$\mu < \tan\theta$ Motion

f_r = up → $a = g\sin\theta - \mu g\cos\theta$ (Down)

f_r = down → $a = g\sin\theta + \mu g\cos\theta$ (Up)

$\mu > \tan\theta$ → REST

$F_c = \sqrt{N^2 + f_r^2}$

$F_c = Mg$

Q. 5. Find acceleration friction and contact force.

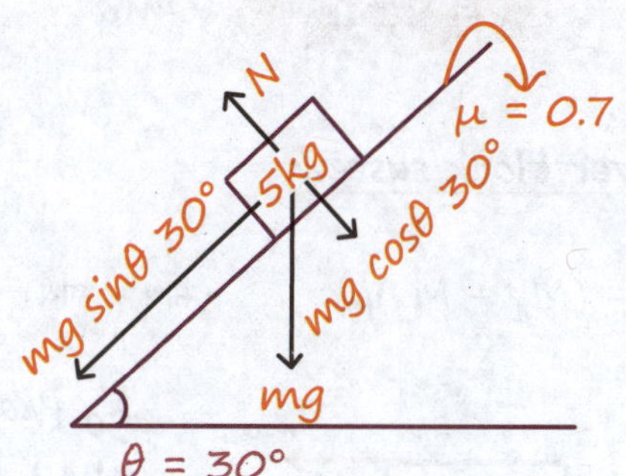

Sol. To determine if the block will slide, compare the parallel component of the gravitational force to the maximum frictional force.

Since μ = 0.7 and tan 30°

$= \frac{1}{\sqrt{3}} \approx 0.577$:

$\mu > \tan\theta$

since, $a = 0$ ($\mu_{limting} > mg\sin\theta$)

friction acting will be static

$f_s = f_{applied}$, $f_s = mg\sin\theta$

$= 5 \times 10 \times \frac{1}{2} = 25N$

$N = mg\cos\theta = 5 \times 10 \times \frac{\sqrt{3}}{2} = 25\sqrt{3}N$

$F_c = \sqrt{f_r^2 + N^2} = \sqrt{(25)^2 + (25\sqrt{3})^2}$

$= \sqrt{(25)^2(1+3)}$

$= 2 \times 25 = 50N$

Max & Min Force req. to move an object on incline:-

When moving the object up the incline, you need to overcome both the gravitational force pulling it down the incline and the frictional force resisting the motion.

- The maximum force required to move the object up:

$F_{max} = mg\sin\theta + \mu mg\cos\theta$ (up)

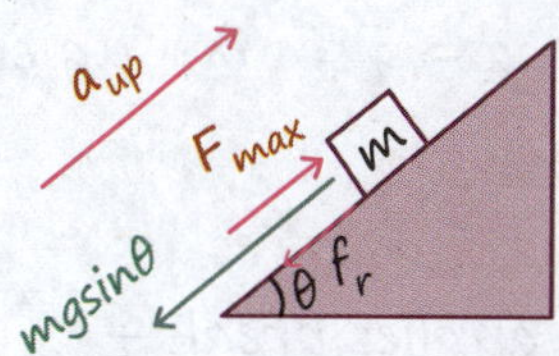

- When moving the object down the incline, the frictional force acts up the incline and you need to apply a force to overcome it.

$F_{min} = \mu mg\cos\theta - mg\sin\theta$ (down)

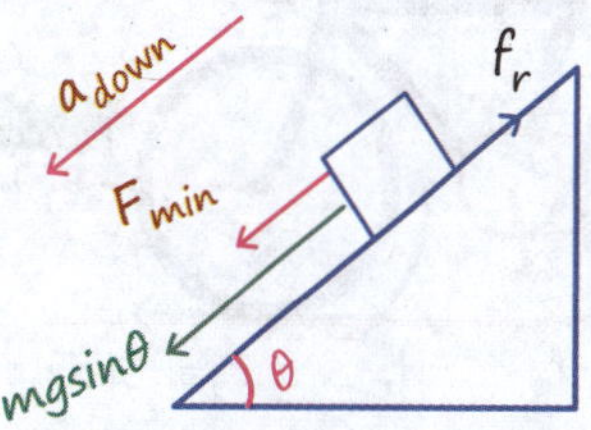

- If ($\mu < \tan\theta$) then force required to keep the object at rest

$F = mg\sin\theta - \mu mg\cos\theta$

+ Object is moving up to inclined plane

$a \text{ (downward)} = g\sin\theta + \mu g\cos\theta$

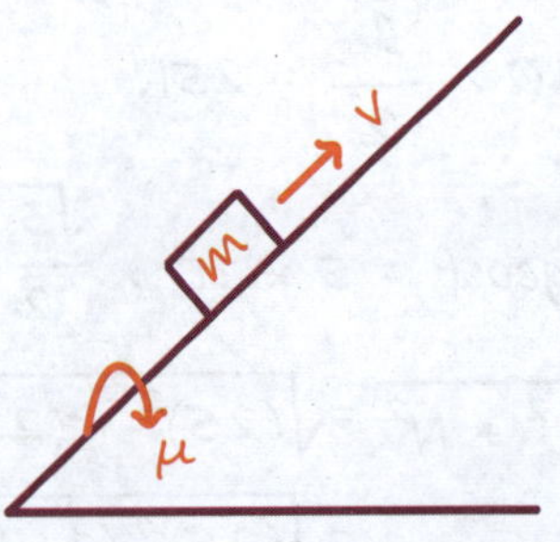

S (Stoping distance on inclined) $= \frac{V^2(\text{intial})}{2(\mu g\cos\theta + g\sin\theta)}$

Note:-

1> Friction kishika saga nai hai uska koi fix dir^n nai h

2> Uska ekhi udeesh hai woh relative motion nai hone dega!

If object sliding downward then acceleration, $a = g\sin\theta - \mu g\cos\theta$

If object is sliding upward then acceleration, $a = g\sin\theta + \mu g\cos\theta$

Note:-

+ When Ramlal applies break:-

f_r will be backward on front tyres.

+ The upper half of incline plane of the inclination is perfectly smooth and the lower half is rough. A block starting from rest at the top of the plane will again come to rest at the bottom if the coefficient of friction between the block and lower half of the plane is given by:-

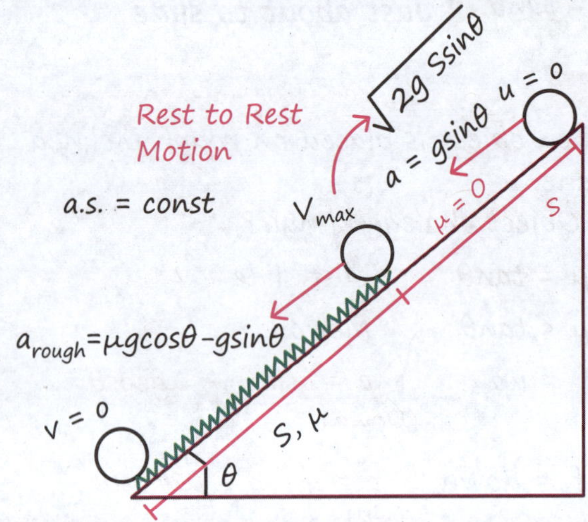

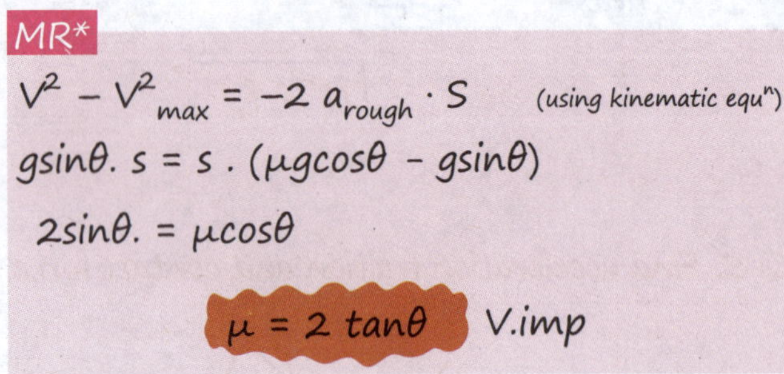

MR*

$V^2 - V^2_{max} = -2\, a_{rough} \cdot S$ (using kinematic equ^n)

$g\sin\theta.\, s = s\,.\, (\mu g\cos\theta - g\sin\theta)$

$2\sin\theta. = \mu\cos\theta$

$\mu = 2\tan\theta$ V.imp

Block over block system

1> $F = (M_1 + M_2)\mu g$ (साथ - साथ)

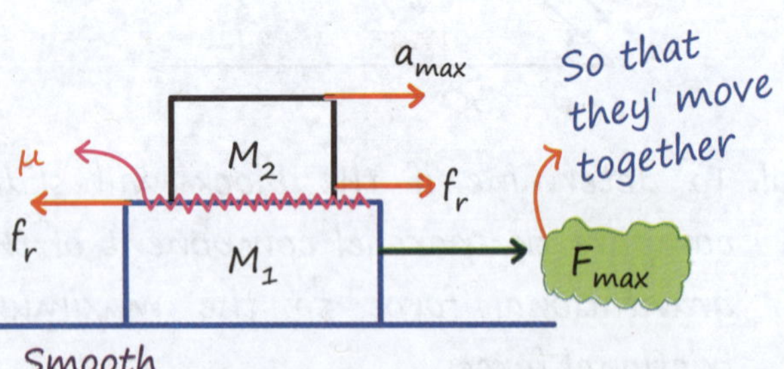

+ The maximum acceleration (a_{max}) at which M_2 can be moved without slipping over M_1 is given by the maximum frictional force divided by M_2:

$a_{max} = \mu g$ } Friction की max औकात है Iss Accl^n से Chalane Ki

- The force required to move both blocks together is

$F_{max} = (M_1 + M_2)a_{max}$

- If the applied force Fapp exceeds the maximum frictional force, M_2 will start sliding over M_1.

For M_2 to slide over M_1, the force F_{app} must be greater than the maximum frictional force that can be applied:

$F_{app} > (m_1 + m_2)\mu g$

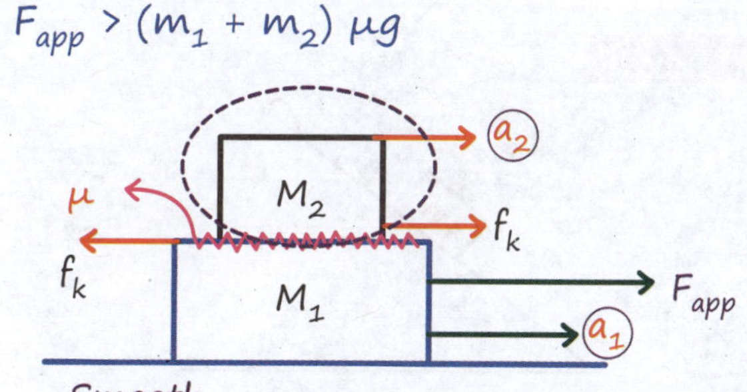

M_2:- Iska maximum acc^n μg hi hoga. Isse jayada nahi ho sakta hai.

"a_1" Ki value "F" पर depend karegi "a_1", ka koi limit nai hai!

$a_2 = \mu g$ } Upar wala issi accl^n से jayega

Kyuki isko friction hi leke jaane wala hai!

$a < a_{max}$	$a = a_{max}$	$a > a_{max}$
Move Together	About to slip!	slip!
f_r = static	f_r = limiting	f_k = kinetic diff. accl^n

$\because a = \frac{F_{app}}{M_{net}}$

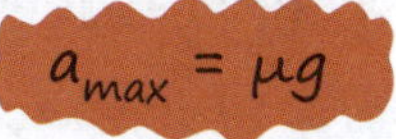

Note:

If F_{app} on upper block:-

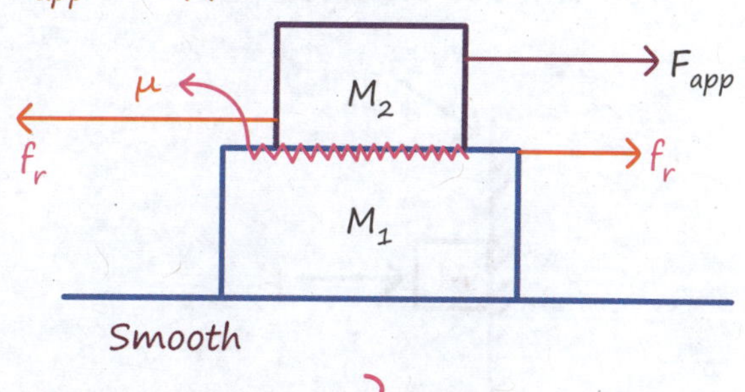

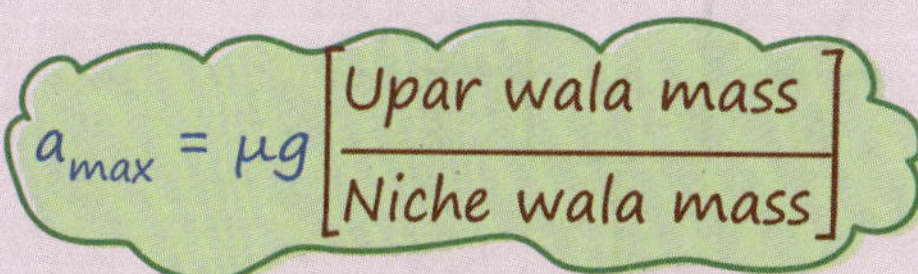

MR*

$a_{max} = \mu g \left[\frac{\text{Upar wala mass}}{\text{Niche wala mass}}\right]$

* An insect crawls up a hemispherical surface very slowly as shown in figure. The coefficient of friction between the insect and the surface is $\frac{1}{3}$. If the line joining the centre of the hemispherical surface to the insect makes an angle α with the vertical, the maximum possible value of α is given?

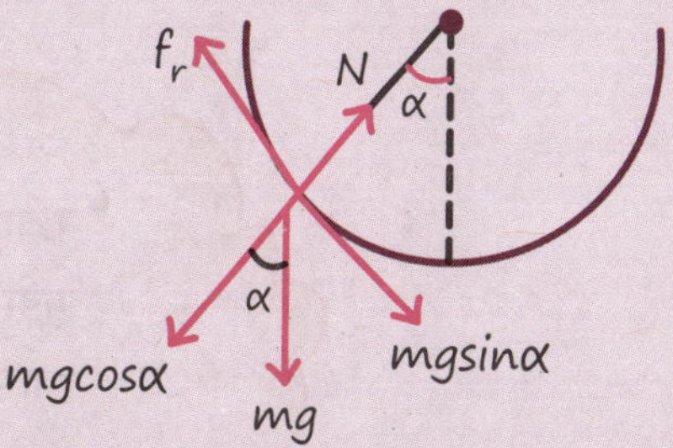

Sol. For insect to be in equilibrium, frictional force should balance the parallel component of the gravitational force.

$f_r = mg\sin\alpha$

$\mu mg\cos\alpha = mg\sin\alpha$

$\mu = \tan\alpha$

Q. 6. If object is at rest then find friction force acting on this object?

Sol. Object is at rest hence static friction force will act on object

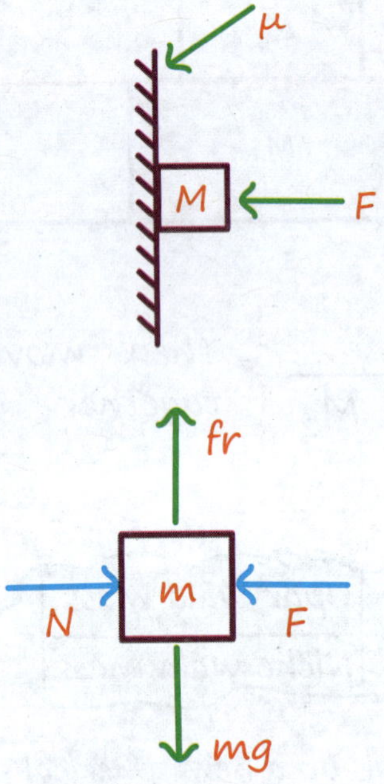

$(Fr)_{static}$ = applied force which create tendency of motion = mg

$f_{limiting} = \mu N = \mu F$ this will act when object about to move

Q. 7. Find acc^n so that block does not slide down?

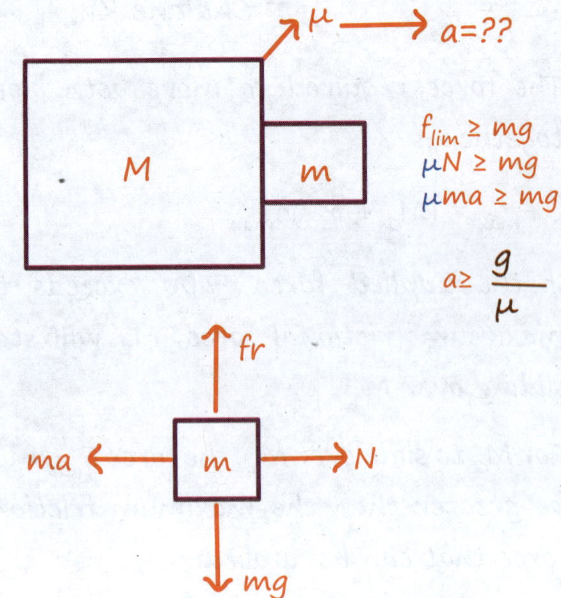

MR* Sawal

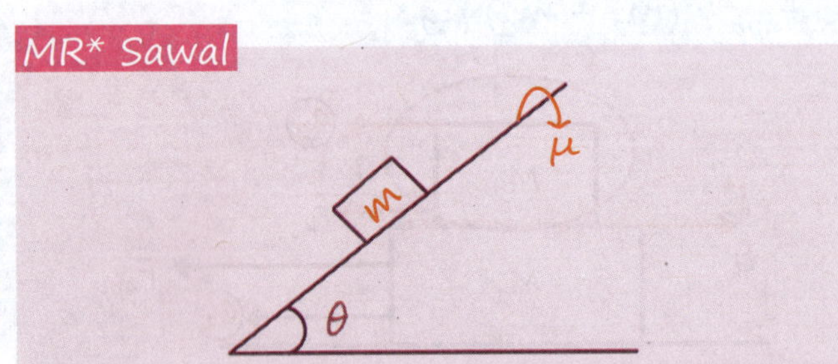

If Angle of Inclination θ increase from 0° then contact force remains constant then decreases.

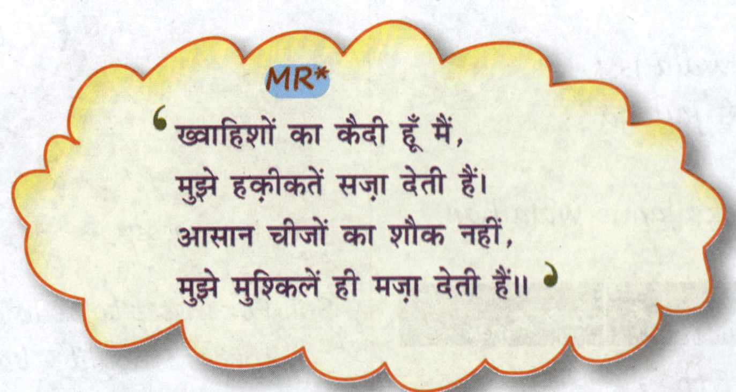

8 Circular Motion

1 Circular motion

When a particle moves in a plane such that its distance from a fixed point (the center) remains constant, this motion is called circular motion. The constant distance from the center is called the radius.

2 General Formula

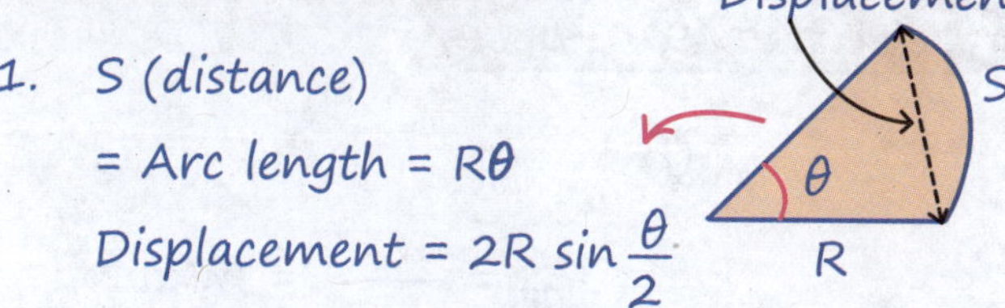

1. S (distance)

   = Arc length = $R\theta$

   Displacement = $2R \sin\frac{\theta}{2}$

   Angular displacement = θ
   - always in Radian
   - Anti-clockwise ⊙ (positive θ)
   - Clockwise ⊗ (negative θ)

   Angular disp^m in π rotation,
   $\theta = \pi(2\pi) = 2\pi^2$

Note:

one complete rotation around a circle corresponds to an angular displacement of 2π radians.

2. V (speed) = $R\omega$

   $\vec{V} = \vec{\omega} \times \vec{R}$

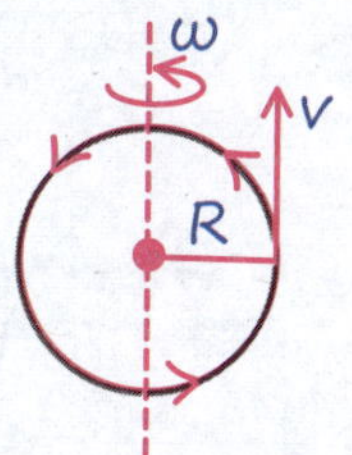

3. Certripetal Accl^n:-

$$a_c = \frac{V^2}{R} = R\omega^2 = \omega V$$

Ye always lagega agar circular motion hai toh.

4. Tangential acceleration:-

$$a_t = \frac{d|v|}{dt} = \alpha R$$

Ye sirf varying speed Ke samay.

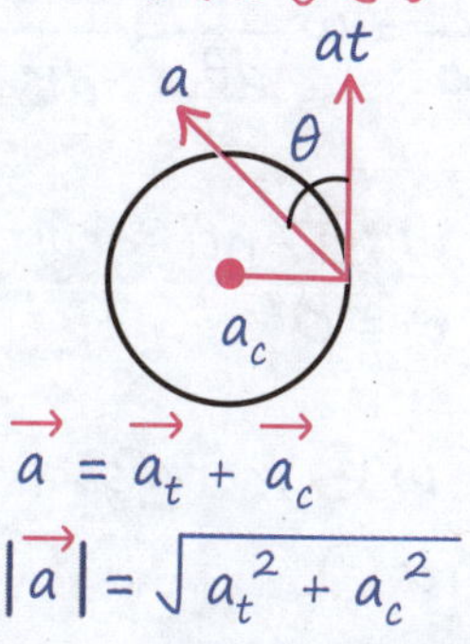

$$\vec{a} = \vec{a_t} + \vec{a_c}$$

$$|\vec{a}| = \sqrt{a_t^2 + a_c^2}$$

Tangential acceleration is the acceleration that occurs when the speed of an object moving along a circular path changes (i.e., the object speeds up or slows down). alpha is the angular acceleration, and R is the radius of the circle.

Net accleration.

5. $|\Delta\vec{\gamma}| = 2r \sin\frac{\theta}{2}$

MR**

$$|\Delta\vec{v}| = 2v \sin\frac{\theta}{2}$$

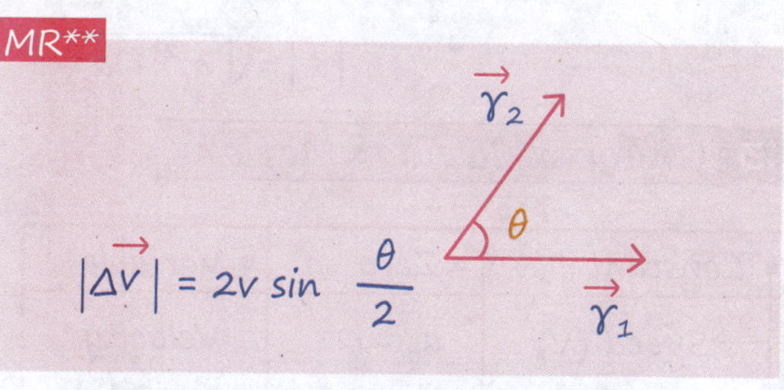

6. $\vec{\omega}_{inst} = \frac{d\theta}{dt}$ $\quad$ $\vec{\omega}_{Avg} = \frac{\Delta\theta}{\Delta t}$
   - axial vector
   - ω(Anti) ⊙
   - ω(clock) ⊗

7. Time, period, $T = \frac{2\pi}{\omega}$

Freqency $= \frac{1}{T}$, $\omega = 2\pi f$

8. Angular acceleration (α)

$$\vec{\alpha}_{Avg} = \frac{\Delta\omega}{\Delta t}$$

(Axial Vector)

$$\vec{\alpha}_{Inst} = \frac{d\omega}{dt} = \omega\frac{d\omega}{d\theta} = \frac{d^2\theta}{dt^2}$$

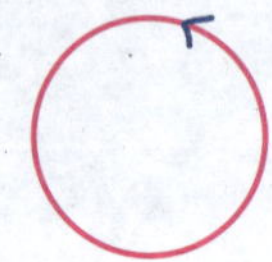

$\omega \odot$ Anti = $cost^n$

$\alpha = 0$

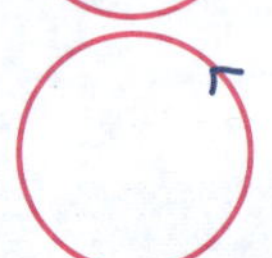

$\omega \odot$ increase ↑

$\alpha \odot$

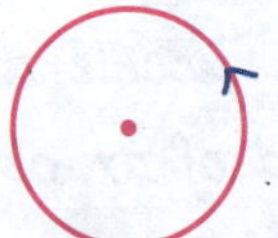

$\omega \odot$ Anti ↓ decreases

$\alpha \otimes$ Clock

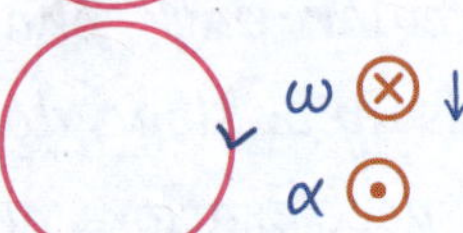

$\omega \otimes \downarrow$

$\alpha \odot$

$\vec{a}$	$\vec{a}_t$	$\vec{a}_c$	$\vec{\alpha}$

$a_t = R\alpha$ $\quad \vec{a}_c = V^2/R \quad \vec{a} = \vec{a}_t + \vec{a}_c$

$$|\vec{a}| = \sqrt{a_t^2 + a_c^2}$$

3 Uniform Circular Motion

• Constant	• Zero	• Variable
→ Speed (V)	$a_t = 0$	Velocity
→ Angular Speed (ω)	$\alpha = 0$	direction
→ K.E.	Work = 0	acc^n
→ $\vec{L}$ (Angular Momentum)	$\tau = 0$	Force Momentum

If Tangential acc, $a_t = 0$

Then, $\vec{a} = \vec{a}_c = \frac{V^2}{R}$ towards Centre

$$(\vec{a}_c)\ Avg = \frac{V^2}{R}\,\frac{\sin\frac{\theta}{2}}{\frac{\theta}{2}}$$

✦ Dir^n of centripetal acc^n is variable but it's magnitude is constant

4 Centripetal Force

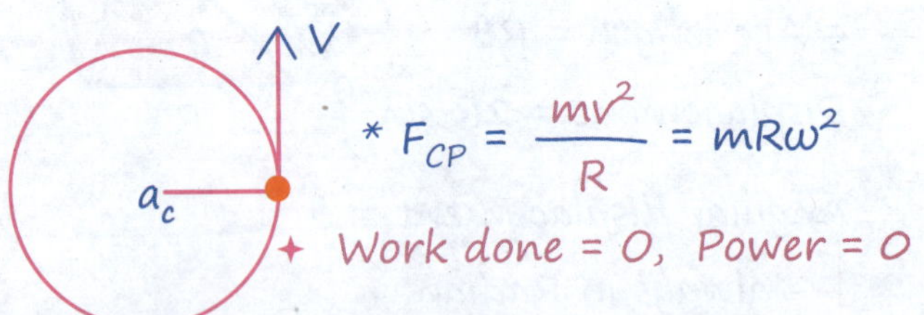

$$* \ F_{CP} = \frac{mv^2}{R} = mR\omega^2$$

✦ Work done = 0, Power = 0

Two Car moving with different speed V_1 & V_2 on circular path of radius r_1 & r_2 with same time period then ratio of angular speed $\omega_1 : \omega_2 = 1 : 1$

5 Non-uniform circular motion

$$\alpha_{avg} = \frac{\vec{\omega}_2 - \vec{\omega}_1}{\partial t} = \frac{\partial\omega}{\partial t}$$

$$a_{net} = \sqrt{a_c^2 + a_t^2}$$

$$a_{net} = \vec{\alpha} \times \vec{R} + \vec{\omega} \times \vec{V}$$

$$\vec{a}_{net} = \vec{a}_t + \vec{a}_c$$

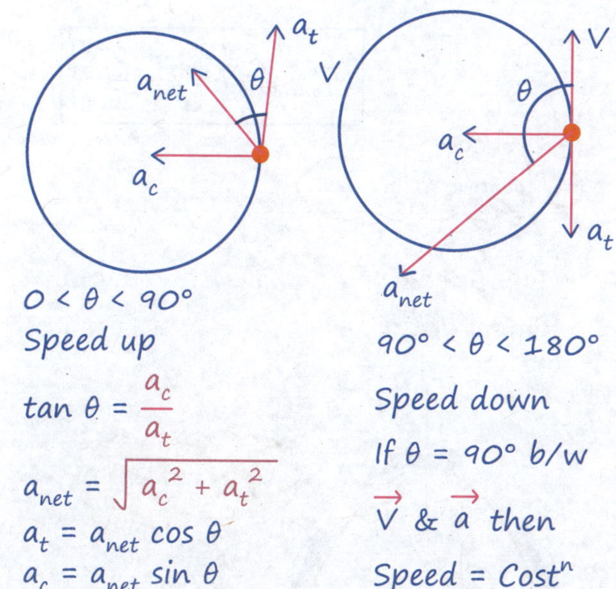

$0 < \theta < 90°$

Speed up

$\tan\theta = \frac{a_c}{a_t}$

$a_{net} = \sqrt{a_c^2 + a_t^2}$

$a_t = a_{net}\cos\theta$

$a_c = a_{net}\sin\theta$

$90° < \theta < 180°$

Speed down

If $\theta = 90°$ b/w $\vec{V}$ & $\vec{a}$ then

Speed = Cost^n

Q.1 Object starts circular motion from rest and tangential acceleration $4\ m/s^2$ find acceleration of object at t = 3 sec if radius is 48m.

Sol. Non uniform circular motion.

$a_t = 4 m/s^2$ $V = u + at$

$V = 4 \times 3 = 12\ m/s$

$$a_c = \frac{V^2}{R} = \frac{12\times12}{48} = 3 m/s^2$$

$$a = \sqrt{a_t^2 + a_c^2} = 5 m/s^2$$

Q.2 A particle moves on circle of 5cm with constant time period 0.2πs then find acceleration.

$$a_c = w^2R = \frac{4\pi^2}{T^2}R = 5 m/s^2$$

6 Circular Motion

Kinematical Equation:-

$\omega = \omega_o + \alpha t$	$V = U + at$
$\theta = \omega_o t + \frac{1}{2}\alpha t^2$	$S = Ut + \frac{1}{2}at^2$.
$\omega^2 - \omega_o^2 = 2\alpha\theta$	$V^2 - U^2 = 2aS$
$\theta_n^{th} = \omega_o + \alpha\left[\frac{2n-1}{2}\right]$	$S_n^{th} = U + a\left[\frac{2n-1}{2}\right]$

MR**

Motion starts from rest with constant Angular acceleration then Angle rotated in 1s, 2s & 3s are in ratio 1 : 4 : 9 & in 1st sec : 2nd sec : 3rd sec = 1 : 3 : 5.

Stopping Angle, $\theta = \frac{\omega_o^2}{2\alpha}$ (α = Angular retardation)

MR** → "F_{pseudo}" is used to validate Newton's law in Non-inertial frame.

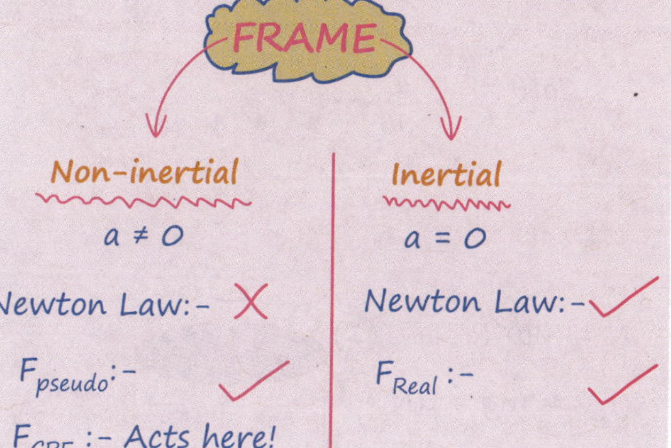

MR*

1. Semi spherical bowl:-

$$\frac{N\sin\theta}{N\cos\theta} = \frac{mv^2}{r\,mg}$$

$$\left[\tan\theta = \frac{V^2}{rg}\right]$$

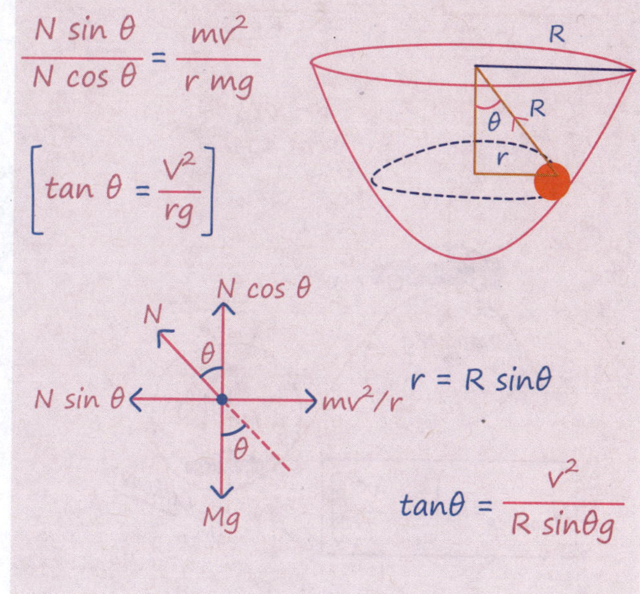

$r = R\sin\theta$

$$\tan\theta = \frac{v^2}{R\sin\theta g}$$

MR*

2. Cone:-
A smooth cone is rotated with an angular velocity ω as shown. A block is placed at height h. The block has no motion relative to the cone. then find valu of h.

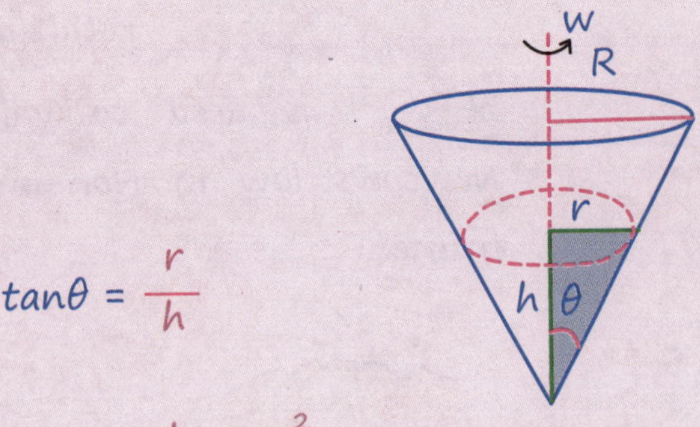

$\tan\theta = \frac{r}{h}$

$\cot\theta = \frac{h}{r} = \frac{v^2}{rg}$

$h = \frac{\omega^2 R^2}{g}$

3. Death well:- Or Rotor

$N = m\omega^2 R$ (1)

$(f_r)s = mg = \mu N$

$mg = \mu\omega^2 Rm$

$\omega^2 = \frac{g}{\mu R}$

4. Car moving on a convex & concave bridge with uniform speed:-

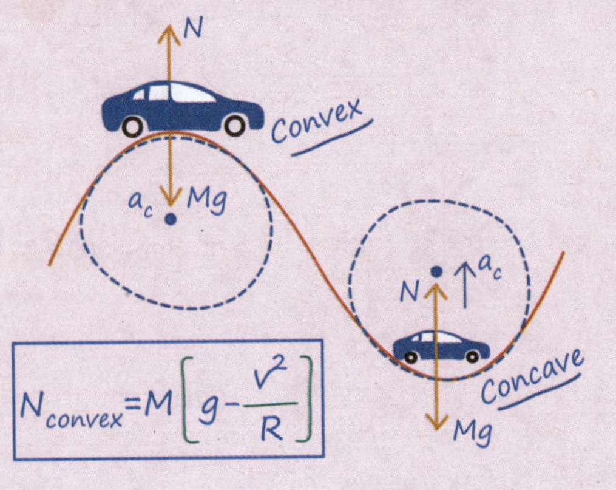

$N_{convex} = M\left[g - \frac{v^2}{R}\right]$

$N_{concave} = M\left[g + \frac{v^2}{R}\right]$

$N_{Concave} > N_{convex}$

5. Max. Speed of a vehicle to move on a convex bridge:-

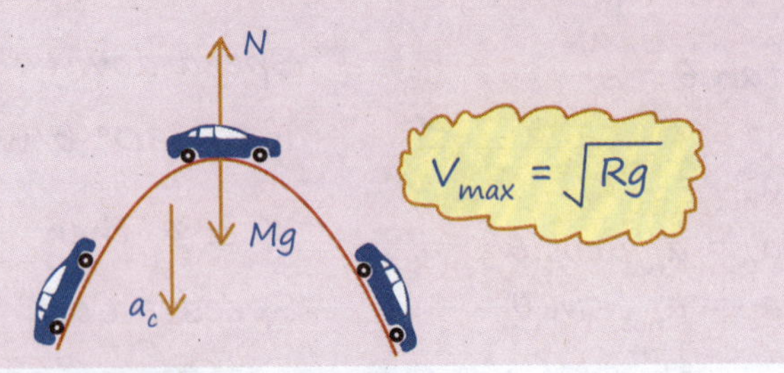

$V_{max} = \sqrt{Rg}$

7 Banking of Roads

Case-I:- Rough horizontal Road:-

(Sirf Friction)

$V_{max} = \sqrt{\mu Rg}$ $\omega_{max} = \sqrt{\frac{\mu g}{R}}$

Case-II:- Smooth Banked Road:-

(Sirf Banking)

$V_{max} = \sqrt{Rg\tan\theta}$

Case-III:- Rough Banked Road:-

(Banking + Friction)

$$V_{max} = \sqrt{Rg\left[\frac{\tan\theta + \mu}{1 - \mu\tan\theta}\right]}$$

$$V_{min} = \sqrt{Rg\left[\frac{\tan\theta - \mu}{1 + \mu\tan\theta}\right]}$$

Safe Ride:- $V_{min} < V < V_{max}$.

✦ Bending of cyclist: $\tan\theta = \frac{v^2}{Rg}$

θ = Angle bend by cyclist from vertical, V = velocity of cyclist, R = Radius of circular path.

8 Conical Pendulum

$\tan\theta = \frac{v^2}{rg}$

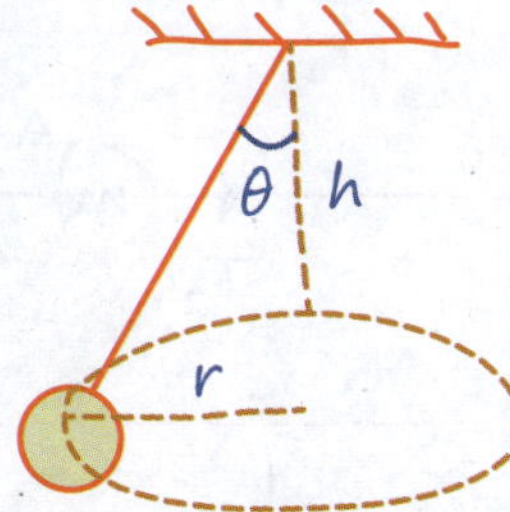

MR* Special

- value of acceleration due to gravity of a planet using Conical Pendulum:-

$V = \sqrt{Rg\tan\theta} = \frac{2\pi r}{T}$ $\quad g = \frac{4\pi^2 r}{T^2 \tan\theta}$

- Time period of Conical Pendulum:-

$T_P = 2\pi\sqrt{\frac{l\cos\theta}{g}}$

$T_P = 2\pi\sqrt{\frac{h}{g}}$

$T_P = 2\pi\sqrt{\frac{(l^2 - r^2)^{1/2}}{g}}$

(Triangle with sides l, h, r and angle θ)

9 Vertical Circular Motion

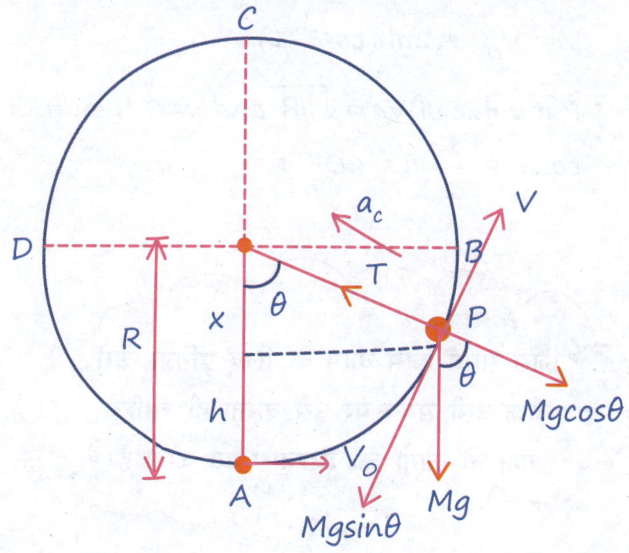

$T_P = mg\cos\theta + \frac{MV^2}{R}$

$V^2 = V_o^2 + 2gR(\cos\theta - 1)$

$T_P = \frac{MV_o^2}{R} - 2mg + 3\,mg\cos\theta$

Max $\quad T_A(\theta = 0) = \frac{MV_o^2}{R} + mg$

$T_B(\theta = 90) = \frac{MV_o^2}{R} - 2mg$

Min $\quad T_C(\theta = 180) = \frac{MV_o^2}{R} - 5mg$

$T_D = \frac{MV_o^2}{R} - 2mg$

$T_A - T_C = 6mg$

$T_A - T_B = 3mg$

$T_A - T_D = 3mg$

$T_B - T_D = 0$

- If reference is at 'A' for potential then total M.E $= \frac{5}{2}$ MgR.
- Work done by tension is zero.
- In case of critical condition $V = \sqrt{5gR}$
- If $0 < V \le \sqrt{2gR}$ → oscillate

  $\theta_{max} \le 90°$ → At extreme $T \ne 0$; $V = 0$

  Ex: $V_o = \sqrt{gR}$ $\quad \theta_{max} = 60°$

  $V_o = \sqrt{2gR}$ $\quad \theta_{max} = 90°$

- If $\sqrt{2gR} < V_o \le \sqrt{5gR}$ Parabolic path at θ_{max} $V \ne 0$, $T = 0$, $90 < \theta_{max} \le 180°$.
- If $V_o = \sqrt{5gR}$ Critical velocity to complete vertical circle.
- If object (bob) connected with massless rod, then $V_o = \sqrt{4gR}$ to complete vertical circular motion.

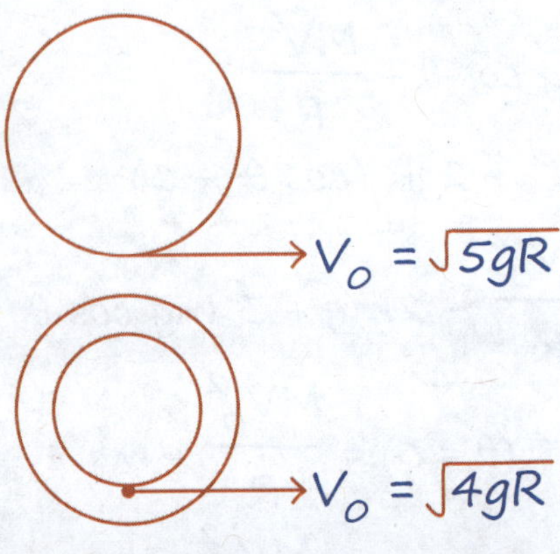

10 Particle Leaving Contact with HEMIsphere

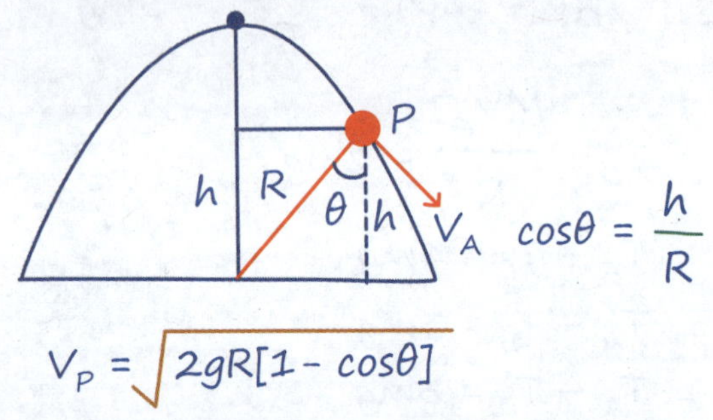

$\cos\theta = \frac{h}{R}$

$V_P = \sqrt{2gR[1 - \cos\theta]}$

- find minimum height h, so that body may succesfully completethe loop of radius R.

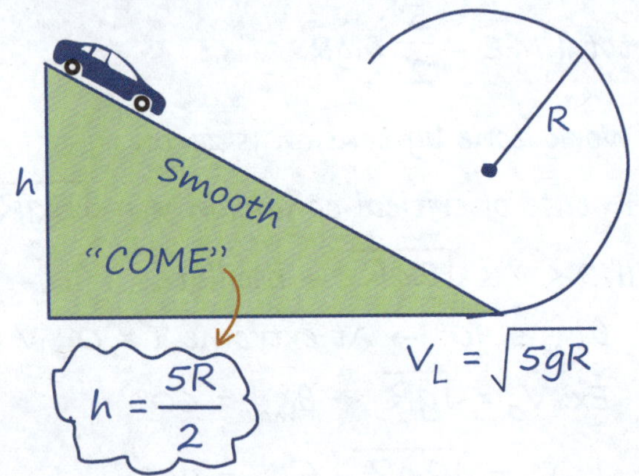

- Concept of motion of light rigid rod.

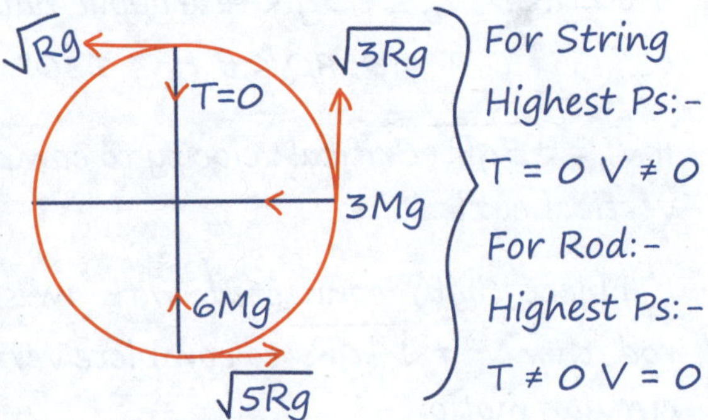

For String
Highest Ps:-
$T = 0 \; V \neq 0$
For Rod:-
Highest Ps:-
$T \neq 0 \; V = 0$

Q.3. Two blocks of mass m connected with string of length l then. Find T_1 & T_2.

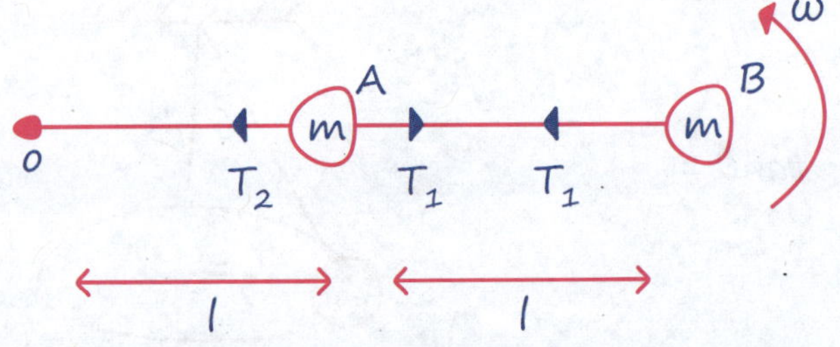

FBD of A and B w.r.t. Non inertial frame.

$T_1 = m\omega^2 (2l)$

$T_2 = T_1 + m\omega^2 (l) = m\omega^2 (2l) + m\omega^2 (l)$

$T_2 = 3\, m\omega^2 l$

Q.4. Object is given velocity $V = \sqrt{3gR}$ at mean position where it will leave circular path.

Sol. V (given) > $\sqrt{2gR}$ hence it will leave circular path where tension becomes zero.

$$T = \frac{mV^2}{l} - 2mg + 3mg\cos\theta$$

Put value of V and T find θ

$$\cos = -\frac{1}{3}$$

Q.5. If V (given) $V > \sqrt{gR}$ then it will leave circular path where velocity becomes zero

$V = V_0^2 + 2gl(\cos\theta - 1)$

Put value of $Vg = \sqrt{gR}$ and V=0 then find

$\cos\theta = \frac{1}{2}$ $\theta = 60°$

MR*
"जिस समय जिस काम के लिए प्रतिज्ञा करो, ठीक उसी समय पर उसे करना ही चाहिये, नहीं तो लोगों का विश्वास उठ जाता है।"

9 Work, Energy, and Power

1 Effects of force

Force

- How long (time): $F \cdot \Delta t = \Delta P$ → When force (F) is applied over a period of time (Δt), it changes the momentum (ΔP) of an object.
- How Far (disp^m, S): $\vec{F} \cdot \Delta \vec{S} = \text{Work}$ → When force ($\vec{F}$) is applied over a displacement (ΔS), it does work.

2 Work Done

By Constant Force:-

$$W = \vec{F} \cdot \vec{S} = FS \cos \theta$$

where,

F = Force, θ = Angle b/w Force and the displacement,

S = displacement of point of application of force

- → Scalar, unit (Joule), Dimension, (ML^2T^{-2})
- → depends on Frame (vary based on the observer's point of view.)

W_{Total} = Scalar sum of all the work by individual force.

By Variable Force:-

$$\int dW = \int F.dS$$

$$W_{Total} = \int_{x_1}^{x_2} F_x d_x + \int_{y_1}^{y_2} F_y d_y + \int_{z_1}^{z_2} F_z d_z$$

Nature of work

How the angle between the force and the displacement affects the work done, and the corresponding effects on speed and kinetic energy (KE) are shown in following table:

Work = +ve	Work = 0	Work = –ve
$0° \le \theta < 90°$ Speed ↑ (Increase) KE ↑ (Increase)	$\theta = 90°$ W = 0 Speed = const^n K.E. = const^n F_c, V	$90° < \theta \le 180°$ Speed ↓ (Decrease) KE ↓ (Decrease)

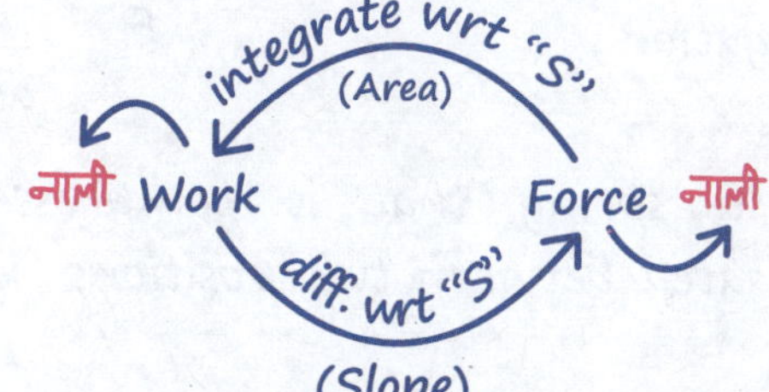

* Man is moving up or down on stairs then work done by normal force is zero.

3 Work done by different forces

Work done By Gravity:-

- ✦ $WD_{gravity} = F_{vertical} (S_{vertical}) = mg\,H$
  $= mg\,(S \sin \theta)$
  $H = S \sin \theta$
- ✦ Path independent.

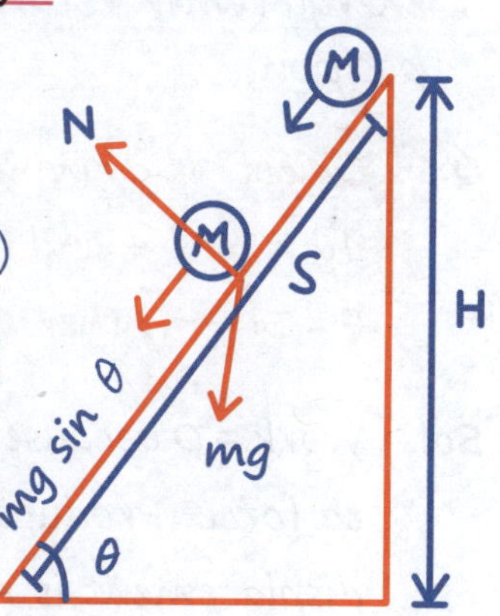

Work Done By Friction:-

$WD = -f_r \cdot S$

(–ve sign indicates force acting opposite to the direction of motion)

Path dependent

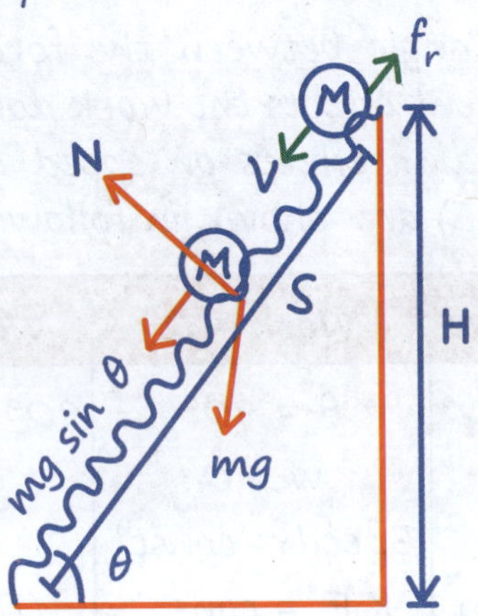

Work Done By Spring Force:-

$$W = -\frac{1}{2}Kx^2$$

here, x = compression or elongation

MP

x_1

F_s

F_{ext}

$\vec{dx}$

- For a spring that is compressed or elongated between two positions x_1 and x_2.

  - $W = -\frac{1}{2}K\left[x_2^2 - x_1^2\right]$

- The spring either stores energy (when compressed or stretched) or releases energy (when returning to its equilibrium position).

Q.1. Object is moving on a straight line $4y = 3x + 4$ and force acting on object is $F = 3\hat{i} - 4\hat{j}$ then work done by this force.

Sol. Work = 0 because disp^m is perpendicular to force. Product of slope of force and displacement is -1

Q.2. Force acting on object $\vec{F} = 2x\hat{i} + 3y^2\hat{j}$ find work when object displace from (0, 2, 3) to (2, 2, 0)

Sol. $dW = F_x dx + f_y dy$

$$\int dW = \int 2x\,dx + \int 3y^2 dx$$

$$\text{Work} = 2\left(\frac{x^2}{2}\right)_0^2 + 3\left(\frac{y^2}{3}\right)_2^2 = (2)^2 + 0 = 4J$$

Q.3. If $\vec{F} = y\hat{i} + x\hat{j}$ then find work when object displace from (1, 2, 3) to (4, 6, 7).

Sol. $dW = F_x dx + F_y dy$

Hint: $ydx + xdy = d(xy)$

$$\int dW = \int (ydx + xdy)$$

$$\text{work} = \int d(xy) = (xy)_{1,2}^{4,6}$$

$$= 6 \times 4 - 1 \times 2 = 22J$$

4 Kinetics energy and Momentum

$$P = mv, \quad KE = \frac{1}{2}mv^2 = \frac{P^2}{2m}$$

P = magnitude of momentum

KE = Energy stored due to motion (Scalar)

$$P = \sqrt{2m\,(K.E)}$$

Statement-1: Two object having same mass and momentum having same K.E. → True

Statement-2: Two object having same mass and K.E. having same momentum → False

- Graphs:-

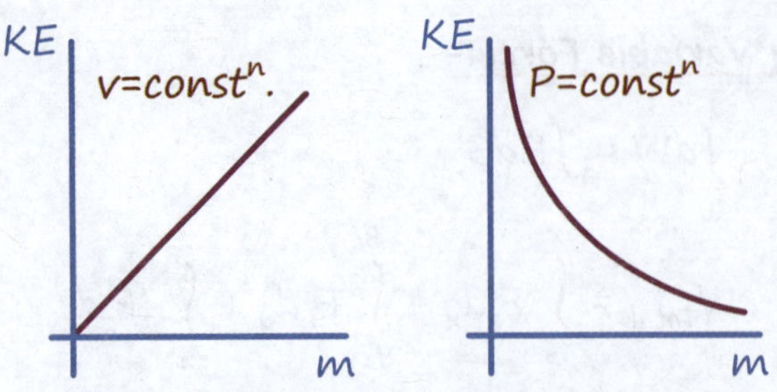

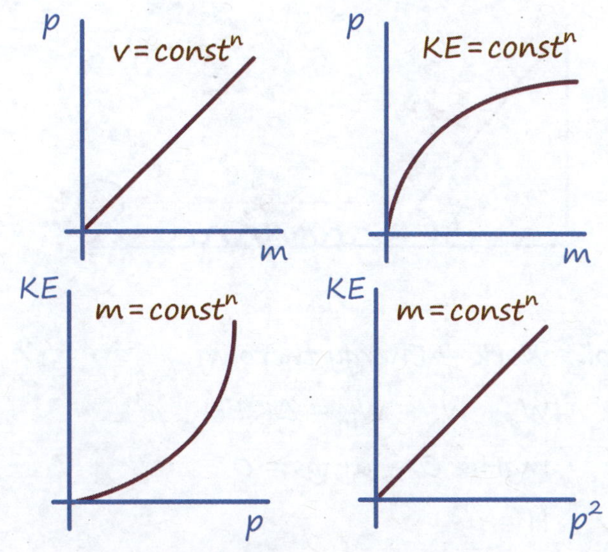

% Change Calculation:-

Small Change: $\lesssim 5\%$	Large Change:-
$\frac{\Delta KE}{KE} \times 100$ $= \frac{2\Delta P}{P} \times 100$	$\%\Delta KE = \frac{K_2 - K_1}{K_1} \times 100$

- For small changes, the percentage change in kinetic energy can be approximated by twice the percentage change in momentum.

5 Work Energy Theorem

$W_{all\ Force} = \Delta KE.$

All means all no except^n at all

$W_{C.F.} + W_{N.C.F.} + W_N + W_T + W_{Pseudo} + W_{etc} = \Delta K.E.$

The total work done by all forces includes:

- Conservative Forces ($W_{C.F.}$): Forces like gravity and spring force, which have potential energy associated with them.
- Non-Conservative Forces ($W_{N.C.F.}$): Forces like friction and air resistance, which dissipate energy.
- Normal Forces (W_N): Perpendicular forces that do not work in the direction of motion.
- Tension Forces (W_T): Forces in strings or ropes.
- Pseudo Forces (W_{pseudo}): Apparent forces in non-inertial reference frames.
- Other Forces (W_{etc}): Any other forces acting on the object.

Special Case:-

$\left.\begin{array}{l} F = \text{constant} \\ S = \text{constant} \end{array}\right\} \rightarrow$ तो Δ KE independent of Mass.

$\Delta KE \propto m^0.$

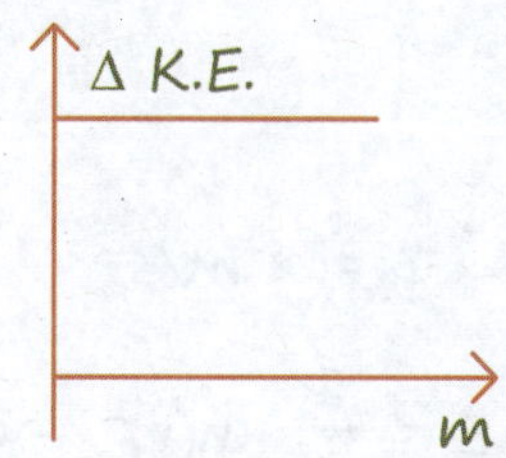

Q.4. If K.E. of a body is increased by 44%, what is the percentage increase in momentum?

Sol. Let initial K.E. is 100% and intial P is 100%

Given in question $K.E_f = 144\%$

$K.E_f = \frac{144}{100}$ $\qquad P = \sqrt{2m\,(K.E)}$

Ignore const term

$P_f = \sqrt{K.E_f}$

$P_f = \sqrt{\frac{144}{100}} = \frac{12}{10}$

for % change

$P_f = \frac{12}{10} \times 100 = 120\%$

Hence increase by 20%

Q.5. If momentum of a body is decreased to 25% then find % change in K.E.

Sol. Let initial K.E. and momentum is 100

Now decrease to 25% hence $P_f = 25\%$

$$P_f = \frac{25}{100} = \frac{1}{4}$$

$$K.E_f = P_f^2 = \left(\frac{1}{4}\right)^2 = \frac{1}{16}$$

for % change $\frac{1}{16} \times 100 = 6.25\%$

[decreased to 6.25%
and decreased by 93.75%]

Q.6. Position of object of mass 2kg $x = \frac{t^2}{3}$ then find work done in first three sec.

Sol. $V = \frac{dx}{dt} = \frac{2t}{3}$

$V_i = 0$

$V_f = \frac{2}{3} \times 3 = 2\ m/s$

$W = \Delta K.E = \frac{1}{2} m(v_f^2 - v_i^2)$

$W = \frac{1}{2} 2\,[(2)^2 - 0] \quad = 4J$

Q.7. Smooth hemispherical surface find velocity of ball at bottom point.

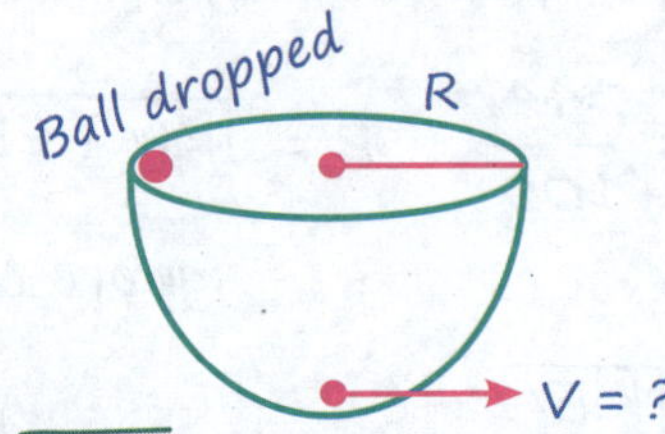

Sol. $V = \sqrt{2gR}$

Q.8. Speed of mass m?

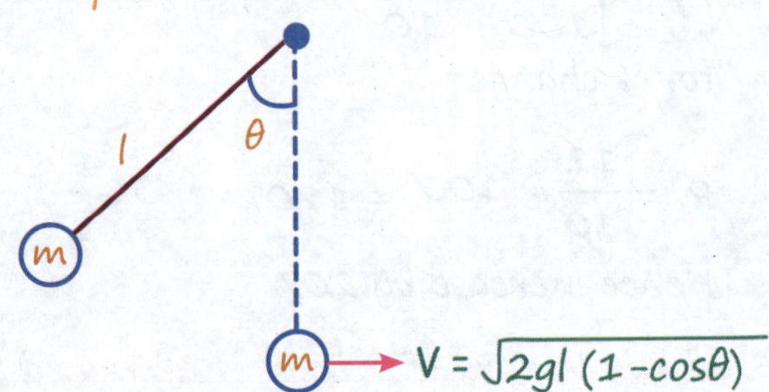

Q.9. Ball is droped from A and comes to at rest at B find horizontal distance moved by Ball on rough surface

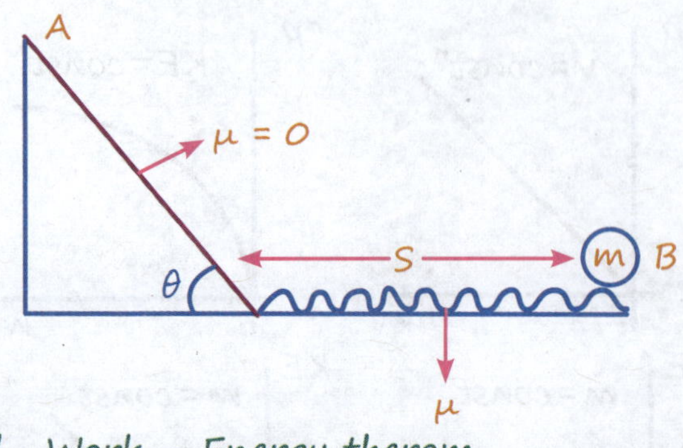

Sol. Work – Energy therom

$W_g + W_N + W_{fr} = \Delta K.E$

$mgH + 0 - \mu mgs = 0$

$H = \mu s$

$S = \frac{H}{\mu}$

6 Conservative and non-conservative force

Force

CONSERVATIVE FORCE (CF)	NON-CONSERVATIVE FORCE (NCF)
✦ Path independent	✦ Path dependent
✦ Concept of Potential is defined here!	✦ Concept of potential is not defined.
✦ W = Reversible	✦ W = Irreversible.
✦ $W = \oint \vec{F} \cdot \vec{dr} = 0$	✦ $W = \oint \vec{F} \cdot \vec{dr} \neq 0$

For a Conservative Force:-

$$\vec{F}_{CF} = -\left(\frac{dU}{dr}\right)$$ Potential Energy gradient (Vector) (Potential Energy is scalar)

$\therefore \quad \Delta U = -W_{C.F.} = -\vec{F}_{C.F.} \cdot \vec{dr}$

Change in potential energy does not depends upon reference, and path.

Potential Energy → depends upon reference not have unique/absolute value.

$$\vec{F}_{CF} = \left(-\frac{\partial U}{\partial x}\hat{i}\right) - \left(\frac{\partial U}{\partial y}\hat{j}\right) - \left(\frac{\partial U}{\partial z}\hat{k}\right)$$

y&z = const^n x & z = const^n x & y = const^n

$$\Delta U = -\int F_x d_x - \int F_y d_y - \int F_z d_z$$

* Apne -Aap Joh Kaam Hota hai Usmein PE Hamesha Ghatega!

MR*

Very Important!

✦ WET:- [Heart of Physics]

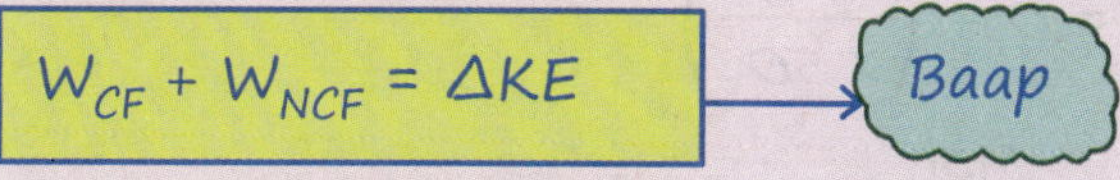

$\Delta U = -W_{CF}$

$W_{CF} = -\Delta U$

Mother always valid

✦ Conservation of mechanical energy

NCF may be acting or not acting but $W_{NCF} = 0$

$(KE + PE)_i = (KE + PE)_f$ Beta

MR*

* Jaha Hum Waha COME nai! (NCF)
* Jaha COME, waha hum nai
* Hum kisi se kam nai

✦ → KE = Const^n [Slow - Slow]

→ NCF = Acting against C. Force.

$\Delta U = W_{NCF}$.   Beti

Q.10. If potential energy $U = x^2y + y^2z + z^2x$ then find force acting on it

Sol. $\left(\frac{dU}{dx}\right)_{y \text{ and } z \text{ constant}} = 2xy + z^2$

$\left(\frac{dU}{dy}\right)_{x \text{ and } z \text{ constant}} = x^2 + 2yz$

$\left(\frac{dU}{dz}\right)_{y \text{ and } z \text{ constant}} = y^2 + 2zx$

$\vec{F} = -(2x\,y + z^2)\,\hat{i} - (x^2 + 2yz)\hat{j} - (y^2 + 2zx)\hat{k}$

Q.11. Work done in bringing object from A to B is 40J then find potential energy at B if potential energy at A is –30 J

Sol. $W_{N.C.F} = \Delta U$

$40\text{ J} = U_B - (-30)$

$40 - 30 = U_B$

$U_B = 10\text{J}$

7 Potential Energy

Energy due to shape, size and position.

✦ Potential energy of Point Objects:-

$U = mg\,h$

$= mg\,S \sin\theta$

$\because\ h = s\sin\theta.$

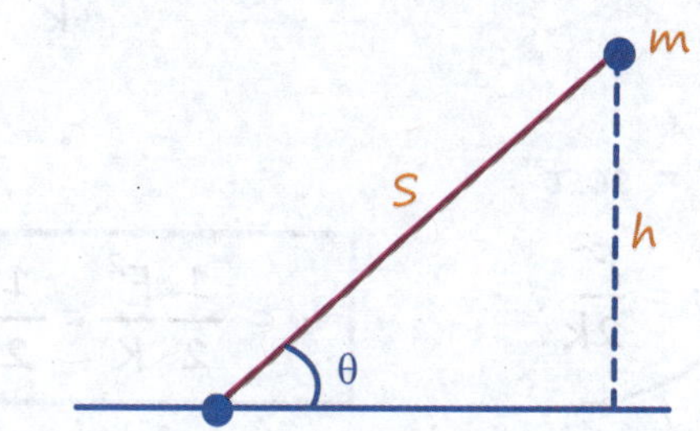

✦ Potential energy of Extended Objects:-

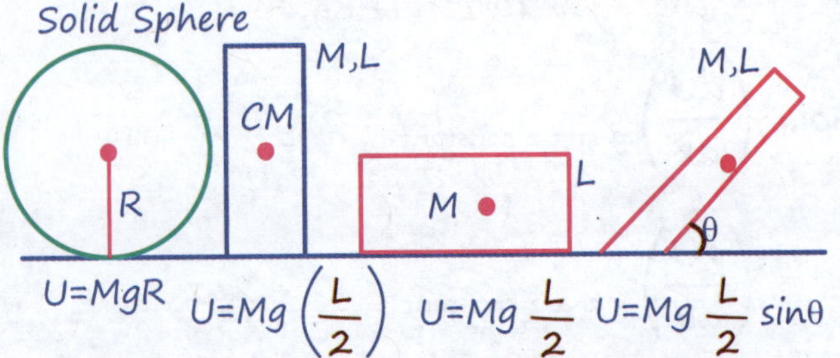

Chain Problem:- (P.E of hanging part)

$$U = \frac{-MgL}{2n^2}$$

Table

$\frac{L}{n}$

✦ Speed of chain when it becomes vertical!

$$V = \sqrt{gL\left[1 - \frac{x^2}{L^2}\right]}$$

x = Initial hanging length

L = Length of Chain.

Spring Force Energy:-

$$\Delta U = U_{x_2} - U_{x_1}$$

$$= \frac{1}{2}K\left[x_2^2 - x_1^2\right]$$

k = Force constⁿ

$$U = \frac{1}{2}Kx^2 \text{ [PE Store]}$$

✦ For a Spring:-

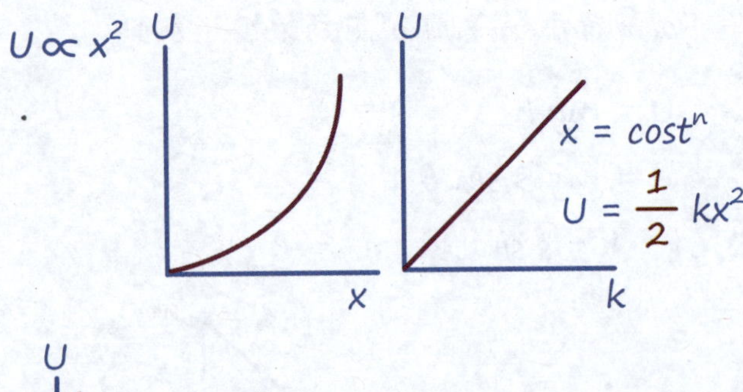

U

F = constⁿ

$$U = \frac{F^2}{2k}$$

k

$$U = \frac{1}{2}\frac{F^2}{K} = \frac{1}{2}Fx$$

Q.12. A chain on a frictionless table one fifth of it's length hanging over edge. If chain has length L Mass M then work done to pull back hanging part on table

Sol. refⁿ taken on the table then

$$U_f = 0$$

$$U_i = -\frac{m}{5}g\frac{l}{10}$$

$$W = \Delta U = U_f - U_i$$

$$= 0 - \left(-\frac{Mgl}{50}\right)$$

$$= \frac{Mgl}{50}$$

Q.13. Surface is smooth then maximum compression in spring.

m V_0 K

Sol. Conservation of energy

$$(K.E + U)_i = (K.E + U)_f$$

$$\frac{1}{2}mV_0^2 + 0 = \frac{1}{2}Kx^2 + 0$$

$$x = \sqrt{\frac{mV_0^2}{K}}$$

Q.14. Initially mass m is held such that spring is in relaxed condition then find elongation in spring, if mass m is

(i) Suddenly released

C.O.M.E applicable

loss in gravitational P.E. = Gain in spring P.E

$$mgx = \frac{1}{2}kx^2 \text{ and } x = \frac{2mg}{K}$$

m

(ii) Slowly released

Non-conservative force (हमारा) is working to released slowly hence COME not applicable

$F = mg = Kx$

$$x = \frac{mg}{K}$$

Q.15. Find minimum mass hanged from spring so that it can just pull up 20 kg object?

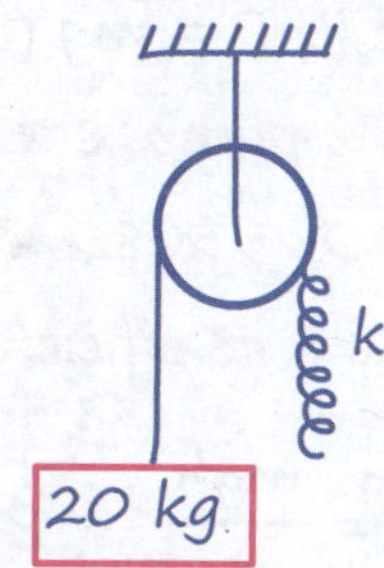

Sol. $m_{mini} = 10$ kg

Spring me suddenly elongation

$$x = \frac{2mg}{K} \text{ hoga.}$$

✦ Ball of mass m, is released with an angle θ from vertical where it'll loose contact!

* $V = \sqrt{2gR(1-\cos\theta)}$

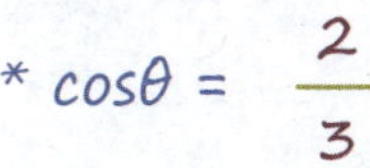

* $\cos\theta = \frac{2}{3}$

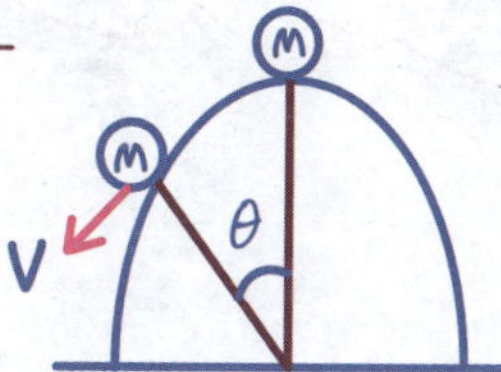

U–x graphs:-

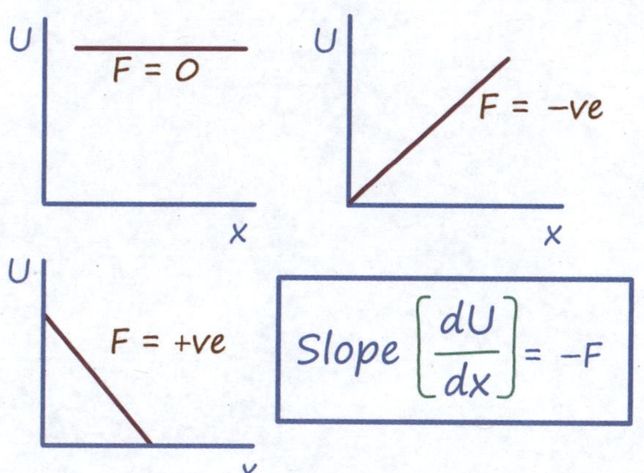

$$\text{Slope}\left[\frac{dU}{dx}\right] = -F$$

8 Equilibrium! and stability

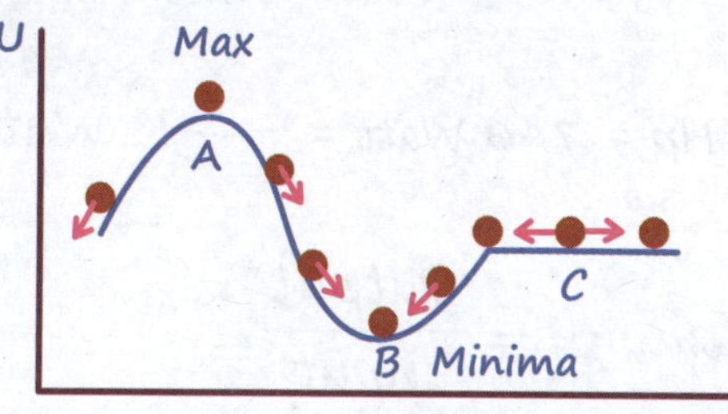

At A	At B	At C
F = O	F = O	F = O
Unstable	Stable	Neutral
$\frac{d^2U}{dx^2} = -ve$ Maxima	$\frac{d^2U}{dx^2} = +ve$ Minima	$\frac{d^2U}{dx^2} = 0$

$$\frac{dF}{dr} = -\frac{d^2U}{dx^2}$$

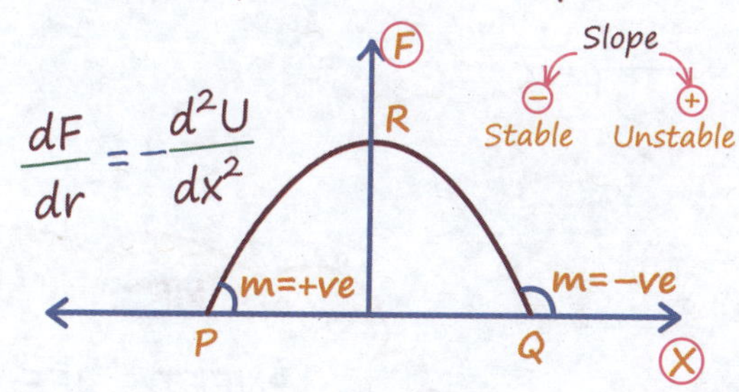

✦ Stable Eq^m = Q $\left[\frac{dF}{dr} = -ve\right]$ ← Slope

✦ Unstable Eq^m = P $\left[\frac{dF}{dr} = +ve\right]$ ← Slope

9 Power

✦ Power is the rate at which work is done or energy is transferred over time.

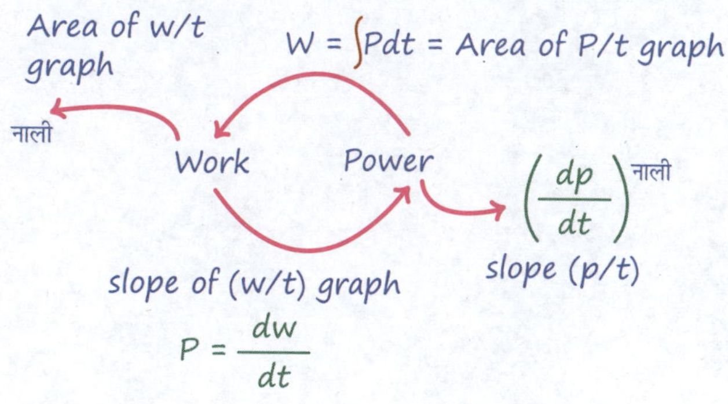

$$P = \frac{dw}{dt}$$

$$P_{avg} = \frac{W_{Total}}{t_{Total}} \qquad P_{inst} = \frac{dW}{dt} = F.v$$

$P_{avg} = \frac{\int P.dt}{\int dt.}$ $\quad P_{inst} = ma.v = Fv\cos\theta$

Unit:- Watt = J/s

$1\ Hp = 746\ Watt = \frac{3}{4} K\ Watt.$

✦ Efficiency (η) = $\frac{Output}{Input}$

* For a = cosnt^n

F = const^n $\quad$ $P \propto t$ $\quad$ $V \propto t$

NOTE: -

- $F = \rho AV^2$ (Pump)
- Power of pump = ρAV^3
- Rate at which K.E Provided to liquid $= \frac{1}{2} \rho AV^3$
- $P = F.v = \rho AV^3$
- When Power Const^n:-

$x \propto v^3 \quad x \propto t^{3/2} \quad v \propto t^{1/2}$

Q.16. Ball projected with speed u at angle θ then power at maximum height and at time 't' by gravitational force

Sol. $P_{inst} = mg\ u_x \cos 90° = 0$

at maximum height velocity and force is perpendicular

at time 't' $P = mg\ (u \sin\theta - gt)$

Q.17. An engine pumps 800 kg water through height 10 m in 80 sec. Find the power of engine if its efficiency is 75%

Sol. $75\%\ P = \frac{mgh}{t}$ $\quad P = \frac{4}{3}$ kW

MR*

'NEET - JEE preparation is not just about being a doctor or engineer. Its always about time management and pressure handling.'

10 Center of Mass

1 Centre of Mass

→ A point where whole mass of the system can be assumed there. COM lies near to heavier obj.

→ It can be inside or outside the body.

→ It always on the axis of symmetry and where two axis of symmetry will cut each other.

→ Position of COM depends upon frame of reference and choice of co-ordinates.

→ Centre of mass does not depend on choice of co-ordinate

COM of discrete particle:-

$$\vec{r}_{cm}\text{(Position of C.O.M.)} = \frac{m_1\vec{r}_1 + m_2\vec{r}_2}{m_1 + m_2}$$

$$\vec{X}_{cm} = \frac{m_1\vec{x}_1 + m_2\vec{x}_2}{m_1 + m_2}$$

$$\vec{Y}_{cm} = \frac{m_1\vec{y}_1 + m_2\vec{y}_2}{m_1 + m_2}$$

motion of the centre of mass

Velocity of C.O.M

$$\vec{V}_{cm} = \frac{m_1\vec{v}_1 + m_2\vec{v}_2}{m_1 + m_2}$$

Acceleration of C.O.M

$$\vec{a}_{cm} = \frac{m_1\vec{a}_1 + m_2\vec{a}_2}{m_1 + m_2}$$

Moment of Mass:

For two objects MR = constant.

✦ This means that the product of mass and distance remains the same even if the individual values of mass and distance change.

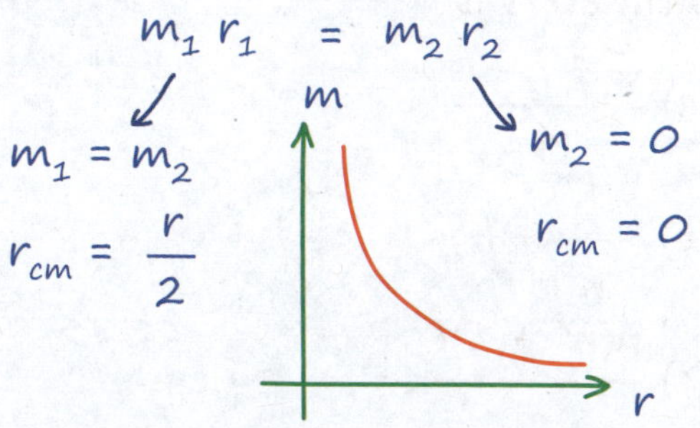

*COM is Closer to Massive body.

* Internal Force की औकात नहीं है की वो COM की $\vec{V}_{c.m}$ change करदे !

Com of two Particle System:-

m_1 —— d_1 | d_2 —— m_2, total distance d

$$d_1 = \frac{m_2 d}{m_1 + m_2} \qquad d_2 = \frac{m_1 d}{m_1 + m_2}$$

Q.1. Find position of C.O.M.?

(a) $\frac{20}{3}$, 8 (b) 8, $\frac{20}{3}$

(c) $\frac{20}{3}$, $\frac{20}{3}$ (d) 8, 4

(0, 10) 8 kg 16 kg (10, 10)

(0, 0) 2 kg 4 kg (10, 0)

Sol: (a)

$$\vec{r}_{cm} = \frac{2 \times 0 + 4(10\hat{i}) + 16(10\hat{i} + 10\hat{j}) + 8 \times 10\hat{j}}{2 + 4 + 8 + 16}$$

$$\frac{40\hat{i} + 160\hat{i} + 160J + 80\hat{j}}{30}$$

Q.2. n particles of masses m, 2m, 3m... 4m grams are placed on the same line at distances l, 2l, 3l, ... 4l c.m. from a fixed point. The distance of centre of mass of the particles from the fixed point in centimeters is

(a) $\frac{(2n+1)l}{3}$

(b) $\frac{l}{n}+l$

(c) $\frac{n(n^2+1)l}{2}$

(d) $\frac{l}{n}-l$

	m	2m	3m	4m.........
(0, 0) →	l	2l	3l	4l.........

Sol: (a)

$$X_{cm} = \frac{ml + 4ml + 9ml + 16ml + \ldots}{m + 2m + 3m + 4m + \ldots}$$

$$= \frac{l(1+4+9+16+\ldots)}{(1+2+3+4+5+\ldots)}$$

MR*

If n = 1	If n = 2	
$x_{cm} = l$	$x = \frac{ml + 4ml}{3m}$	$= \frac{5ml}{3m} = \frac{5l}{3}$

\# If external force is zero then location of centre of mass will not change → False

\# If $f_{ext} = 0$ then state of COM will not change. $[\vec{V}_{cm} = \text{cost}^n]$

\# Shift in C.O.M.

$$\Delta r_{c.m.} = \frac{m_1\vec{\Delta r_1} + m_2\vec{\Delta r_2}}{m_1 + m_2}$$

If C.O.M. does not shift its position then

$$m_1\vec{\Delta r_1} = -m_2\vec{\Delta r_2}$$

2 Centre of mass of Continuous System

Linear Mass Density (λ) for 1-Dimensional Systems:	Surface Mass Density (σ) for 2-Dimensional Systems:	Volume Mass Density (ρ) for 3-Dimensional Systems:
$\lambda = \frac{dm}{dL}$	$\sigma = \frac{dm}{dA}$	$\rho = \frac{dm}{dV}$
finite small mass dm = λdL	$dm = \sigma dA$	$dm = \rho dV$
1-D	2-D	3-D

✦ To find the center of mass (x_{CM}) for a continuous system, we use the following integral formula:

$$X_{CM} = \frac{\int x \cdot dm}{\int dm}, \quad dm = \lambda \cdot dx$$

MR C.O.M of Rod:-**

$X_{CM} = L/2 \quad \lambda = \text{const}^n$

$X_{CM} > L/2 \quad \lambda \propto x$

$X_{CM} = 3L/4 \quad \lambda \propto x^2$

MR*

\# If density of rod varies linearly $\lambda = \lambda_0 x^n$ then position $L/2 < x_{cm} < L$

\# If density of rod $\lambda = \alpha + \beta x$ then COM will be at $x_{cm} = L/2$ if $\beta = 0$

Q.3. Half of the uniform rectangular plate of Length 'L' is made up of material of density 'd_1' and the other half with density d_2. The perpendicular distance of center of mass from AB is

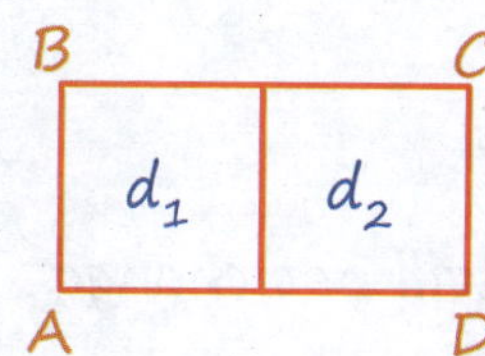

(a) $\frac{2d_1+3d_2}{d_1+d_2} \times \frac{L}{4}$

(b) $\frac{d_1+3d_2}{d_1+d_2} \times \frac{L}{4}$

(c) $\frac{3d_1}{d_1+d_2} \times \frac{L}{4}$

(d) $\frac{3d_2}{d_1+d_2} \times \frac{L}{4}$

MR* Sol. If $d_1=d_2$

C.O.M. at $\frac{L}{2}$ then option (b) correct.

+ If we break cricket bat from c.o.m. then bottom part will have large mass, because c.o.m. divide system in two equal moment of mass.

MR = const.

Q.4. A cricket bat is cut at the location of its centre of mass as shown. Then

(a) The Two pieces will have the same mass

(b) The bottom picece will have large mass

(c) The handle piece will have larger mass

(d) Mass of handle piece is double the mass of bottom piece

Sol. (b)

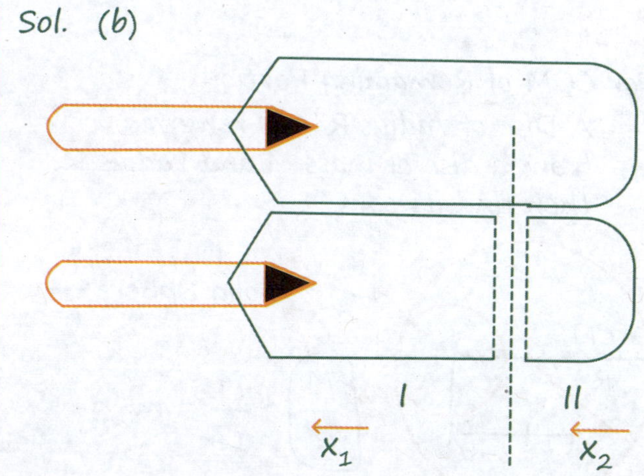

Trick for Com:-

Com of continuous mass system

Semi cirular Ring  $2R/\pi$

Solid Hemi sphere  $3R/8$

Semi cirular Disc  $4R/3\pi$

Hollow Hemi sphere  $R/2$

Triangle / Hollow Cone.  $h/3$

Cone Solid.  $h/4$

COM of Circular ARC:-

$$Y_{CM} = \frac{R \sin \theta/2}{\theta/2}$$

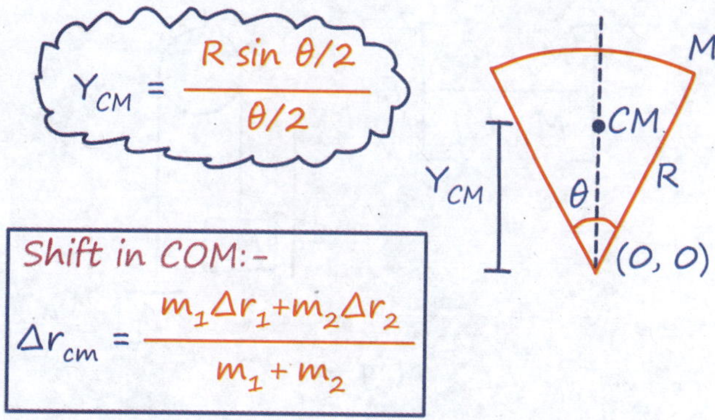

Shift in COM:-

$$\Delta r_{cm} = \frac{m_1\Delta r_1+m_2\Delta r_2}{m_1+m_2}$$

MR**

For COM of Remaining Portion:-

A Disc of radius R/4 is removed from a disc of mass M and radius R. then find its com

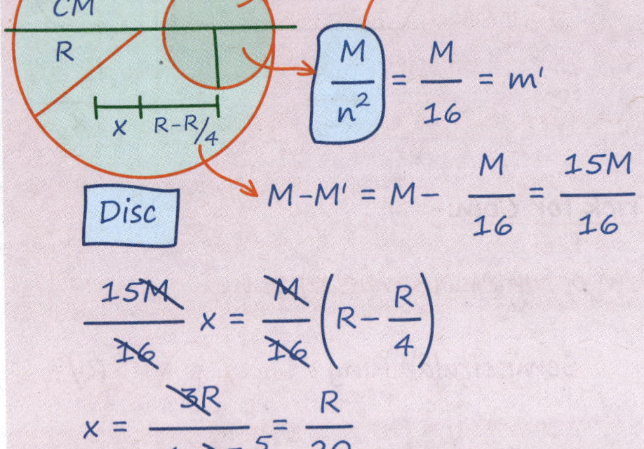

$$\frac{M}{n^2} = \frac{M}{16} = m'$$

$$M - M' = M - \frac{M}{16} = \frac{15M}{16}$$

$$\frac{15M}{16} x = \frac{M}{16}\left(R - \frac{R}{4}\right)$$

$$x = \frac{3R}{4 \times 15} = \frac{R}{20}$$

- find the acceleration of com of both the blocks as shown in fig. 1 and 2 Neglect friction everywhere

$$\vec{a}_{cm} = \frac{m_1 a\hat{i} - m_2 a\hat{j}}{m_1 + m_2}$$

Put $a = \dfrac{m_2 g}{m_1 + m_2}$

Fig-1

MR Ratta*

$$\vec{a}_{cm} = \left[\frac{M_2 - M_1}{M_1 + M_2}\right]^2 g$$

Fig-2

Note:-

Agar $F_{ext} = 0$.

State of COM will not change.

Q.5. Two ball of mass m_1 and m_2 projected with u_1 and u_2 in upward and at 30° from horizontal respectively then acceleration of c.o.m. will be?

Sol. $\vec{a}_{c.m} = g$ (downward) because both have same acc^n g downward.

Aag lage chahe basti mein COM Rahe apne masti mein

Conservation of Linear Momentum of System:

Condition:-

$F_{ext} = 0$. $\vec{P}_{cm}$ = Same.

A stationary object explodes in two unequal part then :-

- External force is zero, hence momentum will be conserved.
- Both part will have equal momentum in opposite direction.
- Smallar mass will have greater kinetic energy.
- Work done by internal force = change in K.E. of system = K.E. of both part.
- For same momentum both have unequal velocity in opposite direction.
- Internal force can change kinetic energy of system → True
- Internal force can change momentum of system → false.
- A body falling vertically downward under gravity breaks in two unequal part then c.o.m. will continoue to do vertical motion does not shift horizontal.
- A shell following parabolic path explode somewhere in many part but c.o.m. will continue parabolic path.

Q.6. A body of mass (4m) is lying in x-y plane at rest. It suddenly explodes into three pieces. Two pieces, each of mass (m) move perpendicular to each other with equal speeds (u). The total kinetic energy generated due to explosion is

Sol.:-

MR**

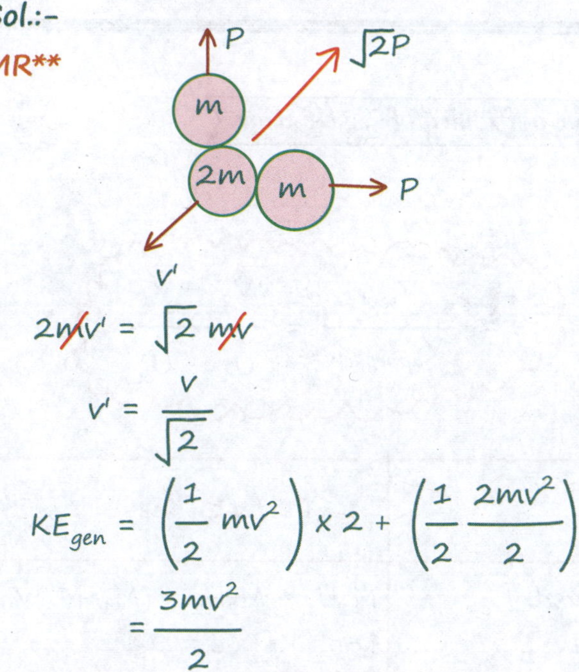

$$v'$$

$$2\cancel{m}v' = \sqrt{2}\,\cancel{m}v$$

$$v' = \frac{v}{\sqrt{2}}$$

$$KE_{gen} = \left(\frac{1}{2}mv^2\right) \times 2 + \left(\frac{1}{2}\frac{2mv^2}{2}\right)$$

$$= \frac{3mv^2}{2}$$

Q.7. A man of mass M stands at one end of a plank of length L which lies at rest on a frictionless surface. The man walks to other end of the plank. If the mass of the plank is M/3, then the distance that the man moves relative to ground is :

Sol. If x = distance moved by Planck, then

$$M(L-x) = \frac{M}{3}, \quad ML = \frac{Mx}{3} + Mx$$

$$x = \frac{ML}{M+M/3} = \frac{ML}{4M/3} = \frac{3L}{4}$$

Displacement of man relative to ground = L−x

$$= L - \frac{3L}{4} = \frac{L}{4}$$

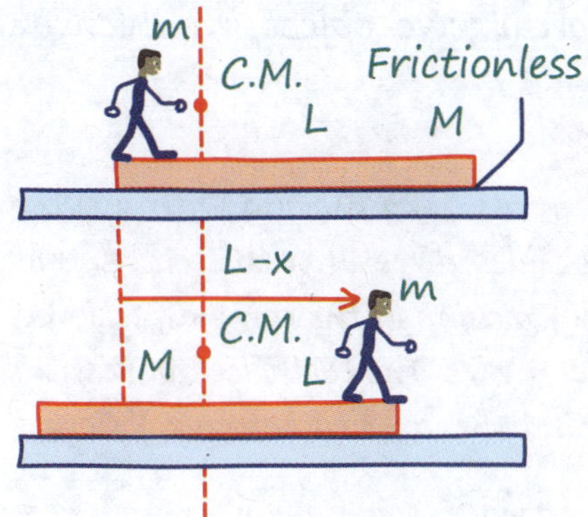

Q.8. A bomb explodes into two parts 6 kg and 8 kg if velocity of 6kg is 10m/s then find K.E. of 8kg will be

Sol. $$K.E = \frac{P^2}{2m} = \frac{6\times6\times10\times10}{2\times8}$$

$$= 225\ J$$

Q.9. A vertical rod is placed on smooth ground and released then path of C.O.M will be

C.M

$\mu=0$

Ans straight line in vertical

Q.10. Object is projected with u at angle θ at maximum height it breaks into two equal part, if one just fall below maximum height then range of other from point of projection.

Sol.

C.O.M ka range तो R हि होगा

$$\vec{r}_{c.m} = \frac{m_1r_1 + m_2r_2}{m_1 + m_2}$$

$$R = \frac{m\,R/2 + m\,r_2}{2m}$$

$$2R = \frac{R}{2} + r_2$$

$$r_2 = \frac{3R}{2}$$

11 Collision

✦ In all Colisions, $\vec{F}_{ext} = 0$; the linear momentum of sytem will be conserved

P = Conserved.

Momentum of individual mass is not conserved but P_{system} = Conserved!

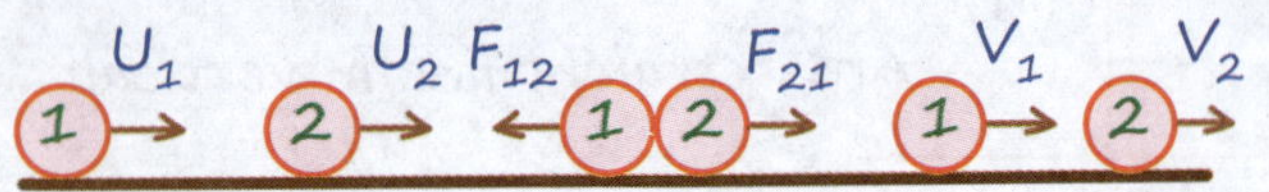

1 Coefficient of Restitution

$$e = \frac{V_{sepr}}{V_{approach}} = \frac{\vec{V}_2 - \vec{V}_1}{\vec{U}_1 - \vec{U}_2}$$

Elastic Collision	Inelastic	Perfectly Inelastic
KE = Conserved e = 1	Not Conserved. $0 < e < 1$	$M_1U_1 + M_2U_2 = (M_1+M_2)V$ $e = 0$
$V_1 = \frac{(m_1 - m_2)\vec{U}_1}{m_1+m_2} + \frac{2m_2\vec{U}_2}{m_1+m_2}$	$\vec{V}_1 = \frac{(m_1 - em_2)\vec{U}_1}{m_1+m_2} + \frac{(e+1)m_2\vec{U}_2}{m_1+m_2}$	$V = \frac{m_1u_1+m_2u_2}{m_1+m_2}$ → V_{com}
$V_2 = \frac{(m_2 - m_1)\vec{U}_2}{m_1+m_2} + \frac{2m_1\vec{U}_1}{m_1+m_2}$	$\vec{V}_2 = \frac{(m_2 - em_1)U_2}{m_1+m_2} + \frac{(e+1)m_1U_1}{m_1+m_2}$	
$e = \frac{\vec{V}_2 - \vec{V}_1}{\vec{U}_1 - \vec{U}_2}$	$\Delta KE_{loss} = \frac{1}{2}\frac{m_1m_2}{m_1+m_2}(U_{rel}^2)(1-e^2)$	$\left(\Delta KE_{max}\right)_{loss} = \frac{1}{2}\frac{m_1m_2}{m_1+m_2}U_{rel}^2$

Elastic collision b/w moving (m) and (nm) at rest

Neet

$$\frac{KE_{Trans}}{KE_{initial}} = \frac{4n}{(1+n)^2}$$

(n) = Kitna bada hai ye matter Krta hai Kon bada hai ye nai.

Fraction of retained K.E. = $\left(\frac{1-n}{1+n}\right)^2$

Note:-

Elastic collsion of two object in which one is at rest.

MR Ek Soch**

Q.1. A ball of mass 2 kg moving with a speed of 5 m/s collides directly with another ball of mass 3 kg moving in the same direction with a speed of 4 m/s. The coefficient of restitution is 2/3. Find the velocities after collision.

→ Formula ek bar he lagana hai dusre object Ki velocity tum direct nikal dena Momentum Conservation after collision से!

$$\vec{V_1} = \left(\frac{2 - \frac{2}{3} \times 3}{5}\right) 5 + \frac{\left(\frac{2}{3} + 1\right) 3 \times 4}{5}$$

$\vec{V_1} = 4$ m/s

$P_i = P_f$ $\quad 10 + 12 = 8 + 3V$

$V = 14/3$

2 Elastic Collision B/N Two Object of Same Mass

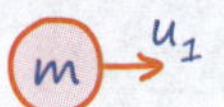

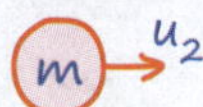

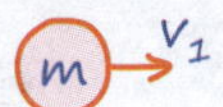

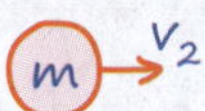

before collision after collision

Woh dono apni velocity interchange Krlenge i.e.

$$\vec{V_1} = \vec{U_2}$$
$$\vec{V_2} = \vec{U_1}$$

Elastic Collision B/N Two Object in Which one Having Mass very Much Greater then other:- (M>>>>m)

Case-I.

∴ Velocity of Ramlal after Collision

= bade ka double – Khud ka velocity

$$\vec{V_2} = 2\vec{U_1} - \vec{U_2}$$

Case-II.

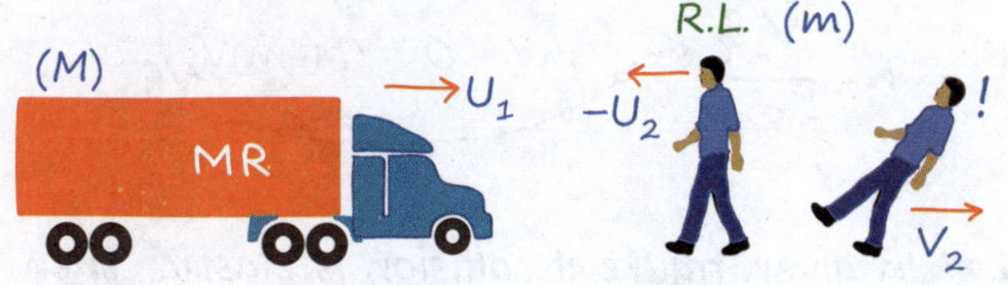

$\vec{V_2}$ = bade ka double + Khud ka velocity

$$\vec{V_2} = 2\vec{U_1} + \vec{U_2}$$

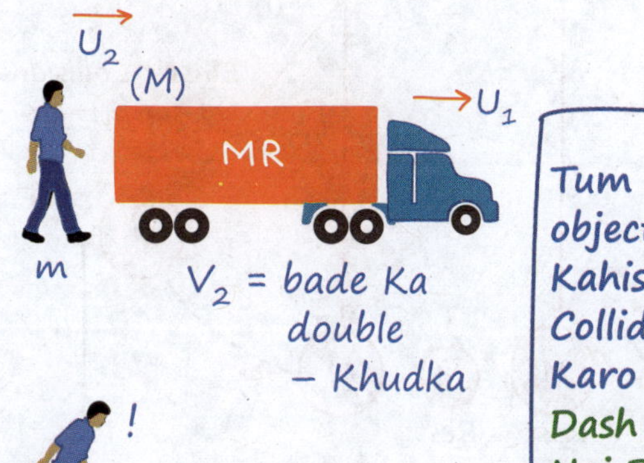

Tum Heavy object Ko Kahise Collide Karo us Ka Dash Farak Nai Padhta'

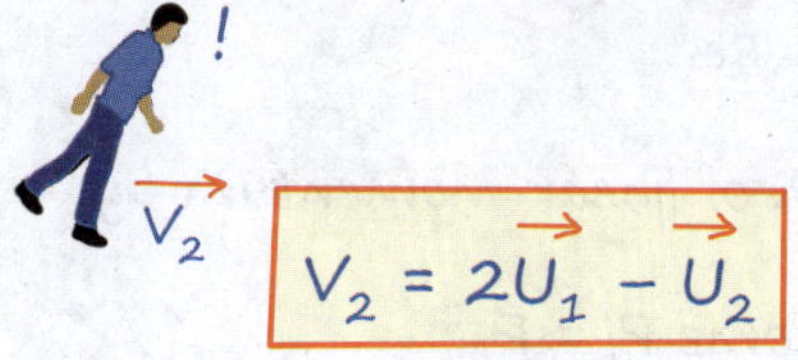

$$V_2 = 2\vec{U_1} - \vec{U_2}$$

MR-Wala Sawaal:-

1> In given figure if collision is perfectly inelastic then find maximum height attain by ball just after collision from mean position.

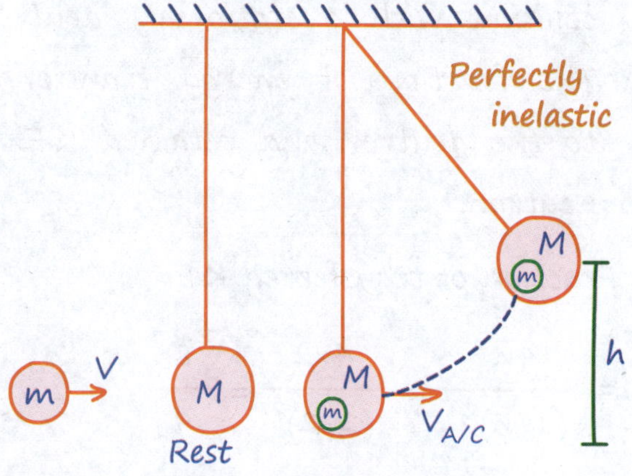

by conservation of linear momentum by come $P_i = P_f$

$$K_i + u_i = K_f + u_f$$

$$h = \frac{V_{A/C}^2}{2g} \leftarrow mV + 0 = (M+m)V_{A/C}$$

2> In given figure if collision is elastic then find maximum height attain by ball just after collision from mean position.

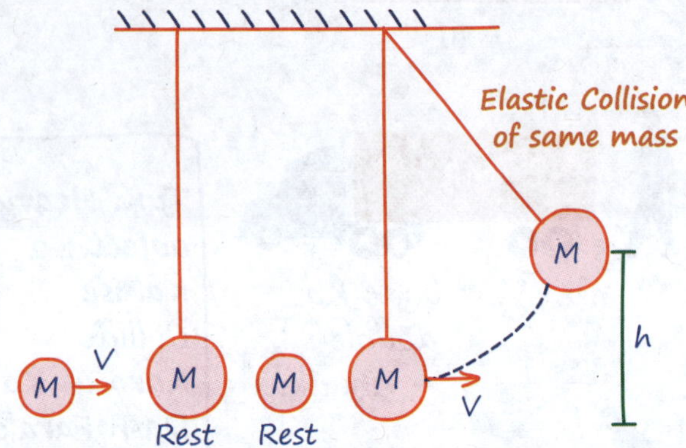

by conservation of linear momentum by

come $P_i = P_f$

$$K_i + u_i = K_f + u_f$$

$$h = \frac{V^2}{2g} \Big\} \text{COME.}$$

Q.2. A neutron makes a head on elastic collision with a stationary deutron. The fraction of energy transferred to the deutron and retained K.E. in neutron.

Sol. Fraction of transferred K.E.

$$= \frac{4n}{(1+n)^2} = \frac{4\times 2}{(1+2)^2} = \frac{8}{9}$$

$$\text{Fraction of retained K.E.} = \left(\frac{1-2}{1+2}\right)^2 = \frac{1}{9}$$

Ball is Drop From Height H:-

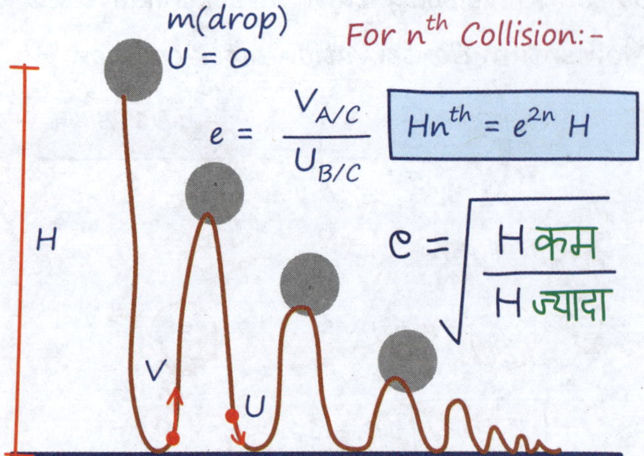

For n^{th} Collision:-

$$e = \frac{V_{A/C}}{U_{B/C}}$$

$$H_{n^{th}} = e^{2n} H$$

$$e = \sqrt{\frac{H\text{ कम}}{H\text{ ज्यादा}}}$$

Ball is Dropped from Height H Then Total Time of Flight and total Distance before coming to Rest.

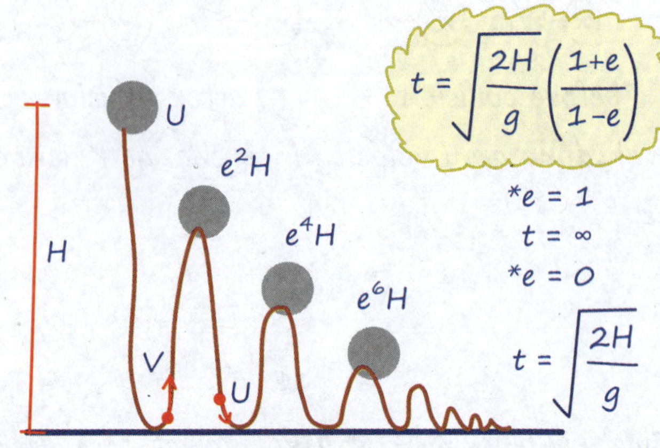

$$t = \sqrt{\frac{2H}{g}}\left(\frac{1+e}{1-e}\right)$$

*e = 1

t = ∞

*e = 0

$$t = \sqrt{\frac{2H}{g}}$$

$$\text{Total distance} = H\left(\frac{1+e^2}{1-e^2}\right)$$

3 Oblique Inelastic Collision

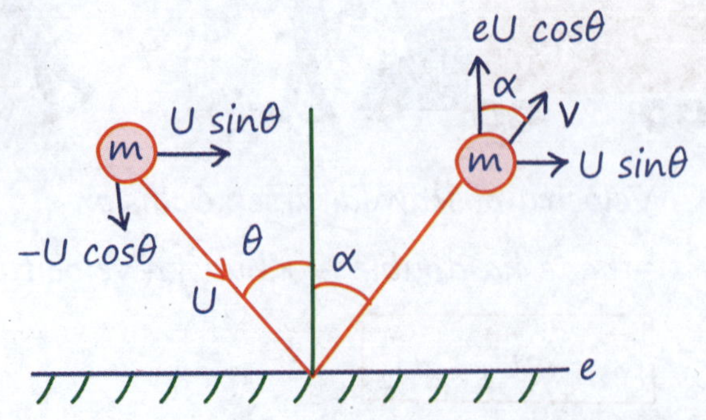

$$V = \sqrt{(U\sin\theta)^2 + (Ue\cos\theta)^2}$$

$$\Delta P = mu\cos\theta\,(1 + e)$$

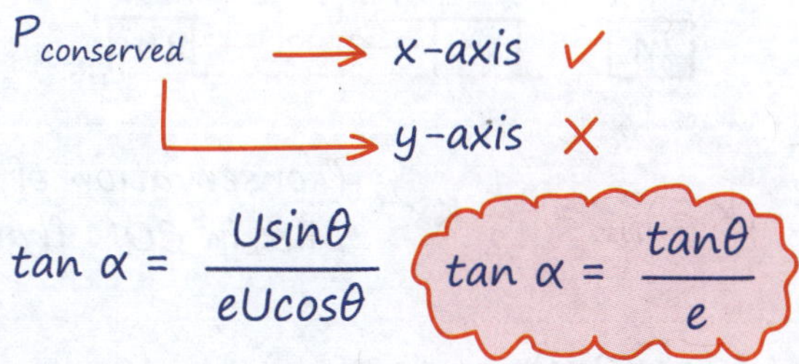

+ Sawaal mein agar "θ" horizontal se liye toh:-

$$*\Delta P = mu\sin\theta + emu\sin\theta$$

$$= mu\sin\theta\,(1 + e)$$

1d Head on Collision:-

$(b = 0)$

2d Oblique Collision:-

$(b > 0)$

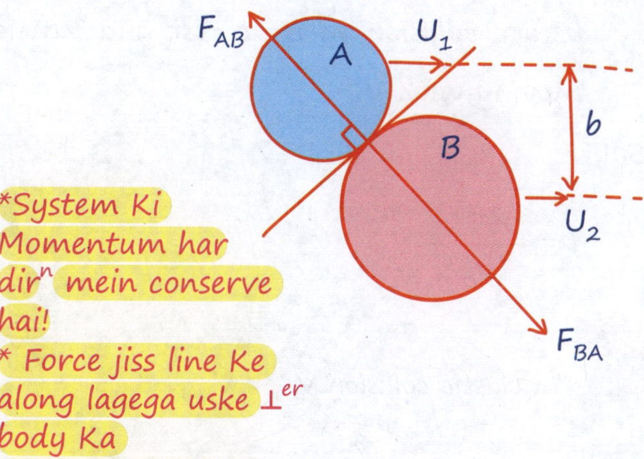

*System Ki Momentum har dir^n mein conserve hai!

* Force jiss line Ke along lagega uske ⊥^{er} body Ka P Conserved hoga!

Q.3. A sphere P of mass m and velocity $\vec{V}$ undergoes an oblique and perfectly elastic collision with an identical Q initially at rest. The angle θ between the velocities of the spheres after the collision shall be

Sol. According to the law of conservation of linear momentum,

$$mv_i + m \times 0 = mv_{Pf} + mv_{Qf}$$

$$\Rightarrow v_i = v_{Pf} + v_{Qf}$$

$$\text{Now, } (v_i \cdot v_i) = (v_{Pf} + v_{Qf})^2$$

$$= v_{Pf}^2 + v_{Qf}^2 + 2v_{Pf}v_{Qf}\cos\theta \quad ...(i)$$

Using conservation of kinetic energy, we get

$$\frac{1}{2}mv_i^2 = \frac{1}{2}mv_{Pf}^2 + \frac{1}{2}mv_{Qf}^2$$

$$\Rightarrow v_i^2 = v_{Pf}^2 + v_{Qf}^2 \quad ...(ii)$$

On comparing (i) and (ii), we get

$$\cos\theta = 0 \Rightarrow \theta = 90°$$

Wait for MR*

Jis direction me collision hoga velocity interchange ho jayega.

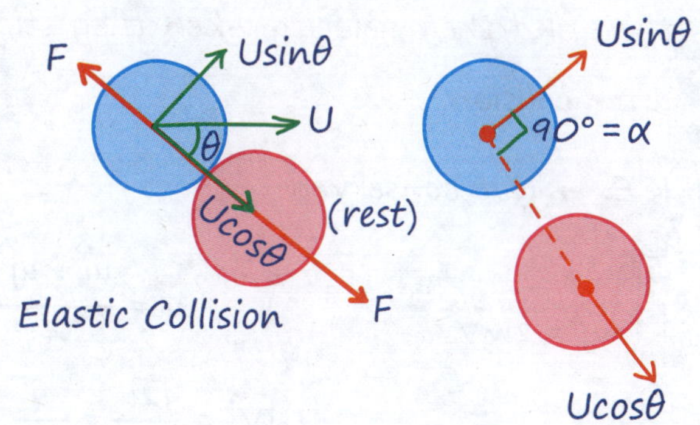

Note:-

+ 2-D Elastic Collision wale sawaal mein:-
KE = Conservation lagao.

+ Perfectly Inelastic wale Sawaal mein:-
P = Conservation lagao.

Q.4. Two sphere A and B of masses m_1 and m_2 respectively collide. A is at rest initially and B is moving with velocity v along x-axis. After collision B has a velocity $\frac{v}{2}$ in a direction perpendicular to the original direction. The mass A moves after collision in the direction. (AIPMT-2012)

Sol.

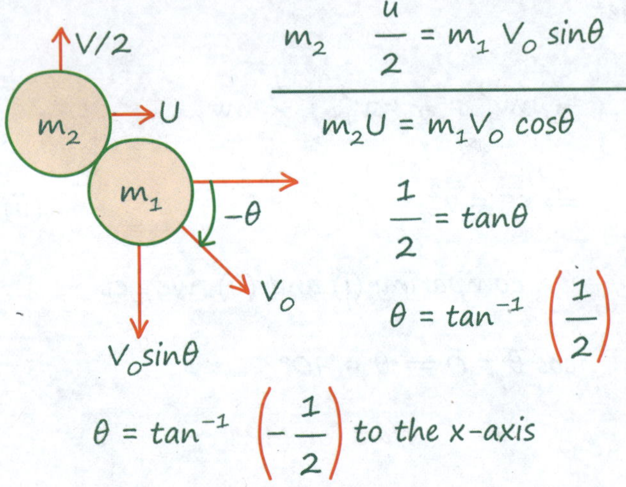

$$m_2 \frac{u}{2} = m_1 V_0 \sin\theta$$

$$m_2 U = m_1 V_0 \cos\theta$$

$$\frac{1}{2} = \tan\theta$$

$$\theta = \tan^{-1}\left(\frac{1}{2}\right)$$

$$\theta = \tan^{-1}\left(-\frac{1}{2}\right) \text{ to the x-axis}$$

Q.5. Two identical block of mass m moving with speed u perpendicular to each other then find their velocity when they stick after collsion.

Sol. K.E. → Not conserved

$P_i = P_f$

$$mu\hat{i} + mu\hat{j} = 2m\vec{V}$$

$$\vec{V} = \frac{u\hat{i} + u\hat{j}}{2}$$

$$|\vec{V}| = \frac{\sqrt{2}u}{2} = \frac{u}{\sqrt{2}}$$

Q.6. A [m] → u, B [m] ─ spring ─ C [2m]; if collision b/w A & B is elastic then maximum compression in spring is?

Sol.

Just after collsion (elastic collsion of same mass)

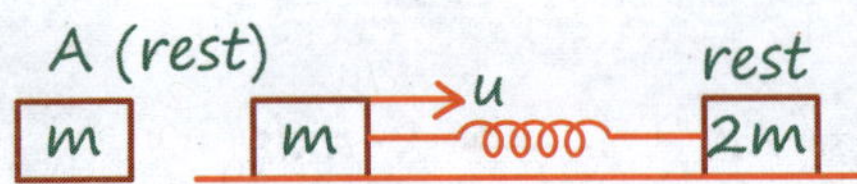

$$(K.E.)_{loss} = \frac{1}{2}Kx^2 \quad \text{(conservation of M.E. in COM frame)}$$

$$\frac{1}{2}\frac{m(2m)}{m+2m}u^2 = \frac{1}{2}Kx^2$$

$$\frac{1}{3}mu^2 = \frac{1}{2}Kx^2$$

$$x = \sqrt{\frac{2mu^2}{3K}}$$

Q.7. Smooth horizontal circular track, as shown in Fig. then find time taken b/w 1st and 2nd collision if collsion is elastic, before collision m is at rest and 2m is moving with V_0

Sol.

$$t = \frac{2\pi r}{V_0}$$

MR*

In elastic collision $V_{sep} = V_{app.}$

(2m) V_0
r
(m) rest

Q.8. Four identical ball placed on horizontal table then find their velocity after collsion.

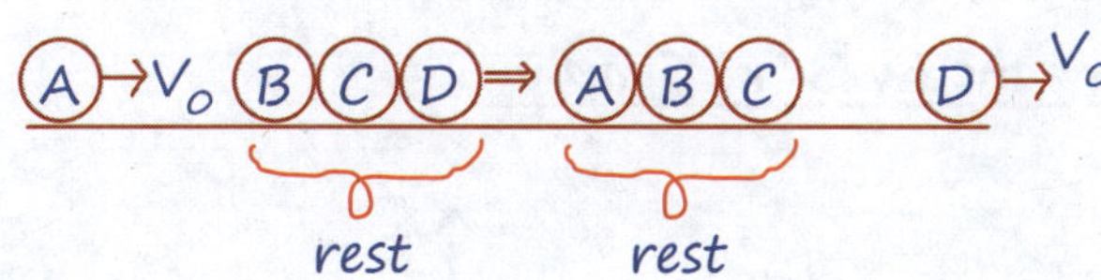

Q.9. Find 'e'

$m_1 \rightarrow 2m/s$ $\leftarrow m_2$ 10m/s $m_1 \rightarrow 1m/s$ $m_2 \rightarrow 4m/s$

B/C A/C

Sol. $e = \frac{4-1}{10+2} = \frac{3}{12} = \frac{1}{4}$

Q.10. Two Identical object moving with velocity 4m/s and 10m/s towards each other find their velocity after collsion if e = 0.5.

Sol. $\vec{V}_1 = \frac{m - 0.5m}{2m} 4 - \frac{(0.5+1)m \times 10}{2m}$

$= \frac{+0.5}{2} \times 4 - \frac{1.5 \times 10}{2}$

$= +1 - 7.5 = -6.5$ m/s

Now Conserved momentum of system

$\vec{P}_i = P_f$

$4m - 10m = -6.5m + mv$

$-6 = -6.5 + v$

$V = 6.5 - 6$

$V = 0.5$ m/s

MR*

"Dusro ke liye kab tak taali bajaoge? Ab aisa karo ki duniya tumhare liye taali bajaye."

12

Rotational Motion

1 Moment of Inertia

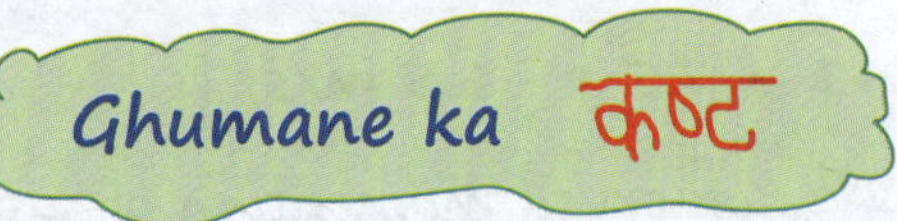

Property of object by which object oppose cause of change in rotational state.

- Unit Kg m^2
- Dimension $[ML^2T^0]$

MR*

$I \propto$ (कितना Mass) × (कितनी Dur)^2

depends more on distance rather than mass

I_1 — 2M, L

I_2 — M, 2L

$I_2 > I_1$

Tensor and Axis of Rotation

→ Tensor. → Calculated from Axis of Rotation.

- **Tensor**: Tensor relates to the property that describes the moment of inertia in different directions.
- **Axis of Rotation (AOR)**: This is the imaginary line around which an object rotates.

Graphs I v/s r Graph:-

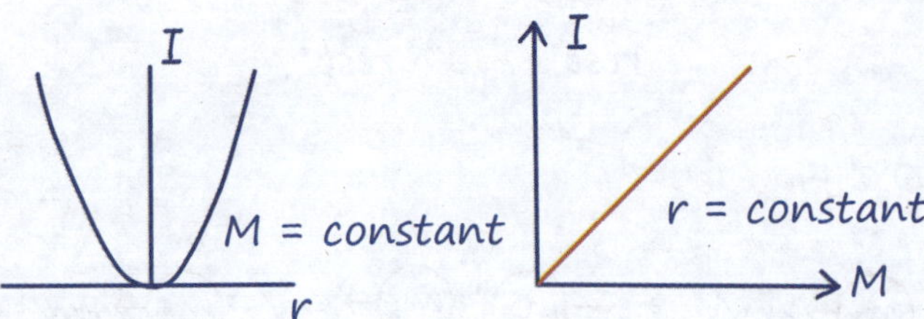

Mass & M.o.I:-

Point Mass:

- For a single point mass at a distance r from the axis of rotation, the moment of inertia

$I = Mr^2$ (r → $\perp^{er}$ dist^n from AOR)

System of Masses:

- For multiple masses at different distances from the axis of rotation

$I = M_1r_1^2 + M_2r_2^2 + ...$

M.O.I. about C.O.M. & $\perp^{er}$ to line:-

- For two masses M1 and M2 connected by a rigid rod with a total length 1: Here, r is the distance from the axis to one of the masses.

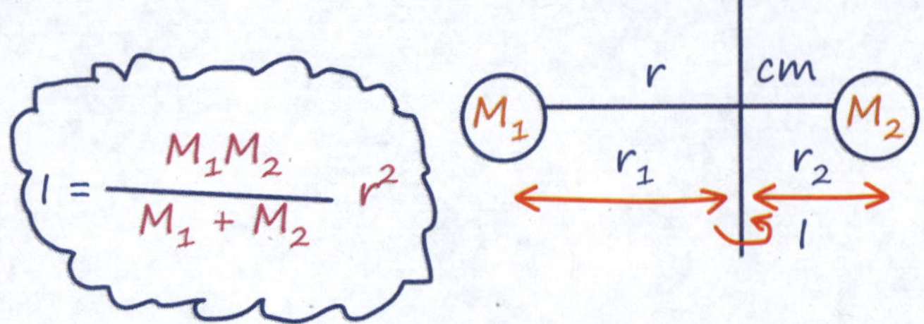

$$I = \frac{M_1M_2}{M_1 + M_2} r^2$$

- Object of mass m is placed at (x, y, z) then moment of

 Inertia about x axis $I_{xx} = m(y^2 + z^2)$

 about y axis $I_{yy} = m(x^2 + z^2)$

 about z axis $I_{zz} = m(y^2 + x^2)$

- Three identical mass placed on the corner of equilateral triangle of side (a).

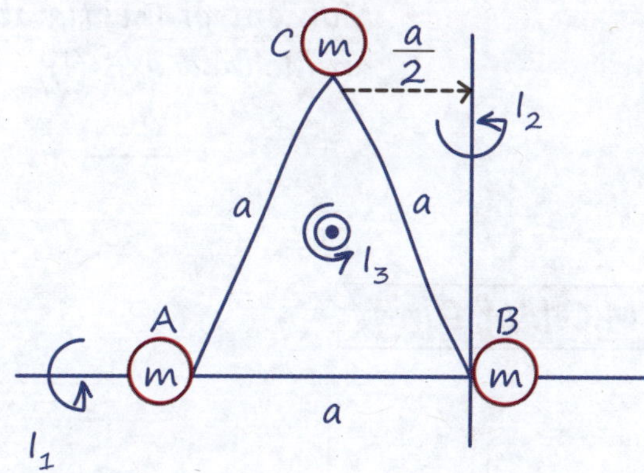

The moment of inertia about an axis passing through one of the sides of the triangle.

Let The coordinate of mass placed at position A is the origin (O,O), then the coordinate of other two masses are at

(a, O) and $\left(\frac{a}{2}, \frac{\sqrt{3}a}{2}\right)$ respectively.

The moment of inertia of masses at A and B would be zero about the line AB.

The moment of inertia of mass m at C about the line AB;

$$\# \; I_1 = m\left(\frac{\sqrt{3}\,a}{2}\right)^2 = \frac{3ma^2}{4}$$

The moment of inertia about an axis passing through one of the masses and perpendicular to the plane of the triangle.

$$\# \; I_2 = ma^2 + \frac{ma^2}{4} = \frac{5ma^2}{4}$$

The moment of inertia about an axis passing through the centroid of the triangle and perpendicular to the plane of the triangle.

$$\# \; I_3 = 3\left[m\left(\frac{a}{\sqrt{3}}\right)^2\right] = ma^2/4$$

4-Point mass placed on the corner of square:-

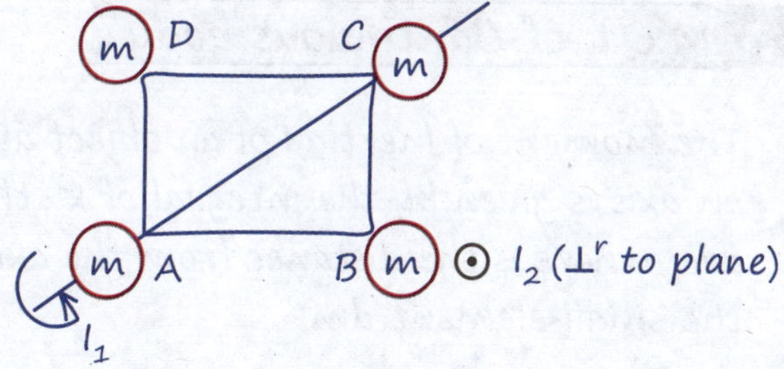

Moment of inertia about an axis passing through the the diagonal of the square:

- Moment of inertia due to mass at A and C would be zero, as they are at axis of rotation.

For masses at B and D:

- The distance from the axis to the mass diagonally opposite is $\frac{a}{\sqrt{2}}$ Therefore moment of inertia;

$$I_1 = 2\left[m\left(\frac{a}{\sqrt{2}}\right)^2\right] = ma^2$$

Moment of Inertia about an axis passing through one of the vertex of square and perpendicular to the plane of the square.

$$I_2 = ma^2+ma^2+m(\sqrt{2}\,a)^2 = 4ma^2$$

2 Theorems

- Parallel axis → valid for all type of body

$I_0 = I_{cm} + Md^2$

d = distance b/w. axis passing through C.O.M. and 'O'

- Perpendicular axis → valid for planer object

$I_z = I_x + I_y$

I_x & I_y → axis parallel to plane

I_z → $\perp_r$ to plane

3 M.O.I. of Continuous body

The moment of inertia I of an object about an axis is given by the integral of x^2 times dm, where is the distance from the axis to the small element dm.

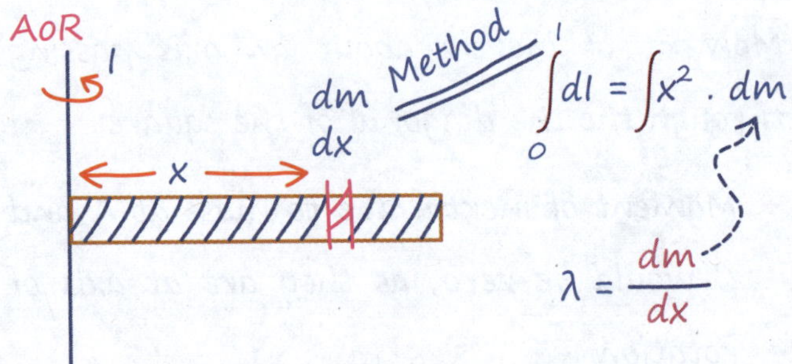

$$\int_0^I dI = \int x^2 \,.\, dm$$

$$\lambda = \frac{dm}{dx}$$

Non Uniform body.

$$I = \int x^2 . dm .... \int x^2 \,.\, \lambda . dx$$

Here, λ (linear mass density) is not constant but, a function of x.

Moi of Rod:-

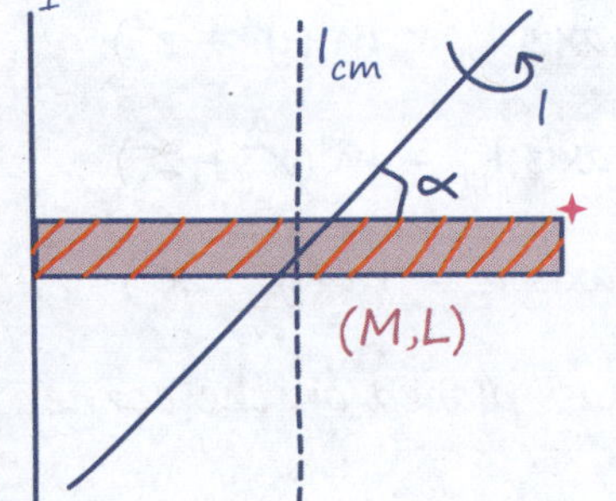

- Moment of Inertia about the Center of Mass:

$$I_{cm} = \frac{ML^2}{12}$$

- Moment of Inertia about one end of the Rod

$$I_1 = \frac{ML^2}{3}$$

- Moment of Inertia about an inclined axis (I)

$$I = \frac{ML^2}{12} \sin^2 \alpha \; *$$

4 M.O.I. of Ring

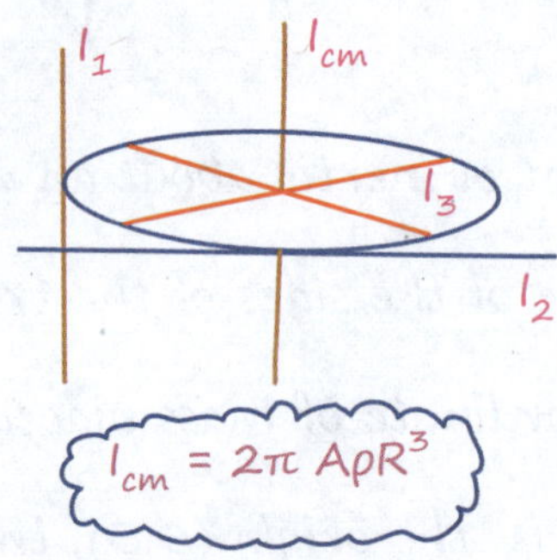

$$I_{cm} = 2\pi A\rho R^3$$

A = cross-section area

(Linear mass density) $\lambda = A\rho$

(ρ = volumetric density)

S. No.	Axis of Rotation	Moment of Inertia
(a)	About an axis passing through its center and perpendicular to its plane	$I_{cm} = MR^2$
(b)	About a tangent perpendicular to its plane	$I_1 = 2MR^2$

(c)	About a tangent in the plane of the ring	$I_2 = \frac{3MR^2}{2}$
(d)	About a diameter	$I_3 = \frac{MR^2}{2}$

M.O.I. of half ring about centre perpendicular to plane:-

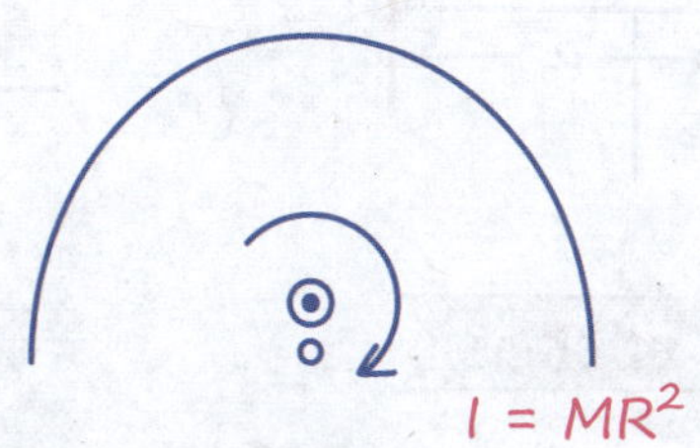

5 M.O.I. of Disc

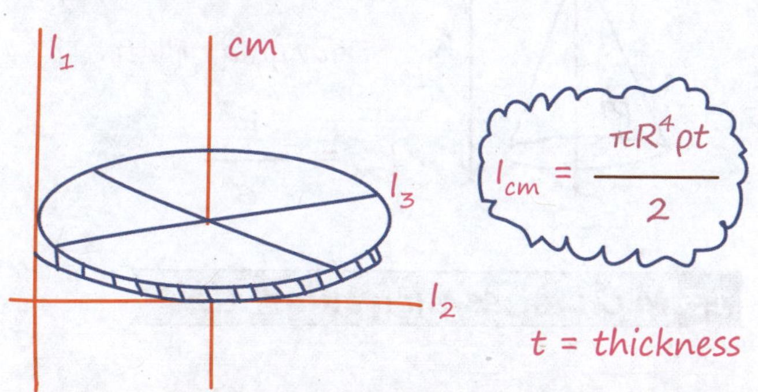

S. No.	Axis of Rotation	Moment of Inertia
(a)	About an axis passing through its center and perpendicular to its plane	$I_{cm} = \frac{MR^2}{2}$
(b)	About a tangent perpendicular to its plane	$I_1 = \frac{3MR^2}{2}$
(c)	About a tangent in the plane of the ring	$I_2 = \frac{5MR^2}{4}$
(d)	About a diameter	$I_3 = \frac{MR^2}{4}$

6 M.O.I. of hollow cylinder

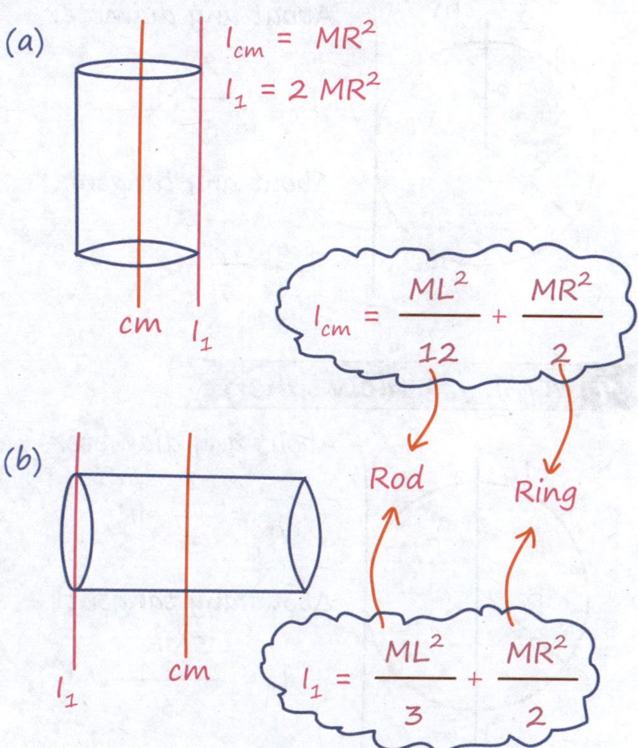

7 M.O.I. of solid cylinder

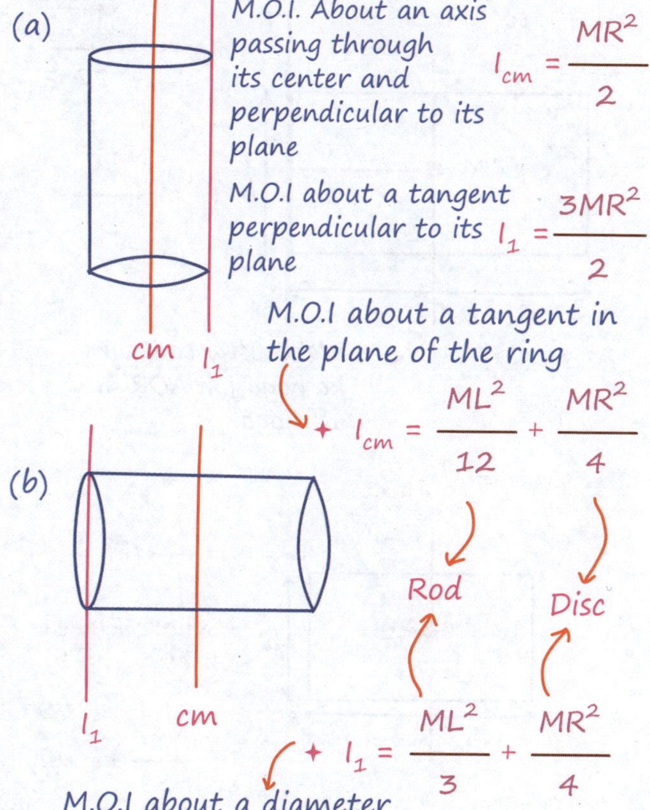

8 M.O.I. of solid sphere

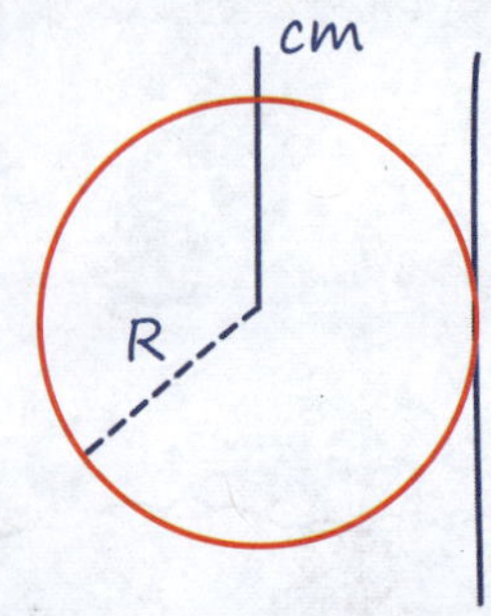

About any diameter

$$I_{cm} = \frac{2}{5} MR^2$$

About any tangent

$$I_T = \frac{7}{5} MR^2$$

9 M.O.I. of hollow sphere

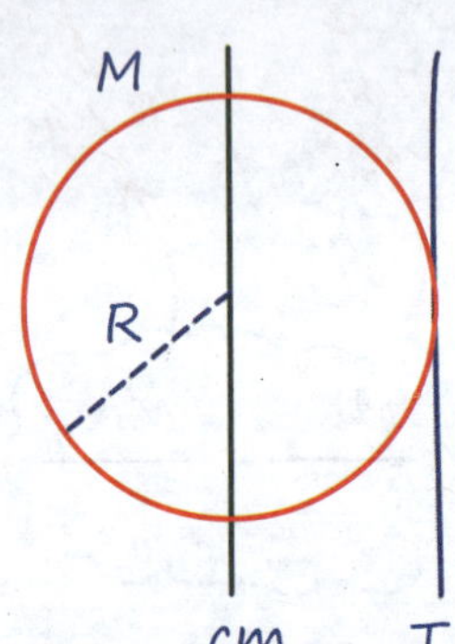

About any diameter

$$+ \; I_{cm} = \frac{2}{3} MR^2$$

About any tangent

$$+ \; I_T = \frac{5MR^2}{3}$$

10 M.O.I. of rectangular plate

(a) IIel to Plane:-

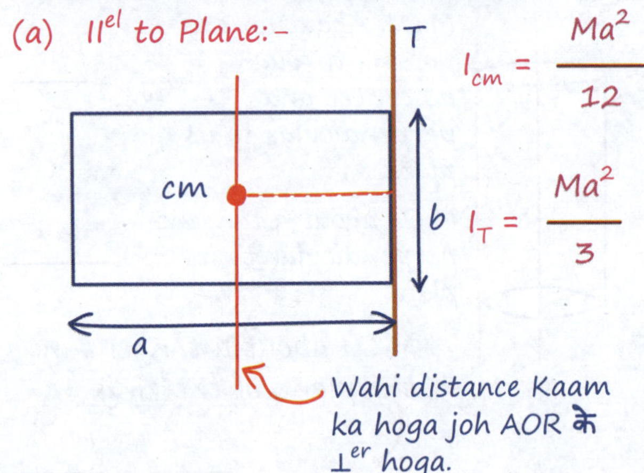

$$I_{cm} = \frac{Ma^2}{12}$$

$$I_T = \frac{Ma^2}{3}$$

Wahi distance Kaam ka hoga joh AOR के ⊥er hoga.

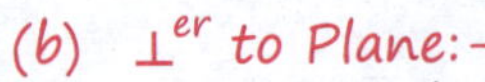

(b) ⊥er to Plane:-

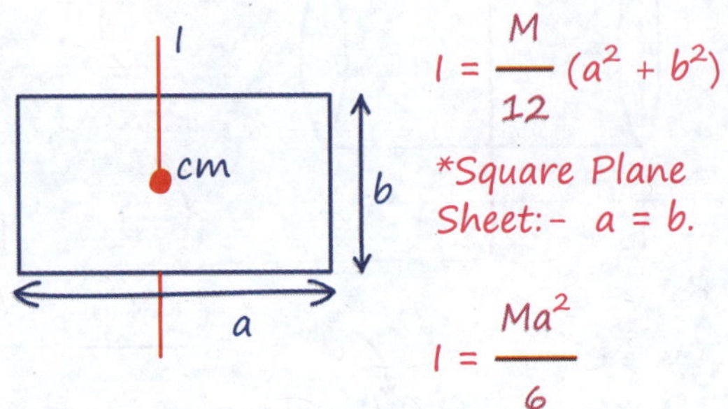

$$I = \frac{M}{12}(a^2 + b^2)$$

*Square Plane Sheet:- a = b.

$$I = \frac{Ma^2}{6}$$

11 M.O.I. of cube

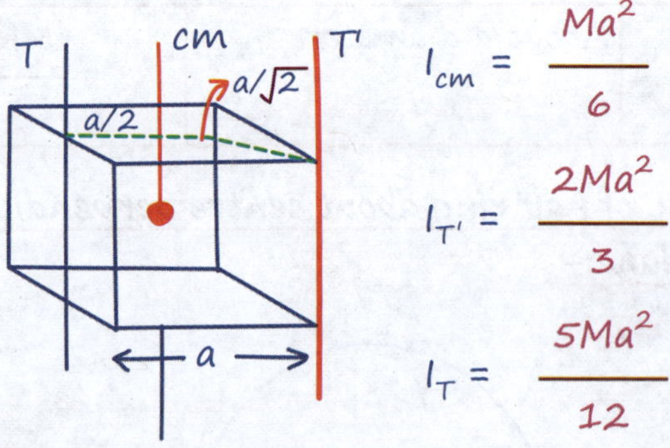

$$I_{cm} = \frac{Ma^2}{6}$$

$$I_{T'} = \frac{2Ma^2}{3}$$

$$I_T = \frac{5Ma^2}{12}$$

12 M.O.I. of cone

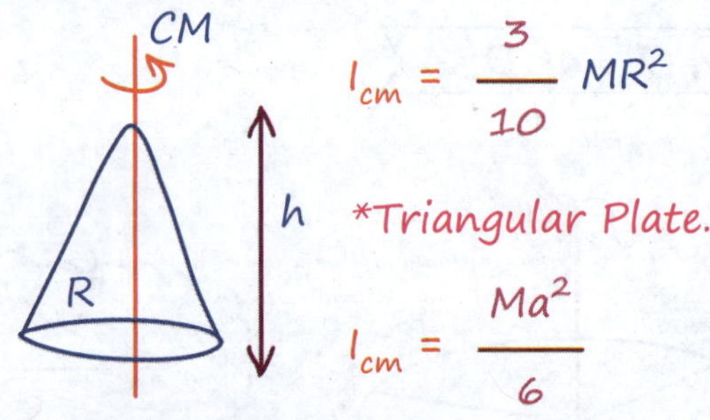

$$I_{cm} = \frac{3}{10} MR^2$$

*Triangular Plate.

$$I_{cm} = \frac{Ma^2}{6}$$

13 M.O.I. of semicircular disc

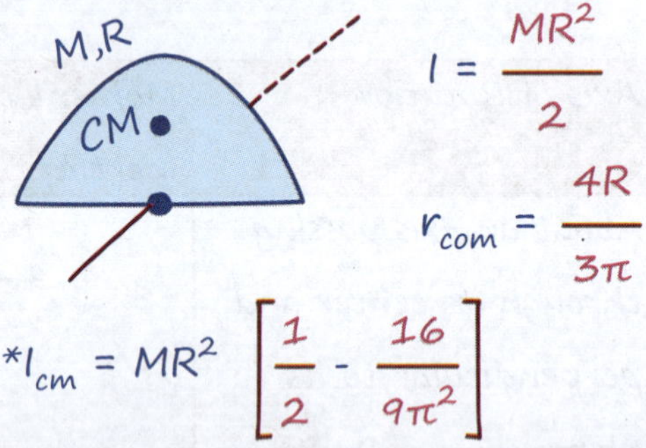

$$I = \frac{MR^2}{2}$$

$$r_{com} = \frac{4R}{3\pi}$$

$$*I_{cm} = MR^2\left[\frac{1}{2} - \frac{16}{9\pi^2}\right]$$

14 Concept of M.O.I. of Cutting Section

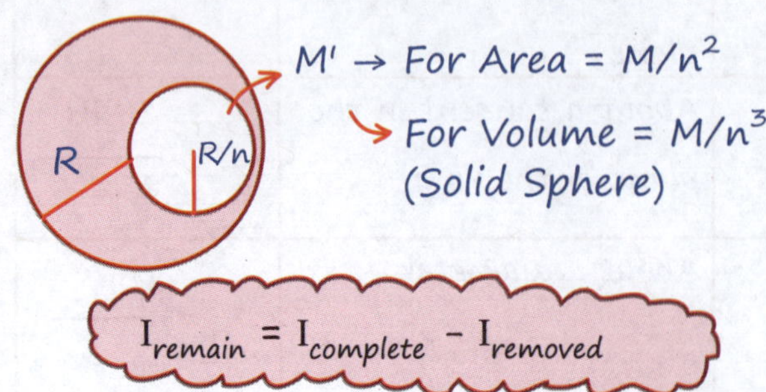

$M' \rightarrow$ For Area $= M/n^2$

For Volume $= M/n^3$ (Solid Sphere)

$$I_{remain} = I_{complete} - I_{removed}$$

15 M.O.I. of Annular disc

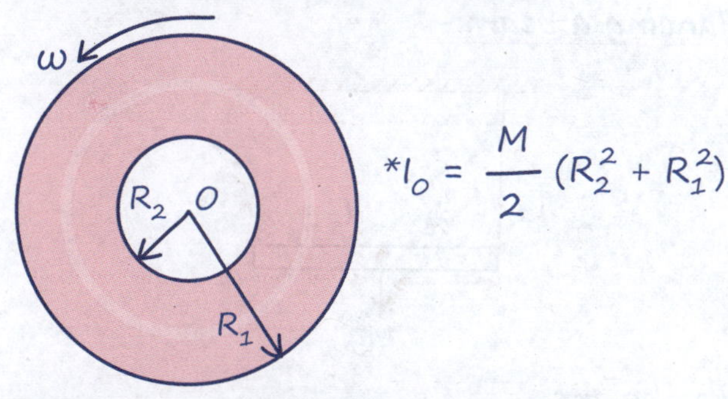

$$^*I_0 = \frac{M}{2}(R_2^2 + R_1^2)$$

16 Radius of Gyration

+ Radius of gyration is denoted by K and gives an equivalent distance at which the entire mass of the body can be thought to be concentrated to produce the same moment of inertia.

The radius of gyration K is defined as:

$$K = \sqrt{\frac{r_1^2 + r_2^2 + ... + r_n^2}{n}}$$

where $r_1, r_2, ..., r_n$ are the distances of the particles of the body from the axis of rotation, and n is the number of particles.

→ COM : not valid. So we use it.

+ The moment of inertia I of a body can be expressed in terms of its radius of gyration K and its mass M:

$$I = MK^2 \text{ or } K = \sqrt{\frac{I}{M}}$$

Q.1. Find the Radius of Gyration K, of a thin Rod about an End:

Sol.

$$K = \sqrt{\frac{I}{M}} = \sqrt{\frac{\frac{ML^2}{3}}{M}} = \sqrt{\frac{L^2}{3}} = \frac{L}{\sqrt{3}}$$

So, the radius of gyration K for the thin rod about an axis through one end is $\frac{L}{\sqrt{3}}$.

+ $I = MK^2$ * Yaad rakhna → Mass dyan से lena.

+ 3 SPHERICAL SHELL (M,R)

$$I = \frac{2}{3}MR^2 + \left(\frac{5}{3}MR^2\right) \times 2$$

$$I = \frac{12}{3}MR^2 = 4MR^2$$

Q.2. Two identical disc placed perpendicular to each other then find radius of gyration about axis passing through centre of disc parallel to one disc.

Sol. $$I = \frac{MR^2}{2} + \frac{MR^2}{4} = \frac{3MR^2}{4} = 2MK^2$$

$$K = \frac{1}{2}\sqrt{\frac{3}{2}}R$$

17 Torque

Torque → Cause of change in rotational state of the body

Torque oppose rotational motion → false

Axial vector

Unit → Nm

Dim^n → $[ML^2T^{-2}]$

$$\vec{\tau} = \vec{r} \times \vec{F} = rF\sin\theta$$

Thumb ($\vec{\tau}$), Four Fingers ($\vec{r}$), Slap ($\vec{F}$)

CW ⊗ Negative

ACW ⊙ Positive

$$\vec{\tau}_{net} = \vec{r}_1 \times \vec{F}_1 + \vec{r}_2 \times \vec{F}_2 + ...$$

$$\vec{\tau} \perp \vec{r} \perp \vec{F}$$

$$\vec{\tau} = I\vec{\alpha} \qquad \vec{\tau} = \frac{d\vec{L}}{dt}$$

Statement:-

If $F_{net} = 0$ then τ_{net} must be zero → False

If $F_{net} \neq 0$ then τ_{net} must be non-zero → False

If $\tau_{net} = 0$ then net force must be zero → False

If $\tau_{net} \neq 0$ then net force must be non-zero → False

MR*

Torque hamesha hinge point के about lagega aur who ek toh body Ko Ghumayega ya toh Ghumte huye object Ka state change Karega.

$\vec{\tau}_{net} = 0$ } Rotational Eq^n.

+ Concept of Rigid body:-

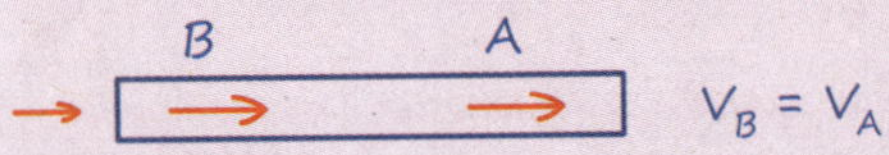

$V_B = V_A$

Imp Que:- in given figure find torque about point 'O'

$\tau = Mg\,(U\cos\theta)t$

At hightest Point

$$\tau' = \frac{Mg\,U^2 \sin 2\theta}{2g}$$

at Hmax

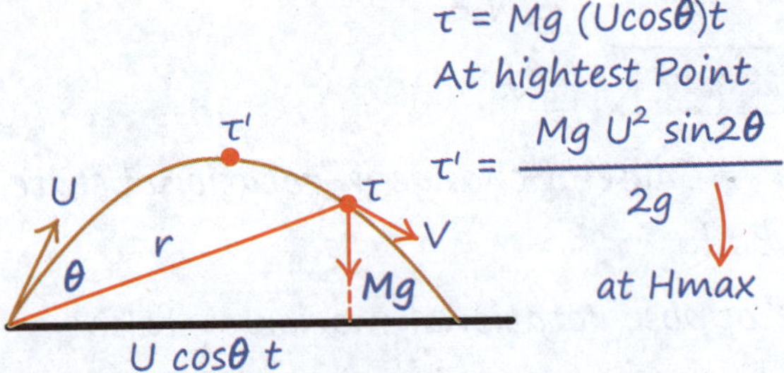

18 Rotational Equilibrium

+ Rotational equilibrium occurs when the net torque acting on an object is zero. This means that there is no unbalanced rotational effect on the object, and it remains in a state of constant rotational motion (either at rest or rotating with constant angular velocity).

+ $\tau_{net} = 0$

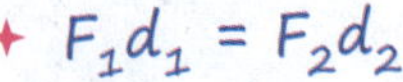

+ $F_1 d_1 = F_2 d_2$

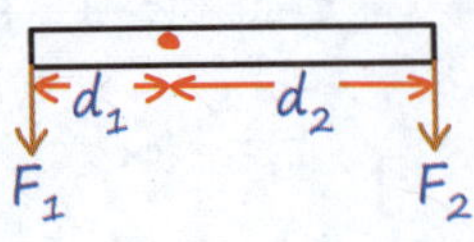

+ Mechanical Advantage $= \frac{F_1}{F_2} = \frac{d_2}{d_1}$ (M.A. > 1)

Balancing a beam:-

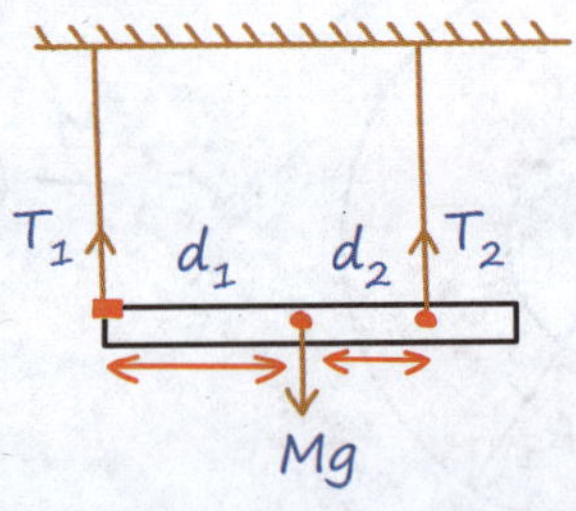

Force balance

$T_1 + T_2 = mg$ ...(i)

Torque balance

$T_1 d_1 = T_2 d_2$ ...(ii)

19 Rotational Kinematics

uniform circular motion

+ $\vec{\omega} = \text{const}^n$ $\quad \alpha = 0 = a_t$.

+ Speed = Const^n $\quad \vec{a_c} = \frac{v^2}{r}$

+ $\theta = \omega t$.

non uniform circular motion

When, α = const^n, Eqn. of Motion:

+ $\omega_2 - \omega_1 = \alpha t$

+ $\theta = \omega_1 t + \frac{1}{2}\alpha t^2$

+ $\omega_2^2 - \omega_1^2 = 2\alpha\theta$

$$\theta = \left(\frac{\omega_2 + \omega_1}{2}\right) t = n2\pi.$$

When α = Variable

$\Delta\theta = \int \omega . dt$ $\quad \Delta\omega = \int \alpha . dt$

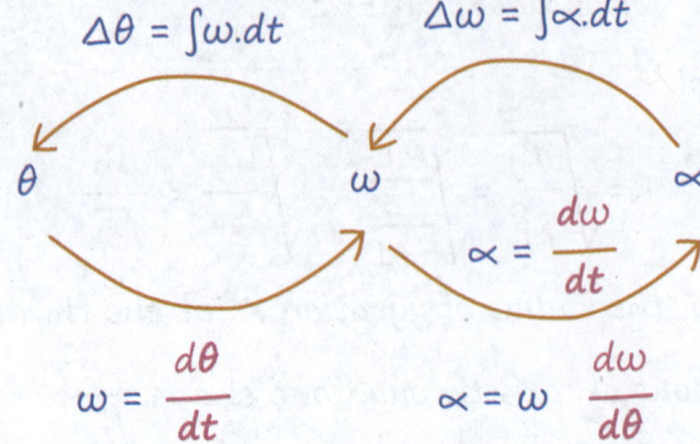

$\omega = \frac{d\theta}{dt}$ $\quad \alpha = \omega \frac{d\omega}{d\theta}$

✦ Rest to Rest Motion:-

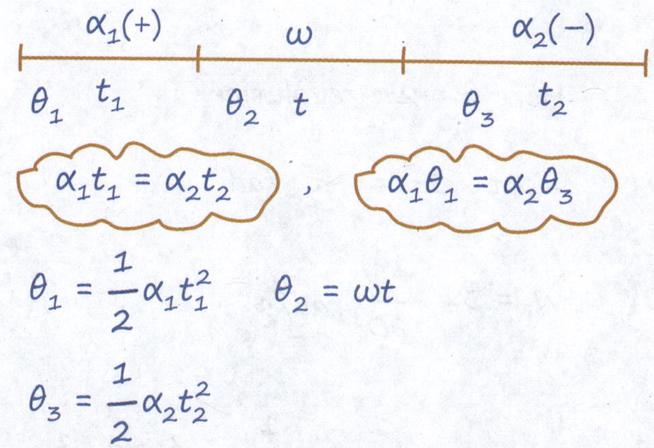

$\theta_1 = \frac{1}{2}\alpha_1 t_1^2 \qquad \theta_2 = \omega t$

$\theta_3 = \frac{1}{2}\alpha_2 t_2^2$

✦ The time at which two particle with different speed, start moving from same position Meet?

$V_{relative} = \frac{2\pi R}{T}$

✦ Pure Motion:-

1> Rotational Motion:-

$\omega_1 = \omega_2 = \omega_3$

$\alpha_1 = \alpha_2 = \alpha_2$

$v_1 \neq v_2 \neq v_3$

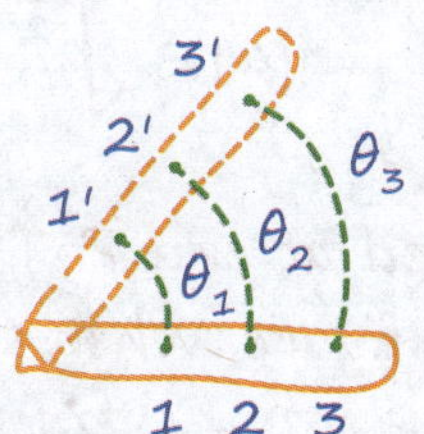

2> Translational Motion:-

$v_1 = v_2 = v_3$

$1 \to v_1$

$2 \to v_2$

$3 \to v_3$

20 Analogy

Translation	Rotational
S	θ
$V = \frac{dx}{dt}$	$\omega = \frac{d\theta}{dt}$
$a = \frac{dv}{dt}$	$\alpha = \frac{d\omega}{dt}$

$F = \frac{dP}{dt} = ma$	$\tau = \frac{dL}{dt} = I\alpha$
$P = mV$	$L = I\omega$
$W = F.S.$	$W = \vec{\tau}.\vec{\theta}$
$KE = \frac{1}{2} mv^2$	$KE = \frac{1}{2} I\omega^2$
$P = \vec{F}.\vec{v}.$	$P = \vec{\tau}.\vec{\omega}$
Impulse = m.V.	Impulse = Iω

21 Analogy

Translation	Rotational
* $S = ut + \frac{1}{2}at^2$	$\theta = \omega t + \frac{1}{2}\alpha t^2$
* $V = U + at$	$\omega = \omega_0 + \alpha t$
* $V^2 - U^2 = 2as$	$\omega^2 - \omega_0^2 = 2\alpha\theta$
* $S_{nth} = U + \frac{a}{2}(2n-1)$	$\theta_{nth} = \omega_0 + \frac{\alpha}{2}(2n-1)$
* $S = \left(\frac{U+V}{2}\right)T$	MR* $\theta = \left(\frac{w_i+\omega_f}{2}\right)T$
* $S = \frac{1}{2}\left(\frac{\alpha\beta}{\alpha+\beta}\right)T^2$	$S = \frac{1}{2}\left(\frac{\alpha\beta}{\alpha+\beta}\right)T^2$
* Stoping distⁿ. $= \frac{U^2}{2a}$	* $\theta = \frac{\omega_0^2}{2\alpha}$
$V_{max} = \left(\frac{\alpha\beta}{\alpha+\beta}\right)T$	$\omega_{max} = \left(\frac{\alpha\beta}{\alpha+\beta}\right)T$

Q.3. A solid sphere (M, R) hinged about centre and free to rotate then find angular accⁿ.

Sol.

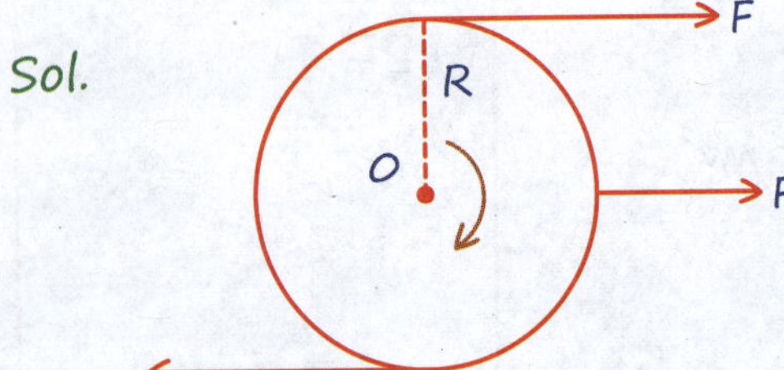

$\tau_o = I\alpha$

$2FR = \frac{2}{5}MR^2\alpha \Rightarrow \alpha = \frac{5F}{MR}$

Q.4. A solid cylinder of mass 2 kg and radius 4 cm rotating about its axis at the rate of 3 rpm. The torque required to stop after 2π revolution is

Sol. Using Work Energy Theorem

$W = \frac{1}{2}I(\omega_f^2 - \omega_i^2)$

Here θ = 2π revolution

$= 2\pi \times 2\pi = 4\pi^2$ rad

$W_i = 3 \times \frac{2\pi}{60}$ rad/s

$\Rightarrow -\tau\theta = \frac{1}{2} \times \frac{1}{2} mr^2(0^2 - \omega_i^2)$

$\Rightarrow -\tau = \frac{\frac{1}{2} \times \frac{1}{2} \times 2 \times (4\times10^{-2})^2 \left(-3\times\frac{2\pi}{60}\right)^2}{4\pi^2}$

$\Rightarrow \tau = 2 \times 10^{-6}$ Nm

Note:-

$a_t = R\alpha \qquad a_c = R\omega^2 = \frac{V^2}{R}$

$|\vec{a}| = \sqrt{a_t^2 + a_c^2}$

Q.5. A rod PQ of mass M and length L is hinged at end P. The rod is kept horizontal by a massless string tied to point Q as shown in figure. When string is cut, the initial angular acceleration of the rod is (NEET 2013, AIP MT – 07/11/ IIT-4 X).

$[\tau_P = I_P\alpha_P]$ hinged point/axis of Rotⁿ

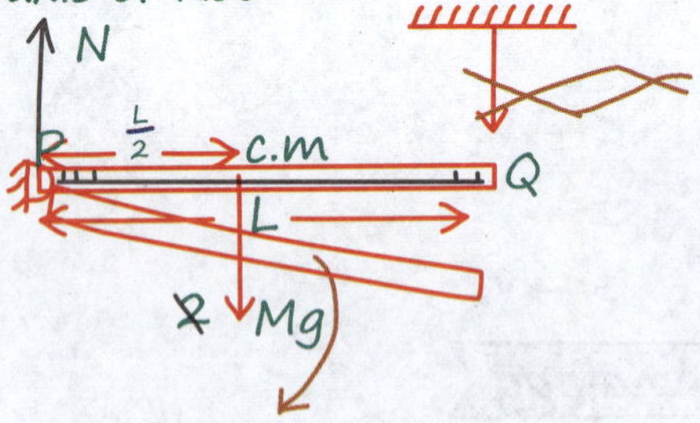

(a) $\frac{2g}{L}$

(b) $\frac{2g}{2L}$

(c) $\frac{3g}{2L}$

(d) $\frac{g}{L}$

$N \times 0 + mg\frac{L}{2} = \frac{ML^2}{3}\alpha$

$\frac{g}{2} = \frac{L\alpha}{3}$

Angular accⁿ of Rod (every point of Rod) about 'p'

$* \; \alpha = \frac{3g}{2L}$

Find liner accⁿ of C.O.M.??

$a_t = r(\alpha)$ diffⁿ for all Points

$a_{c.m} = \frac{L}{2}\left(\frac{3g}{2L}\right) = \frac{3g}{4}$ m/s²

(Ask 4X in IIT) (Most Imp. Que.)

in given figure find the acceleration of block.

MR*

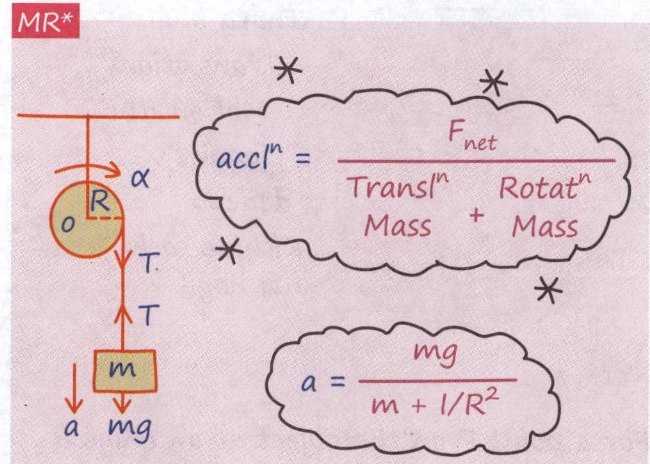

✦ **Relation b/n ω_1 & ω_2:-**

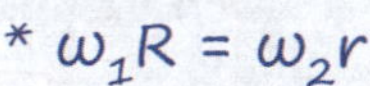

* $V_1 = V_2$
* $\omega_1 R = \omega_2 r$
* $a_{t_1} = a_{t_2}$
* $\alpha_1 R = \alpha_2 r$

✦ Two bodies of masses m and m are attached to the two ends of a string. The string passes over a pulley of mass m and radius R. If $m_1 < m_2$, the acceleration of the system will be-

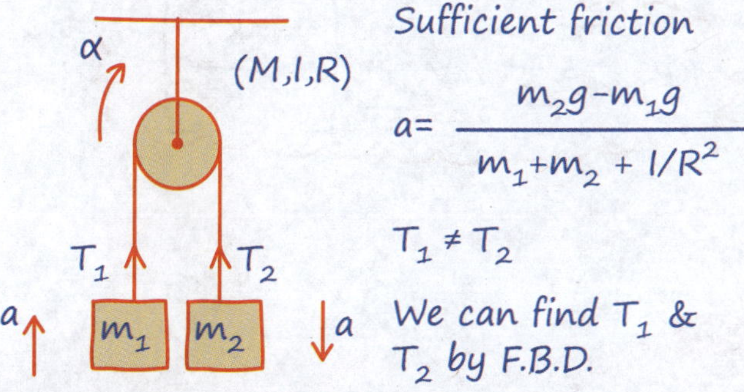

Sufficient friction

$$a = \frac{m_2 g - m_1 g}{m_1 + m_2 + I/R^2}$$

$T_1 \neq T_2$

We can find T_1 & T_2 by F.B.D.

22 Application of Com in Pure Rotational Motion

$$KE = \frac{1}{2} I\omega^2 = \frac{L^2}{2I}$$

Rod release from vertical position

Rod is released from vertical position Find ω when rod becomes horizontal.

Applying Energy conseration

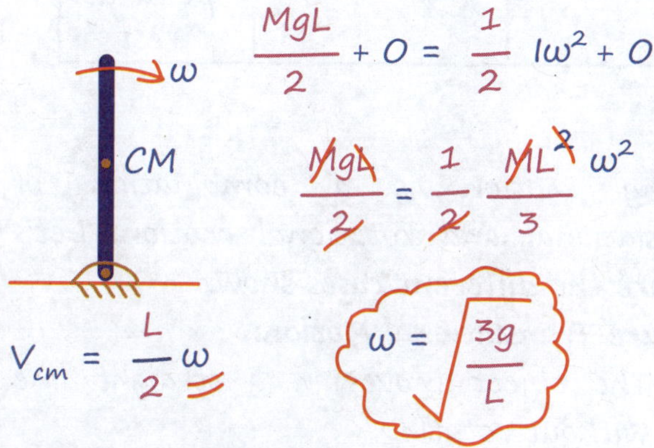

$$\frac{MgL}{2} + 0 = \frac{1}{2} I\omega^2 + 0$$

$$\frac{MgL}{2} = \frac{1}{2}\frac{ML^2}{3}\omega^2$$

$$V_{cm} = \frac{L}{2}\omega$$

$$\omega = \sqrt{\frac{3g}{L}}$$

23 Rolling Motion

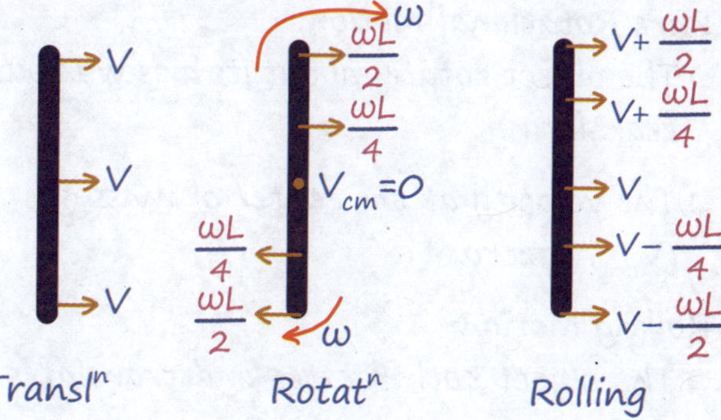

$$*KE = \frac{1}{2} mV_{cm}^2 + \frac{1}{2} I\omega^2. *$$

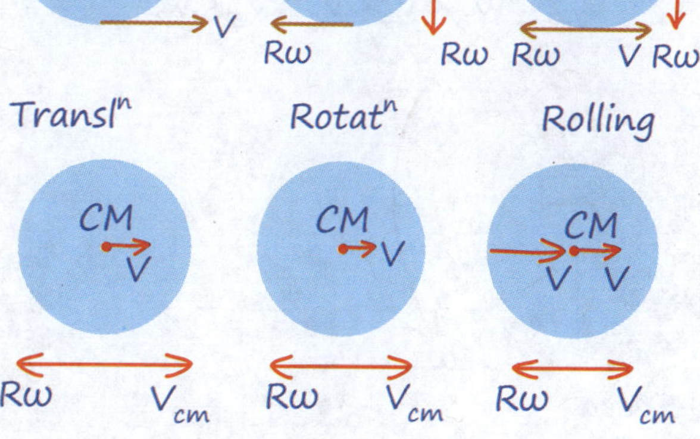

Forward Sliping	Backward Sliping	No Sliping
$V_{cm} > R\omega$	$V_{cm} < R\omega$	$V_{cm} = R\omega$
f_k = Back	f_k = Front	Lowest point at rest

Note:-

Rolling motion is a combination of translational and rotational motion. Let's explore the different cases shown:

1. **Pure Translational Motion:**
   - The object moves in a straight line without rotating.
   - Velocity (V) is constant for all points on the object.
2. **Pure Rotational Motion:**
   - The object rotates about its axis without translating.
   - The velocity at the center of mass (V_{cm}) (V_{cm}) is zero.
3. **Rolling Motion:**
   - The object both rotates and translates.
   - The velocity of the center of mass (V_{cm}) and the rotational velocity combine.
   - The speed at the top of the rolling object is $V_{cm} + R\omega$.
   - The speed at the bottom (contact point) is $V_{cm} - R\omega$.

$*V_3 < V_4 < V_1 < V_5 < V_2*$

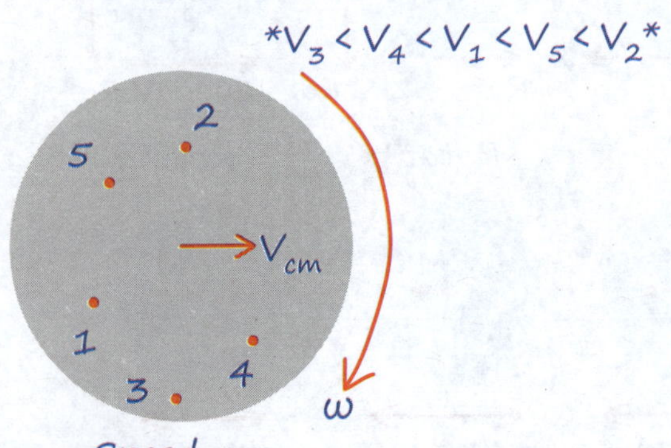

Speed.

#

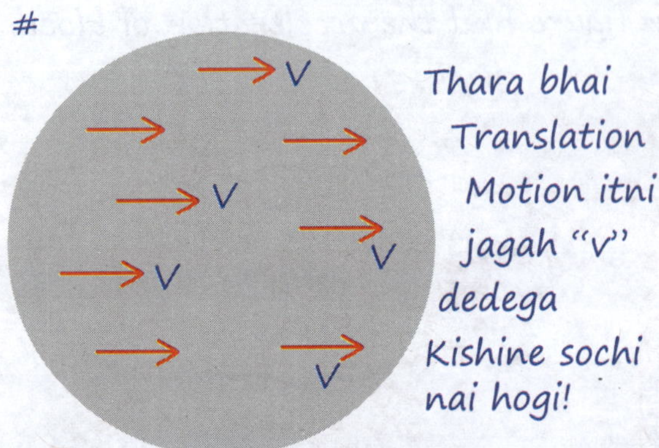

Thara bhai Translation Motion itni jagah "v" dedega Kishine sochi nai hogi!

Note:-

For a point P on the object at an anlge θ.

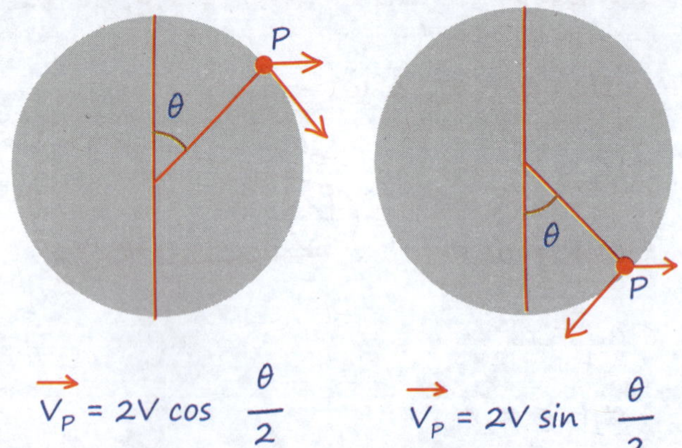

$$\vec{V}_P = 2V\cos\frac{\theta}{2} \qquad \vec{V}_P = 2V\sin\frac{\theta}{2}$$

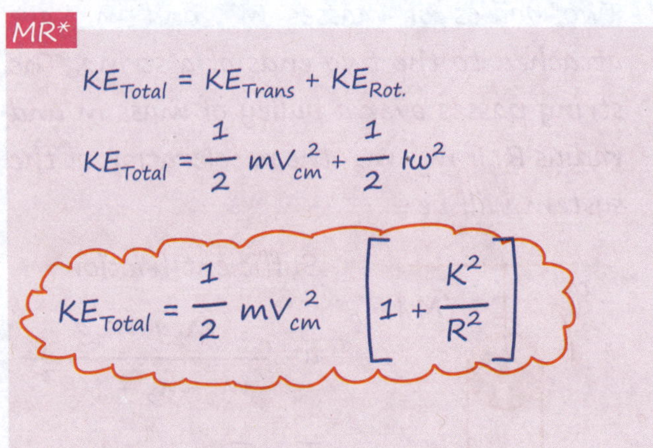

MR*

$$KE_{Total} = KE_{Trans} + KE_{Rot.}$$

$$KE_{Total} = \frac{1}{2}mV_{cm}^2 + \frac{1}{2}I\omega^2$$

$$KE_{Total} = \frac{1}{2}mV_{cm}^2\left[1 + \frac{K^2}{R^2}\right]$$

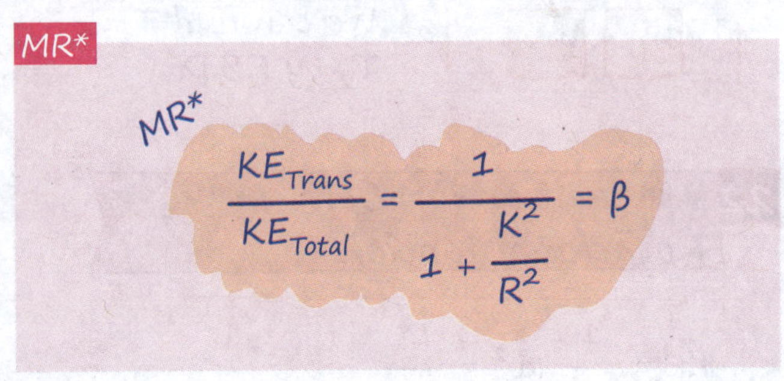

MR*

MR*

$$\frac{KE_{Trans}}{KE_{Total}} = \frac{1}{1 + \frac{K^2}{R^2}} = \beta$$

MR * TABLE (THE PRO VERSION)

Physical quantity	Ring hollow cylinder	Hollow sphere	Disc, solid cylinder	Solid sphere
β	$0.5 = \frac{1}{2}$	$0.6 = \frac{3}{5}$	$0.66 = \frac{2}{3}$	$0.71 = \frac{5}{7}$
KE_{Trans}/KE_{Total}	0.5 = 50%	0.6 = 60%	0.66 = 66%	0.71 = 71%
KE_{Rot}/KE_{Total}	0.5 = 50% (1/2)	0.4 = 40% (2/5)	0.33 = 33% (1/3)	0.28 = 28% (2/7)
KE_{Trans}/KE_{Rot}	1 : 1	3 : 2	2 : 1	5 : 2
Accl^n on inclined → $g\sin\theta\,\beta$	$\frac{g\sin\theta}{2}$	$\frac{3}{5}g\sin\theta$	$\frac{2}{3}g\sin\theta$	$\frac{5}{7}g\sin\theta$
Time req. to come down →	$t = \sqrt{\frac{2S}{g\sin\theta\beta}}$ →	→	→	
Velocity at bottom of inclined $V = \sqrt{2gh\beta}$	$V = \sqrt{gH}$	$V = \sqrt{\frac{6gH}{5}}$	$V = \sqrt{\frac{4gH}{3}}$	$V = \sqrt{\frac{10gH}{7}}$
H_{max} attained by particle	$H = \frac{V_{cm}^2}{2g\beta}$ $H = \frac{V_{cm}^2}{g}$	$H = \frac{5V_{cm}^2}{6g}$	$H = \frac{3V_{cm}^2}{4g}$	$H = \frac{7V_{cm}^2}{10g}$
Friction on inclined	$f_r = Mg\sin\theta(1-\beta)$ $f_r = Mg\sin\theta/2$	$f_r = \frac{2}{5}Mg\sin\theta$	$f_r = \frac{Mg\sin\theta}{3}$	$f_r = \frac{2}{7}Mg\sin\theta$
μ_{min} to start pure rolling	$\mu = (1-\beta)\tan\theta$ $\mu = \tan\theta/2$	$\mu_s = \frac{2}{5}\tan\theta$	$\mu = \frac{\tan\theta}{3}$	$\mu = \frac{2}{7}\tan\theta$

* Jahan "g" wahi "β" * Konse bhi sawaal mein Rolling aayega toh β lagado.*

Rolling on Smooth inclined plane:-

- $a = g \sin\theta$ → independent of Mass, Shape, Size.
- $V = \sqrt{2gh}$ → Velocity at bottom.
- $H = \dfrac{V_{cm}^2}{2g}$ → "H_{max}"

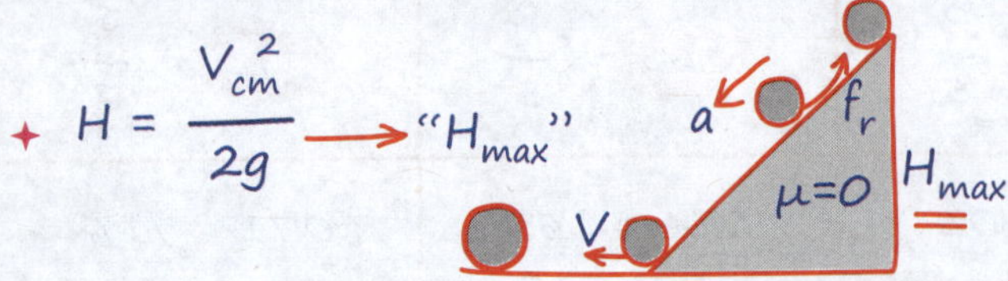

Rolling Motion on Rough inclined plane:-

$V = \sqrt{2gh\beta}$ $\qquad a = \beta g \sin\theta$

$$H = \frac{V_{cm}^2}{2g\beta}$$

Object Upar jaye ya niche

*f_r always acts upwards!

24 Pure rolling on a horizontal plane

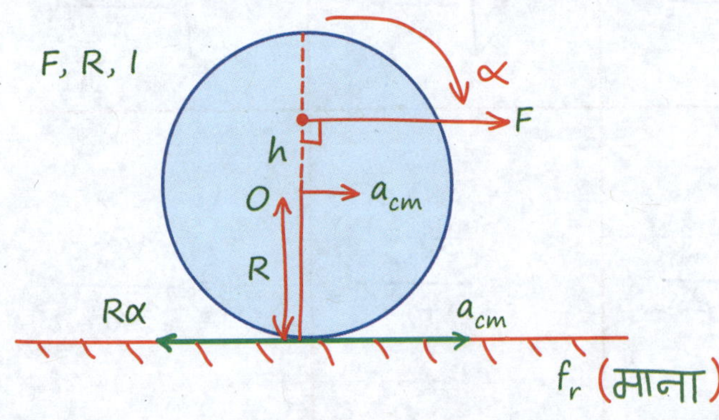

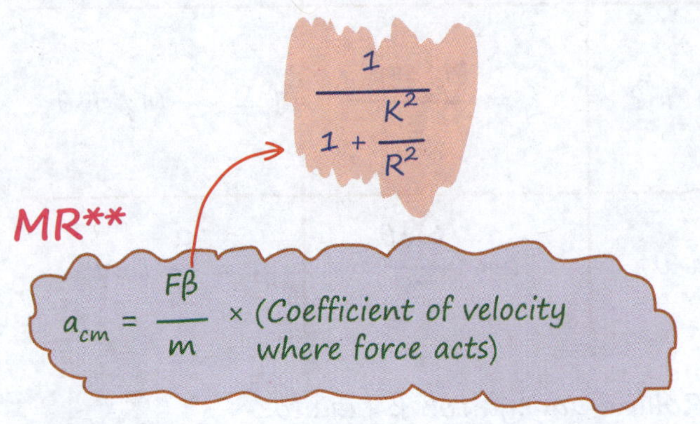

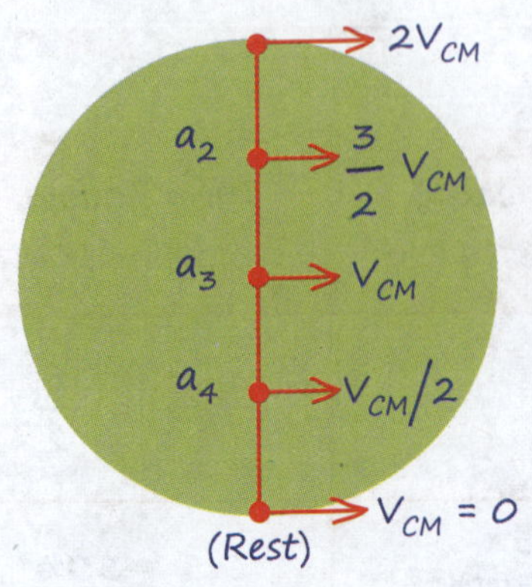

$a_1 = \dfrac{F\beta(2)}{m}$ $\qquad a_2 = \dfrac{F\beta}{m}\left(\dfrac{3}{2}\right)$

$a_3 = \dfrac{F\beta(1)}{m}$ $\qquad a_4 = \dfrac{F\beta}{m}\left(\dfrac{1}{2}\right)$

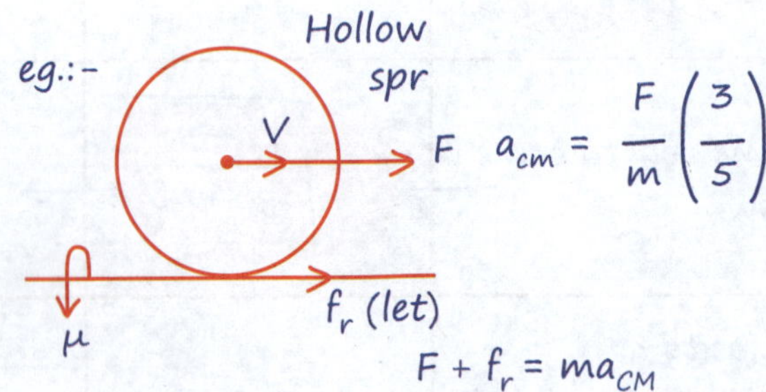

$F + f_r = ma_{CM}$

$f_r = \dfrac{3F}{5} - F = -\dfrac{2F}{5}$ (backward)

Note:-

* Pure rolling motion can start on smooth horizontal surface

25 Angular Momentum

The moment of linear momentum is called angular momentum. It is denoted by L.

- Depends on Frame of Ref.

$L = rmv \sin\theta$

$L = rP \sin\theta$

$\vec{L} = \vec{r} \times \vec{P}$ (Axial Vector)

θ = Angle b/n $\vec{r}$ & $\vec{P}$

→ When object is moving on straight line with const. velocity then:- L = same.

Relation to Torque:

Torque (τ) is the rate of change of angular momentum.

$$\tau = \frac{dL}{dt}$$

Example: Object is projected with speed "U" at an angle "θ" with horizontal then find angular momentum in projectile about point 'O' (orrin):-

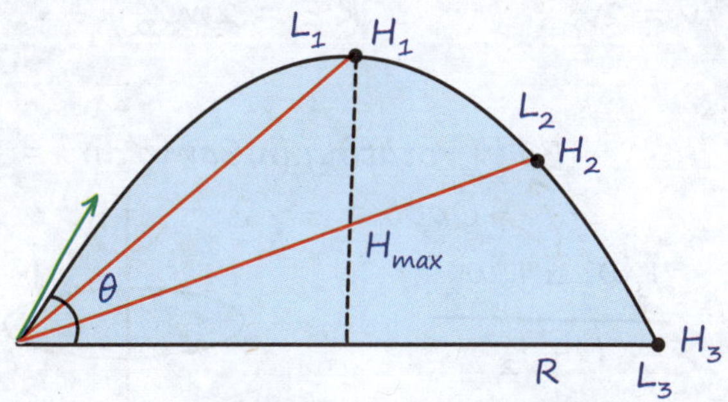

$$L_1 = mu\cos\theta \times H_{max} = mu\cos\theta . \frac{U^2\sin^2\theta}{2g}$$

$$L_3 = 4L_1 = mu\cos\theta . \frac{2U^2\sin^2\theta}{g}$$

L when objects about to Collide = 4 L at H_{max} *

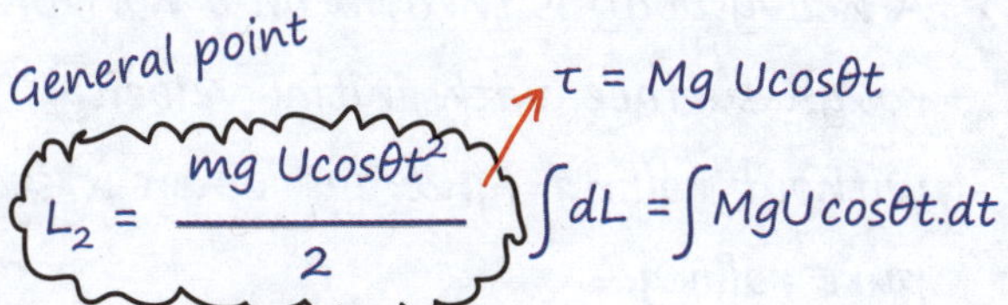

✦ Pure Rotational Motion:-

$$L = I\omega \qquad KE = \frac{1}{2}I\omega^2 = \frac{L^2}{2I}$$

✦ A rolling body is rolling without slipping on a moving plank.

$$V_1 - R\omega = V$$

$$V_1 - V = R\omega.$$

✦ Angular momentum of object w.r.t. origin:-

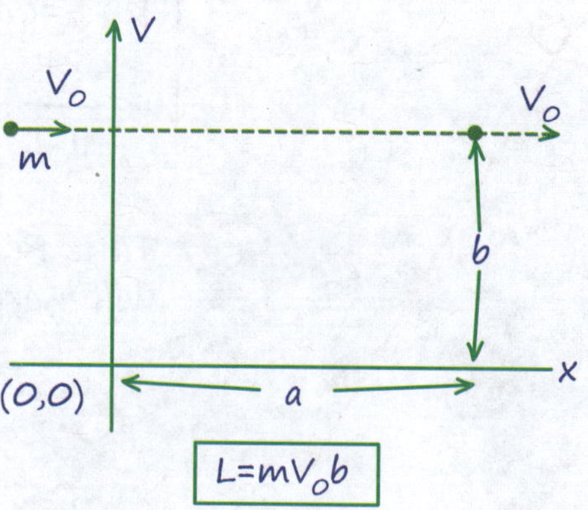

$L = mV_0 b$

Q.6. A solid sphere is rotating with angular speed ω then angular momentum about given axis:-

⇒ $L = I\omega$

$$L = \frac{2}{5}MR^2\omega \qquad L = \frac{7}{5}MR^2\omega$$

Q.7. in given figure find angular momentum about point A ,B and origin.

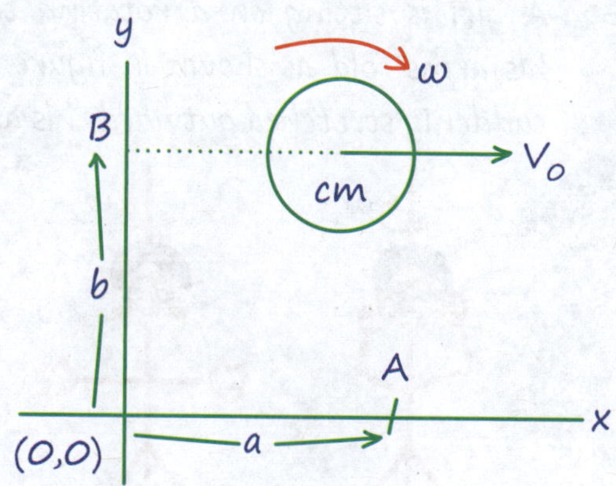

⇒ $L_A = L(0,0) = mV_0 b + I_{cm}\omega$

$L_B = I_{cm}\omega$

26 Conservation of Angular Momentum

If there is no external torque then the angular momentum is conserved.

$\tau = 0, \quad \vec{L} = const^n = I\omega$

Therefore,

$I_1\omega_1 = I_2\omega_2$

1>

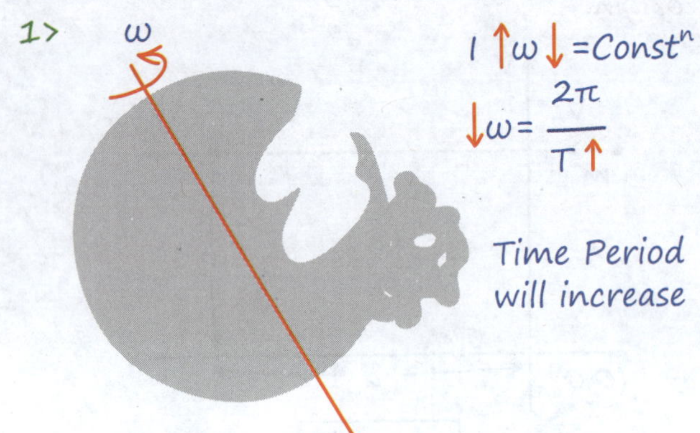

$I\uparrow \omega\downarrow = Const^n$

$\downarrow\omega = \frac{2\pi}{T\uparrow}$

Time Period will increase

2> what happen when ice on pole will melt

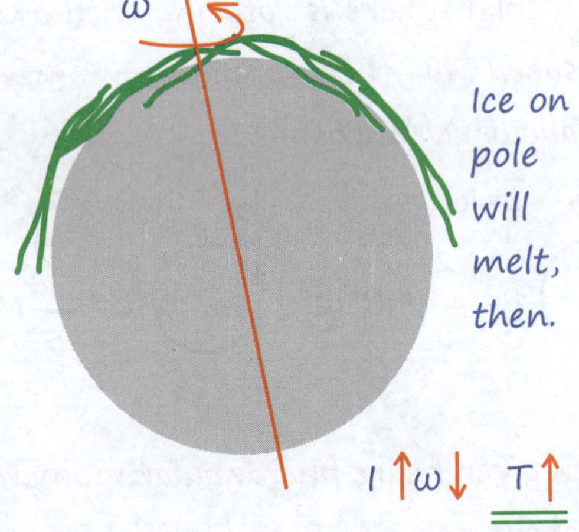

Ice on pole will melt, then.

$I\uparrow \omega\downarrow \quad T\uparrow$

3> A girl is sitting on a rotating table with his arms fold as shown in figure. When he suddenly stretched outwards his arms, then

Q.8. 'm' is moving on circular path of radius r with speed v_o, then find its speed when radius becomes $\frac{r}{2}$ by increasing 'T'.

Sol. $L_o = cos^n$

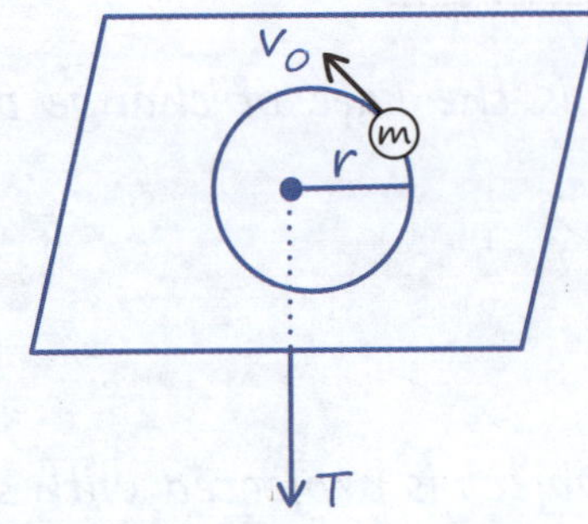

$mv_o r = mv\frac{r}{2} \qquad K.E_i = \frac{1}{2}mv_o^2$

$v = 2v_o \qquad K.E_f = 2mv_o^2$

Imp:-

Body rotating in Same dir^n = +

Oppo = −

$$\omega = \frac{I_1\omega_1 \pm I_2\omega_2}{I_1 + I_2}$$

Same = −

Oppo = +

$$\underset{loss}{\Delta KE} = \frac{1}{2}\frac{I_1 I_2}{I_1 + I_2}\omega_{rel}^2$$

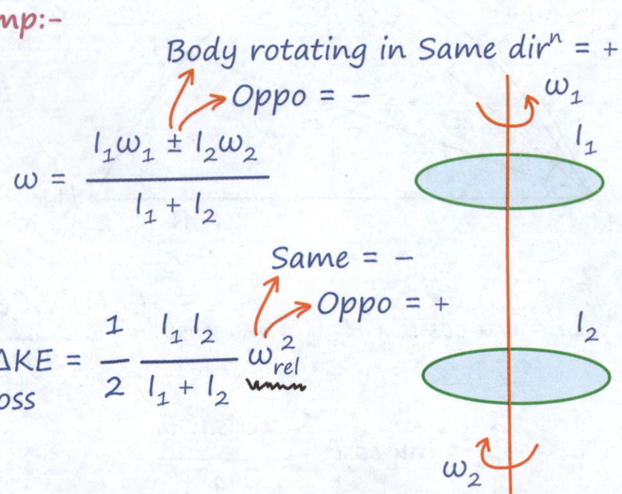

Imp. Model Problems MR*

1> A heavy body is thrown on a horizontal rough surface with initial velocity "U" without rolling. Find "V" when it start pure rolling:-

MR*

$V = \beta U \qquad \beta = \frac{1}{1 + \frac{K^2}{R^2}}$

3> A Rotating body with "ω_0" Placed on rough surface then find Angular velocity when it starts pure rolling motion:-

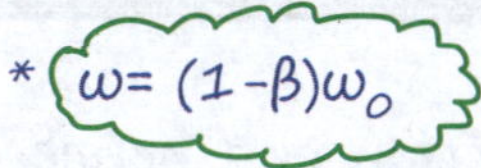

4> A body rolls on horizontal floor. Find W (work) to stop it:-

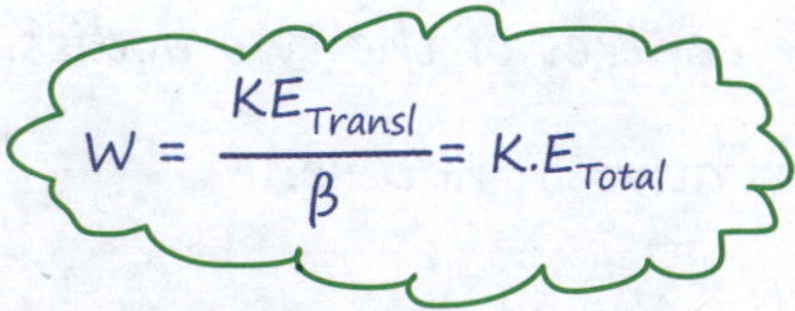

27 Toppling

MR*

$\tau_O = 0$

$Mg \dfrac{\cancel{L}}{2} = F.\cancel{L}$

F, L, CM, L, O, Mg

Don't take $\tau_{CM} = 0$ take

$\tau_O = 0$

Toppling में Normal Ko Shift KrKe Object apne appko palatne se bachata hai!

Q.9. In which case probability of toppling is high:-

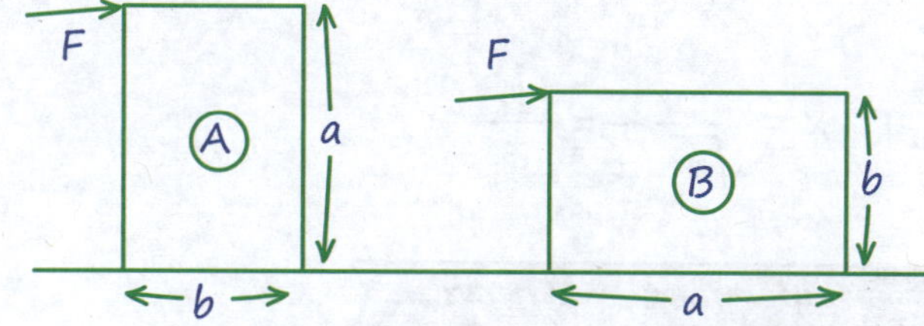

Sol. In A probability of toppling is high

Q.10. What will be the value of maximum acceleration of the truck in the forward direction so that the block kept on the back does not topple:-

Sol. ma_{truck}

h/2

a/2 mg

Block will not topple if

$\tau_{ma_{truck}} \le \tau_{mg}$

$$\frac{ma_{truck}h}{2} \le mg\,\frac{a}{2}$$

$$a_{truck} \le \frac{ag}{h}$$

$$\therefore\ a_{truck} = \frac{ag}{h}$$

Q.11. Ring, solid sphere and disc of mass M and radius R rotating with same angular speed ω, then work to stop it is :-

Sol. $W_{ring} > W_{Disc} > W_{solid\ sphere}$

MR*

"Koi kam Sahi galat nahi hota, bas ush kam ko krne ka samay sahi galat hota hai."

13 Gravitation

1 Gravitation

Every object in the universe pulls on every other object. This pulling force is called gravitation.

Gravity:-

Gravity is the force by which the Earth pulls objects towards its center. It's what makes things fall when you drop them.

2 Newton's law of gravitation

Valid for point and spherical object.

$$F = \frac{Gm_1m_2}{\gamma^2} \qquad G = 6.67 \times 10^{-11}\ kg^{-1}m^3s^{-2}$$

$$= M^{-1}L^3T^{-2}$$

$$\vec{F} = \frac{Gm_1m_2}{|\gamma^3|}\vec{\gamma}$$

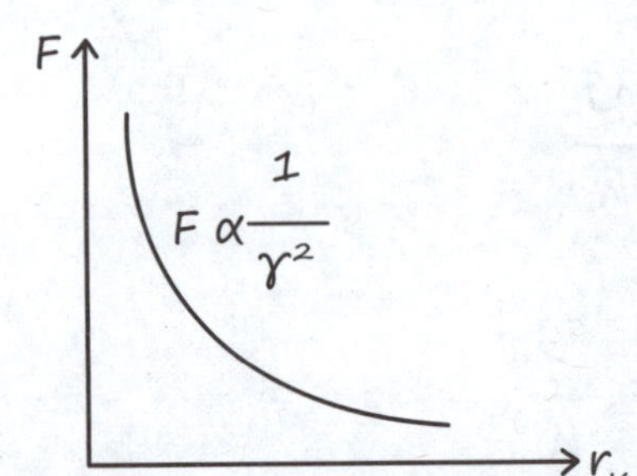

Properties of Gravitational Force:-

- It is always attractive.
- It is independent of nature of intervening medium.
- It is independent of the presence or absence of other bodies.
- Forms action and reaction pair, i.e., the gravitational forces are equal in magnitude and opposite in direction and hence obey Newton's third law of motion.
- It is a central force as it acts along the line joining the centres of the two bodies.
- It is conservative in nature.

Neutral point:-

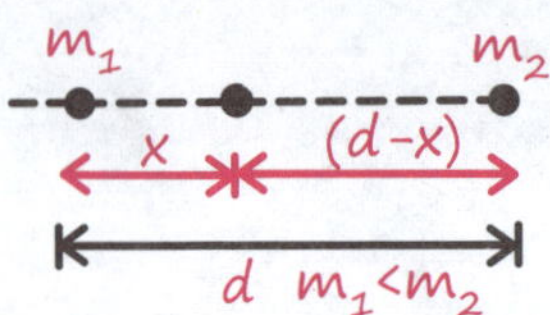

$$x = \frac{d}{\sqrt{n} + 1}$$

Where, x: Distance from the smaller mass to the neutral point.

d: Total distance between the two masses.

n: Ratio of the larger mass to the smaller mass.

Q.1. Find x so that field at that point will be zero

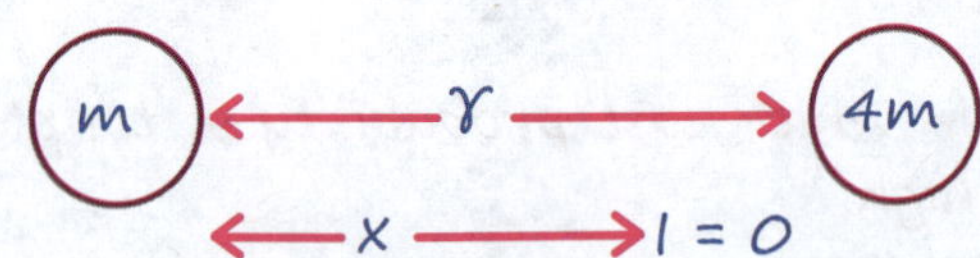

Sol. $x = \frac{r}{\sqrt{4} + 1} = \frac{r}{3}$

3 Superposition Theorem

- Net force on one object is a vector sum of all other forces acting on it due to other masses.

$$\vec{F} = \vec{F}_1 + \vec{F}_2 + \vec{F}_3 + \ldots \vec{F}_n$$

Q.2. In Given figure find net force on mass m placed at corner.

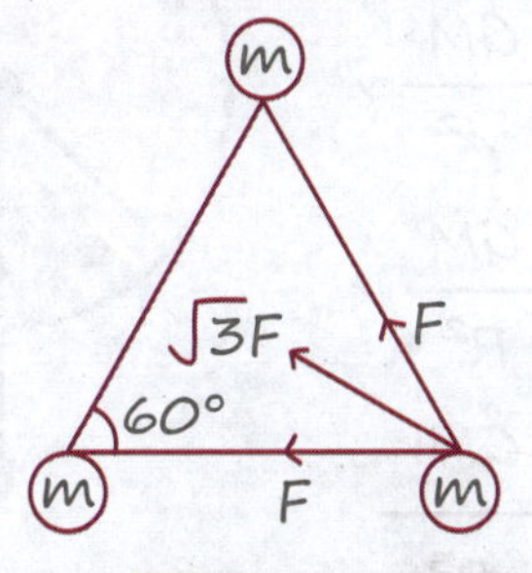

Sol. $F_{net} = \sqrt{F^2 + F^2 + 2FF\cos 60°}$

$$F_{net} = \sqrt{3}F = \frac{\sqrt{3}Gm^2}{a^2}$$

Q.3. In Given figure find net force on mass m placed at corner.

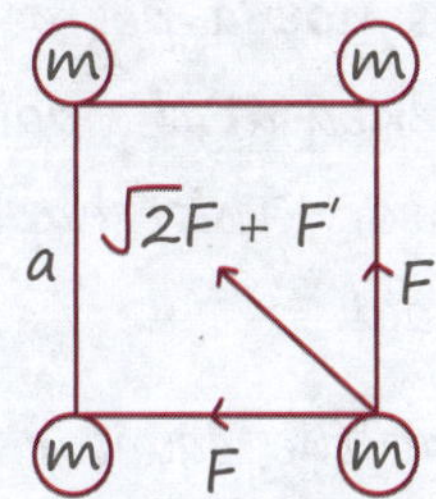

Sol. Each side of the square is of length a.

The net force on one mass (Fnet) due to the other two masses is

$$F_{net} = \frac{Gm^2}{a^2}\left(\frac{2\sqrt{2}+1}{2}\right)$$

- A particle with mass m is located a distance d from one end of a rod with mass M and length L.

The gravitational force exerted by the rod on the particle is

$$F_{net} = \frac{GMm}{d(d+L)}$$

due to rod on particle

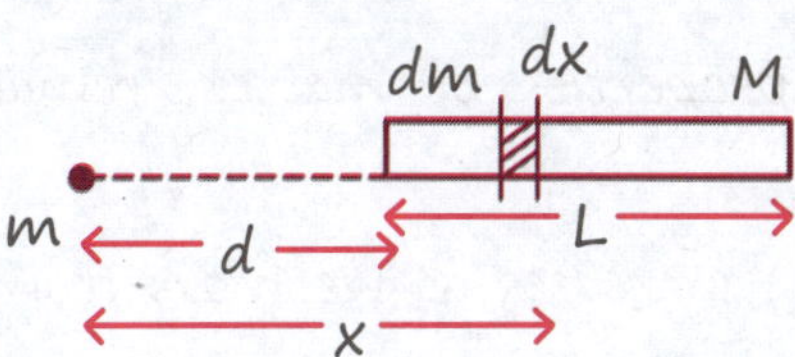

4 Gravitational Field

A gravitational field is a region around a mass where other masses experience a force of attraction.

Gravitational Field intensity

This is how strong the gravitational field is at a certain point.

It is represented by E_g or I.

$\vec{I} = \frac{\vec{F}}{m}$ Unit =N/Kg,

Intensity due to a Point Mass:-

At a distance r from the center of a body with mass M, the intensity of the gravitational field is given by: $I = \frac{GM}{r^2}$

Here, G is the gravitational constant, M is the mass of the body (like Earth), and r is the distance from the center of the mass.

Note:-

The intensity of the gravitational field is a vector quantity, meaning it has both magnitude and direction.

- The direction of the gravitational field intensity is always towards the center of the mass creating the field (e.g., towards the center of the Earth).

Gravitational filed intensity due to ring:-

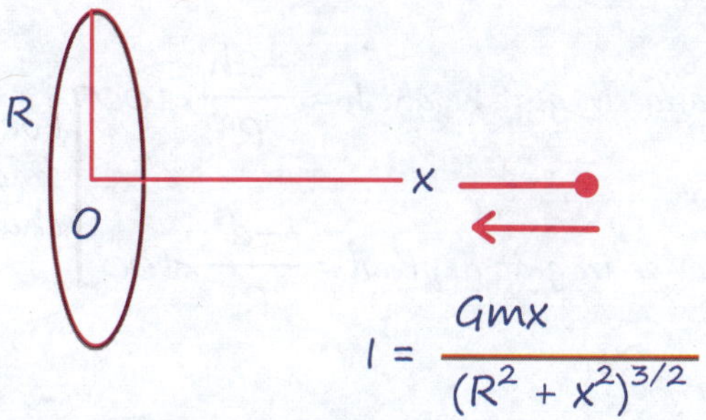

$$I = \frac{Gmx}{(R^2 + x^2)^{3/2}}$$

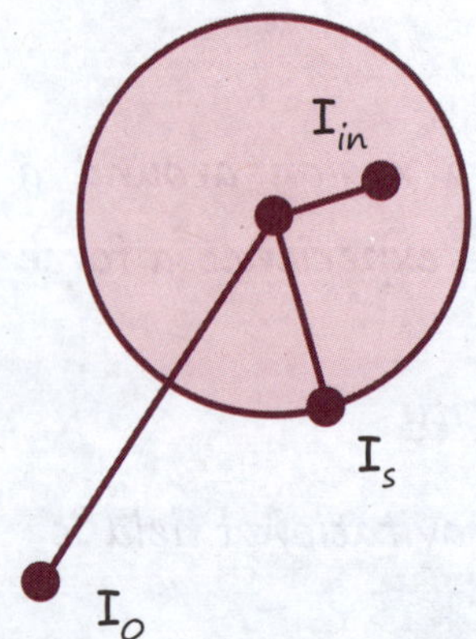

HOLLOW SPHERE.

$$I_{out} = -\frac{GM}{\gamma^2}\hat{\gamma}$$

$$I_{sur} = -\frac{GM}{R^2}\hat{R}$$

$$I_{inside} = 0$$

SOLID SPHERE.

$$I_{out} = -\frac{GM}{\gamma^2}\hat{\gamma}$$

$$I_{sur} = -\frac{GM}{R^2}\hat{R}$$

$$I_{in} = \frac{GMr}{R^3}$$

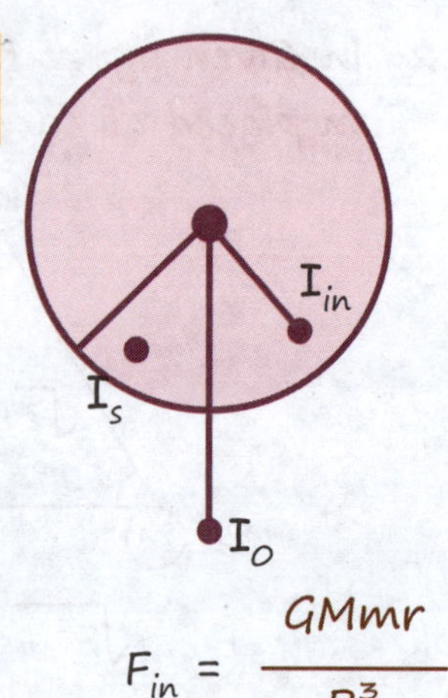

$$F_{in} = \frac{GMmr}{R^3}$$

5 Acceleration due to gravity

On Surface of earth:

$$\rho_{earth} = \frac{M}{\frac{4}{3}\pi R_e^3} = 5.5 \times 10^3\ kg/m^3$$

$$g_0 = \frac{GM_e}{R_e^{\ 2}} = \frac{4}{3}\pi RG\rho$$

$M = const^n$: $g_0 \propto \frac{1}{R^2}$

$\rho = const^n$: $g_0 \propto R$

Above the Surface of the earth:

$$g_h = \frac{g_0 R^2}{(R+h)^2}, \qquad g_h = g_0\left[1 - \frac{2h}{R}\right]$$

Below the Surface of the earth:

$$g_d = g\left[1 - \frac{d}{R}\right]$$

- % change in g at height h = $\frac{-2h}{R} \times 100$
- % change in g at depth d = $\frac{-d}{R} \times 100$

(For small change)

Variation of "g" due to shape of Earth:-

Shape of the Earth:

- The Earth is not a perfect sphere; it is slightly flattened at the poles and bulging at the equator. This shape is called an oblate spheroid.
- The radius at the equator (R_e) is greater than the radius at the poles (R_p) by about 21 km. Therefore, g is minimum at equator and maximum at pole

Effect on Gravity (g):

- Gravity at the poles (g_p) is greater than gravity at the equator (g_e).
- This is because gravity is inversely proportional to the square of the radius. Since the radius at the equator is larger, the gravity there is slightly weaker compared to the poles.

i.e. $R_e = R_p + 21$ km, $g_p > g_e$

Note:-

Mass = properties of matter, remains same everywhere.

W(weight) increases as we move from equator to pole.

Variation of "g" due to rotation of earth about its own axis:-

$g_{eff} = g_o - R\omega^2\cos^2\theta$ Equator pr accl^n due to gravity 0.34% kam hoti hai pole se!

- At pole, $\theta = 90°$, $g_{eff} = g_o$ (maximum)
- At equator, $\theta = 0°$, $g_{eff} = g_o - r\omega^2$ (minimum)

Where,

g_o = standard gravitational acceleration ($9.8\ m/s_2$).

R is the radius of the Earth.

ω, is the angular velocity of Earth's rotation.

θ is the latitude (measured from the equator).

Special Points:-

Rotational speed and weightlessness

T= 84.6 min This inticates that If earth rotates 17 times its present rotational speed so body at equator feels weightless.

Increase in Angular Velocity (ω)

- With increased angular velocity, the centrifugal force acting on objects on Earth's surface increases.
- This would lead to a decrease in the effective gravity (g) experienced by objects at all places except the poles.
- At the poles, the centrifugal force is zero because the poles are on the axis of rotation, so gravity remains unaffected there.
- If $\omega\uparrow$, then g = decrease at all place except pole

Direction of Earth's Rotation and Rocket Launch

- The Earth rotates from west to east, which means if a rocket is launched in the same direction (W → E), it benefits from the Earth's rotational speed.
- Launching a rocket in the direction of the Earth's rotation (west to east) provides it with additional velocity due to the Earth's rotational speed.
- This can be beneficial for launching rockets into orbit as it reduces the amount of fuel needed to reach the required orbital speed.

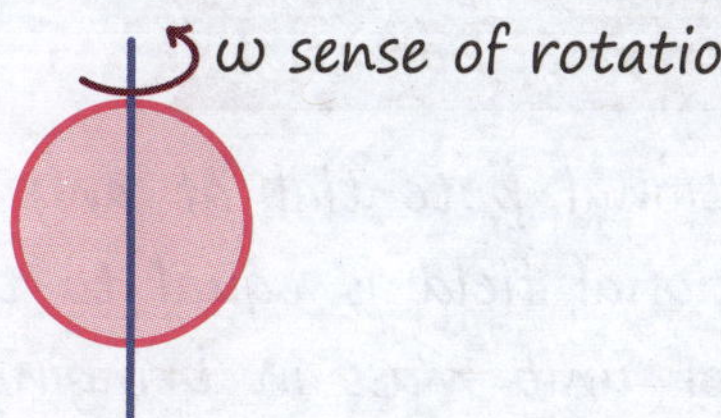

Q.4. The depth 'd' at which the value of acceleration due to gravity becomes $\frac{1}{n}$ times the value at the earth's surface is (R radius of earth)

Sol. Acceleration due to gravity at depth d under the surface $g' = g_s\left(1 - \frac{h}{R}\right)$

Given : $g' = \frac{g_s}{n}$

$$\therefore \frac{g_s}{n} = g_s\left(1 - \frac{d}{R}\right)$$

$$\text{or } \frac{1}{n} = \left(1 - \frac{d}{R}\right)$$

$$\Rightarrow d = R\left(\frac{n-1}{n}\right)$$

Q.5. The height from earth's surface at which acceleration due to gravity becomes is $\frac{g}{4}$ (where g is acceleration due to gravity on the surface of earth and R is radius of earth).

Sol. $g_h = g_0 \frac{R^2}{(R+h)^2}$

$$\frac{g_0}{4} = g_0 \frac{R^2}{(R+h)^2}$$

$$\frac{R}{R+h} = \sqrt{\frac{1}{4}} = \frac{1}{2}$$

$$2R = R + h$$

$$h = R$$

6 Gravitational Potential

- Gravitational potential at any point in gravitational field is equal to the work done per unit mass in bringing a very light body from infinity to that point.

Gravitational potential, $V_g = \frac{W}{m}$

- SI unit is J/kg and it is a scalar quantity.
- Dimensional formula is $[L^2T^{-2}]$.

7 GravitationaL PE

Gravitational potential energy is the energy possessed by an object due to its position in a gravitational field. It is the work done to move an object from one point to another in the field.

$$W_{CF} = -\Delta U \qquad U_B = -[W_{A\to B}]_{GF}$$

- Here, W_{CF} represents the work done by the conservative force.
- ΔU is the change in gravitational potential energy.

8 Gravitational potential due to point mass at a distn "γ"

$$U = -\frac{GMm_o}{\gamma} \;\&\; V = \frac{U}{m_o} = -\frac{GM}{\gamma}$$

- Since, work W is negative, the gravitational potential is always negative.

Note:-

Gravitational potential energy per unit mass

$$V = \frac{U}{M} \quad \vec{I} = \frac{\vec{F}}{M} \quad \vec{I} = -\frac{dv}{dr} \quad \vec{F}_{cF} = -\frac{dU}{dr}$$

$$\vec{I} = -\left[\frac{\delta v}{\delta x}\hat{i} + \frac{\delta v}{\delta y}\hat{j} + \frac{\delta v}{\delta z}\hat{k}\right]$$

$$\Delta V = -\int \vec{I}.\vec{dr} \qquad \Delta U = -\int \vec{F}.\vec{dr}$$

9 Gravitational potential due to combination of point mass

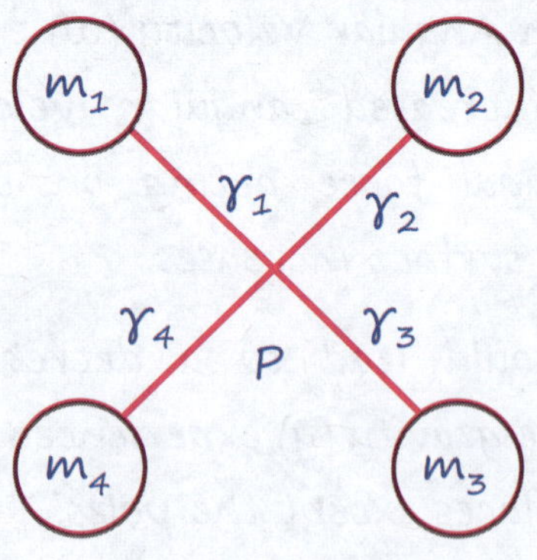

$$V_p = -\frac{Gm_1}{\gamma_1} - \frac{Gm_2}{\gamma_2} - \frac{Gm_3}{\gamma_3} - \frac{Gm_4}{\gamma_4}$$

$U_p = -\frac{GMm}{\gamma}$ Total no. of pairs $= \frac{N(N-1)}{2}$

N is the total no. of masses.

10 Gravitational potential due to

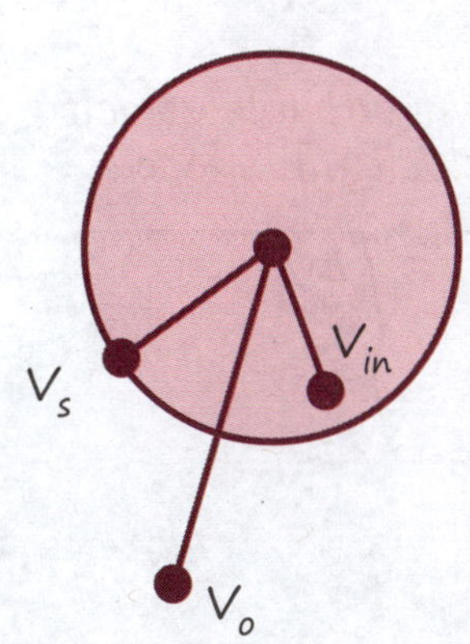

a> Uniform thin shell	b> Uniform Solid Sphere
$V_{out} = -\frac{GM}{\gamma}$	$V_{out} = -\frac{GM}{r}$
$V_{surface} = -\frac{GM}{R}$	$V_{surface} = -\frac{GM}{R}$
$V_{in} = -\frac{GM}{R}$	$V_{in} = -\frac{GM}{2R^3}[3R^2 - \gamma^2]$
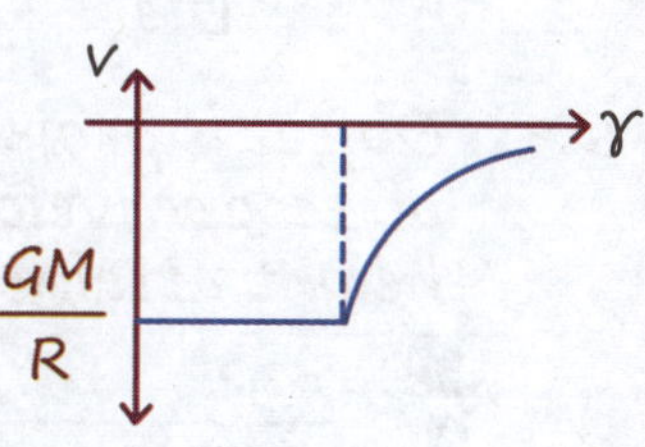	$V_{centre} = -\frac{3}{2}\frac{GM}{R}$

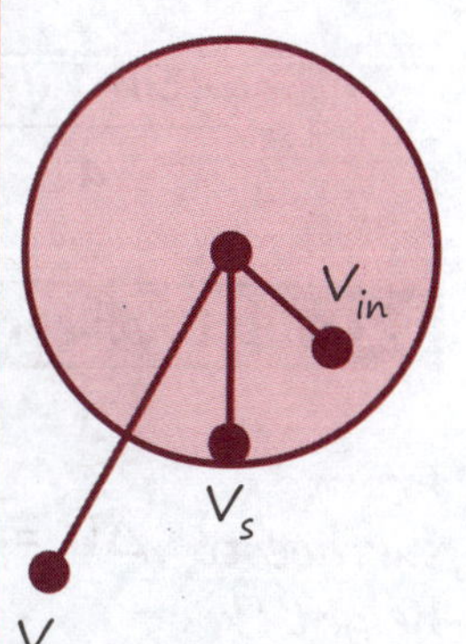

11 Gravitational potential due to uniform ring

At a axial point:-

$$V_p = -\frac{GM}{\sqrt{x^2+R^2}}$$

At center (x = 0)

$$V = -\frac{GM}{R}$$

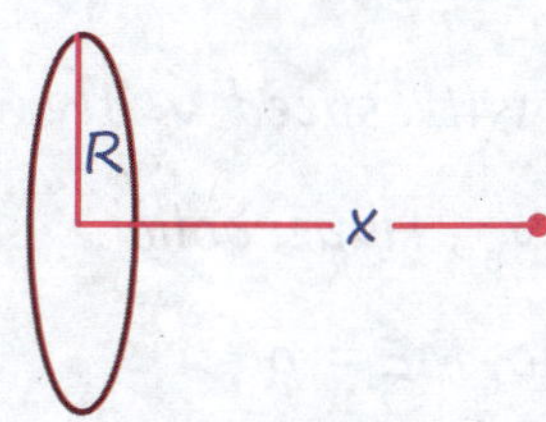

At centre of half ring $V_o = -\frac{GM}{R}$

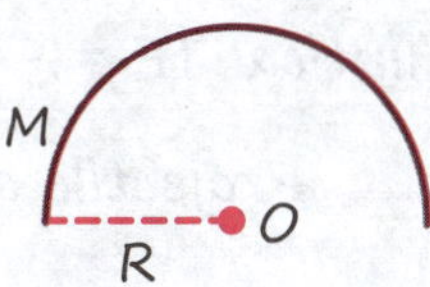

Ex.: find potential energy of the given system

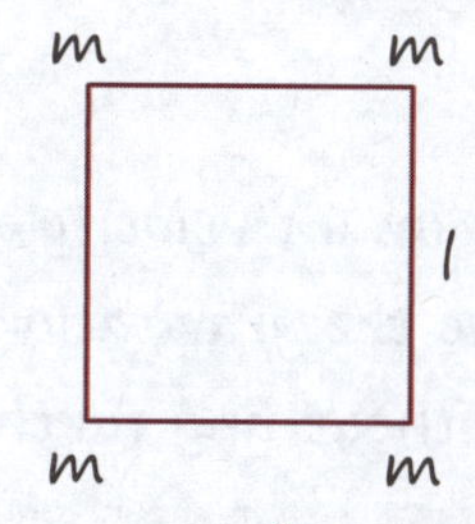

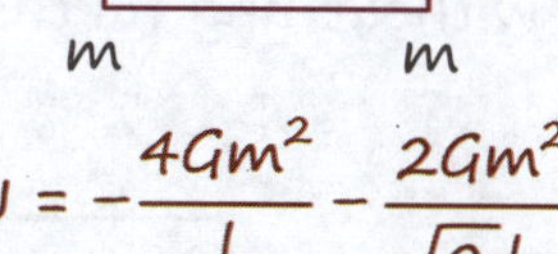

$$U = -\frac{4Gm^2}{l} - \frac{2Gm^2}{\sqrt{2}\,l}$$

Ex:

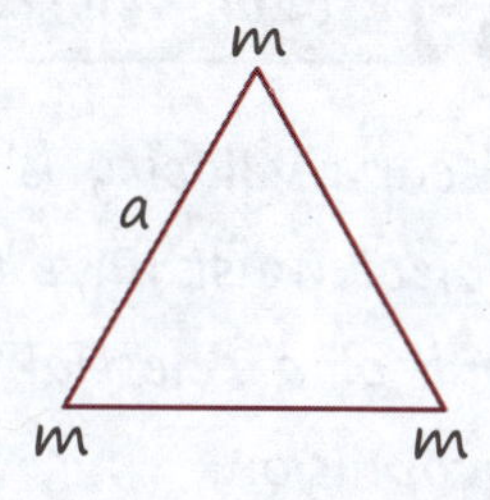

$$U = -\frac{3Gm^2}{a}$$

Q.6. Four point mass m placed at cornor of square of side a, find work done in bringing 5th mass m from infinity to centre of square.

Sol.

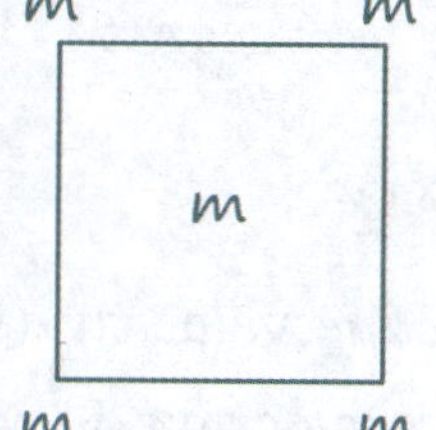

$U_i = U_o$ (Let)

$W = \Delta U = U_f - U_i \qquad U_f = U_o - \dfrac{4Gm^2}{a/\sqrt{2}}$

$= -\dfrac{4Gm^2\sqrt{2}}{a}$

12 G. p.E. of earth-mass system

From surface to height "h":- $\Delta U = \dfrac{mgh}{(1+\frac{h}{R})} = \dfrac{GMmh}{R^2\left[1+\frac{h}{R}\right]}$

Q.7. Object of mass m raised to height h = 2R from earth surface change in its potential energy.

Sol. $\Delta U = \dfrac{mgh}{1+\frac{h}{R}} = \dfrac{mg2R}{1+\frac{2R}{R}} = \dfrac{2}{3}mgR$

13 Escape Velocity

Escape velocity is the minimum velocity an object must have to escape the gravitational pull of a celestial body without any further propulsion.

$V_e = \sqrt{\dfrac{2GM}{R}}\ V_e = \sqrt{2gR}, V_e = \sqrt{\dfrac{8}{3}G\rho\pi R^2}$

→ Escape velocity interns of density.

$\rho = const^n \qquad V_e \propto R$

$M = const^n \qquad V_e \propto 1/\sqrt{R}$

Note:-

Escape velocity does not depend upon angle of projection & mass of object.

14 velocity of object at ∞ when projected with vg > Vea

$V_\infty = \sqrt{V_g^2 - V_e^2}$

Q.8. Object is projected with double velocity of escape then its velocity in space

Sol. $V_{space} = \sqrt{V_g^2 - V_e^2} = \sqrt{4V_e^2 - V_e^2}$

$= \sqrt{3}V_e$

If Vg < Ve THEN hmax = ?

$h_{max} = \dfrac{R}{\left[\dfrac{V_e^2}{V_g^2} - 1\right]}$

Q.9. Object is projected with one-fourth of escape velocity then maximum height attained?

Sol. $h = \dfrac{R}{\left(\dfrac{V_e}{V_g}\right)^2 - 1} = \dfrac{R}{16-1} = \dfrac{R}{15}$

15 Escape velocity from height "h" from surface

$V = \sqrt{\dfrac{2GM}{R+h}}$

For earth to become a black hole

R = 1.48 mm

If object is projected with speed V_g then,

$V_g > V_e$ ⟶ ME = + V_e, Hyperbolic

$V_g = V_e$ ⟶ parabolic, ME = o

$V_o < V_g < V_e$ ⟶ Elliptical close, ME = $-V_e$

$V = V_o$ ⟶ Circular TE = – ve

$V < V_o$ ⟶ Elliptical TE = – ve

$V <<<<< V_o$ ⟶ Projectile open path

16 Orbital velocity

$$V_o = \sqrt{\frac{GM}{R+h}} \qquad V_e = \sqrt{2}\,V_o$$

Satellite near surface of earth:-

$$V_o = \sqrt{gR} = \sqrt{\frac{4}{3}\pi G\rho R^2}$$

Note:-

$V_{satellite}$ inc. by $\sqrt{2}$ times or 41.4 %

then it'll escape to ∞

Q.10. The radii of the circular orbits of two satellites A and B of the earth are 4R and R, respectively. If the speed of satellite A is 3v, then the speed of satellite B will be

Sol. $V \propto \frac{1}{\sqrt{R}}$

$$\frac{3V}{V'} = \sqrt{\frac{R}{4R}}$$

$$V' = 3V \times 2 = 6V$$

17 Energies of satellite

Potential energy, $U = -\frac{GMm}{R+h}$

Kinetic energy $K = \frac{GMm}{2(R+h)}$

Total energy $TE = -\frac{GMm}{2(R+h)}$

Binding energy $BE = \frac{GMm}{2r}$

$U : K : T : BE : = -1 : \frac{+1}{2} : \frac{-1}{2} : \frac{+1}{2}$

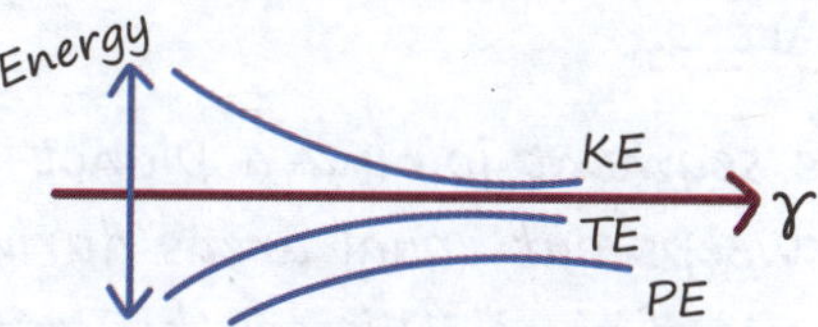

✦ B.E. of a particle placed on earth:-

$$BE = \frac{GMm}{R}$$

18 Time period of satellite

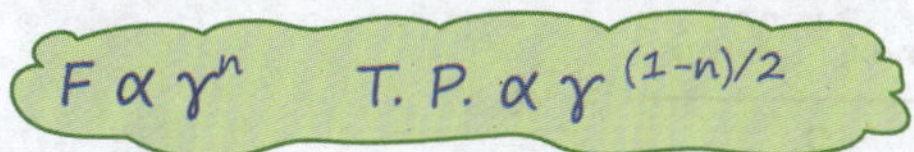

$$T = \frac{2\pi\gamma}{V_o} = 2\pi\sqrt{\frac{(R_e+h)^3}{GM_e}}$$ independent of mass of satellite

$T^2 \propto \gamma^3$

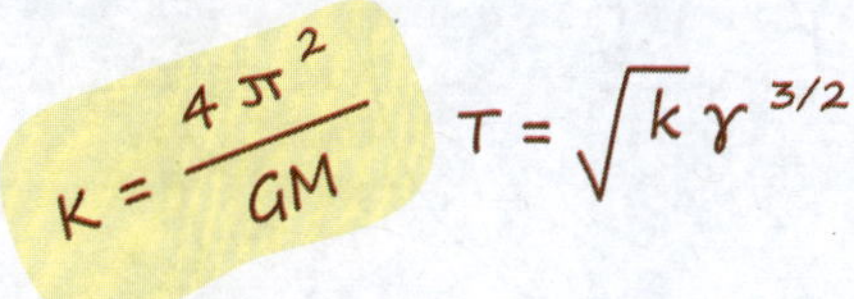

$$T = \sqrt{k}\,\gamma^{3/2}$$

19 Kepler's law

Law of Orbits:

✦ Planets move in elliptical orbits with the Sun at one focus.

✦ $\frac{V_{max}}{V_{min}} = \frac{r_{max}}{r_{min}} = \frac{a(1+e)}{a(1-e)}$

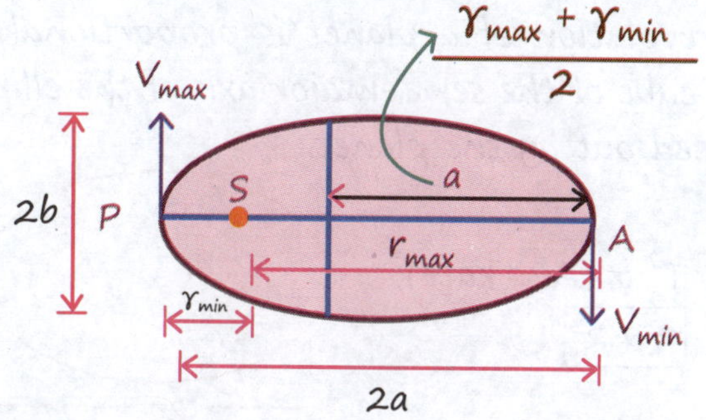

✦ $\gamma_{max} + \gamma_{min} = 2a$

Law of Areas:

- A line segment joining a planet and the Sun sweeps out equal areas during equal intervals of time. This can be written as:

$$\frac{dA}{dt} = const^n$$

$$\gamma V = const^n$$

- $\frac{L}{2m} = const^n$

- $\gamma_{mean} = \frac{\gamma_{max} + \gamma_{min}}{2} = a$

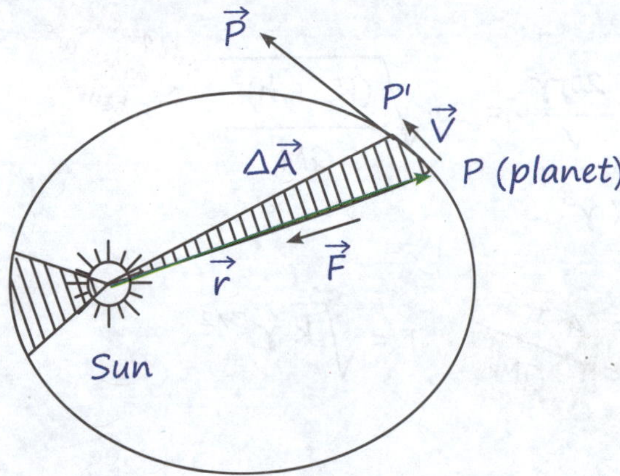

Law of Periods:

- The square of the orbital period (T) of a planet is directly proportional to the cube of the semi-major axis (a) of its orbit.

- $T^2 \alpha\ a^3 \alpha \left[\frac{\gamma_{max} + \gamma_{min}}{2}\right]^3$

Semi major axis

Law of Periods: The square of the time period of revolution of a planet is proportional to the cube of the semi-major axis of the ellipse traced out by the planet.

$T^2 \alpha$ (Avg rate)^3

$\boxed{T^2 \alpha\ R^3}$

Sun

2a

$a_{as} = a$

20 Satellite on circular path

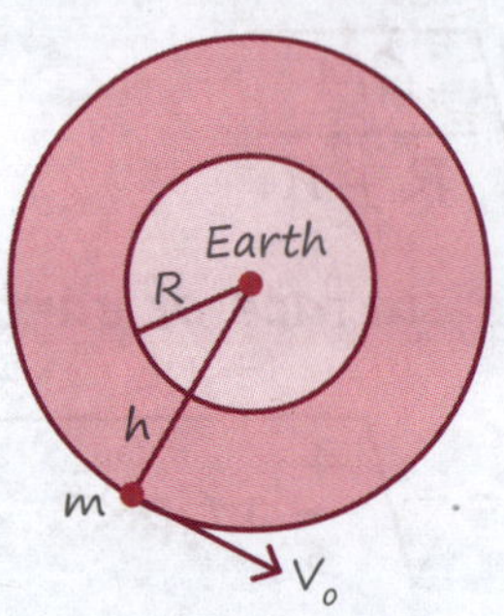

$|\vec{\gamma}| = const^n$

All energies = $const^n$

$L = const^n \qquad \vec{\tau} = o$

$\vec{P}$ = variable

$\vec{v}$ = variable (due to dir^n)

$\omega = o \qquad V = const^n$

$\tau = \frac{dL}{dt} \quad F = \frac{dP}{dt}$

Note:-

- Total energy = $-\frac{GMm}{(\gamma_{max} + \gamma_{min})}$

escape energy = +

$$\frac{GMm}{R_e}$$

29> $V_g = KV_e$

h_{max} from Surface of Earth = $\frac{RK^2}{1-K^2}$

If $F \alpha \frac{1}{\gamma^n}$

$\therefore V_o \alpha \frac{1}{\gamma^{n-1/2}}$

$\therefore T.P. \alpha\ \gamma^{n+1/2}$

Q.11. A satellite of mass m moving around planet of mass M in circular path radius r_1 then energy required to shift this path to radius r_2.

Sol. $E_{given} = E_{total\ final} - E_{total\ initial}$

$$= -\frac{GMm}{2r_2} - \left(-\frac{GMm}{2r_1}\right)$$

$$= \frac{GMm}{2}\left(\frac{1}{r_1} - \frac{1}{r_2}\right)$$

Q.12. The time period of a satellite in a circular orbit of radius R is T. The period of another satellite in a circular orbit of radius 4R is

(i) 4T (ii) $\frac{T}{2}$

(iii) 8T (iv) $\frac{T}{8}$

Sol. (iii) $T^2 \propto R^3$

$$= \left(\frac{T}{T^1}\right)^2 = \left(\frac{R}{4R}\right)^3$$

$$= \left(\frac{T}{T^1}\right)^2 = \frac{1}{64}$$

$$= \frac{T}{T^1} = \frac{T}{\sqrt{64}}$$

$$= \boxed{8T = T^1}$$

Q.13. The kinetic energies of a planet in an elliptical orbit about the Sun, at positions A, B and C are KA, KB and KC respectively. AC is the major axis and SB is perpendicular to AC at the position of the sun S as shown in the figure. Then

(i) $K_A < K_B < K_C$ (ii) $K_A < K_B < K_C$

(iii) $K_B < K_A < K_C$ (iv) $K_B < K_A < K_C$

Sol. (ii)

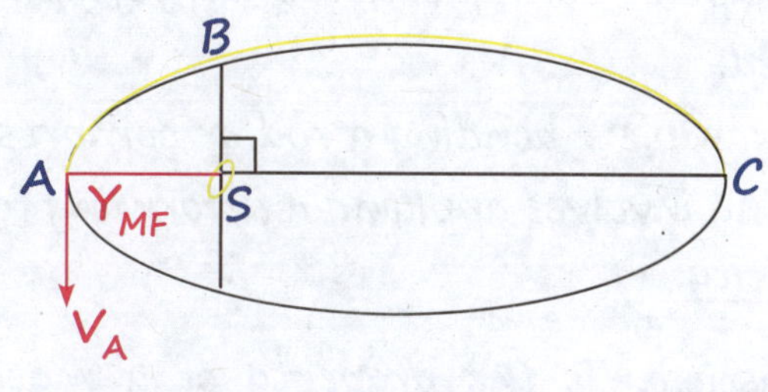

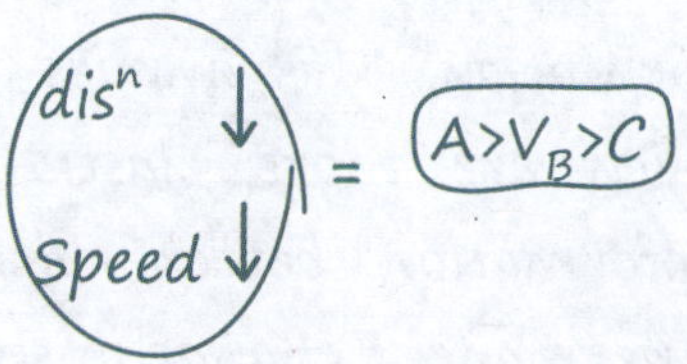

Q.14. The figure shows elliptical orbit of a planet m about the sun S. The shaded area SCD is twice the shaded area SAB. If t_1 is the time for the planet to move from C to D and t_2 is the time to move from A to B then

(i) $t_1 = 4t_2$ (ii) $t_1 = 2t_2$

(iii) $t_1 = t_2$ (iv) $t_1 > 4t_2$

sol. (ii)

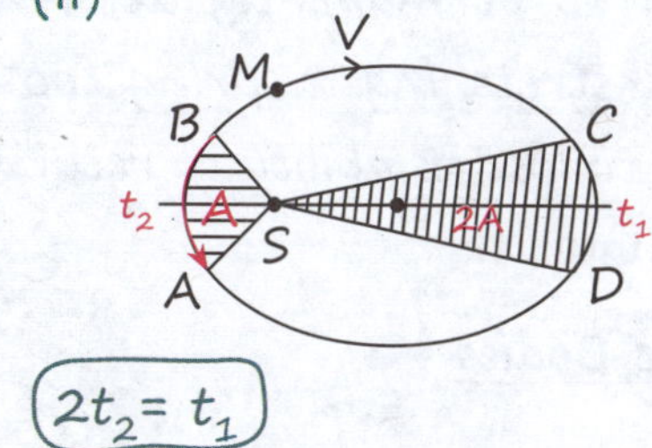

$\boxed{2t_2 = t_1}$

MR*

"Question krte time Galat option ko dhoondo, sahi apne aap mil jayenge."

14 Elasticity

1 Deforming Force

A deforming force is any force that changes the shape or configuration of an object when applied.

For example:- bending a rod or compressing a spring involves applying a deforming force.

Elasticity:

- Elasticity is the property of a material to resist deformation. This implies that materials which are more elastic can withstand deformation better without undergoing a permanent change in shape.

Example:- **Steel is more elastic than rubber**. This means steel returns to its original shape more efficiently after the deforming force is removed compared to rubber.

Elastic Limit

- The elastic limit is the maximum extent to which a material can be deformed and still return to its original shape once the force is removed. Beyond this limit, the material will be permanently deformed. For example, stretching a spring too far may cause it to lose its ability to return to its original shape.

Perfectly Elastic Bodies

- These are ideal materials that can completely return to their original shape immediately after the deforming force is removed. Examples include quartz and phosphor bronze.
- Ex. Quartz is near approach to perfect elastic body.

$\gamma = \infty \qquad e = 0.$

Perfectly Plastic Bodies:

- These materials do not regain their original shape at all after the deforming force is removed. They remain deformed. Examples include putty, paraffin, and wax.

2 Stress

$$\frac{F_{restoring}}{\text{Area}} = \text{stress}$$

Stress is not a vector quantity unlike a force, the stress cannot be assigned a specific direction. Force acting on the portion of a body on a specified side of a section has definite direction.

Types of Stress

Normal Stress

- Deals with forces applied perpendicular to the surface, causing changes in length.

Volumetric Stress

- Deals with forces applied uniformly over the entire surface, causing changes in volume.

Tangential Stress

- Deals with forces applied parallel to the surface, causing changes in shape.

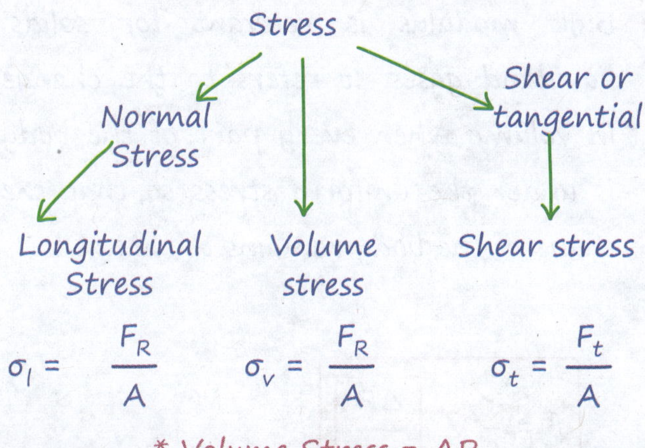

$$\sigma_l = \frac{F_R}{A} \qquad \sigma_v = \frac{F_R}{A} \qquad \sigma_t = \frac{F_t}{A}$$

* Volume Stress = ΔP

3 Strain

- It's the fractional change in configuration due to an applied force.

$$\text{Strain} = \frac{\text{Change in size the body}}{\text{Original size of the body}}$$

type of strain

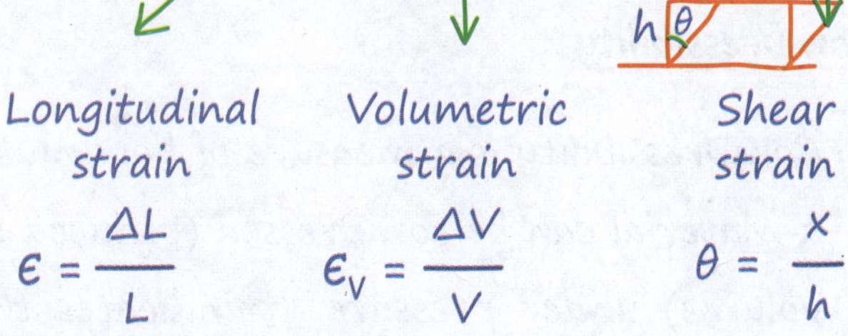

$$\epsilon = \frac{\Delta L}{L} \qquad \epsilon_V = \frac{\Delta V}{V} \qquad \theta = \frac{x}{h}$$

Strain is unitless and dimensionless

- **Longitudinal Strain** occurs when there is a change in length.
- **Volumetric Strain** occurs when there is a change in volume.
- **Shearing Strain** occurs when there is a change in shape due to layers sliding over each other.

4 Hooke's Laws

Within the elastic limit, the stress applied to a material is directly proportional to the strain it produces.

i.e.,

Stress $\propto$ Strain

$\sigma \propto \epsilon$

$\sigma = E\epsilon$

$$E = \frac{\sigma}{\epsilon}$$

Where, E = modulus of elasticity, depends on the nature of the material of the body and is independent of its dimensions (i.e. length, volume etc.)

Material Prop.

$$\text{Modulus of Elasticity (E)} = \frac{\text{Stress}}{\text{Strain}}$$

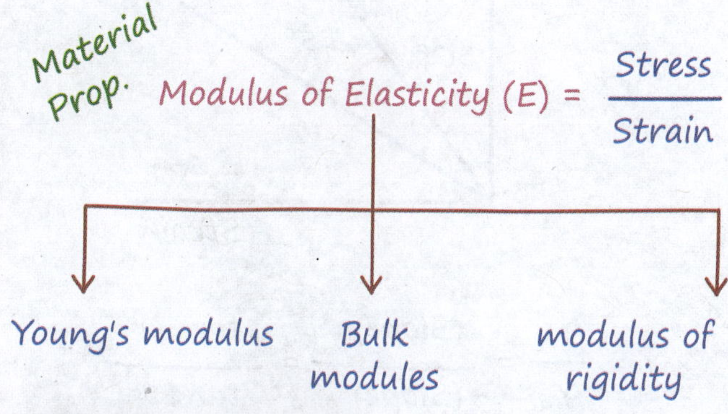

5 Young's Modulus

Young's modulus is the ratio of normal stress to the longitudinal strain.

$$Y = \frac{\text{Longitudinal Stress}}{\text{Longitudinal Strain}}$$

$$Y = \frac{FL}{A\Delta L}$$

- Hooke's law is valid only in the linear part of stress-strain curve.
- The Young's modulus and shear modulus are relevant only for solids since only solids have lengths and shapes.
- Metals have larger values of Young's modulus than alloys and elastomers. A material with large value of Young's modulus requires a large force to produce small changes in its length.

- Slope of stress strain graph is Young's modulus.

Q.1. Find ratio of young's modulus

Sol. Slope of stress-strain graph is young's modules.

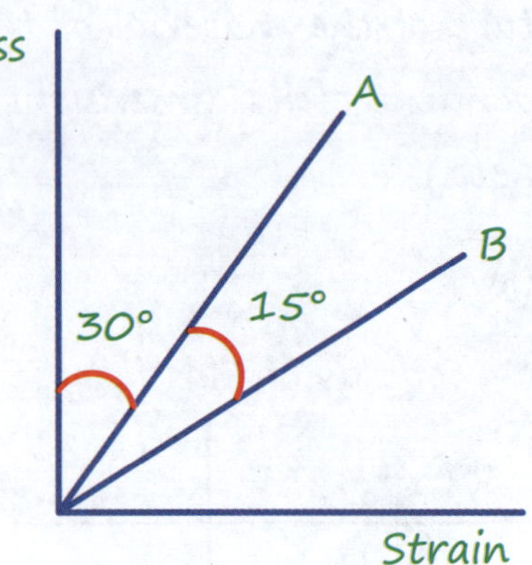

$$Y = \frac{Y_A}{Y_B} = \frac{(\text{Slope})_A}{(\text{Slope})_B} = \frac{\tan 60°}{\tan 45°}$$

$$= \sqrt{3} : 1$$

- Young Modulus $\propto \frac{1}{\text{Temprature}}$

e.g.

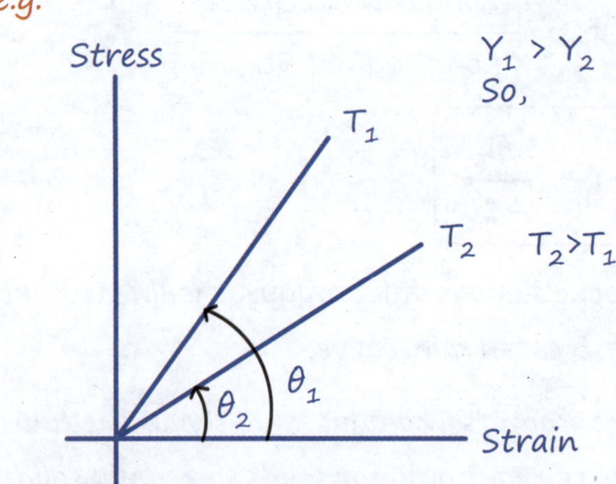

6 Bulk's Modulus

- The bulk modulus is the ratio of volumetric stress to the volumetric strain.

$$\beta = \frac{\text{volume Stress}}{\text{volume strain}}$$

- Bulk modulus is relevant for solids, liquidand gases. It refers to the change in volume when every part of the body is under the uniform stress so that the shape of the body remains unchanged.

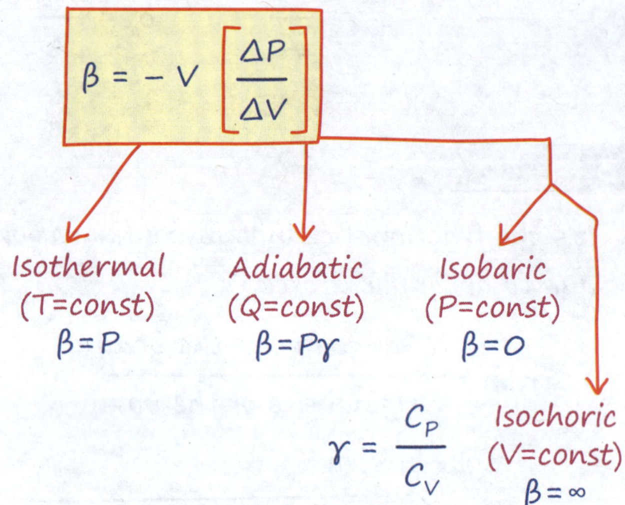

$$\gamma = \frac{C_P}{C_V}$$

Isochoric (V=const) $\beta = \infty$

- $B_{soild} < B_{liquid} > B_{gas}$

Compressibility

- Compressibility is a measure of how much a material can be compressed (reduced in volume) under pressure. It indicates the relative volume change of a material in response to a pressure change.
  - Compressibility (C) of a material is the reciprocal of its bulk modulus of elasticity (K)
  - Mathematically, it is expressed as

$$C = \frac{1}{\beta} = \frac{1}{-V}\frac{\Delta V}{\Delta P}$$

  - Density of Compressed Liq. ↑

$$\rho' = \rho\left[1 + \frac{\Delta P}{K}\right] = \rho\,[1 + C\Delta P]$$

7 Shear Modulus

✦ Shear modulus or modulus of rigidity is the ratio of tangential stress to the shearing strain.

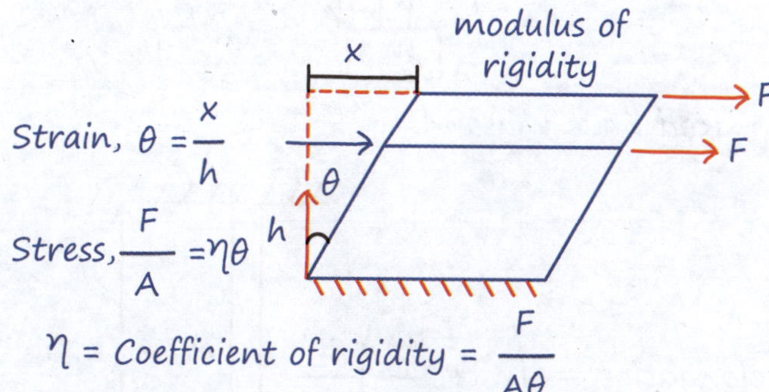

Strain, $\theta = \frac{x}{h}$

Stress, $\frac{F}{A} = \eta\theta$

η = Coefficient of rigidity = $\frac{F}{A\theta}$

Potential Energy Stored in Wire:-

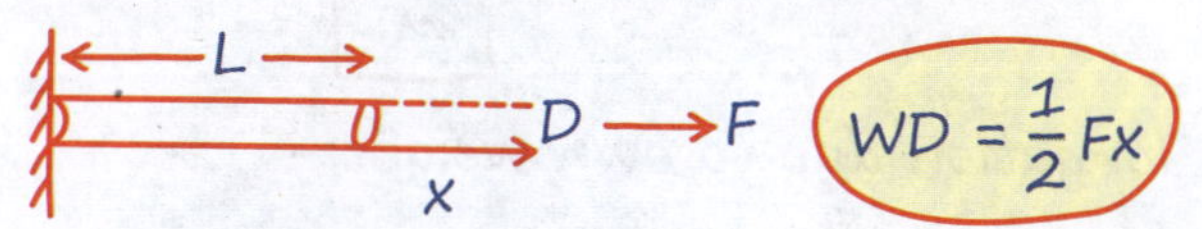

$WD = \frac{1}{2}Fx$

x = elongation
work done in elongation x
w = Fx

○ Energy stored = $\frac{1}{2}Fx = U$

$U = \frac{1}{2}\frac{A\gamma x^2}{L}$

using Hooks chacha ka law

Potential Energy Stored in wire unit volume:-

$$\frac{U}{V} = \frac{1}{2}\gamma(\text{Strain})^2 = \frac{1}{2}\frac{(\text{Stress})^2}{\gamma}$$

$$= \frac{1}{2}\text{Stress} \times \text{Strain}.$$

Breaking Stress (P) :- Breaking stress does not depends upon area.
Material Property.

$$P = \frac{\text{Breaking Force}}{\text{Area}}$$

✦ For a Particular Material:- B.Force ∝ A

Stress-strain curve

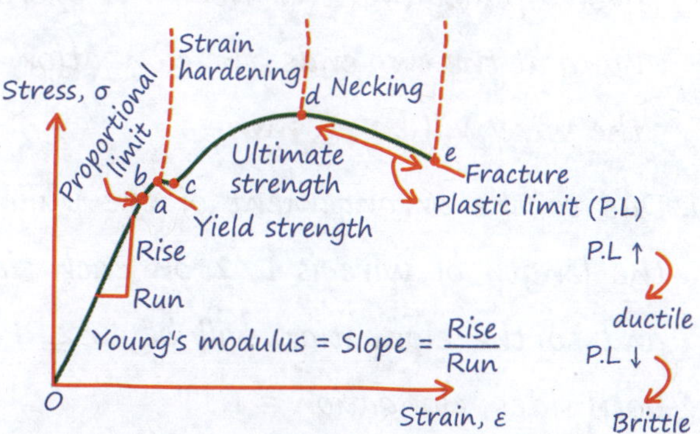

Elastic region, O–a:-

✦ Hooke's law Valid. (Stress ∝ Strain)

$F_{elastic}$ Conservative.

Completely regain its shape.

O–b:-

✦ Conservative limit.

Stress # Strain

Almost Regain its shape.

$F_{elastic}$ Conservative.

Plastic region b–c:-

✦ $F_{elastic}$ is not Conservative

Non-linear region where permanent deformation occurs.

Fracture Point (D)

✦ The point where the material breaks.

Q.2. If the longitudinal strain of a iron rod is 0.01 and its poisson's ratio of 0.2, then the lateral strain will be

Sol. Since,

$$\text{Poisson's ratio} = \frac{\text{Lateral Strain}}{\text{Longitudinal Strain}}$$

⇒ Lateral strain = (Poisson's ratio) × (Longitudinal strain)

⇒ Lateral strain = (0.2) (0.01) = 0.002

Q.3. A wire elongates by L mm when a load W is hanged from it. If the wire goes over a pulley and two weighs W each is hung at the two ends, the elongation of the wire will be (in mm).

Sol. Due to the arrangement of the pulley, the length of wire is L/2 on each side and so the elongation will be L/2. For both sides, elongation = L

Poisson's Ratio

$$\sigma = \frac{\text{Lateral Strain}}{\text{Longitudinal Strain}}$$

$$\sigma = -\frac{\Delta D/D}{\Delta L/L}$$

Volume strain = 2 × lateral strain + longitudinal strain

$\sigma = 0.5$ V = constant.

Q.4. If the elastic potential energy density store in a material is $3 \times 10^4 J/m^3$ due to the application of longitudinal stress of $1 \times 10^{11} N/m^2$ then, the strain developed in it would be

Sol. $U_e = \times \frac{1}{2}$ stress × strain

$$\text{strain} = \frac{2Ue}{\text{stress}}$$

$$\text{strain} = \frac{2 \times 3 \times 10^4}{10^{11}} = 6 \times 10^{-7}$$

Special Case:-

✦ Elongation in massless rod due to attatched block of mass m:

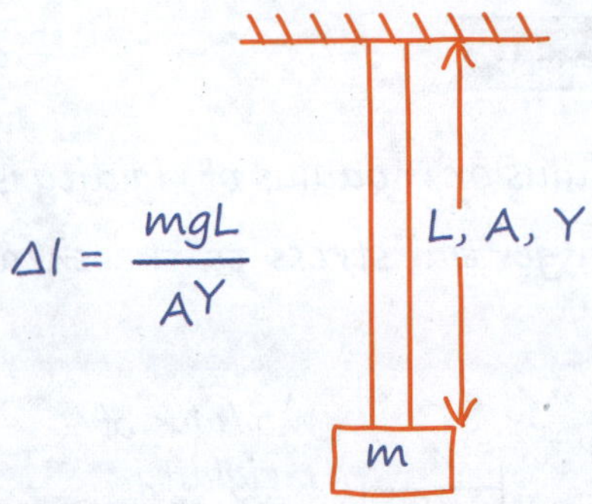

$$\Delta l = \frac{mgL}{AY}$$

✦ Rod have mass M:

$$\text{elongation} = \left(m + \frac{M}{2}\right)\frac{gL}{AY}$$

M, A, L, Y

m

✦ Elongation due to own weight:-

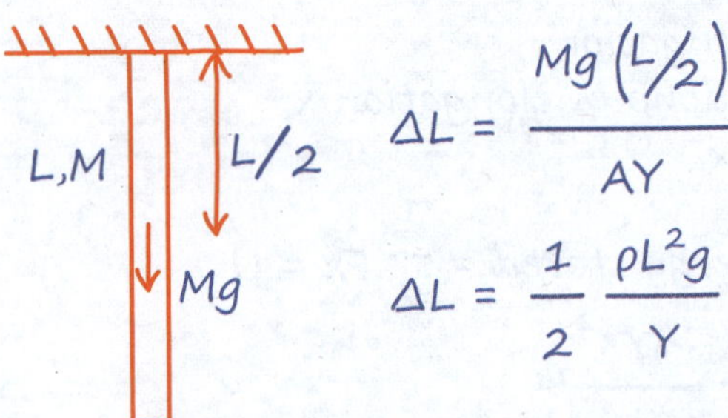

$$\Delta L = \frac{Mg\left(L/2\right)}{AY}$$

$$\Delta L = \frac{1}{2}\frac{\rho L^2 g}{Y}$$

✦ l_{max} of a wire which can hang under his own weight:-

$$l_{max} = \frac{P}{\rho g}$$

✦ Linear Strain = $\alpha \Delta T$

Linear Stress = $Y.\alpha\Delta T$

Q.5. Two wires are made of the same material and have the same volume. However, wire 1 has a cross-sectional area A and wire 2 has cross- sectional area 3A. If the length of the wire 1 increases by Δx on applying force F, how much force is needed to stretch wire 2 by the same amount?

Sol. For the same material, Young's modulus is the same and it is given that the volume is the same and the area of the cross-section for the wire L_1 is and that of L_2 is 3A

$V = V_1 = V_2$

$V = A \times L_1 = 3A \times L_2 \rightarrow L_2 = L_1/3$

$Y = (F/A)/(\Delta L/L)$

$F_1 = YA(\Delta L_1/L_1)$

$F_2 = Y3A(\Delta L_2/L_2)$

Given $\Delta L_1 = \Delta L_2 = \Delta x$ (for the same extension)

$F_2 = Y3A(\Delta x/(L_1/3)) = 9.(YA\Delta X/L_1) = 9F_1 = 9F$

✦ Fractional Change in Radius of Sphere:-

β = Bulk Modulus. $\quad \dfrac{dR}{R} = \dfrac{F}{3A\beta} = \dfrac{Mg}{3A\beta}$

✦

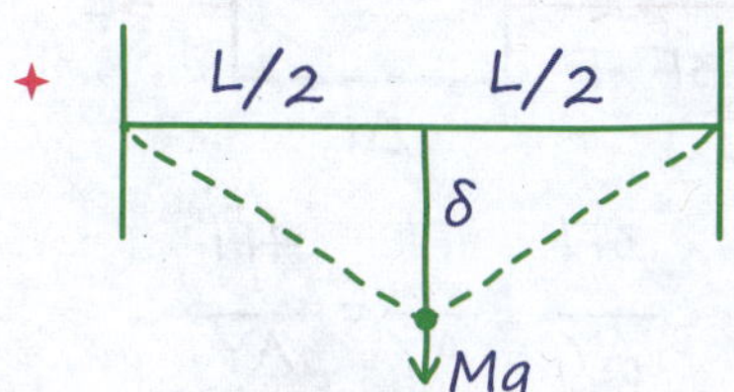

$$\delta = \frac{L}{2}\left[\frac{Mg}{AY}\right]^{1/3}$$

✦ Relation between shear angle and angle of twist.

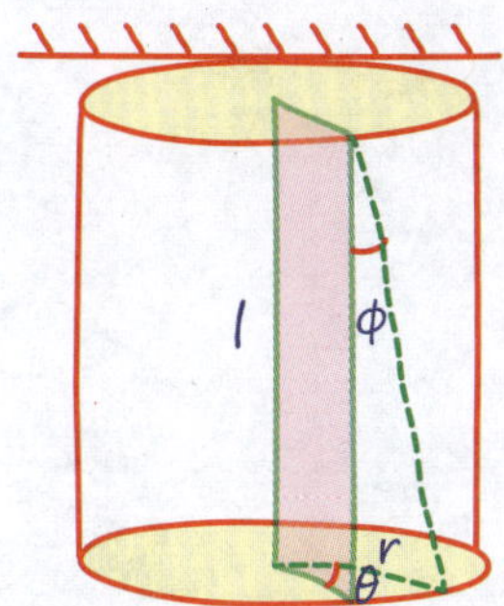

θ = Twist Angle

ϕ = Shear Angle

$$PE = \frac{1}{2}C\theta^2 = \frac{1}{2}Ke^2$$

$$r\theta = l\phi$$

✦ Relation among Y, K, η & σ :-

$$Y = \frac{9\eta K}{3K + \eta} \qquad Y = 2\eta\,[1 + \sigma]$$
$$Y = 3K\,[1 - 2\sigma]$$

$$Y = \frac{3K - 2\eta}{6K + 2\eta}$$

✦ If length of wire l_1 at tension T_1 and l_2 at tension T_2, then find original length $T_1 \rightarrow l_1$ $\Rightarrow T_2 \rightarrow l_2$

$$L = \frac{T_1 l_2 - T_2 l_1}{T_1 - T_2} = \frac{T_2 l_1 - T_1 l_2}{T_2 - T_1}$$

Q.6. When a block of mass M is suspended by a long wire of length L, the length of the wire becomes (L+l). The elastic potential energy stored in the extended wire is:

Sol. Using Formula

$$U = \frac{1}{2} \times \text{stress} \times \text{strain} \times \text{volume}$$

$$\text{stress} = \frac{\text{Force}}{\text{Area}} = \frac{Mg}{A}$$

$$\text{strain} = \Delta L/L = l/L$$

$$\text{volume} = \text{area} \times \text{length} = A \times L$$

$$\text{Hence, } U = \frac{1}{2} \times \frac{Mg}{A} \times \frac{l}{L} \times A \times L$$

$$U = \frac{1}{2} Mgl$$

Stress-Strain graph for Elastomer:- e.g., Tissue of Aorta,

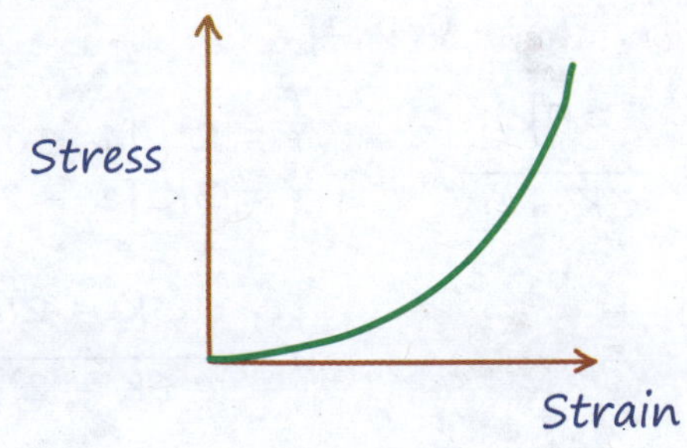

✦ Parallel & Series Combination of Young's Modulus:-

1> Parallel Combination:-

In both wire have same elongation but different stress

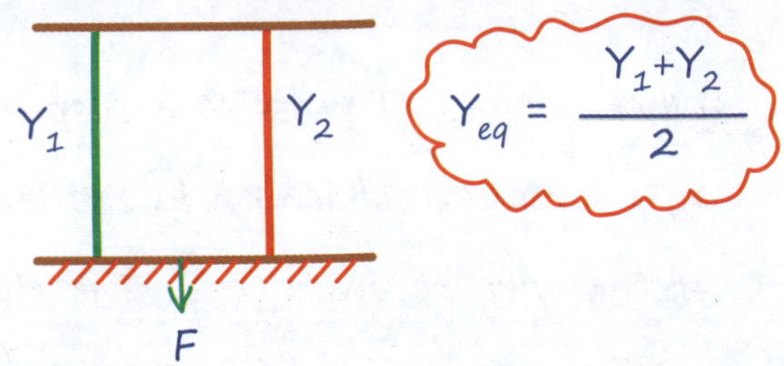

$$Y_{eq} = \frac{Y_1+Y_2}{2}$$

2> Series Combination:-

In both wire have same stress but different elongation.

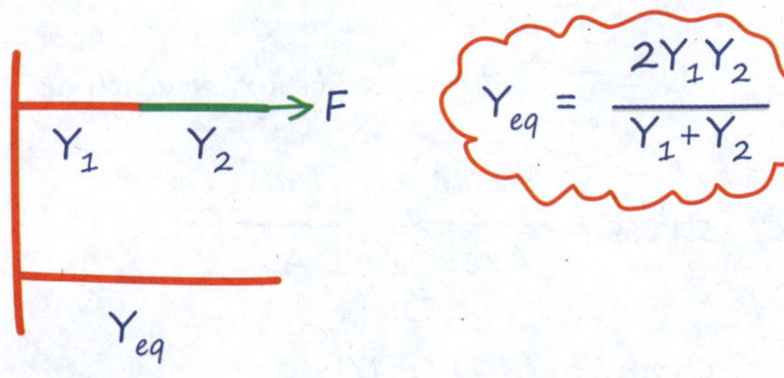

$$Y_{eq} = \frac{2Y_1Y_2}{Y_1+Y_2}$$

NOTE:-

✦ On removing deforming Force Solid regain its shape by this graph.

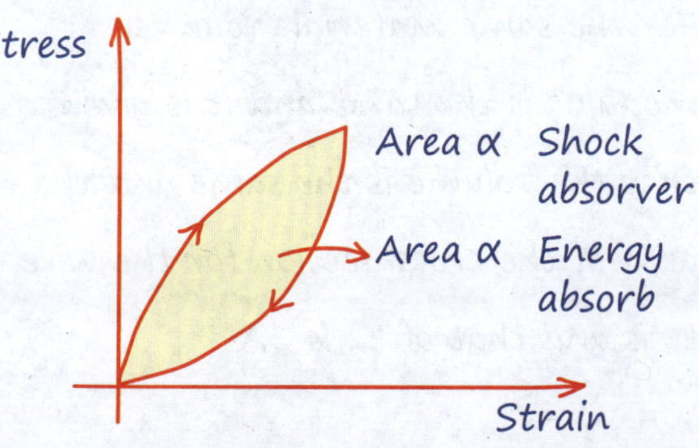

Special Type Questions:-

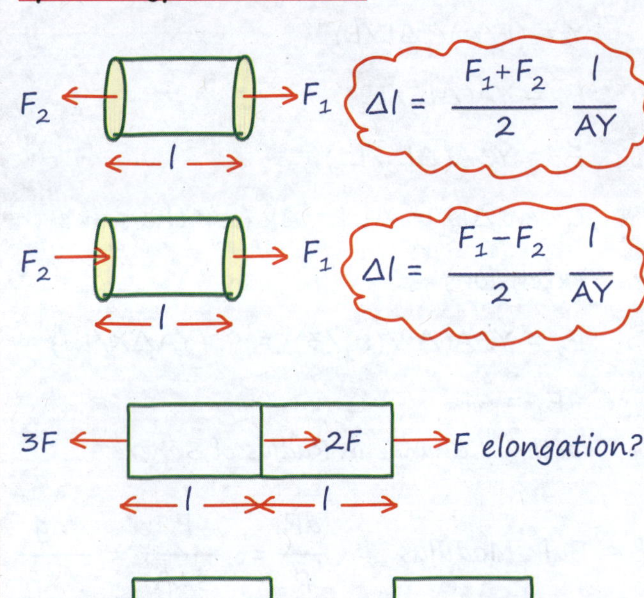

$$\Delta l = \frac{F_1+F_2}{2}\ \frac{l}{AY}$$

$$\Delta l = \frac{F_1-F_2}{2}\ \frac{l}{AY}$$

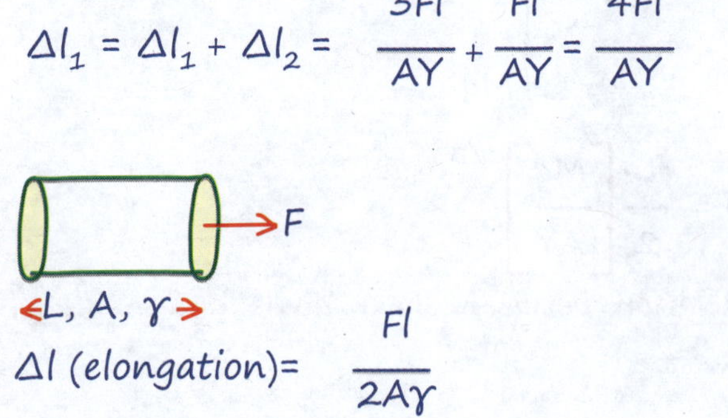

$$\Delta l_1 = \Delta l_1 + \Delta l_2 = \frac{3Fl}{AY} + \frac{Fl}{AY} = \frac{4Fl}{AY}$$

F

L, A, γ

$$\Delta l \text{ (elongation)} = \frac{Fl}{2A\gamma}$$

MR*
"Push Yourself, because No One Else is going to do it for you."

15 Fluid Mechanics

1 Fluid

Substances that can flow when an external force is applied. This includes both liquids and gases.

Types of Fluids

Liquids:

- Have a definite volume but take the shape of their container.
- Examples: Water, oil, and milk.

Gases:

- Do not have a definite volume or shape and expand to fill their container.
- Examples: Air, helium, and carbon dioxide.

✦ Relative density (R.D):-

- Relative density is the ratio of the density of a substance (object) to the density of water.

$$R.D. = \frac{\rho_{obj.}}{\rho_{water}} = \text{Unitless.}$$

$\rho_w = 1\ gm/cm^3$ (Water)

$\rho_{oil} = 0.8\ gm/cm^3$ (Oil)

$\rho_{Hg} = 13.6\ gm/cm^3$ (Mercury)

$\rho_{milk} = 1.04\ gm/cm^3$

Example 1: Oil

$$R.D. = \frac{\rho_{oil.}}{\rho_w} = \frac{0.8\ g/cm^3}{1\ g/cm^3} = 0.8$$

The relative density of oil is 0.8, meaning oil is less dense than water.

Example 2: Mercury

$$R.D. = \frac{\rho_{Hg}}{\rho_w} = \frac{13.6\ g/cm^3}{1\ g/cm^3} = 13.6$$

The relative density of mercury is 13.6, meaning mercury is much denser than water.

Note:-

✦ Objects with a relative density less than 1 will float in water.

✦ Objects with a relative density greater than 1 will sink in water.

2 Density

Density of an object is the ratio of its mass to its volume. i.e.,

$$\text{Density} = \frac{\text{Mass}}{\text{Volume}}$$

✦ it is a scalar quantity with dimensional formula

$[ML^{-3}]$

✦ Density of water = $1\ g/cm^3$ or $10^3\ kg/m^3$

Density of mixture:-

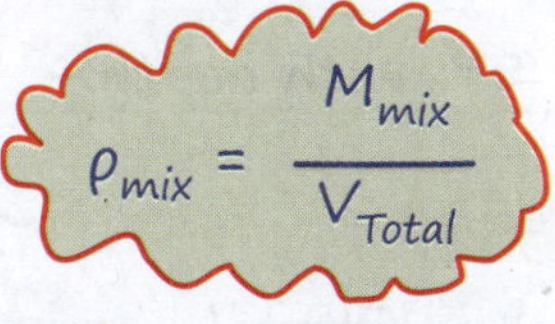

$$\rho_{mix} = \frac{M_{mix}}{V_{Total}}$$

Same Mass, Different Density

m = same ρ = different

$$\rho_{mix} = \frac{2m}{\frac{m}{\rho_1} + \frac{m}{\rho_2}} = \frac{2\rho_1\rho_2}{\rho_1 + \rho_2}$$

Same volume, different density

V = same ρ = different

$$\rho_{mix} = \frac{\rho_1 V + \rho_2 V}{2V} = \frac{\rho_1 + \rho_2}{2}$$

3 Hydrostatics or static fluid

- This branch deals with fluids at rest.
- It studies how fluids exert pressure and the effects of this pressure.
- Example: Water in a tank or a swimming pool.

Thrust:-

- Thrust is the total normal force exerted by a liquid at rest on a given surface.
- SI Unit: Newton (N).

Pressure (P):-

$$P = \frac{F}{A}$$

- It is a scalar quantity
- Unit: N/m^2
- Dimension: $[ML^{-1}T^{-2}]$
- 1 atm = 1.01×10^5 N/m^2

Variation of "P" with depth:-

$\Delta P = \rho g h$

ρ = density of liquid.

h = depth.

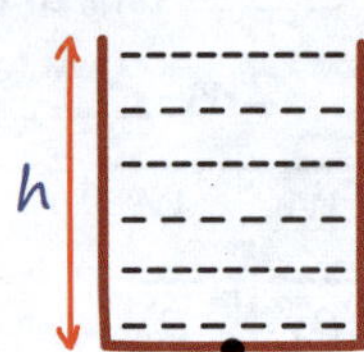

Note:-

- ΔP does not depend upon amount of liquid, shape of container.
- The mean pressure is given by:

$$\frac{h\rho g}{2}$$

This means that on average, the pressure exerted by the liquid on the walls of the container is half of the pressure at the bottom of the container.

4 Pascal's Law

Static liquid, "P" on horizontal level must be same.

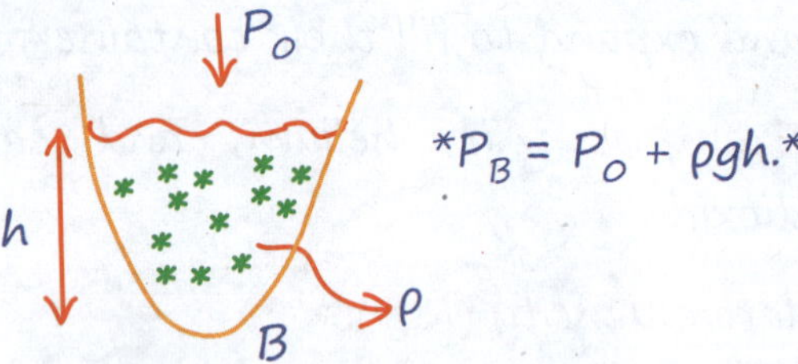

Relation between Absolute & Gauge Pressure:-

$P_{absoluble} = P_{net}$

$P_{gauge} = P_{due\ to\ liq.\ only.}$

$$P_{absolute} = P_o + P_{gauge}$$

Moving Container "P" Calculation:-

(Be "g_{eff}" lelena hai!)

1. Lift is moving up:- (with "a")

   $P_{net} = \rho(g + a)H$

2. Lift is moving down:- (with "a")

   $P_{net} = \rho(g - a)H$

3. Lift is in free fall:- (a = g)

   $P_{net} = 0$

1>

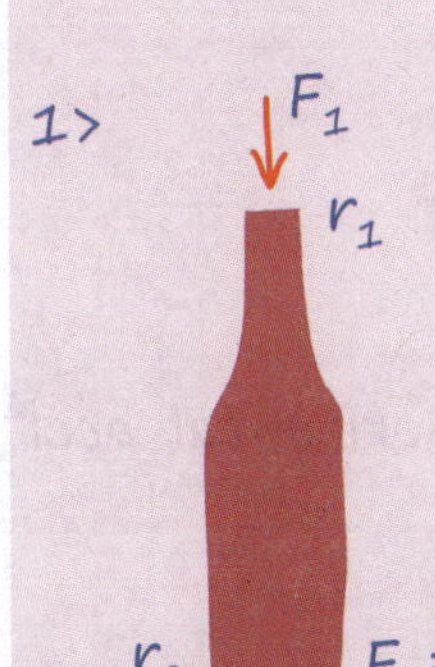

Pascal चाचा :-

Static fluid me pressure balance kerte hai

$$\frac{F_1}{\pi r_1^2} = \frac{F_2}{\pi r_2^2}$$

✦ pressure is the same at any two points at the same level in the fluid. The shape of the container does not matter

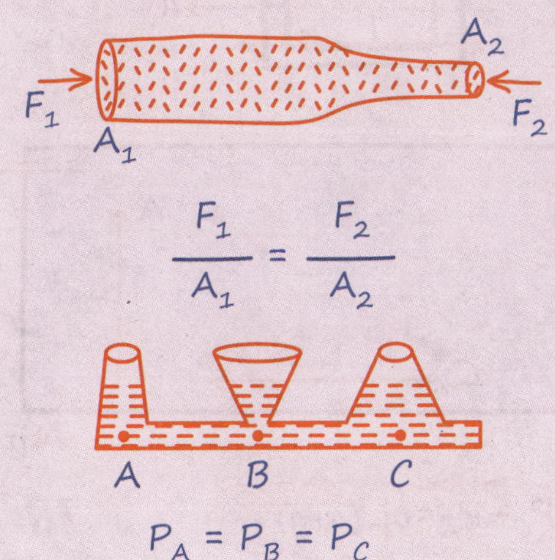

$$P_A = P_B = P_C$$

Q.1. In a hydraulic jack as shown, mass of the car W = 800 kg, A_1 = 10 cm^2, A_2, = 10m^2. The minimum force F required to lift the car is

(a) 1N (b) 0.8 N

(c) 8 N (d) 16 N

Sol.

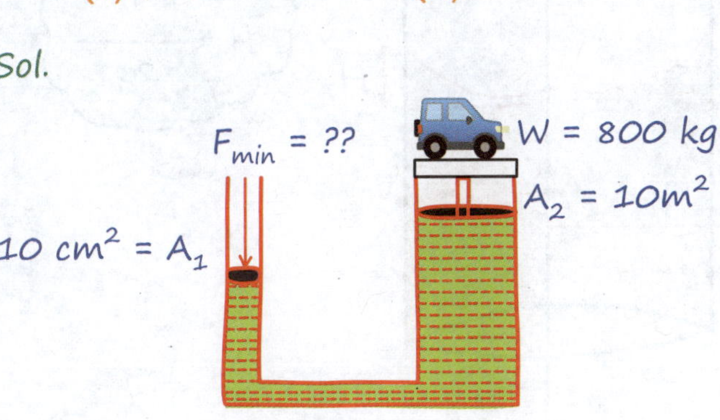

$$\frac{F_{min}}{10\ cm^2} = \frac{800 \times 10}{10 \times 10^4\ cm^2}$$

$$f_{min} = \underline{0.8\ N}$$

MR*

✦ When air bubble move from bottom to surface of lake then $p_1V_1 = p_2V_2$.

2> MR**

Agar Koi Cheez linearly vary Krti hai toh uska Avg:-

$$X_{avg} = \frac{X_i + X_f}{2}$$

5 Measurement of atmospheric Pressure

1. Barometer:- "Torricilli"

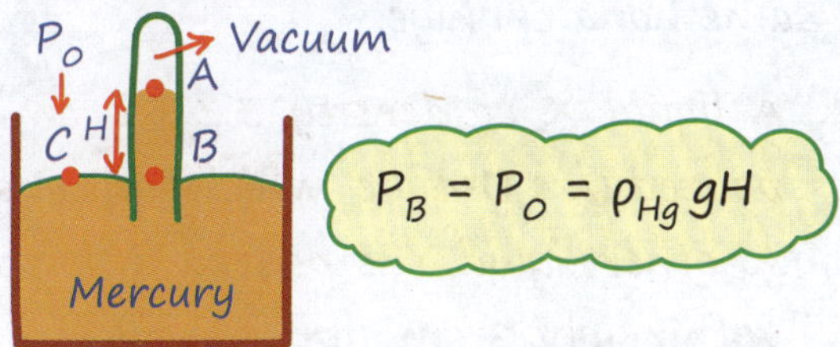

$$P_B = P_o = \rho_{Hg} gH$$

H = 76 cm of Hg = 1 atm

The Garib Ramlal Exp:-

Height of water column in Barometer:-

$$H = 10.1\ m$$

6 Horizontal Accelerating Lift

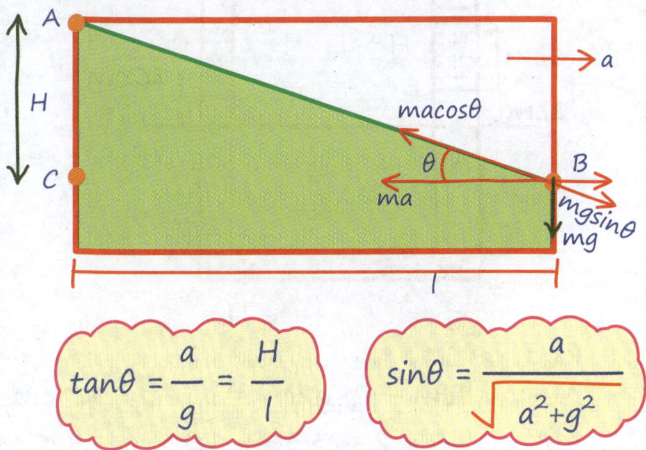

$$\tan\theta = \frac{a}{g} = \frac{H}{l}$$

$$\sin\theta = \frac{a}{\sqrt{a^2+g^2}}$$

7 U-Tube

Initialy one liquid is on same horizontal level, other liquid is put in left arm of container then 1st liquid moves x down from initial level in left arm of U-tube.

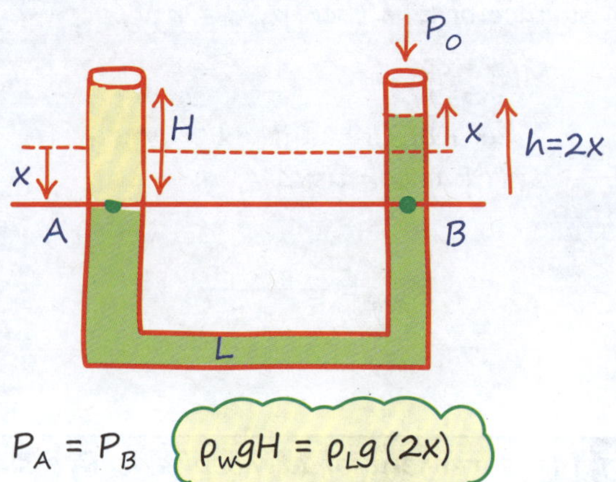

$P_A = P_B$ $\quad \rho_w g H = \rho_L g (2x)$

ek aaise line select Karo jiske about "P" same hona chahiye.

Q.2. A liquid X of density 3.36 g cm^{-3} is poured in a U-tube, which contains Hg. Another liquid Y is poured in left arm with height 8 cm, upper levels of X and Y are same. What is density of Y? (given - height of x = 10cm)

(a) 0.8 gcc^{-1} (b) 1.2 gcc^{-1}

(c) 1.4 gcc^{-1} (d) 1.6 gcc^{-1}

Sol.

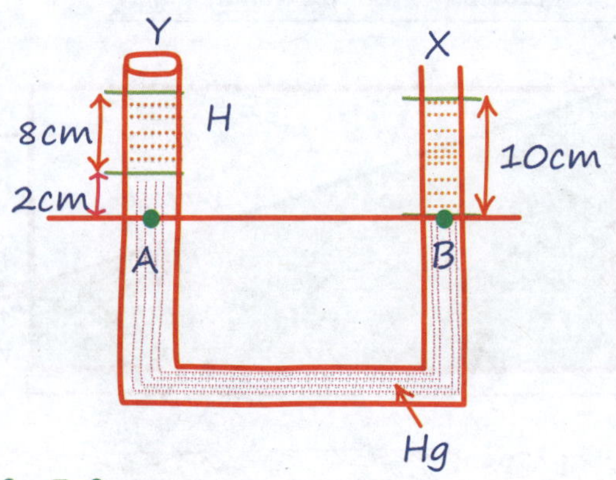

$\rho_A = \rho_B$

$\rho_0 + \rho_y g 8cm + \rho_{mg} g 2cm = \rho_0 + \rho_x g\ 10\ cm$

$\rho_y 8 + 13.6 \times 2 = 3.36 \times 10$

$8\rho_y + 27.2 = 33.6$

$8\rho_y = 6.4$

$\rho_y = \frac{6.4}{8}$

Important Case

1> When U-tube is given horizontal accl^n

(a) rise is liquid column:-

$$\frac{a}{g} = \frac{H}{l} = \tan\theta$$

$$H = al/g$$

2>

$$P_C - P_A = \rho L (a+g)$$

$$P_B - P_A = \rho g L \qquad P_C - P_B = \rho a L$$

3> A vessel is rotated about vertical axis. Find rise in water "H":-

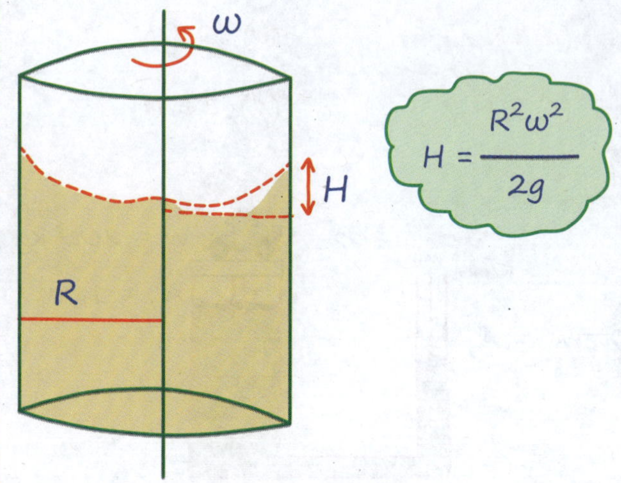

8 Open-Tube Manometer

Gas Container:

- Contains the gas whose pressure we want to measure.

U-Shaped Tube:

- One end is connected to the gas container, and the other end is open to the atmosphere.

Liquid Column:

- The liquid inside the U-shaped tube moves in response to the pressure differences.

I >

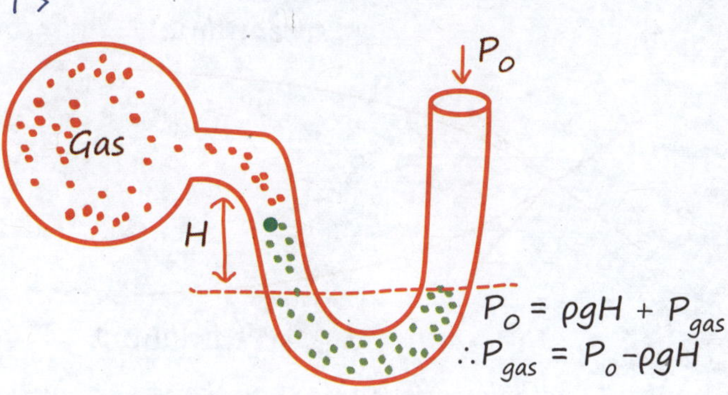

$P_0 = \rho g H + P_{gas}$

$\therefore P_{gas} = P_0 - \rho g H$

II>

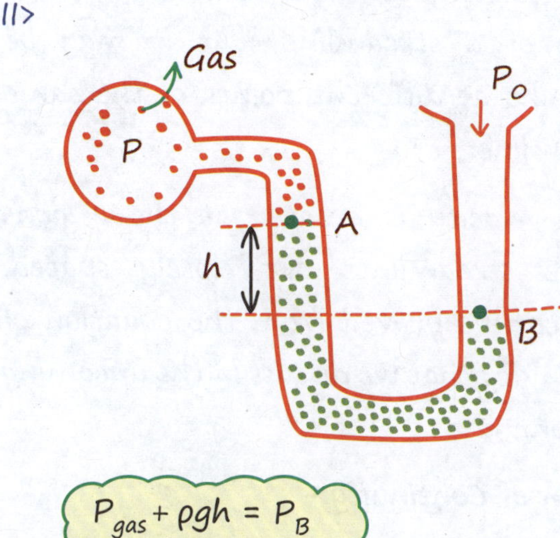

$$P_{gas} + \rho g h = P_B$$

9 Archimedes Principle

F_B (Buoyant Force) = Weight of displaced liquid.

→ depends on density of liquid, volume of solid submerged in liquid.

* does not depends on density, location of object inside liquid.

Volume of displaced liq. = Volume of Solid in liq.

Volume of solid inside liquid

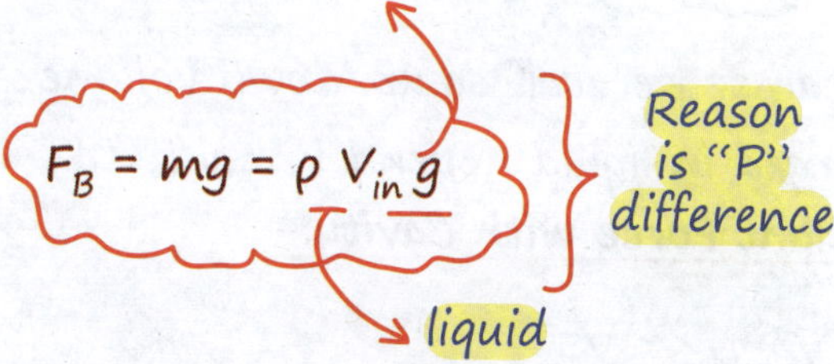

* Apparent Weight:-

σ = object ρ = Liquid

$\sigma > \rho$	$\sigma = \rho$	$\sigma < \rho$
Sink!	Submerge & Float.	Float on Surface
$N = mg_{eff}\left[1 - \frac{\rho}{\sigma}\right]$	N = 0	N = 0
$N = mg_{eff}\left[1 - \frac{\text{Chota}}{\text{bada}}\right]$	Object Remains where it's Placed!	$\frac{V_{in}}{V_T} = \frac{\sigma}{\rho}$
$N = mg_{eff}\left[1 - \frac{1}{RD}\right]$		$\frac{V_{in}}{V_T} = \frac{\text{Chota}}{\text{bada}}$

*Object of density "σ" is released then find accl^n of object inside liquid:-
($\sigma > \rho$)

$$a = g\left[1 - \frac{\rho}{\sigma}\right]$$

*A ball of density "D" is immersed in liquid of density "d" to a depth "h" below the surface of liquid & then released. Upto what height will be ball jump out of the liquid:-

Condition = $d > D$

$$H = h\left[\frac{d}{D} - 1\right]$$

*A container is at rest then inside volume is V_{in} and when accl^n up with a_0 the inside volume is V_{in}' of an object then:-

$$V_{in} = V_{in}'$$

Container Ke accl^n se up/down volume emerged in liquid depend hi nai Krta.

Buoyant Force with Cavity:-

$$V_{matter} = V_T - V_C$$

$$*N = \sigma[V_T - V_C]g - \rho V_T g$$

To Find V_{cavity} :-

1> $N = mg - F_B$

$F_B = \rho_L V_T g$

$V_{matter} = \frac{m}{\sigma}$

$V_C = V_T - V_M$

} Problem Solving Strategy

*** Rise / Fall of liq:-**

When ICE placed on liquid will Melt.

$\rho_L > \rho_0$ = Rise

$\rho_L < \rho_0$ = Fall

$\rho_L = \rho_W$ = Same.

ρ_0 = density of water.

ρ_L = Surr. Liquid.

10 Hydrodynamics or fluid dynamics

+ This branch deals with fluids in motion.

Example: Water flowing in a river or air flowing around an airplane wing.

Ideal Fluid:-

+ Non-viscous
+ Incompressible
+ $\rho = const^n$
+ Streamline & IRROTATIONAL Flow.

11 Properties of Streamlines

+ The tangent at any point of a streamline gives the direction of flow of fluid particle at that point.
+ In a steady flow, no two streamlines can cross each other, for if they do so, two tangents can be drawn at the point of intersection. It means the oncoming fluid particle can go either one way or the other. Thus, the flow would not be steady

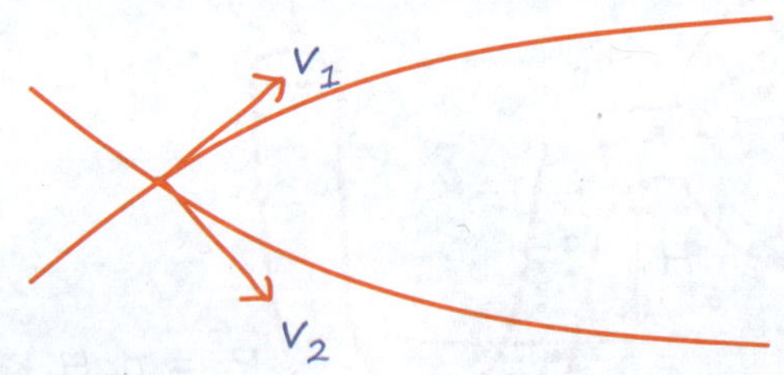

+ Fluid velocity remains constant at any point of a streamline, but it may be different at different points of the same streamline.
+ Fluid velocity is greater at the regions where streamlines are closely spaced. This can be proved from the equation of continuity that we discuss in the upcoming section.

Equation of Continuity:-

(Conserv^n of Mass)

*Area × Velocity = Const^n

Rate of Volume Flow remains Const^n.

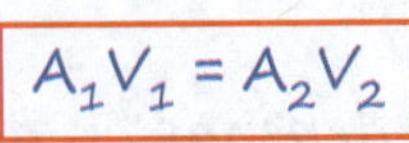

$$A_1V_1 = A_2V_2$$

water flowing from the tap.

Q.3. An incompressible liquid is flowing through a horizontal pipe as show in figure. Find the value of v.

Sol. $(AV)_{incoming} = (Av)_{outgoing}$

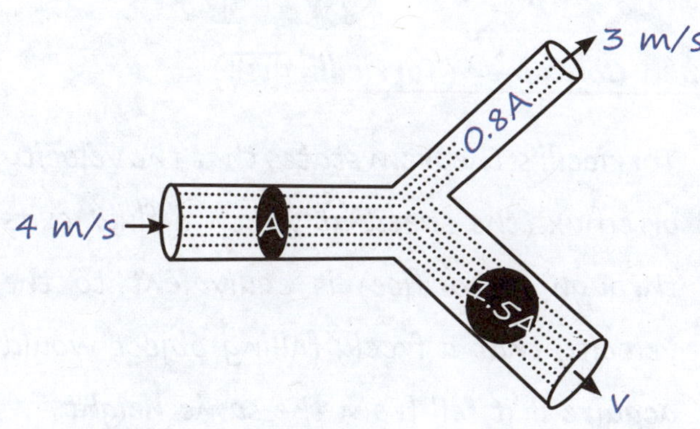

$$4A = 3(0.8A) + \frac{3}{2}A\,v$$

$$4 = 2.4 + 1.5v$$

$$4 - 2.4 = 1.5v$$

$$1.6 = 1.5v$$

$$V = \frac{1.6}{1.5} = 1.1 \text{ m/s}$$

Freely falling liquid:-

$$V^2 = 2gh\left[\frac{A_o^2}{A_o^2 - A^2}\right]$$

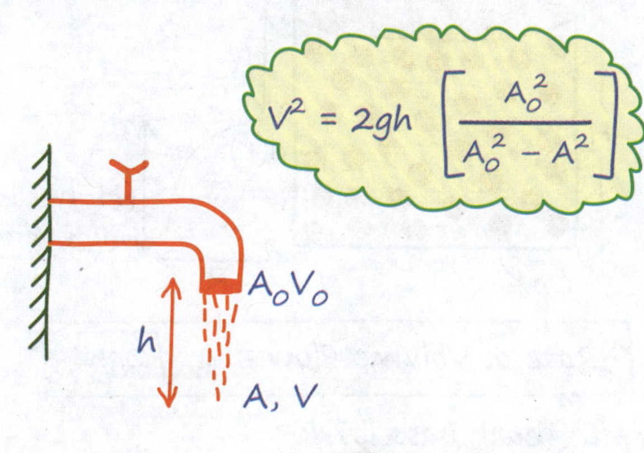

$$\text{Rate of Volume Flow} = AA_o\sqrt{\frac{2gh}{A_o^2 - A^2}}$$

12 Bernouli's equation

- Bernoulli's Theorem states that for an ideal fluid (non-viscous, incompressible, and irrotational) flowing in a streamline manner, the total energy (pressure energy, kinetic energy, and potential energy) per unit volume of the fluid remains constant along a streamline.

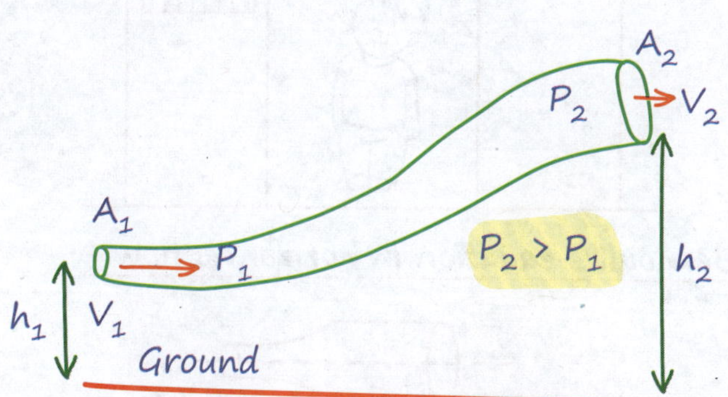

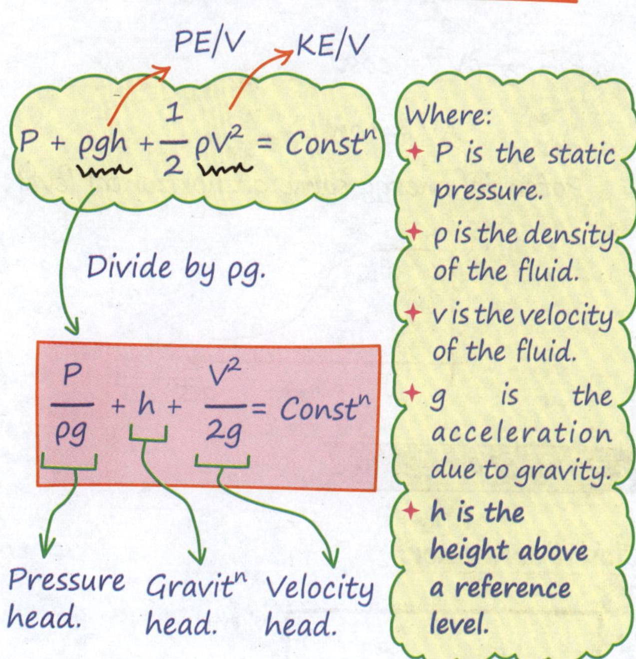

Where:
- P is the static pressure.
- ρ is the density of the fluid.
- v is the velocity of the fluid.
- g is the acceleration due to gravity.
- h is the height above a reference level.

Ramlal House

Wind is flowing outside then pressure inside P_1 and outside house P_2

$$P_2 < P_1$$

Hence upthrust force will act on roof

$$F = (P_1 - P_2) \times \text{Area}$$

- Another way to express this force using the wind speed (V) and the density of air (ρ) is:

$$F = \frac{1}{2}\rho V^2 A$$

Bernoulli's equation in horizontal flow:-

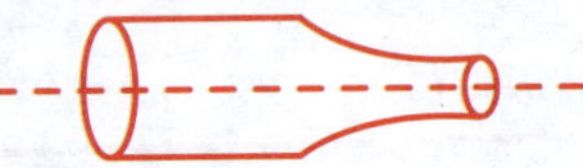

$$P + \frac{1}{2}\rho V^2 = const^n$$

$$AV = const^n$$

Potential energy same on horizontal level

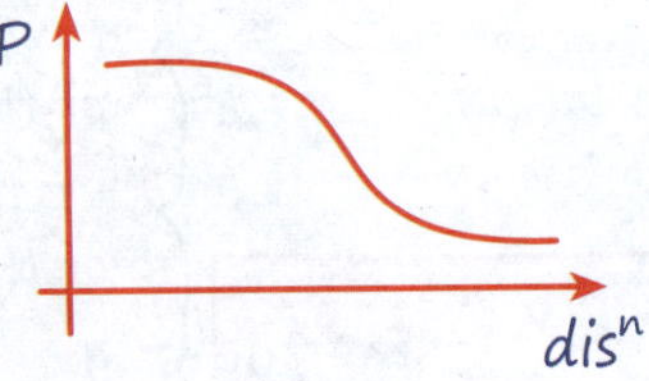

13 Velocity of Efflux

Closed container:

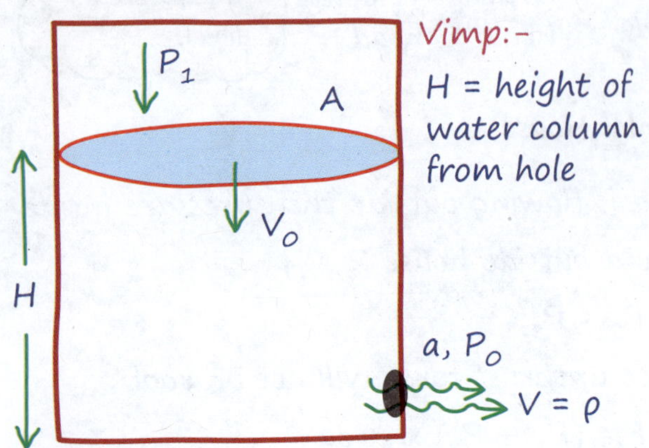

Vimp:- H = height of water column from hole

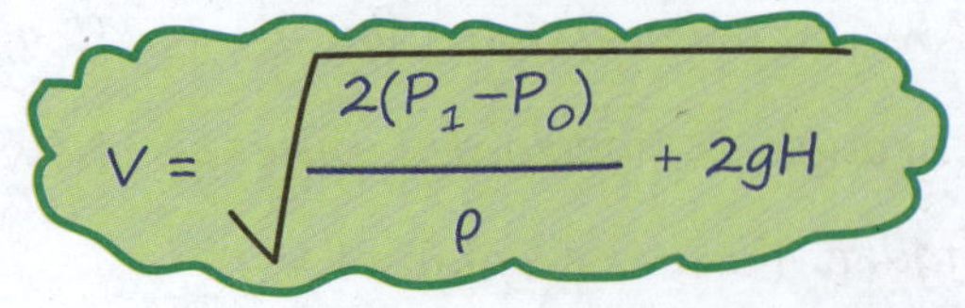

$$V = \sqrt{\frac{2(P_1 - P_0)}{\rho} + 2gH}$$

$P_1 = P_{Top}$ $\quad$ ρ = liquid

P_0 = atm P. $\quad$ $AV_0 = av$

Open Container (Torricilli चाचा)

- Torricelli's Theorem states that the velocity of efflux (the speed at which a fluid exits through an orifice) is equivalent to the velocity that a freely falling object would acquire if it fell from the same height.

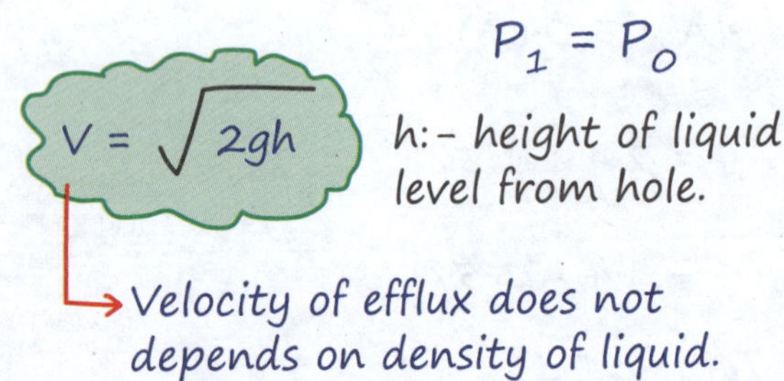

$P_1 = P_0$

$$V = \sqrt{2gh}$$

h:- height of liquid level from hole.

→ Velocity of efflux does not depends on density of liquid.

14 Range, Heights, Time of Flight of Efflux

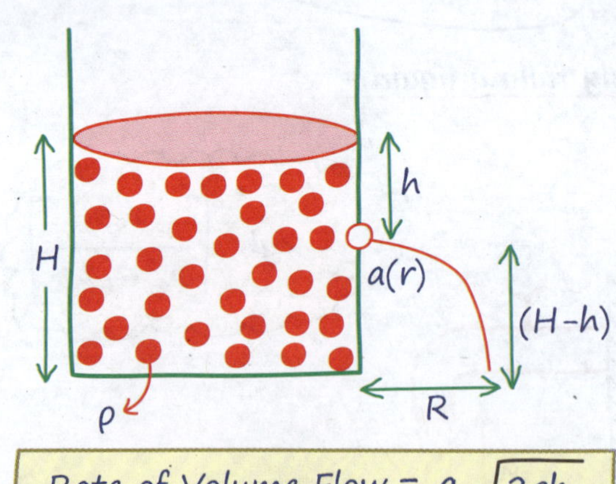

Rate of Volume Flow = $a\sqrt{2gh}$

Time to reach base level:-

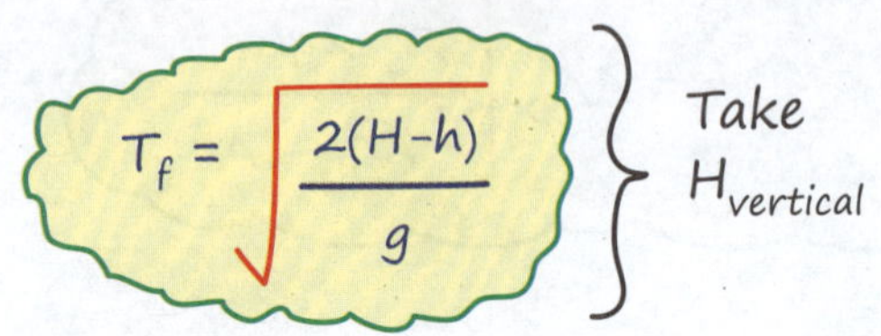

$$T_f = \sqrt{\frac{2(H-h)}{g}}$$ Take $H_{vertical}$

Horizontal range:-

$$R = U_x T_f = \sqrt{2gh}\sqrt{\frac{2(H-h)}{g}}$$

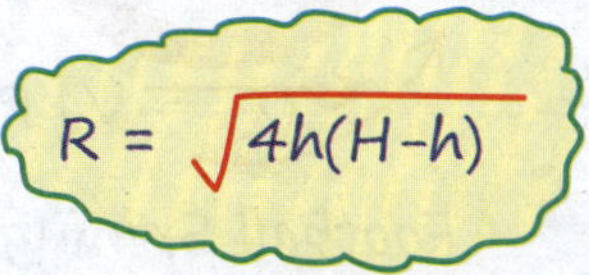

Time to Empty the Tank:

✦ If the orifice is at the bottom of the tank, the time required to empty the tank is:

$$t = \frac{A}{a}\sqrt{\frac{2H}{g}}$$

Maximum horizontal range

✦ The horizontal range of the liquid stream is maximum when the orifice is located at half the height of the liquid column.i.e.,

$$h = \frac{H}{2}$$:- R = max?

Therefore,

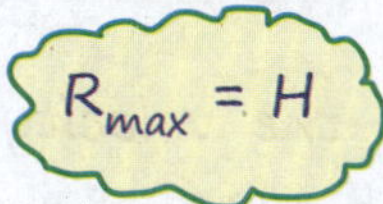

Just like. Motion in 1-D.

Angles and Range:-

✦ At 45°, the range is maximized, illustrating the symmetry in projectile motion.

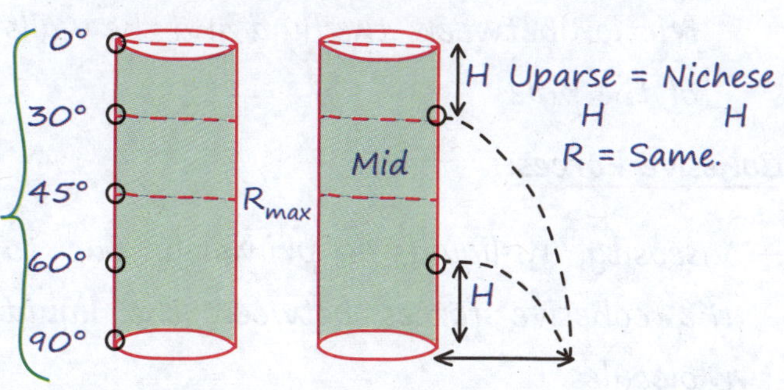

Force on Container due to efflux:-

$$F = \rho a V^2$$

$$* \; F = 2\rho agH \; * \qquad \because \; v = \sqrt{2gH}$$

ρ = density of liq.

H = Height of liquid column from Hole.

a = Area of orifice.

For massless container, "μ_{min}" to keep container at rest:-

$$\mu mg = 2\rho agh \qquad \begin{matrix} a = \text{hole} \\ A = \text{Container Area.} \end{matrix}$$

$$\mu_{min} = \frac{2a}{A}$$

Time Taken to Move Liquid from Height H2 to H1

$$V_0 = \frac{a}{A}\sqrt{2gy}$$

$$t = \frac{A}{a}\sqrt{\frac{2}{g}}\left(\sqrt{H_2} - \sqrt{H_1}\right)$$

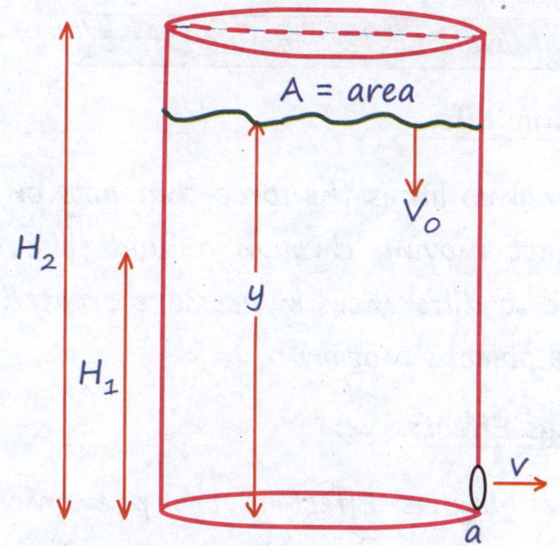

Time to Fall down:-

✦ To bottom :- $t = \frac{A}{a}\sqrt{\frac{2H}{g}}$

✦ Ratio from h → $\frac{h}{2}$ & $\frac{h}{2}$ → bottom

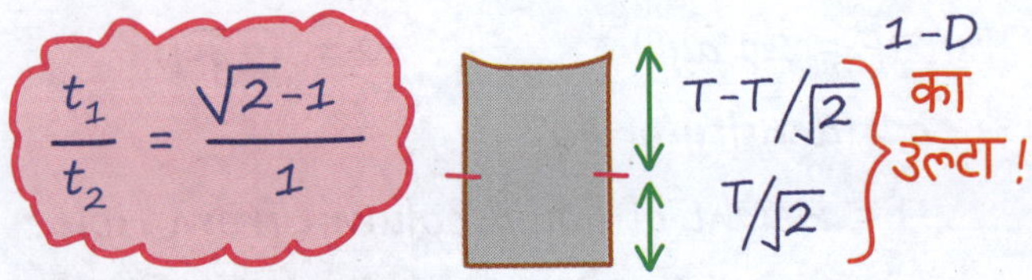

15 Venturimenter

✦ Measure rate of Volume Flow

✦ Based on Bernoulli's Principle.

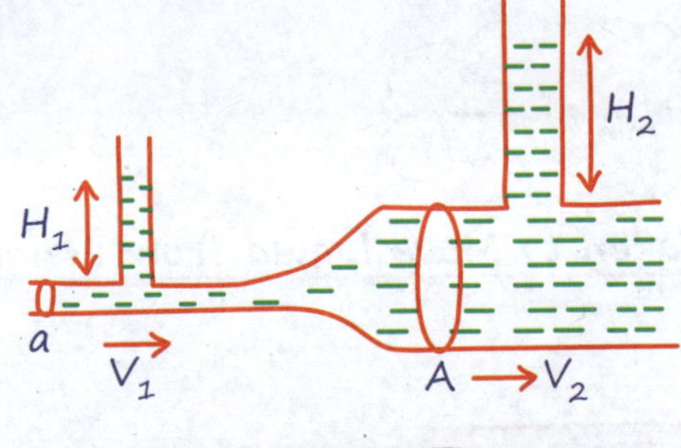

$$V_2^2 = \frac{2g\,(H_2 - H_1)\,a^2 A^2}{A^2 - a^2}$$

* $aV_1 = AV_2 = V$ = Rate of Volume Flow:

$$P_1 + \frac{1}{2}\rho V_1^2 + 0 = P_2 + \frac{1}{2}\rho V_2^2 + 0$$

16 Dynamic Lift & Magnus Effect

Dynamic Lift:

✦ Dynamic lift is the force that acts on an object moving through a fluid (like air) due to differences in pressure created by the object's motion.

Magnus Effect:

✦ The Magnus Effect is the phenomenon where a spinning object moving through a fluid creates a force perpendicular to the direction of motion, due to pressure differences caused by the spin.

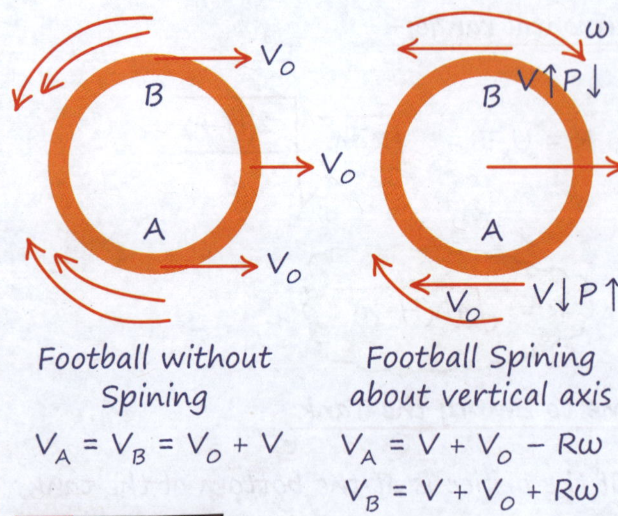

Football without Spining

$V_A = V_B = V_O + V$

Football Spining about vertical axis

$V_A = V + V_O - R\omega$

$V_B = V + V_O + R\omega$

17 Viscocity

✦ Viscosity is a measure of a fluid's resistance to flow. It describes how thick or sticky a fluid is.

✦ **"1 Poissulle = 10 Poise."**

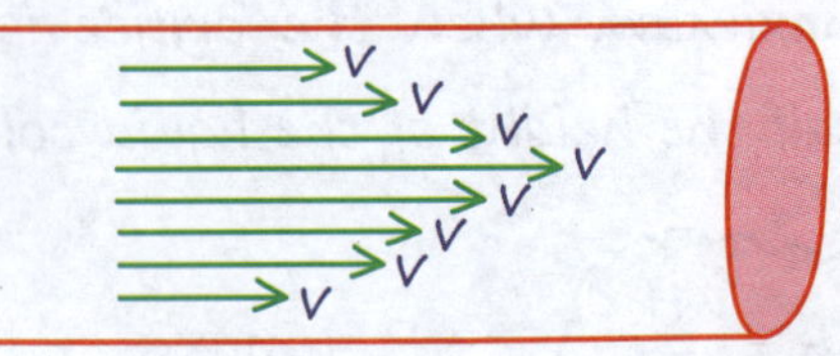

- The diagram shows a fluid flowing through a pipe.
- The arrows represent the velocity (v) of the fluid particles at different layers within the pipe.
- Near the center of the pipe, the fluid moves faster, while near the walls of the pipe, the fluid moves slower due to friction between the fluid and the walls of the pipe.

Cohesive Forces:

✦ Viscosity in liquids is primarily due to the cohesive forces between the liquid molecules.

✦ In gases, viscosity is due to the diffusion of molecules.

Force between two liq:-

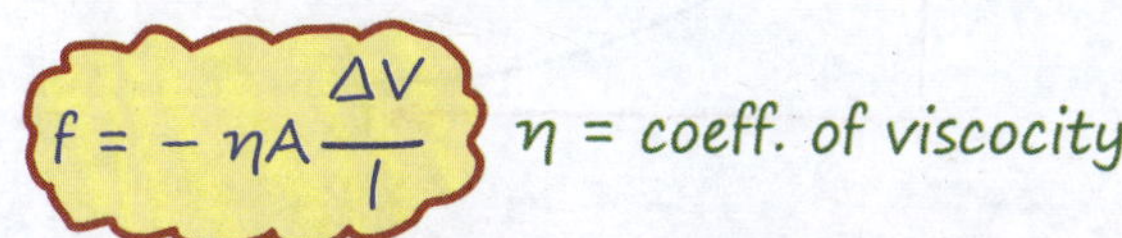

η = coeff. of viscocity.

$$\eta = \frac{f}{A\left(\frac{\Delta V}{\Delta x}\right)} = \frac{\text{Shear Stress}}{\text{Velocity gradient}} = \frac{\text{Shear Stress}}{\text{Strain Rate}}$$

Q.4. A metal block of base area $0.2 m^2$ is connected to a 0.02 kg mass via a string that passes over an ideal pulley as shown in figure. A liquid film of thickness 0.6 mm is placed between the block and the table. When released the block moves to the right with a constant speed of 0.17 m/s. The coefficient of viscosity of the liquid is: (2017-Gujarat)

(a) $3.45 \times 10^3 \times$ Pa-s

(b) $3.45 \times 10^{-2} \times$ Pa-s

(c) $3.45 \times 10^{-3} \times$ Pa-s

(d) $3.45 \times 10^3 \times$ Pa-s

Sol. (c) $F_r = T = mg$

$$\eta A \frac{\Delta v}{d} = mg$$

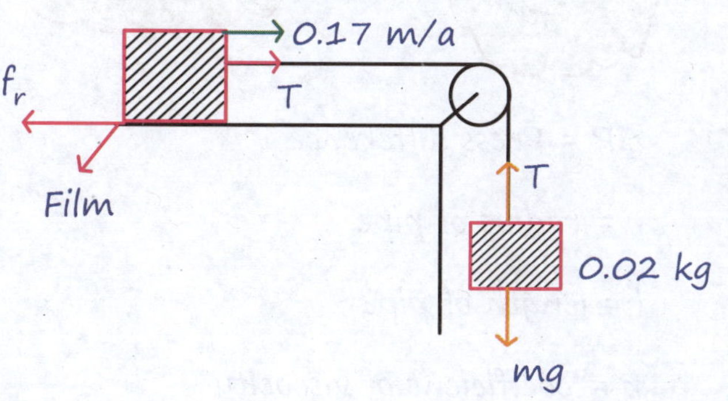

$$\eta = \frac{mgd}{A\Delta v} = \frac{2 \times 10 \times 6 \times 10^{-3}}{\frac{2}{10} \times \frac{17}{100}}$$

$$= \left(\frac{60}{17}\right) \times 10^{-3} \text{ Pa-s}$$

Difference Viscosity and Solid Friction;-

	Viscosity		(Solid Friction) NLM friction
(a)	The viscous force is directly proportional to the surface area of contact of fluid layers.	(a)	It is independent of the area of contact.
(b)	The viscous force is directly proportional to the relative velocity between the two layers in contact.	(b)	It is independent of the relative velocity of the contact surface.
(c)	The viscous force does not depend on the normal reaction between the two layers of the fluid.	(c)	It is directly proportional to the normal reaction between the surfaces in contact.
(d)	The viscous force on a solid moving in a fluid depends on the shape and size of the solid.	(d)	It is independent of the shape and size of the bodies in contact.

18 Stoke's Law

Stoke's law describes the force of viscosity acting on a small spherical object moving through a fluid.

Only For Sphere.

Jab Force between Solid-liq:-

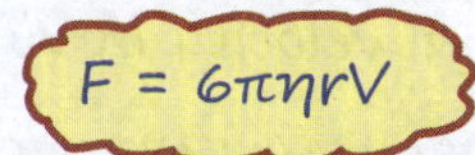

Where:

η = coefficient of viscosity of the fluid.

r = radius of the spherical body.

V = terminal velocity of the body.

Terminal Velocity:-

The maximum constant velocity acquired by a body while falling through a viscous medium is called its terminal velocity. It is usually denoted by V_T

$$mg = F_B + f_r$$

$$V_T = \frac{2}{9}\frac{r^2}{\eta}(\sigma - \rho)g.$$

Where,

r = radius of object, σ = density of object, ρ = density of liquid.

Graph:-

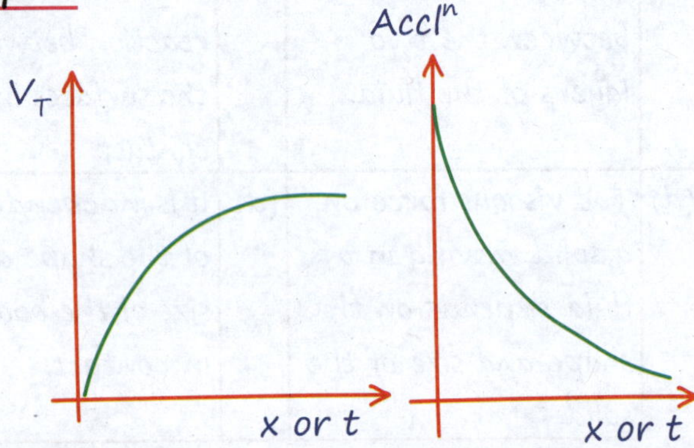

- $V_T \propto r^2$:- The terminal velocity is directly proportional to the square of the radius of the body. That is why bigger rain drops fall with a larger velocity as compared to the smaller drops.

- Ball is thrown down with velocity greater than terminal velocity in viscous liq. then variation of "v" v/s "t"? (NEET-2022)

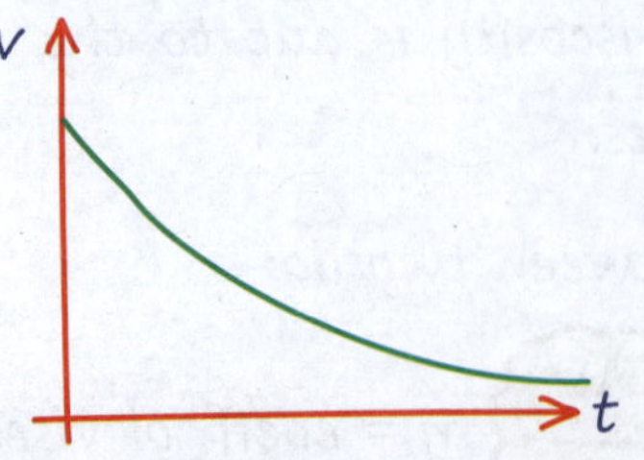

- Coalesce of Drops:-

- If V_T is the terminal velocity of the smaller drops and V'_T is the terminal velocity of the larger drop, then

$$V_T' = n^{2/3} V_T \ * \ R = n^{1/3} r$$

- Temperature dependence of η:-

For liquid

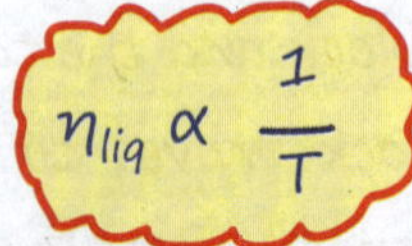

For gas

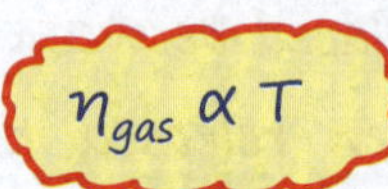

Poiseuille Equn:-

↳ Used to calculate the flow rate of Viscous liq.

↳ Bernoulli's theorem is not valid

Woh sirf ideal liquid keliye.

Fluid Current:-

$$Q = \frac{\Delta P}{\frac{8\eta l}{\pi r^4}} = \frac{\Delta P}{R_{Fluid}}$$

$$R_{fluid} \text{ of pipe} = \frac{8\eta l}{\pi r^4}$$

ΔP = Press difference

r = radius of pipe.

l = length of pipe.

η = coefficient of viscocity

Series

Q = same

ΔP = Diff.

$$Q = \frac{\Delta P}{R_{eq}}$$

$R_{eq} = R_1 + R_2$

$\Delta P = P_1 - P_2$

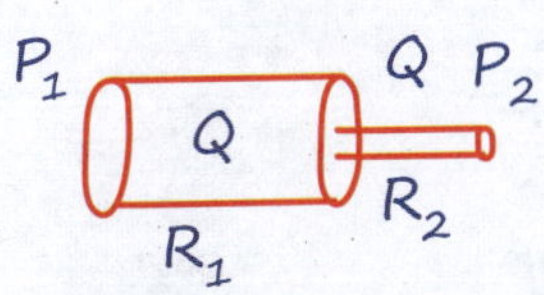

Parallel

Q = Diff.

ΔP = same

$$Q = \frac{\Delta P}{R_{eq}}$$

$$R_{eq} = \frac{R_1 R_2}{R_1 + R_2}$$

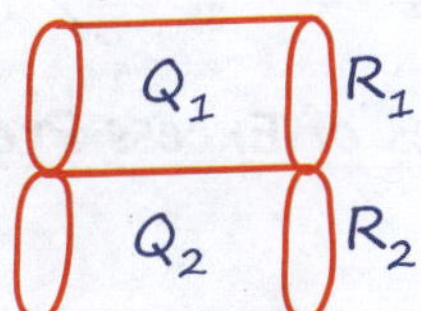

19 Surface Tension

Excess force required for lift

- Surface Tension is a property of liquid at rest by virtue of which a liquid surface gets contracted to a minimum area and behaves like a stretched membrane.
- Surface Tension of a liquid is measured by force per unit length on either side of any imaginary line drawn tangentially over the liquid surface, force being normal to the imaginary line
- It is a scaler quantity and depends only on nature of liquid but independent of area of surface.

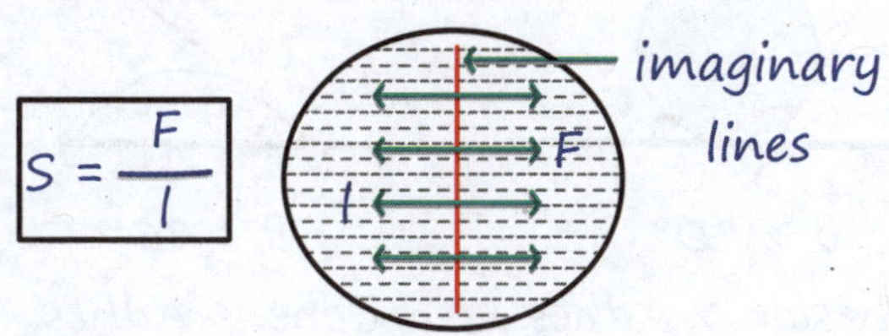

Q.5. A liquid film is formed over a frame ABCD as shown in figure. Wire CD can slide without friction. The mass to be hung from CD to keep it in equilibrium is

(a) $\frac{Tl}{g}$ (b) $\frac{2Tl}{g}$

(c) $\frac{g}{2Tl}$ (d) $T \times 1$

Sol. $m = \frac{2sl}{g}$

$mg = sl(2)$

$$\text{mass} = \left(\frac{F}{l} \times 1\right) = \frac{acc^n}{mass}$$

A B l liquid film C D Mg

Excess force required to lift: a

-Rod :- $S = \frac{F}{2l}$

-Disc :- $S = \frac{F}{2\pi r}$

-Ring :- $S = \frac{F}{2\pi r + 2\pi R}$

-Annular Disc :- ↑

Rod :-

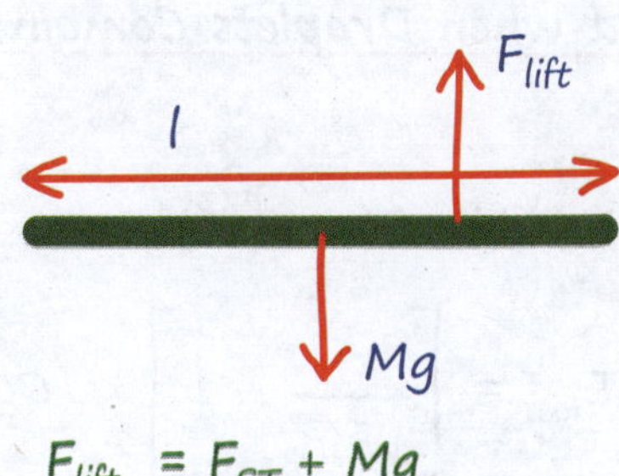

$$F_{lift} = F_{ST} + Mg$$
$$= 2Sl + Mg$$

20 Surface Energy

* $E = S.A.$

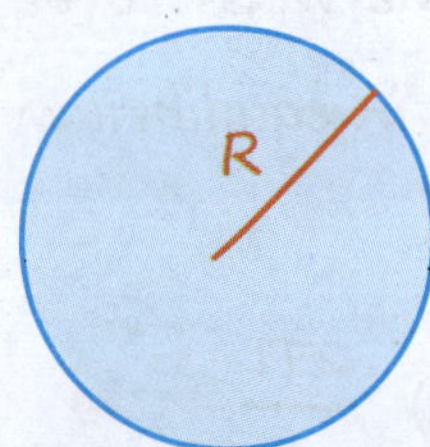

Sphere.

Surface area $= 4\pi R^2$

Volume $= \frac{4}{3}\pi R^3$

Circum ference $= 2\pi R$

21 Splitting of Drops into Droplets

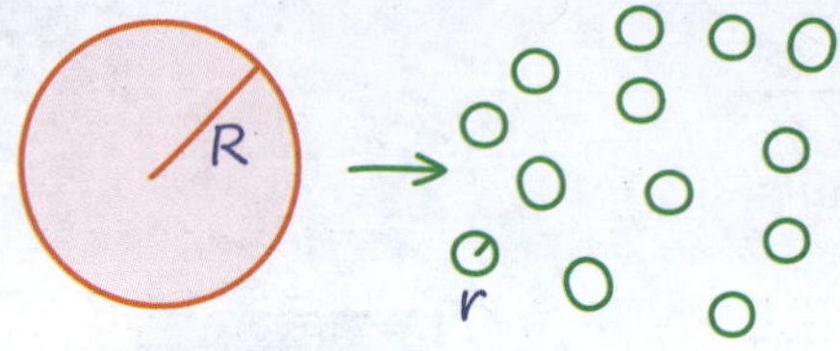

$R = n^{1/3} r$

$\Delta A = 4\pi\,[nr^2 - R^2]$; $\Delta A = 4\pi R^2 [n^{1/3} - 1]$

$$\Delta A = 4\pi R^2 \left[\frac{R}{r} - 1\right]$$

Bubble **Drop**

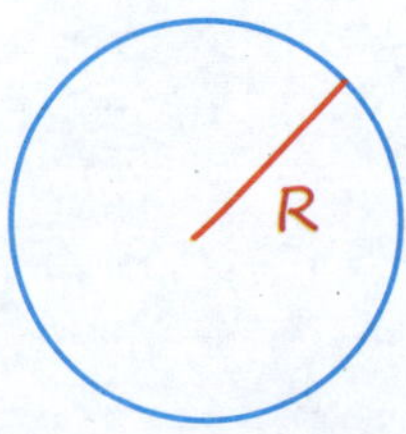

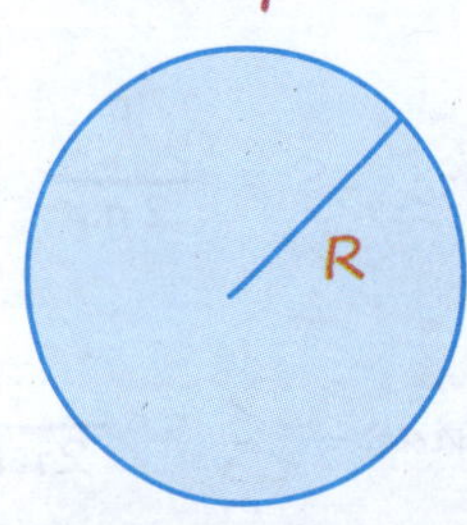

$A = 4\pi R^2 \times 2$ $A = 4\pi R^2$

Film:- Take Area double.

Energy Released when Droplets Combine to Form a Drop :-

$$\Delta E = E(n - n^{2/3}) = S4\pi r^2 (n - n^{2/3})$$

$$\frac{E_f}{E_i} = \frac{1}{n^{1/3}},\ \%\,E_{loss} = \left[\frac{1}{n^{1/3}} - 1\right] \times 100$$

$$\Delta E = n4\pi r^2 - 4\pi R^2$$

$$\Delta E = 3VT\left[\frac{1}{r} - \frac{1}{R}\right]$$ V = Vol of larger drop

Pressure is always high in concave side

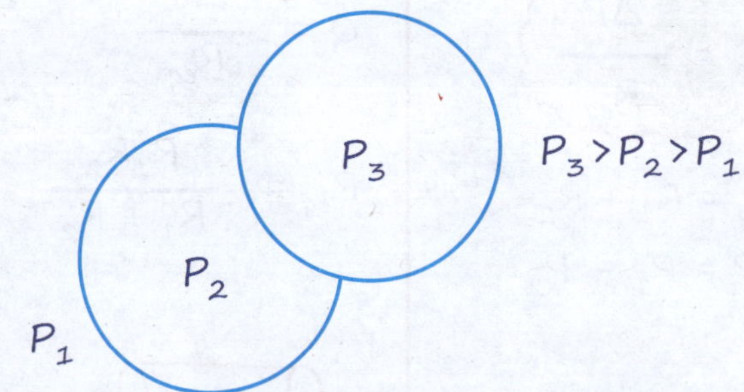

Values of Excess Pressure :-

Drop

$$\Delta P = P_{in} - P_{out} = \frac{2S}{R}$$

$$P_{in} = P_{out} + \frac{2S}{R}$$

Bubble

$$\Delta P = P_{in} - P_{out} = \frac{4S}{R}$$

$$P_{in} = P_{out} + \frac{4S}{R}$$

Radius of Coalesce :-

Two drops of radius r_1 & r_2 coalesce under isothermal condition.

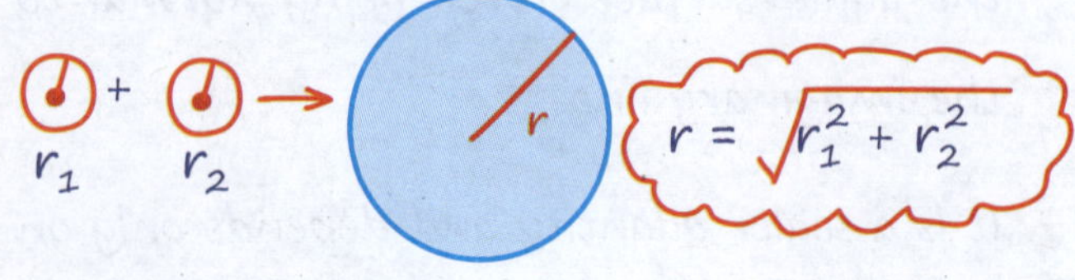

$$r = \sqrt{r_1^2 + r_2^2}$$

Angle of Contact :-

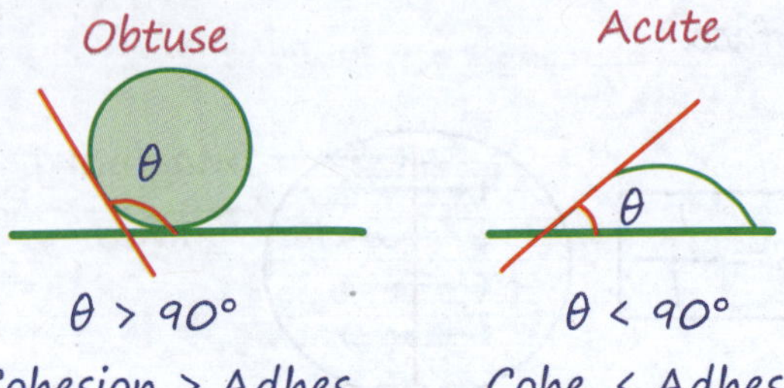

$\theta > 90°$ — Cohesion > Adhes.

$\theta < 90°$ — Cohe. < Adhes.

22 Capillary Tube

Rise	Fall
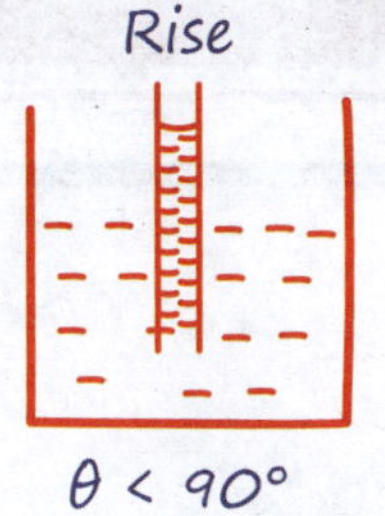	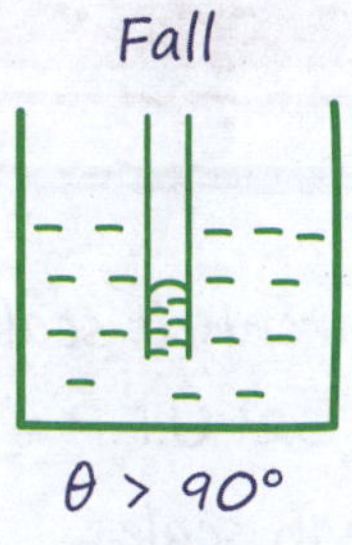
$\theta < 90°$	$\theta > 90°$
Concave Meniscus	Convex Meniscus

#

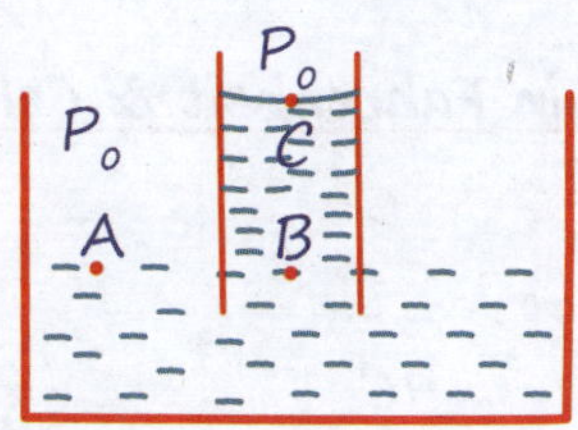

"$P_A = P_B = P_O$ ...(1)

Due to excess pressure

$$P_O - \frac{2S}{R} = P_C \quad ...(2)$$

Due to pressure variation in liquid

$$P_C + \rho gh = P_B = P_O$$

Putting value of P_C from (2)

#

$$P_O - \frac{2S}{R} + \rho gh = P_O$$

$$\frac{2S}{R} = \rho gh \Rightarrow \quad h = \frac{2S}{\rho gR}$$

Here, R is the radius of meniscus

If, r = radius of tub, then, $r = R\cos\theta$

Therefore, $h = \dfrac{2S\cos\theta}{\rho gr}$

+ **Height of liq. rising in Capillary Tube**

$$h = \frac{2S\cos\theta}{\rho rg}$$ → "g_{eff}"

$$M = \rho Ah = \frac{A2S\cos\theta}{rg}$$

"Mass of liq. in tube"

+ Rise of liq. in tube of insufficient length :-

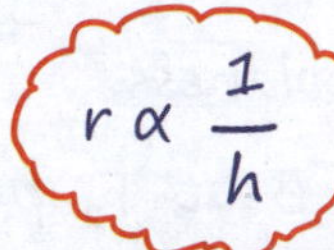

$r \propto \dfrac{1}{h}$ don't overflow

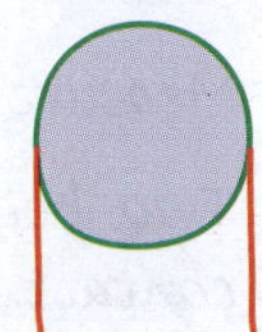

Radius of Interface

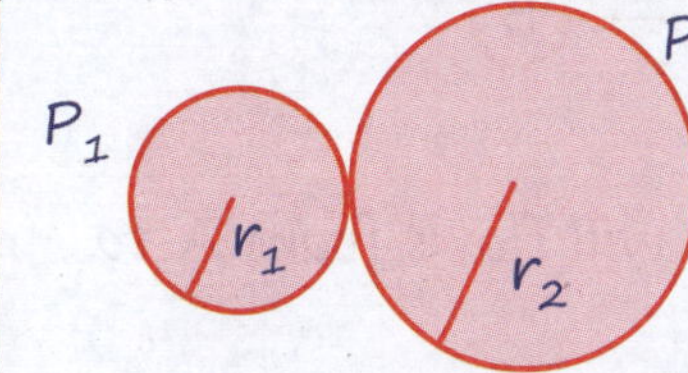

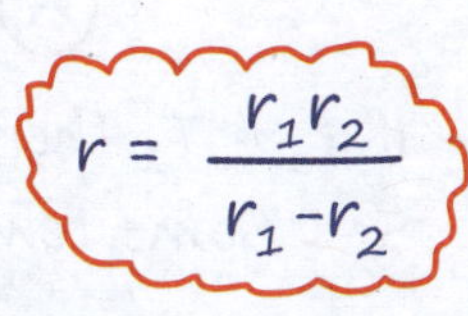

$$r = \frac{r_1 r_2}{r_1 - r_2}$$

+ Height/Depth of liquid $\propto 1/r$
+ Mass of liquid $\propto r$
+ Potential energy of liquid $\propto r^o$
+ If container with capillary in a freely falling lift, liquid rise upto complete length and does not overflow.

$$h \propto \cos\theta \propto \frac{1}{\theta}$$

+ For two different liquid if h, S and r same then find relation b/w density and contact angle.

$$h = \frac{2S\cos\theta}{\rho gr}$$

$$\cos\theta \propto \rho$$

$$\theta \propto \frac{1}{\rho}$$

MR* ‘धीमे कदमों पर यूं मायूस ना हो,
पहाड़ कभी भाग कर नहीं चढ़े जाते!’

16 Thermal Properties of Matter

1 Temperature

- Measure of hotness and coldness.
- Two body A at T_1 and B at T_2 put in contact.

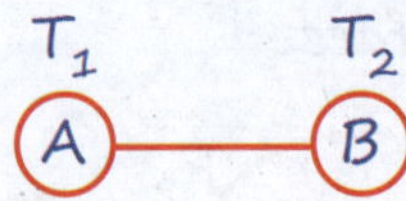

If $T_1 > T_2$ then

S-1 Some temp^r will flows from A to B ⇒ False

S-2 Some heat will flows from A to B ⇒ True

S-3 Heat will increase in B ⇒ False

S-4 Tempr of A will decrease ⇒ True

- Heat can not be stored it can flows from Body A to B.

2 Measurement of temperature

$$\frac{\text{Temp}^r - \text{L.F.P}}{\text{U.F.P} - \text{L.F.P}} = \text{Constant}$$

$$\frac{°C - 0}{100 - 0} = \frac{F - 32}{212 - 32} = \frac{K - 273}{373 - 273}$$

$$= \frac{\text{MR} - \text{L.F.P}}{\text{U.F.P} - \text{L.F.P}}$$

- Where
  - LFP = lower fixed point
  - UFP = upper fixed point

Note:-

For celcius scale

L.F.P = 0, U.F.P = 100

For Fahrenheit scale

L.F.P = 32, U.F.P = 212

For kelvin scale

L.F.P = 273, U.F.P = 373

Relation in Fahrenheit & Celcius:-

$$F = \frac{9C}{5} + 32$$

$$C = \frac{5F}{9} - \frac{160}{9}$$

$$\frac{\Delta C}{5} = \frac{\Delta F}{9}$$

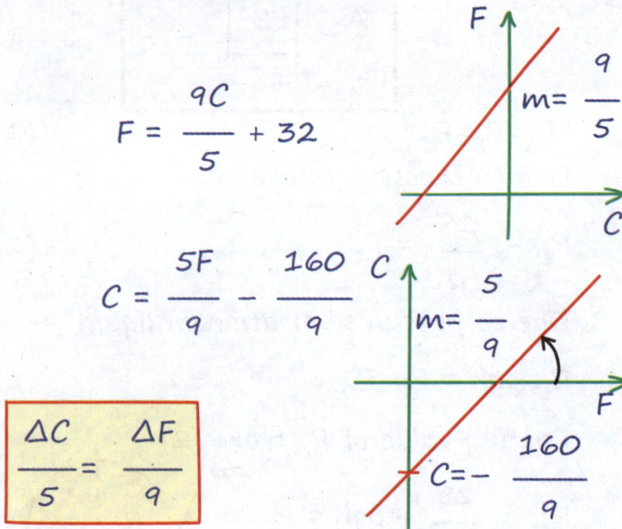

Relation between Kelvin & Celcius:-

$$C = K - 273$$

$$K = C + 273$$

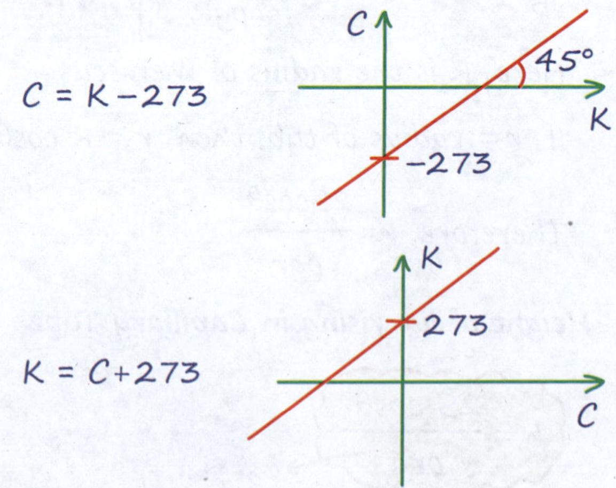

- Relation between °F & Kelvin :-

$$F - 32 = \frac{9}{5}(K - 273)$$

✦ Change in Temperature

$$\Delta C = \Delta K = \frac{5}{9}\Delta F$$

Q.1. The freezing point on MR* scale is 20° and boiling point 150°. A temperature of 60° thermometer will be read as.

Sol. Let x is reading at 60°C

$$\frac{x-20}{150-20} = \frac{60-0}{100-0}$$

[°CX]

3 Construction of thermometer

$$\frac{\text{Change in physical quantity}}{\text{per unit raise in temperature}} = \text{constant}$$

Resistance Thermometer:-

$$t = \frac{R_t - R_0}{R_{100} - R_0} \times 100$$

Pressure thermometer:-

$$t = \frac{P_t - P_0}{P_{100} - P_0} \times 100$$

Volume Thermometer:-

$$t = \frac{V_t - V_0}{V_{100} - V_0} \times 100$$

Length Thermometer:-

$$t = \frac{L_t - L_0}{L_{100} - L_0} \times 100$$

Q.2. Length of rod at 20°C is 10m and at 80°C is 40 m then find temperature when its length is 30 m.

Sol. Change in length per unit raise temperature = const

$$\frac{40\text{ m} - 10\text{ m}}{80°\text{C} - 20°\text{C}} = \frac{\ell - 10\text{ m}}{t - 20}$$

$$\frac{1}{2} = \frac{30 - 10}{t - 20}$$

$t = 60°C$

4 Thermal Expansion

Linear expansion:-

$\Delta L = L_0 \alpha \Delta \theta$; $L_2 = L_0(1 + \alpha \Delta \theta)$

Areal expansion:-

$\Delta A = A_0 \beta \Delta \theta$; $A_2 = A_0(1 + \beta \Delta \theta)$

Volume expansion:-

$\Delta V = V_0 \gamma \Delta \theta$; $V_2 = V_0(1 + \gamma \Delta \theta)$

Important relations:-

Bulk modulus and thermal coefficient of volume expansion.

$$\Delta P = \beta \frac{\Delta V}{V} = \beta \gamma \Delta \theta$$

β = Bulk Modulus of Elasticity

$$\frac{\alpha}{1} = \frac{\beta}{2} = \frac{\gamma}{3}$$

$\alpha : \beta : \gamma = 1 : 2 : 3$

✦ For anisotropic crystal

$\gamma = \alpha_x + \alpha_y + \alpha_z$

For Isotropic $\alpha_x = \alpha_y = \alpha_z$

$\gamma = 3\alpha$

✦ For Two Rod

$\alpha_1\ l_1$ $\quad$ $\alpha_2\ l_2$

* If difference in length of these two rod is independent of Temperature then $l_1\alpha_1 = l_2\alpha_2$

Ramlal Ne socha winter me gold buy karunga, summer me sold karunga tab length Increase ho Jayga ☺

→ Ramalal ko koi benefit nahi hoga

5 Cavity Problems

Photographic Enlargement.

T↑ → L↑ l↑

T↓ → L↓ l↓

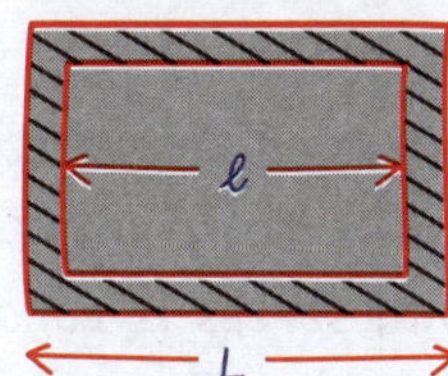

6 Bimetallic Strip

- A bimetallic strip is composed of two different metals with different coefficients of thermal expansion (α_1 and α_2).

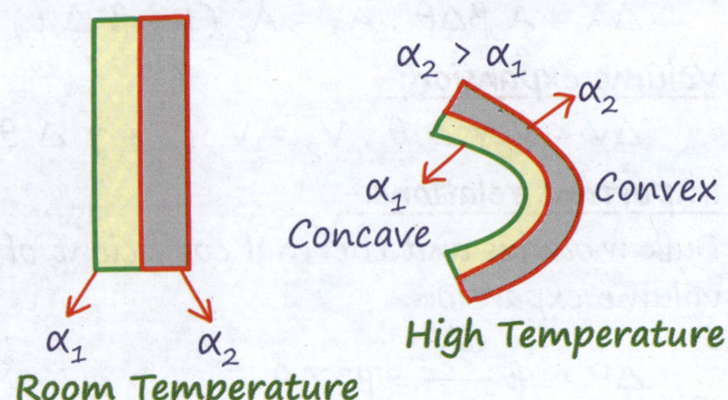

Behavior at room temperature:

At room temperature, the strip remains straight because both metals expand or contract equally.

Behavior at High Temperature:

- When the temperature increases, the metal with the higher coefficient of thermal expansion (α_2) expands more than the metal with the lower coefficient (α_1).
- This differential expansion causes the bimetallic strip to bend.
- The side with the metal that has a higher expansion coefficient (α_2) becomes convex (curves outward).
- The side with the metal that has a lower expansion coefficient (α_1) becomes concave (curves inward).

7 Thermal Stress

$$T.S. = \frac{F}{A} = \alpha Y \Delta\theta$$

Y :- Young's Modulus

A rod of length ℓ and area A placed on smooth surface then due to to increase in temprature thermal stress is zero.

Pendulum Clock:-

Note :-

Temperture↑ = Clock Slow = Time loss

Temperture↓ = Clock Fast = Time Gain

Loss or gain in time, $\frac{\Delta T}{T} = \frac{1}{2}\frac{\Delta l}{l} = \frac{1}{2}\alpha\Delta\theta$

Variation of density with temperature

$$\rho_2 = \rho_0(1 - \gamma\Delta\theta)$$

Anomalous Behaviour of Water:-

Water shows unique behavior near 4°C, where its volume decreases to a minimum and density increases to a maximum.

- **Volume vs. Temperature**: Minimum volume at 4°C.
- **Density vs. Temperature**: Maximum density at 4°C.

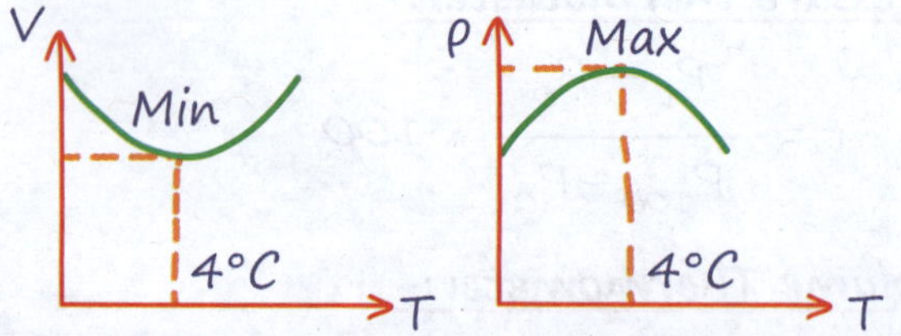

8 Apparent Coefficient of Volume Expansion of Liquid

The apparent coefficient of volume expansion of a liquid is the observed expansion when the liquid is heated in a container.

Formula

$$\gamma_{app} = \gamma_{real} - \gamma_{container}$$

$$\gamma_l' = \gamma_{liquid} - \gamma_{soild}$$

- If $\gamma_l = \gamma_s$ $\quad \gamma_l' = 0$

  Level Unchanged
- If $\gamma_l > \gamma_s$ $\quad \gamma_l' = +ve$

  The liquid overflows because it expands more than the container.
- If $\gamma_l < \gamma_s$ $\quad \gamma_l' = -ve$

  The liquid level goes down because the container expands more than the liquid.

9 Effect of Expansion on Apparent Weight in Liquid

The apparent weight (N) of an object submerged in a liquid is the weight of the object minus the buoyant force exerted by the liquid.

$$N = mg - \rho_l V_s g$$

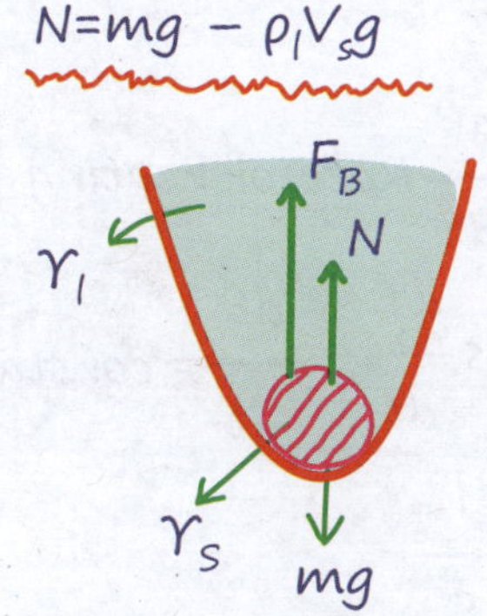

Scenarios Based on Coefficients of Expansion:

Case 1: $\gamma_l = \gamma_s$:

- If the coefficient of volume expansion of the liquid is equal to that of the solid, the apparent weight remains unchanged (N = same).

Case 2: $\gamma_l = \gamma_s$:

- If the coefficient of volume expansion of the liquid is less than that of the solid, the apparent weight decreases (N = dec.), and the submerged volume increases ($V_s \uparrow$).

Case 3: $\gamma_l > \gamma_s$:

- If the coefficient of volume expansion of the liquid is greater than that of the solid, the apparent weight increases (N = inc.), and the submerged volume decreases ($V_s \downarrow$).

Note:-

1> Effect of tempture on angular velocity

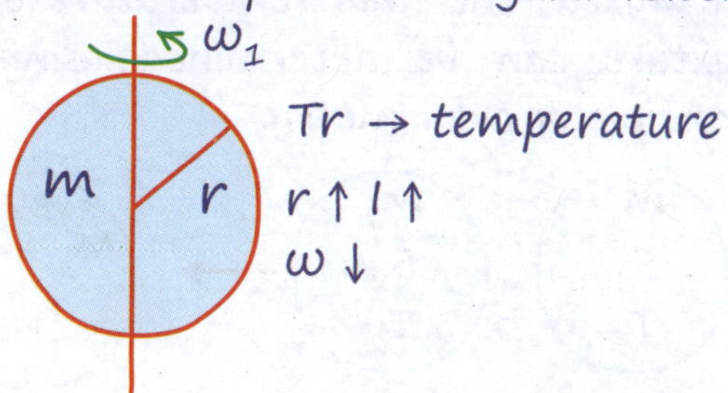

10 Meter Scale

- The true length of a rod can be determined using a meter scale by accounting for any thermal expansion.

Only one type will be asked :-

True length of Rod = Reading taken + ΔL

$$\Delta L = L\alpha\Delta\theta$$

- where ΔL is the change in length due to thermal expansion.

11 Calorimetry

Calorimetry is the study of measuring the amount of heat exchanged

Specific Heat Capacity:-

Heat required to raise the temperature by unit degree °C of unit mass

$Q = mS\Delta\theta$

raise T^r of "m" Kg by $\Delta\theta$

Unit of S = J/kg kelvin

- $S = \dfrac{\phi}{m\Delta T}$

If $\Delta T = 0$, $S = \infty$ (isothermal)

- $S = \dfrac{\phi}{m\Delta T}$

If $\phi = 0$, then $S = 0$ (for adiabatic process)

Heat Capacity (C):-

Heat required to raise temperature by unit of m mass.

$C = mS$ $(\Delta\theta = 1°C)$

raise T^r of object by 1°C

MR*

→ Joh Jaldi Garam hoga woh jaldi Thanda hoga. Uska specific heat capacity kam hai.

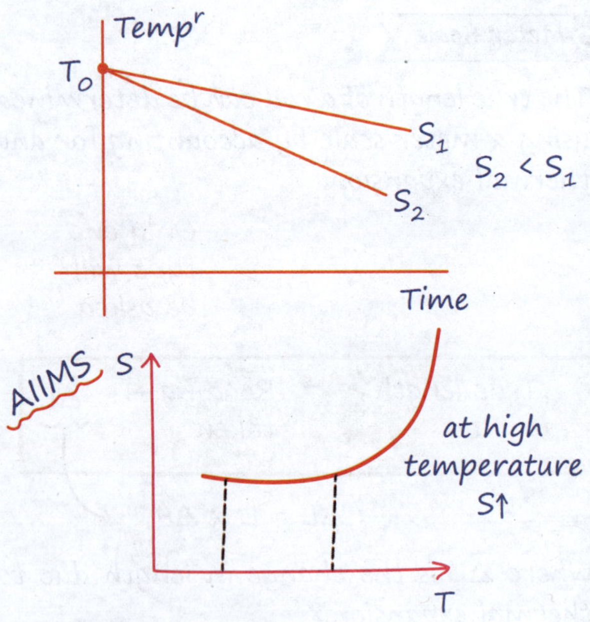

Molar heat Capacity [C]:-

$Q = nC\Delta\theta$

raise Temp^r of 1 mole Sub^s by 1°C

Note :-

$S_w = 1$ cal/gm°C = 4200 J/KgK

$S_{ice} = 0.5$ cal/gm°C = 2100 J/KgK

$S_{steam} = 0.5$ cal/gm°C = 2100 J/KgK

1 cal = 4.2 J

Latent Heat:-

$Q = mL$

heat req. to change state.

$L_{ice} = 80$ cal/gm

$L_{steam} = 540$ cal/gm

12 Temp.-time graph

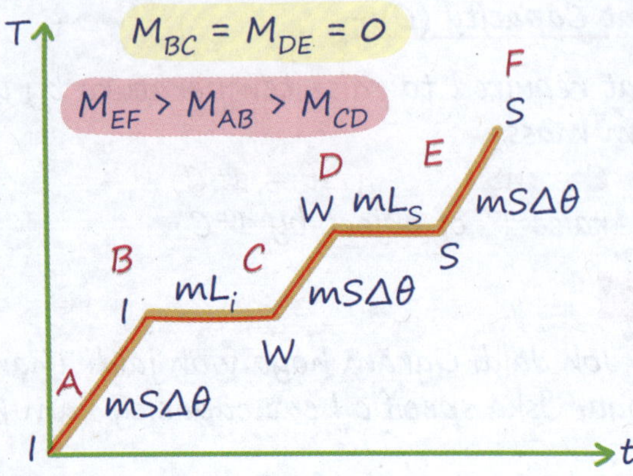

MR* feel

- Water Can't exist below 0°C. Ice can't exist above 0°C. at 0°C both can exist. Mixture of (Ice + water) only Possible at 0°C. Zero se upar gya matlb sab pani ho gya, zero se Niche sab Ice.
- Steam + water mixture only exist at 100°C.
- Slope $= \frac{d\theta}{dT}$ Rate of heating
- $\frac{d\theta}{dt} = ms\frac{dT}{dt}$, $\frac{d\theta}{dt} =$ constant

  $\frac{dT}{dt} \propto \frac{1}{S}$

Q.3. 10 gm ice at 0°C mixed with 10 gm water at 40°C then mass of water in mixture?

Sol. $Q_{ice} = mL = 800$

$Q_{water} = wt = 10 \times 40 = 400$

only 5 grm ice will melt

m (water) = 15 gm.

13 Principle of Calorimetry

- The principle of calorimetry states that in an isolated system, the heat lost by hot objects is equal to the heat gained by cold objects.

  Heat Loss = Heat Gain

14 Mixture

Final Temperature of Mixture:-

- When two liquids of different temperatures are mixed, the final temperature of the mixture can be determined using the principle of calorimetry.

m_1, S_1, T_1 + m_2, S_2, T_2 → $M_1 + M_2$, T_{mix}

Final Temperature of Mixture when two liquid of mass m_1 & m_2 of specific heat capacity S_1 and S_2 at temperature T_1 and T_2 mixed.

$$T_{mix} = \frac{m_1S_1T_1 + m_2S_2T_2}{m_1S_1 + m_2S_2}$$

Mixture of ice & water:-

MR*

- m gram Ice at 0°C mixed with W gram water at T°C
- Bring both ice & water at common Temp^r and same phase.

Required :- Q = mL (m gram Ice melt into m gram water at 0°C)

Supply :- $Q = WT = mS\Delta T$ (When W gram water cool down from T to 0°C)

1> mL = WT

$T_{mix} = 0°C$ $\qquad (amt)_{ice} = 0$

$(amt)_w = W + m$

2> mL > WT

$T_{mix} = 0°C$ $\qquad (amt)_{ice} = m - m'$

$(amt)_W = W + m'$

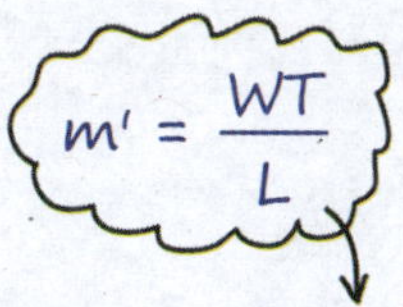

Amt. of ice converted to water.

3> mL < WT

$$T = \frac{WT - mL}{m + W}$$

*Paani hi Rahega.

Mixture of ice & steam-

Required :- Q = mL = 80 m

Supply :- $Q = WL + WS\Delta\theta = 640\ W$

1> $Q_{Supply} = Q_{Req.}$

$M_{ice} = 8W$

$$\frac{M_{ice}}{W_{steam}} = \frac{8}{1}$$

$T_{mix} = 0°C.$

2> Ice 0°C $\xrightarrow{80m}$ water 0°C $\xrightarrow{100m}$ Water 100°C

Steam 100°C $\xrightarrow{540W}$ Water 100°C

180m = 540W

$$\frac{M_{ice}}{W} = \frac{3}{1}$$

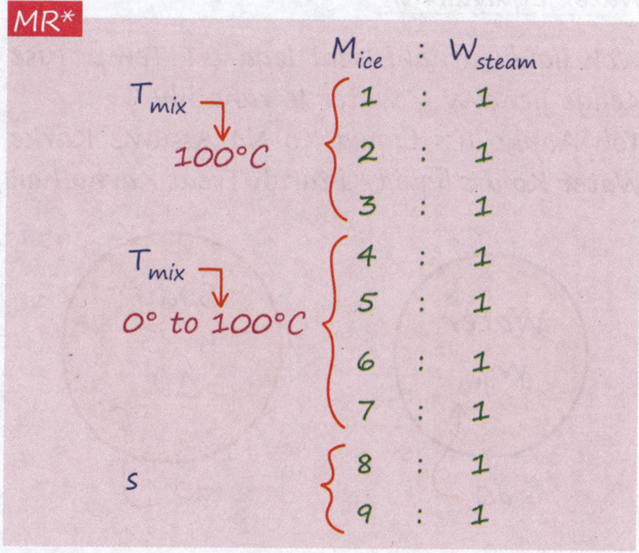

Q.4. 200 grm ice at –20°C is mixed with 500 grm water it 20°C, then find temperature of mixture and amount of water, ice in mixture

Sol. MR* Dono ko kisi ek phase me same temperature par le ke aao.

$Q_1 = ms\,\Delta T + mL = 2000\text{ cal} + 16000\text{ cal}$

heat given to ice to melt Δ°C.

$Q_2 = ms\Delta t = 10000\text{ cal.}$

Heat given by water when it fall from 20° to 0°C

$Q_1 > Q_2$ hence complete ice will not melt, out of 10000 cal heat given by water 2000 cal used to increase temperature and 8000 cal use to melt ice

$m'L = 8000$

$m' = 100$ ice will melt

$T_{mix} = 0°$ because (ice + water) mixture

$M_{water} = 600$ grm

$M_{ice} = 100$ grm

Q.5. 5 gm water at 30°C and 5 gm ice at −20°C mixed then temperature of mixture.

Sol. $T_{mix} = 0°$ Requirement > supply

Water Equivalent:-

Woh liquid utnahi Heat lega ΔT Temp. Rise Keliye jitna W_{gm} water le raha hh!
Toh Aapko uss Liquid Ko Na Assume Karke Water Ko uss liquid Ki tarah Treat Karna hai!

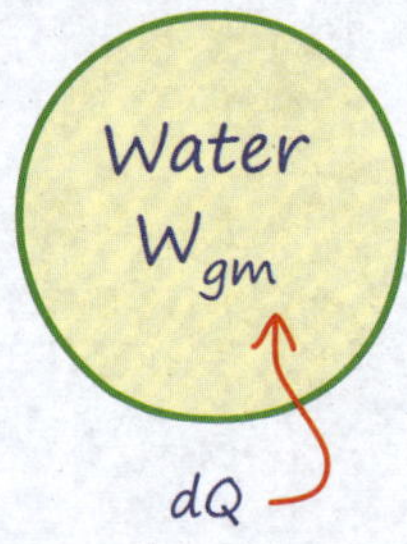

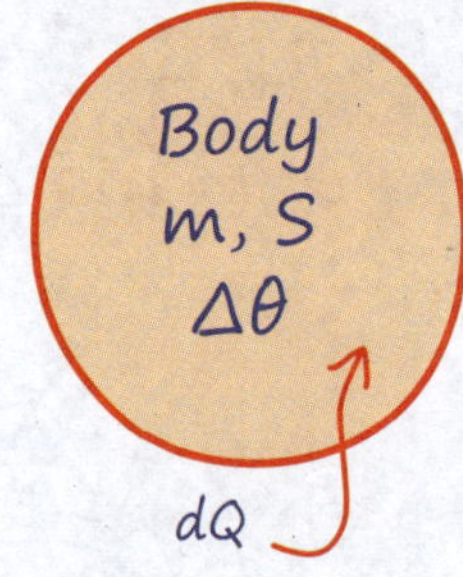

$\therefore\ mS\Delta\theta = W \times 1 \times \Delta\theta$

$W = mS$

eg :- Water equivalent = 55g at 40°C.

$WT = 55 \times 40 = 2200$ cal.

V. Imp Question :-

*A bullet of mass "m" moving with "u" hits an ice block of "M" gm Kept on a frictionless floor & gets stuck in it. How much ice will melt if x% of the lost KE goes to ice ? (initial temp^r of block & bullet = 0°C).

$$\left[\frac{1}{2}\ \frac{mMu^2}{M+m}\right] x\% = mL$$

✦ Ice Formation :-

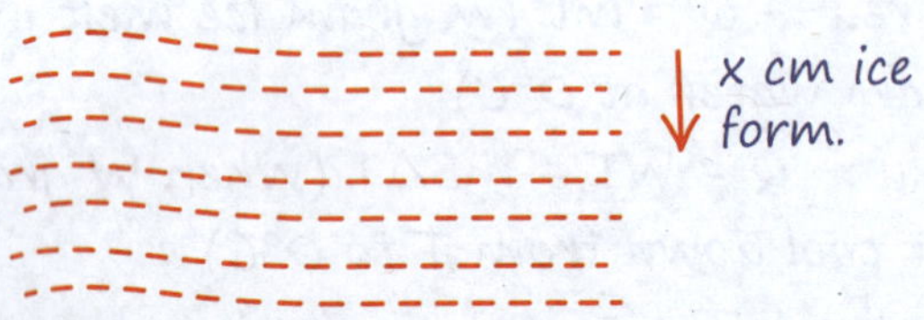

Formation of ice :-

0 to x : x to 2x : 2x to 3x

t : 3t : 5t

0 to x : 0 to 2x : 0 to 3x

t : 4t : 9t

$t \propto x_2^2 - x_1^2$

MR*

"It's okay to feel up and down Its normal and natural, don't overthink, move forward and work hard"

17 Heat Transfer

1 Conduction

Heat flows from hot end to cold end, medium required but particles of medium simply oscillate but do not leave their position.

- Slow process
- Takes places in solid
- Path may be zig-zag
- Temperature of medium increases.

2 Convection

- Medium required, each particle of medium absorbs heat and moves from hot end to cold end.
- Slow process
- Occurs in fluid not in solid
- Temp^r of medium increases.

3 Radiation

- Heat flows in the form of electromagnetic waves.
- Medium is not required
- Path straight line
- No change in temp^r of medium.

4 Law of Thermal Conductivity

l

A = Area

T_1 T_2

$$H = \frac{Q}{t} = \frac{KA\Delta T}{L}$$

K = Coefficient of Thermal Conductivity (Material Property)

Q = Heat.

- $\frac{Q}{t} = KA\frac{dT}{dx}$

Q.1. Three rods made of the same material and having the same cross section have been joined as shown in the figure. Each rod is of the same length. The left and right ends are kept at 0°C and 90° respectively.

The temperature of the junction of the three rods will be

(a) 45°C (b) 60°C

(c) 30°C (d) 20°C

Sol.

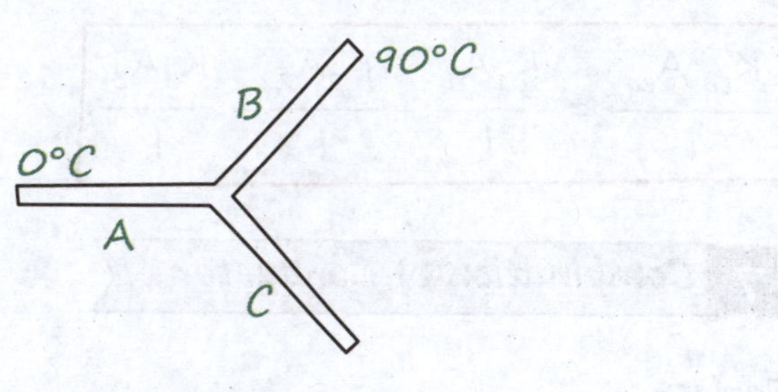

$$T = \frac{T_1 + T_2 + T_3}{3}$$

$$T = \frac{0 + 90 + 90}{3} = \frac{180}{3} = 60$$

5 Heat Current

$$H = \frac{\Delta T}{R_T}, \quad R_T = \frac{l}{KA}$$

R_T :- Thermal Resistance

6 Combination of Rod

Series Combination:-

"R" add होगा

$$R_{eq} = R_1 + R_2 + R_3 \ldots\ldots$$

$$\boxed{\frac{l_{eq}}{K_{eq}} = \frac{l_1}{K_1} + \frac{l_2}{K_2} + \frac{l_3}{K_3}}$$

Junction Temp:-

Rate of heat flow same in series combination.

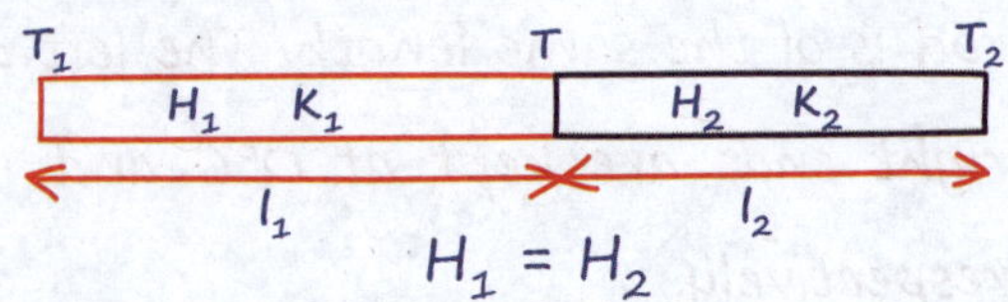

$$H_1 = H_2$$

$$\frac{K_1(T_1 - T)}{l_1} = \frac{K_2(T - T_2)}{l_2}$$

Parallel Combination:-

$$\frac{1}{R_{eq}} = \frac{1}{R_1} + \frac{1}{R_2} + \frac{1}{R_3} + \ldots$$

$$\boxed{\frac{K_{eq}A_{eq}}{l} = \frac{K_1A_1}{l} + \frac{K_2A_2}{l} + \frac{K_3A_3}{l}}$$

7 Combination of Conductor

Series:-

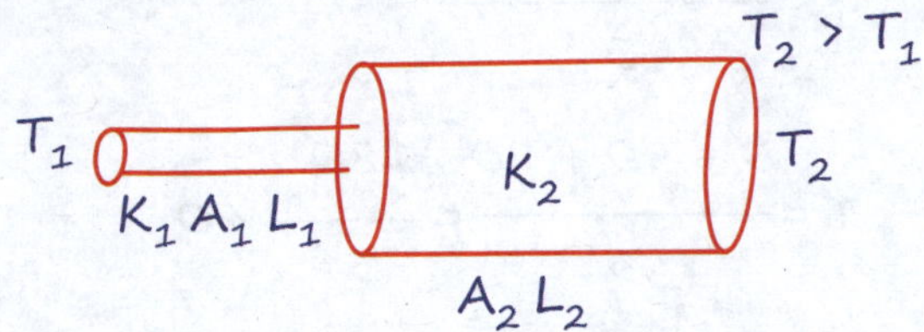

$$R_{eq} = R_1 + R_2 = \frac{L_1}{K_1A_1} + \frac{L_2}{K_2A_2}$$

$$\frac{H}{dt} = \frac{\Delta T}{R_{eq}} = \frac{T_2 - T_1}{R_{eq}}$$

$$K_{eq} = \frac{L_1 + L_2}{\frac{L_1}{K_1} + \frac{L_2}{K_2}}$$

If $L_1 = L_2$

Then

$$K_{eq} = \frac{2K_1K_2}{K_1 + K_2}$$

$$T_{mid} = \frac{\frac{K_1A_1T_1}{L_1} + \frac{K_2A_2T_2}{L_2}}{\frac{K_1A_1}{L_1} + \frac{K_2A_2}{L_2}}$$

Parallel:-

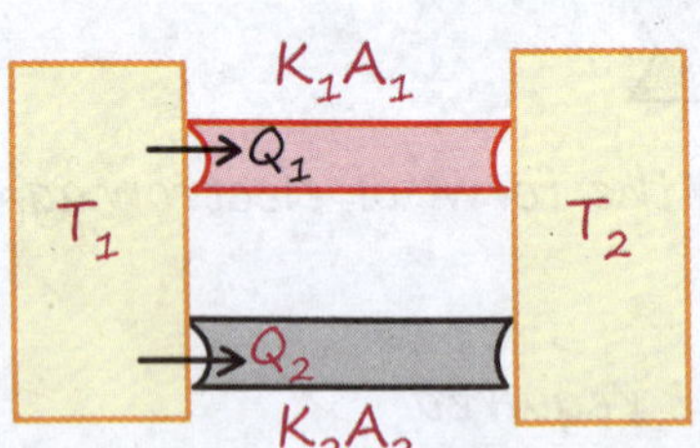

$T_1 > T_2$

Length Should be same

$L_1 = L_2 = L$

$$\frac{1}{R_{eq}} = \frac{1}{R_1} + \frac{1}{R_2}$$

$$\frac{K_{eq}[A_1 + A_2]}{l} = \frac{K_1A_1}{l} + \frac{K_2A_2}{l}$$

$$K_{eq} = \frac{K_1A_1 + K_2A_2}{A_1 + A_2}$$

✦ If $A_1 = A_2 = A$

Then

$$K_{eq} = \frac{2K_1K_2}{K_1 + K_2}$$

$$H_{eq} = Q_1 + Q_2$$

$$= \frac{K_1(T_1 - T_2)A_1}{L} + \frac{K_2(T_1 - T_2)A_2}{L}$$

Wheat stone Bridge :-

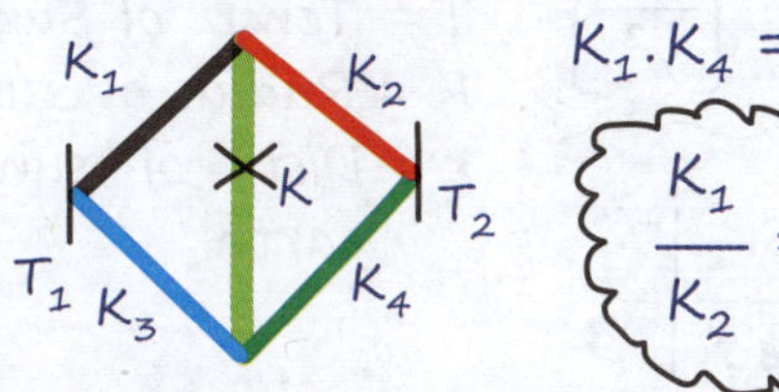

$$K_1.K_4 = K_3.K_2$$

$$\frac{K_1}{K_2} = \frac{K_3}{K_4}$$

8 Radiation

Black Body :-

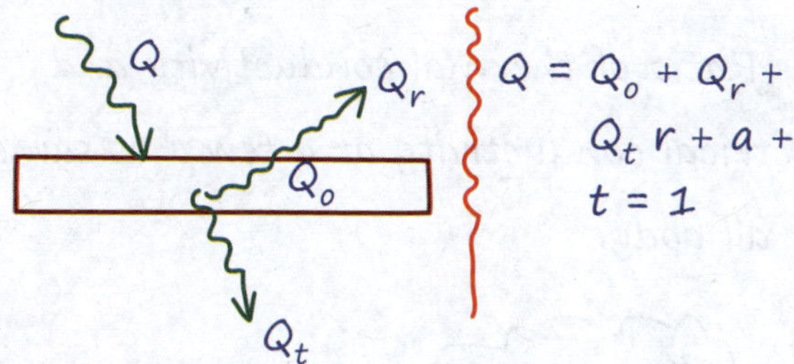

$Q = Q_o + Q_r + Q_t$

$r + a + t = 1$

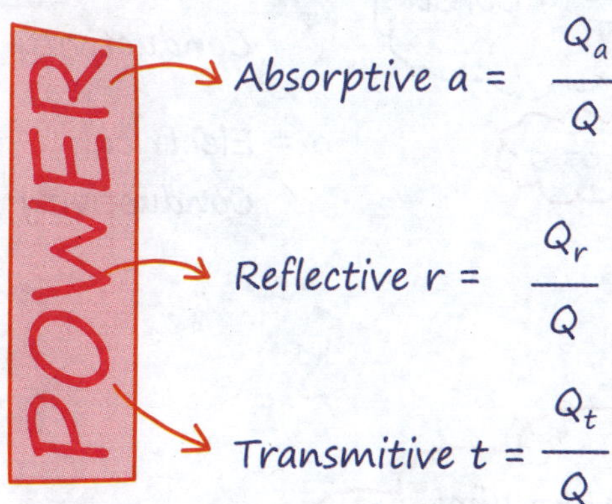

Absorptive $a = \frac{Q_a}{Q}$

Reflective $r = \frac{Q_r}{Q}$

Transmitive $t = \frac{Q_t}{Q}$

✦ Emissive Power [Intensity] :-

$$E = \frac{Q}{At} = \frac{J}{m^2s} = \text{Watt/m}^2$$

Stefan's Law :-

$E = \sigma T^4$ $\qquad$ T = Kelvin.

$$\sigma = 5.67 \times 10^{-8} \frac{\text{Watt}}{m^2K^4}$$

Emissivity :-

$$e = \frac{\text{Emissive Power of Normal body } (\varepsilon)}{\text{Emissive Power of Black body } (E)}$$

For Black Body (e) = 1.

Emissive Power of Normal Body (ε):-

$$\varepsilon = eE = e\sigma T^4$$

$$P = \frac{Q}{t} = Ae\sigma T^4$$

Stefan-Boltzmann's Law:-

$$P_{loss} = P_{emit} - P_{absorb}$$

$$P_{loss} = \sigma eA\,[T^4 - T_0^4]$$

T_0 = Surr. Temp^r, T = body Temp^r

Q.2. If the temperature of a boody is increased by 50%, then the increase in the amount of radiation emitted by it will be

Sol. ∵ Percentage increase in the amount of radiations emitted

$$\therefore \frac{E_1 - E_2}{E_1} \times 100 \frac{(1.5T_1)_4 - T_1^4}{T_1^4} \times 100$$

$$\Rightarrow \frac{E_1 - E_2}{E_1} \times 100 = [(1.5)^4 - 1] \times 10$$

$$\frac{E_1 - E_2}{E_1} \times 100 = 400\%$$

Wien's Law :-

$\lambda_m T = b$

$\lambda_3 > \lambda_2 > \lambda_1$

$T_1 > T_2 > T_3$

$b = 2.9 \times 10^{-3}$ mK

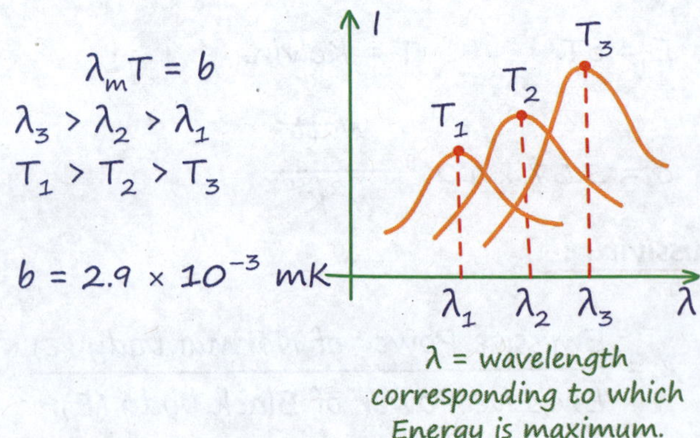

λ = wavelength corresponding to which Energy is maximum.

Newton's Law of Cooling :-

- Time 't' taken to fall temperature T_1 to T_2 where T_0 is the temperature of surrounding.

$$\frac{T_1 - T_2}{t} = K\left(\frac{T_1 + T_2}{2} - T_0\right)$$

- Rate of cooling ∝ Temperature difference
- 90°C $\xrightarrow[t_1]{}$ 80°C $\xrightarrow[t_2]{}$ 70°C $\xrightarrow[t_3]{}$ 60°C

  Time $t_1 < t_2 < t_3$

  (Time taken to fall temp^r for every 10°C)

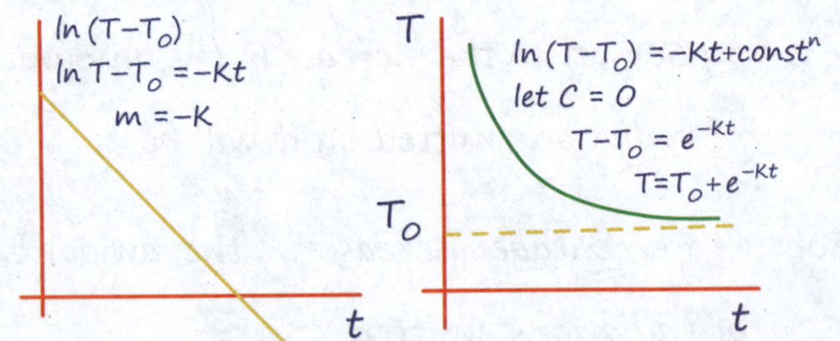

Kirchoff's Law :-

A good absorber is a good emitter!

$$\Rightarrow \frac{E_\lambda}{a_\lambda} E_\lambda \text{(constant)}$$

Solar Constant:-

Total Thermal Energy falling per unit area per sec.

$$S = e\sigma T^4 \left[\frac{R^2}{r^2}\right]$$

e = emissivity of sun.

σ = Stephan's const^n

T = Temp. of Sun.

R = Radius of Sun.

r = Dist^n of sun & earth.

$$T_{sun} \propto \left[\frac{Sr^2}{e\sigma}\right]^{1/4}$$

The surface temp of sun found to be 5742K

Weisman-Fraz Law:-

Ratio of thermal conductivity and electrical conductivity at a temp^r is same for all body.

$$\frac{K}{\sigma T} = Const^n$$

$$K \propto \sigma$$

K = Thermal Conductivity

σ = Electrical Conductivity

MR*

"Jeet ke khatir junoon chahiye, ho ubal ayesa khoon chahiye, Aasman bhi ayega zameen par bas irado main jeet ki goonj chahiye."

KTG & Thermodynamics

1 Thermodynamics

System:

Open → Mass exchange/Energy energy exhange
Closed → Mass not exchange/Energy energy exhange
Isolated → Mass/energy not exchange

Thermal Equilibrium

If Heat transfer between 2 system is not in process i.e. their temperature is same then the 2-system are said to be in thermal equalibrium.

2 Zeroth Law of Thermodynamics

- This law states that if two systems (A and B) are each in thermal equilibrium with a third system (C), then A and B are in thermal equilibrium with each other. This means that the temperature of A, B, and C are equal:

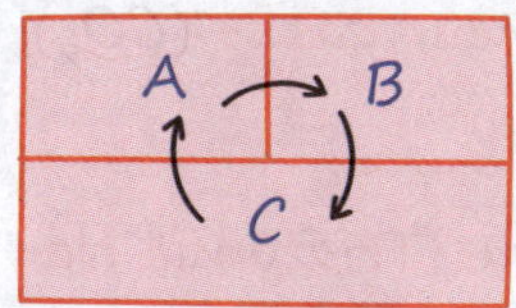

$T_A = T_B = T_C$ (They are in thermal contact)

3 Work done in thermodynamics

$W = P.\Delta V = \int P.dV$

Work done By gas
- Vol. increase, expansion
- $W = +ve$

On gas
- Vol decrease compression
- $W = -ve$

4 P-V Graph → Indicator Diagram

- This graph shows the work done in a cyclic process. The area enclosed by the loop in a P-V graph represents the work done.

Work = Area of Close loop in PV Graph

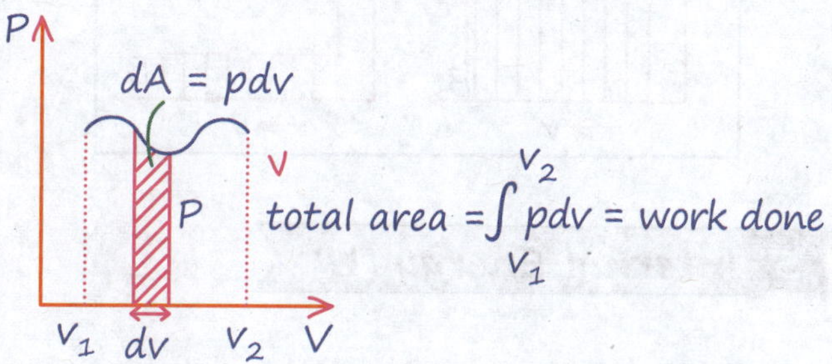

total area $= \int_{V_1}^{V_2} pdv$ = work done

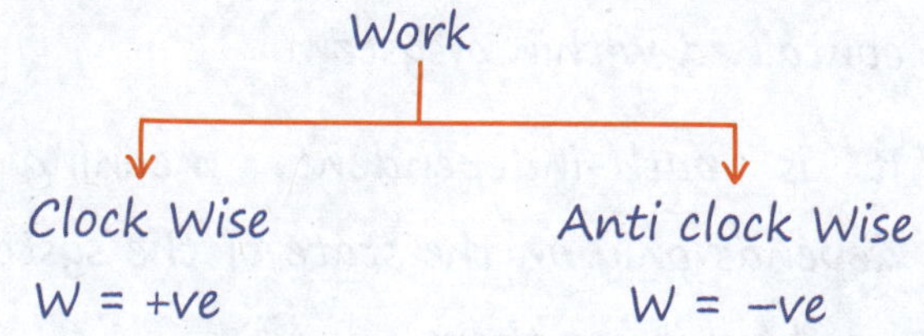

- Expansion → $\Delta V = +ve$ W ↑
- Compression → $\Delta V = -ve$ W ↓

Q.1. In given fig. find workdone

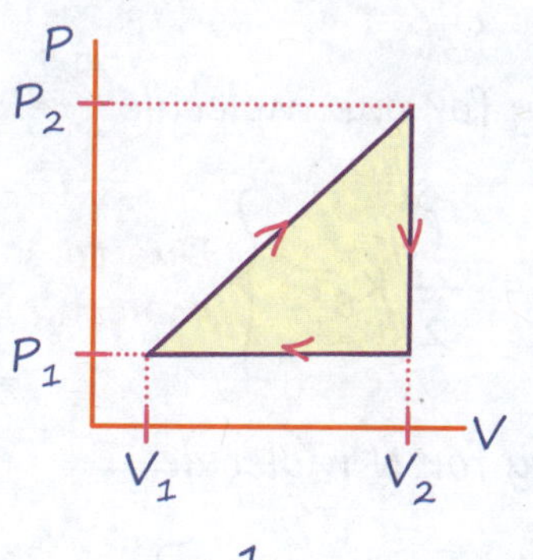

$$Work = +\frac{1}{2}(V_2 - V_1)(P_2 - P_1)$$

Q.2. In the following figure two indicator diagrams are shown. If the amounts of work done in them are W_1 and W_2 respectively, then

(a) $W_1 > W_2$ (b) $W_1 < W_2$

(c) $W_1 = W_2$ (d) $W1 = W_2/4$

Sol. (a)

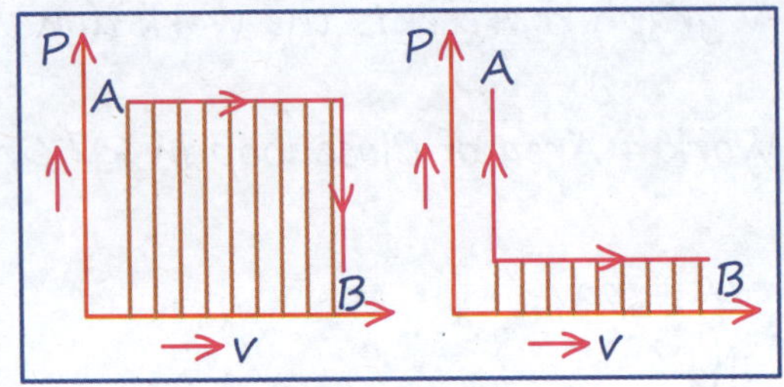

5 Internal Energy (U)

- Internal Energy (U) is the total energy contained within a system.
- It is path-independent, meaning it depends only on the state of the system, not how it got there.

$$U = KE + PE.$$

- It is **temperature-dependent**, meaning it changes with the temperature.

$$U \propto T$$

- Internal energy for one molecule

$$U = \frac{1}{2} K_B T = \frac{f}{2} K_B T \quad \Big\} \text{ Due to 1 molecule}$$

- Internal energy for N molecule

$$U = \frac{f}{2}\frac{R}{N_A} T \times N = \frac{nfRT}{2} \quad \Big\} \text{ N molecule}$$

Where,

f = degree of freedom.

N = No. of molecules.

n = No. of moles

$N_A = 6.023 \times 10^{23}\ mol^{-1}$

$K_B = \frac{R}{N_A}$ = Boltzmann Constant

$K_B = 1.38 \times 10^{-23}\ m^2 kg\ s^{-2} k^{-1}$ → JK^{-1}

- $dU = nCvdT$
- Where, n = no. of mole, $\gamma = \frac{C_p}{C_v}$

6 Degree of Freedom

- Degree of freedom is the number of independent ways in which molecules is free to move

1> Monoatomic Gas :- (Ne, He, Ar)

f = 3T + OR

2> Diatomic gas :- (N_2, O_2, H_2)

f = 3T + 2R

3> Triatomic Linear gas :- (CO_2)

f = 3T + 2R

4> Polyatomic gas/Triatomic Non-linear gas:- (SO_2, CH_4, NH_3)

f = 3T + 3R

When Considering Vibrational Motion:-

- No effect for Monoatomic gas
- Diatomic = f = 3T + 2R + 2V = 7
- Triatomic
  - → Linearf = 3T + 2R + 2V = 7
  - → Non-Linear f = 3T + 3R + 2V = 8

7 Heat Capacities

Specific heat Capacity (S) :-

$$S = \frac{dQ}{m\,dt}$$

Heat Capacity (C) :-

$$C = \frac{dQ}{dt}$$

Molar heat Capacity :-

$$C_m = \frac{dQ}{n\,dt}$$

Relation between Specific & Molar specific. Heat capacity :-

* $C_m = M_{wt} S$ $\quad C_m = \frac{W}{n} S$

✦ C_P & C_V denote the Specific Heat per unit mass of an ideal gas of Mwt "M" then :-

$C_P - C_V = R \rightarrow$ Molar Sp. heat

$S_P - S_V = \frac{R}{M}$ (Specific heat per unit mass)

Molar Heat Capacity:-

Const^n P

$$C_P = \frac{dQ}{dt\,n}$$

Const^n V

$$dQ = dU$$

$$C_V = \frac{fR}{2}$$

Note:-

$$\frac{C_P}{C_V} = \gamma$$

$$C_P - C_V = R$$

$$C_V = \frac{R}{\gamma - 1}$$

$$\gamma = 1 + \frac{2}{f}$$

MR* Table

Gas	DOF (f)	$C_V = \left(\frac{Rf}{2}\right)$	$C_P = (C_V + R)$	$\gamma = \left(\frac{C_P}{C_V}\right)$
Monoatomic	3	$\frac{3R}{2}$	$\frac{5R}{2}$	$\frac{5}{3} = 1.66$
Diatomic	5	$\frac{5R}{2}$	$\frac{7R}{2}$	$\frac{7}{5} = 1.44$
Triatomic Linear	5	$\frac{5R}{2}$	$\frac{7R}{2}$	$\frac{7}{5} = 1.4$
Triatomic Non-linear	6	3R	4R	$\frac{4}{3} = 1.33$
Diatomic at high Temperature	7	$\frac{7R}{2}$	$\frac{9R}{2}$	$\frac{9}{7}$

8 Gas Mixture

$$(C_V)_{mix} = \frac{n_1 C_{V1} + n_2 C_{V2}}{n_1 + n_2}$$

$$f_{mix} = \frac{n_1 f_1 + n_2 f_2}{n_1 + n_2}$$

$$(C_P)_{mix} = \frac{n_1 C_{P1} + n_2 C_{P2}}{n_1 + n_2}$$

$$\gamma_{mix} = \frac{(C_P)_{mix}}{(C_V)_{mix}}$$

$$T_{mix} = \frac{n_1 f_1 T_1 + n_2 f_2 T_2}{n_1 f_1 + n_2 f_2}$$

$$\frac{n_1 + n_2}{\gamma_{mix} - 1} = \frac{n_1}{\gamma_1 - 1} + \frac{n_2}{\gamma_2 - 1}$$

9 1st Law of Thermodynamics

$dQ = dU + dW$

Based on Energy Conservation

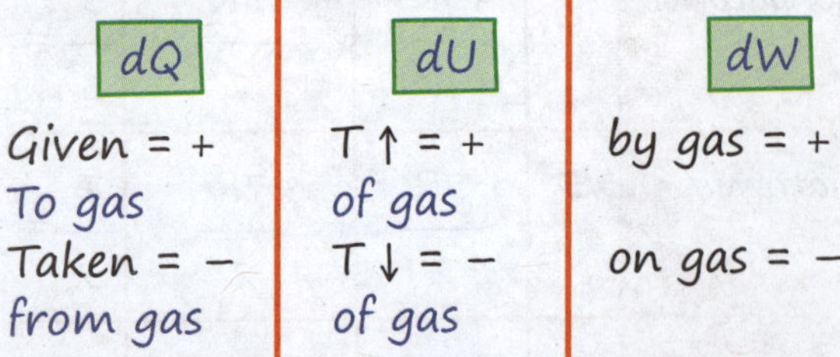

dQ	dU	dW
Given = + To gas Taken = − from gas	T ↑ = + of gas T ↓ = − of gas	by gas = + on gas = −

10 Gas laws and themodynamic process

Gay-Lusaac Law:-

At constant volume, the pressure on a given substance is directly proportional to absolute temperature, i.e,.

$V = Const^n \ (P \propto T)$

Isochoric Process !

- Work done = 0
- Heat added, $Q = \Delta U = nC_V \Delta T$
- For a monatomic ideal gas, the heat added can also be expressed as $Q = \dfrac{nfR\Delta T}{2}$

Graph:-

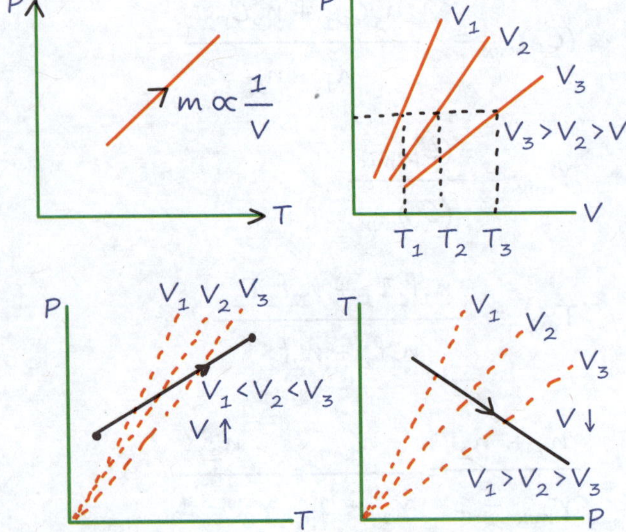

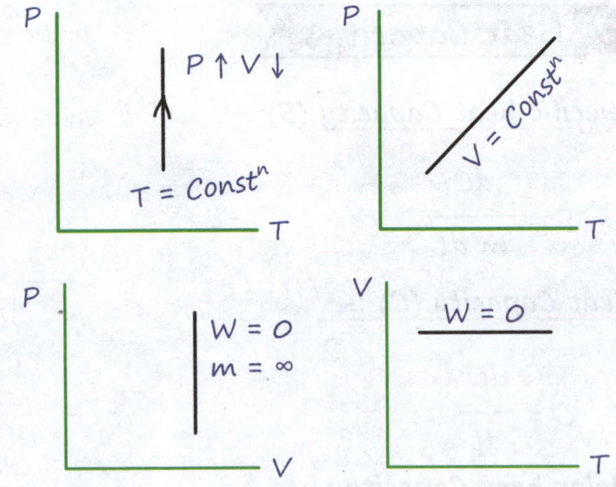

Charles Law:-

Charles' Law states that the volume of a given mass of gas is directly proportional to its absolute temperature, provided the pressure remains constant.

i.e., $P = Const^n$ $(V \propto T)$

Isobaric Process !

When the pressure is constant, the thermodynamic process is isobaric and the work done by the gas in this process is given by the

$W = P\Delta V = P(V_2 - V_1)$

Fraction of Heat goes in Work :-

In an isobaric process, not all the heat added to the system goes into doing work. Some of it goes into increasing the internal energy of the gas.

- **Fraction of Heat that goes into Work done**

$$\frac{W}{Q} = 1 - \frac{1}{\gamma}$$

- **Fraction of Heat that goes into Internal Energy:**

$$\frac{dU}{dQ} = \frac{1}{\gamma}$$

Graph:-

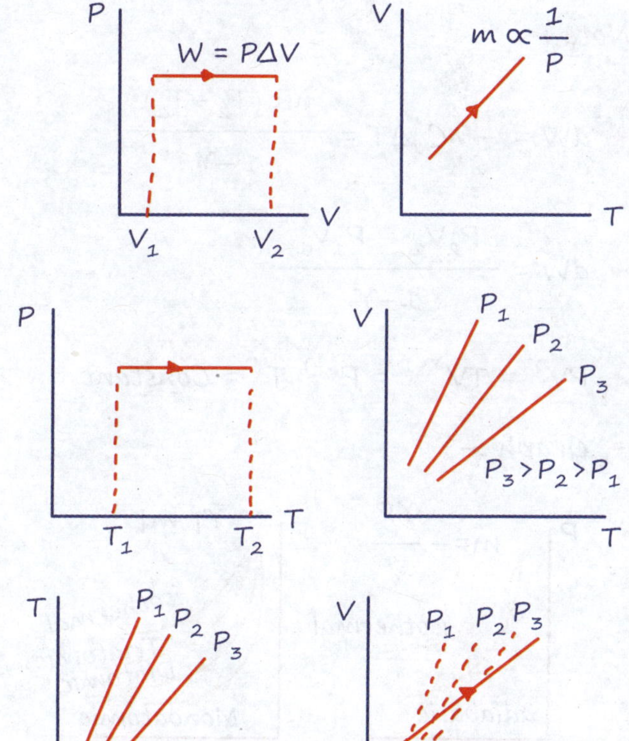

Boyle's Law:-

- Boyle's Law states that the pressure of a given mass of gas is inversely proportional to its volume, provided the temperature remains constant.

  - T = Const^n PV = Const^n

Isothermal Process !

- An isothermal process is a thermodynamic process in which the temperature of the system remains constant.

Characteristics:-

- This process is **reversible**, meaning it happens very slowly to ensure equilibrium at all stages.
- The heat added to the system is entirely used for doing work.
- $\Delta U = 0$ → Hum jitna Kam Karenge sab heat mein jayega!
- Molar heat capacity for an isothermal process is theoretically infinite because any amount of heat can be added without changing the temperature. $C_m = \frac{dQ}{n\Delta T} = \infty$

Work done in Isothermal process

$$* \text{Work} = nRT\left[\ln\left(\frac{V_2}{V_1}\right)\right] *$$

$$W = 2.303\, nRT\left[\log_{10}\left(\frac{V_2}{V_1}\right)\right] = dQ$$

$$W = 2.303\, nRT\left[\log_{10}\left(\frac{P_1}{P_2}\right)\right] = dQ$$

Graph:-

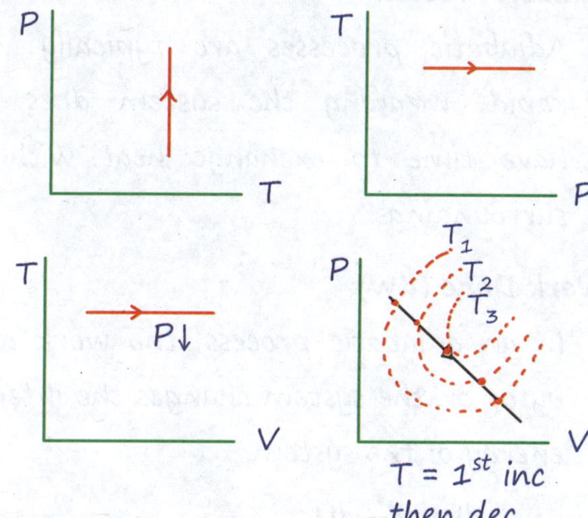

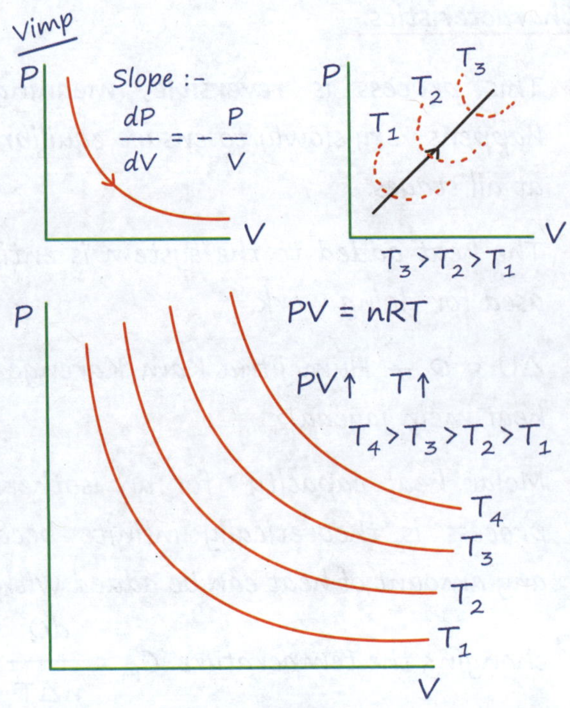

11 Adiabatic Process

- In an adiabatic process, the system is perfectly insulated, so no heat enters or leaves the system. I.e.,
  - $\Delta Q = 0$ $\quad PV_{\gamma} = $ Constn
  - γ = Adiabatic Coefficient = $\frac{C_P}{C_V}$

Process charecterstics:-

Sudden Process:

- Adiabatic processes are typically very rapid, meaning the system does not have time to exchange heat with its surroundings.

Work Done (dW):

- In an adiabatic process, the work done by or on the system changes the internal energy of the system.

$$dW = -dU$$

Example: Tyre burst

Expansion :- V↑ W = +ve $\quad$ U = −ve T↓

Compression :- V↓ W = −ve $\quad$ U = +ve T↑

- $C_m = \frac{dQ}{n\Delta T} = 0$
- Entropy, S = 0

Note :-

- $dW = -nC_V\Delta T = \frac{nR(T_2 - T_1)}{1-\gamma}$
- $dW = \frac{P_2V_2 - P_1V_1}{1-\gamma}$
- $PV^{\gamma} = TV^{\gamma-1} = P^{1-\gamma}T^{\gamma}$ = Constant
- Graph :-

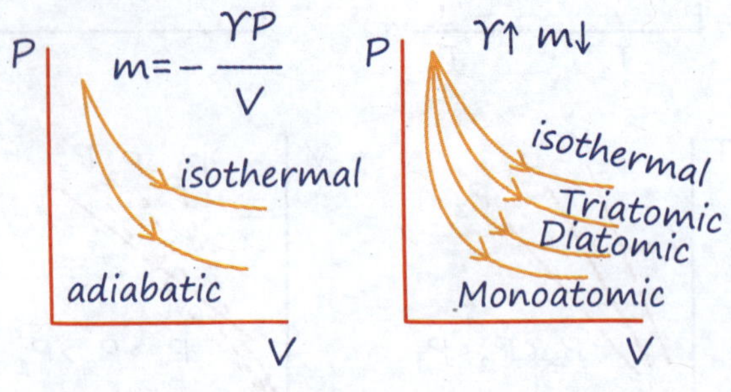

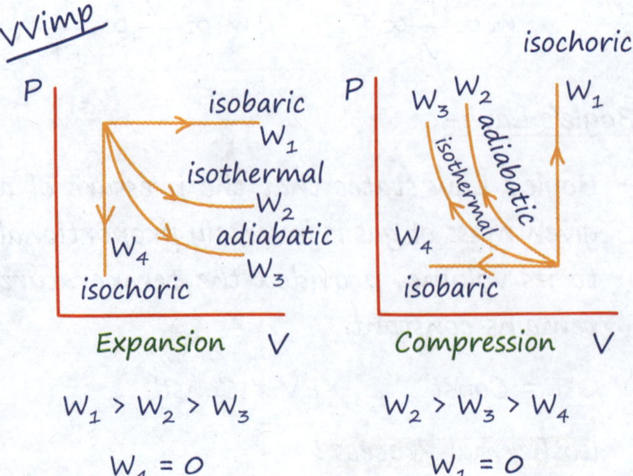

$W_1 > W_2 > W_3$ $\qquad$ $W_2 > W_3 > W_4$

$W_4 = 0$ $\qquad$ $W_1 = 0$

Note :-

Adiabatic Elasticity = Bulk Modulus = γP

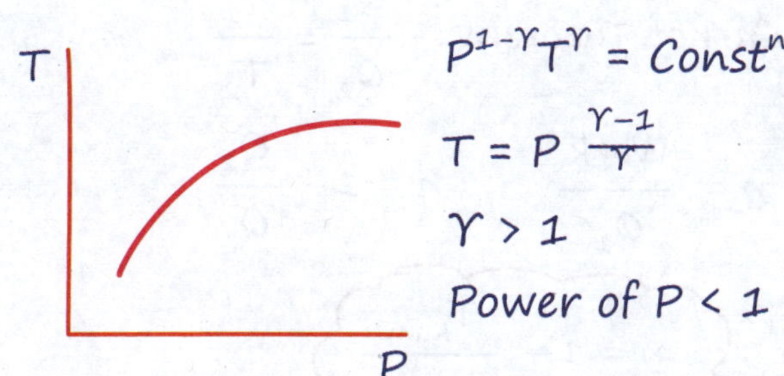

$P^{1-\gamma}T^{\gamma} = Const^n$

$T = P^{\frac{\gamma-1}{\gamma}}$

$\gamma > 1$

Power of P < 1

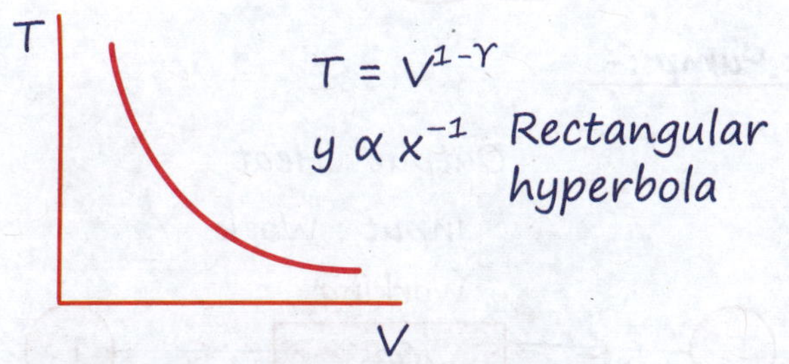

$T = V^{1-\gamma}$

$y \propto x^{-1}$ Rectangular hyperbola

12 Polytropic Process

$PV^x = TV^{x-1} = P^{1-x}T^x = const^n$

$$W = \frac{nR(T_2 - T_1)}{1-x} \qquad C_m = C_V + \frac{R}{1-x}$$

$$C_m = \frac{R}{\gamma - 1} + \frac{R}{1-x}$$

Graph Conversions :-

With the help of P-V diagram draw P-T and T-V diagram

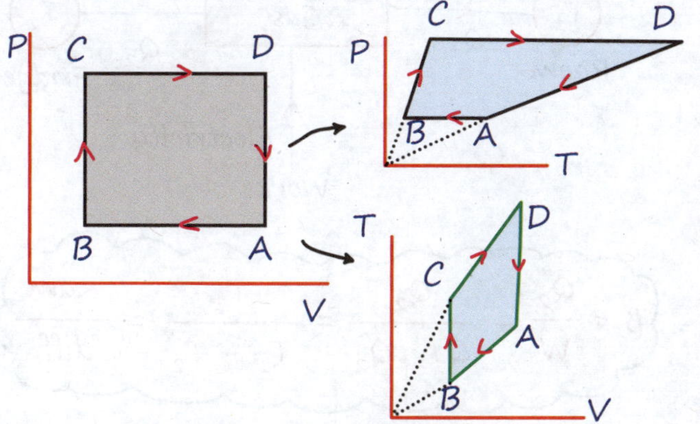

$A \to B = P = Const^n$ V↓ T↓

$B \to C = V = Const^n$ P↑ T↑

$C \to D = P = Const^n$ V↑ T↑

$D \to A = V = Const^n$ P↓ T↓

With the help of P-V diagram draw P-T

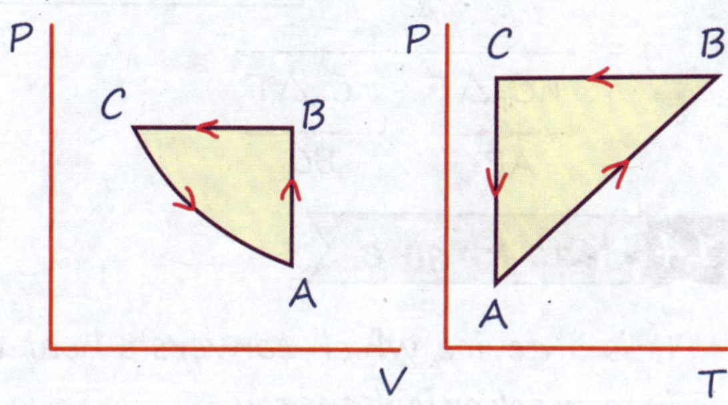

$A \to B$	P↑ T↑	$V = Const^n$
$B \to C$	V↓ T↓	$P = Const^n$
$C \to A$	P↓ V↑	$T = Const^n$

With the help of P-V diagram draw P-T

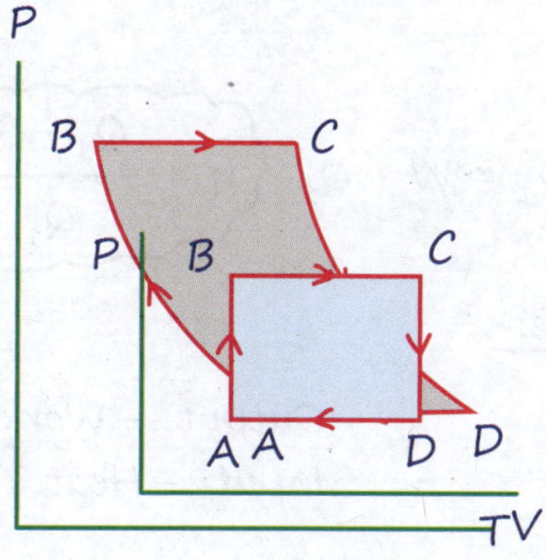

$A \to B$	$T = Const^n$ P↑ V↓
$B \to C$	$P = Const^n$ T↑ V↑
$C \to D$	$T = Const^n$ P↓ V↑
$D \to A$	$P = Const^n$ T↓ V↓

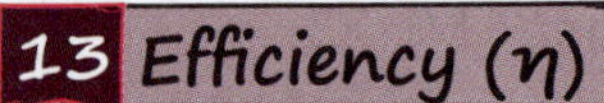

13 Efficiency (η)

$$\eta = \frac{\text{Out put}}{\text{Input}}$$

$$\eta = \frac{WD}{Q_{given}}$$

$$\eta = \frac{A_{req}}{Q_1 + Q_2}$$

P, V, A, B, C, D, $Q_1 = +ve$, $Q_2 = +ve$, $Q_3 = -ve$, $Q_4 = -ve$

$$\eta = \frac{A_{req}}{\underbrace{nC_p\Delta T}_{AB} + \underbrace{nC_V\Delta T}_{BC}}$$

14 Heat Engine

✦ It is a device which converts heat engine into mechanical energy.

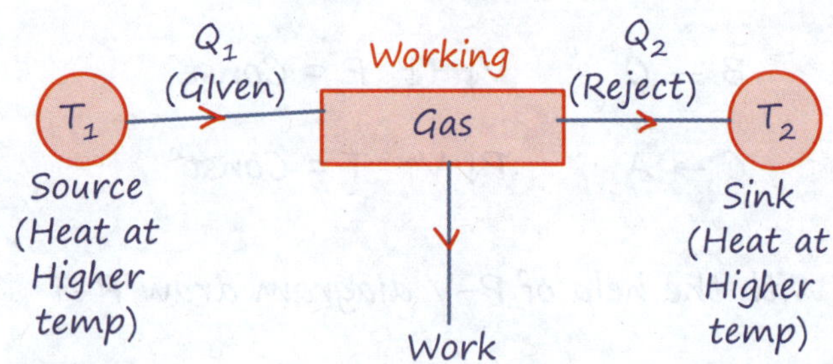

$$Q_1 = W + Q_2 \qquad \eta = \frac{Q_1 - Q_2}{Q_1}$$

Carnot Engine:-

Output :- Work

Input :- Heat

P, V, A, B, C, D, Q_1, Q_2

AB = Isothermal Exp.

BC = Adiabatic Exp.

CD = Isothermal = Comp

DA = Adiabatic = Comp

Carnot Theorem :- $\frac{Q_1}{Q_2} = \frac{T_1}{T_2}$

$$\eta = \frac{Work}{Q_1} \qquad \eta = 1 - \frac{Q_2}{Q_1}$$

$$\therefore \eta = 1 - \frac{T_{Kam}}{T_{Jyada}}$$

Heat Pump :-

Output : Heat

Input : Work

Working Gas, T_1 Source, Q_1, Q_2, T_2 Sink, Work

$Q_1 = W + Q_2$

Pump Engine से उलटा Kaam Krta hai !

$$\eta = \frac{Q_1}{W} = \frac{1}{\eta_{engine}}$$

Refrigerator :-

Output :- Heat

Input :- Work

• Same as Pump.

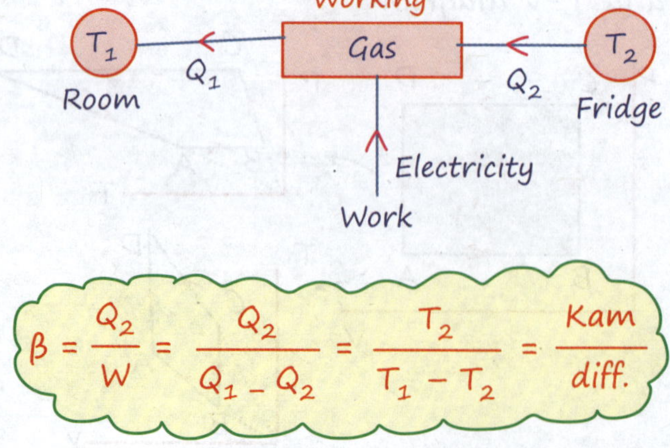

$$\beta = \frac{Q_2}{W} = \frac{Q_2}{Q_1 - Q_2} = \frac{T_2}{T_1 - T_2} = \frac{Kam}{diff.}$$

$$\beta = \frac{1-\eta}{\eta} = \frac{1}{\eta} - 1$$ Vimp

Note:-

Heat engines and refrigerator are deleted from the syllabus

Q.3. The efficiency of a Carnot heat engine is $\frac{1}{3}$ Find the coefficient of performance of Carnot refrigerator when both heat engine and refrigerator are working between similar source and sink.

Sol. Efficiency of heat engine is, $\eta = \frac{1}{3}$

The relation between β & η when same Carnot engine is used,

$$\Rightarrow \beta = \frac{1-\eta}{\eta}$$

$$\therefore \beta = \frac{1-\frac{1}{3}}{\frac{1}{3}} = 2$$

Q.4. A carnot engine works as a refrigerator in between 250 K and 300 K. If it acquires 750 calories from heat source at low temperature, then what is the heat generated at higher temperature. (in calories)?

Sol. $\eta = \frac{T_1-T_2}{T} = \frac{Q_1-Q_2}{Q_1}$

$$= \frac{300-250}{300} = \frac{Q-750}{d}$$

$$\Rightarrow \frac{50}{300} = 1 - \frac{750}{Q}$$

$$\Rightarrow \frac{750}{Q} = 1 - \frac{1}{6} = \frac{5}{6}$$

$$\therefore Q = \frac{750 \times 6}{5}$$

$$= 900$$

Q.5. A system is taken from state A to state B along two different paths 1 and 2. If the heat absorbed and work done by the system along these two paths are Q_1, Q_2 and W_1, W_2 respectively, then

(a) $Q_1 = Q_2$

(b) $W_1 = W_2$

(c) $Q_1 - W_1 = Q_2 - W_2$

(d) $Q_1 + W_1 = Q_2 + W_2$

Sol. Internal energy does not depends on path. Heat & work depends (c) $Q_1 - W_1 = Q_2 - W_2$ is correct.

Q.6. In a given process, $dW = 0$, $dQ < 0$, then for the gas:

(a) Temperature increases

(b) Volume decreases

(c) Pressure decreases

(d) Pressure increases

Sol. $dW = 0 \quad dQ < 0$

$dU + dW = dQ$

$dU = dQ$

$dU < 0$

Temper decrese

Volume constant

$P \propto T$

Pressure decreases

Q.7. If 32 gm of O_2 at 27°C is mixed with 64 gm of O_2 at 327°C in an adiabatic vessel, then the final temperature of the mixture will be:

Sol. For adiabatic process, gain of heat loss of heat =

or, $m_1S(T-T_1) = m_2S(T_2-T)$

or, $m_1(T-T_1) = m_2(T_2-T)$

or, $32 \times (T-27) = 64 \times (327-T)$

or, $T-27=2(327-T)$

or, $T-27\ 654-2T$

or, $3T=681$

or, $T=227°C$

Q.8. If W_1 is the work done in compressing an ideal gas from a given initial state through a certain volume isothermally and W_2 is the work done in compressing the same gas from the same initial state through the same volume adiabatically, then:

(a) $W_1 = W_2$ (b) $W_1 < W_2$

(c) $W_1 > W_2$ (d) $W_1 = 2W_2$

Sol. (b)

Q.9. During an experiment and ideal gas is found to obey an additional law VP^2 = constant. The gas is initially at a temperature T and volume V. When it expand to a volume 2V, the temperature becomes.

Sol. Ideal gas: $V P^2$ = constant

Again PV = nRT from equation of state,

Hence, $VP \times P$ = constant i.e. $nRT \times P$ = constant

Again, $P = \frac{nRT}{V}$

$\therefore \frac{(nRT)^2}{V}$= constant

$\frac{T^2}{V}$ = constant

Thus volume V when expanded to 2V, temperature T_2

$T_2 = \sqrt{\frac{2V}{V}} = 2\sqrt{T_1}$

Q.10. A carnot engine having an efficiency of $\frac{1}{10}$ th of heat engine, is used as a refrigerator. If the work done on the system is 10 J, the amount of energy absorbed from the reservoir at lower temperature is:

Sol. Coefficient of performance of a refrigerator

$\beta = \frac{1-\eta}{\eta}$

$\beta = \frac{1-\frac{1}{10}}{1/10} = 9$

Also $\beta = \frac{Q_L}{W}$ (where W is the work done)

or

$Q_L = \beta \times W = 9 \times 10 = 90J$

Q.11. During an adiabatic process, the pressure of a gas is found to be proportional to the cube of its temperature. The ratio of C_P/C_V for the gas is equal to:

Sol. $P \propto T^3$,

$PV = nRT$

$P \propto T^3$

$P \propto (PV)^3$

$P^2V^3 = \text{constant}$

$PV^{\frac{3}{2}} = \text{constant}$

$\gamma = \frac{3}{2}$

Q.12. A monoatomic gas at pressure p_1 and V_1 is compressed adiabatically to $\frac{1}{8}$th its original volume. What is the final pressure of the gas?

Sol. Correct option is (a)

It is given that $\frac{C_P}{C_V} = \gamma = \frac{5}{3}$

For an adiabatic process,

$P_1V_1^{\gamma} = P_2V_2^{\gamma}$

$$\Rightarrow \frac{P_2}{P_1} = \left(\frac{V_1}{V_2}\right)^{5/3}$$

$$= \left(\frac{8}{1}\right)^{5/3} = 32$$

15 Kinetic Theory of Ideal Gas

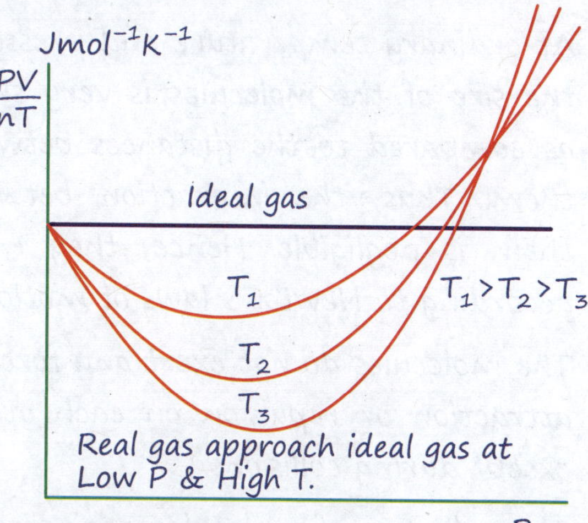

Ideal Gas Behavior:

- For an ideal gas, the relationship between pressure (P), volume (V), and temperature (T) is given by the ideal gas law: PV = nRT
- In the graph, this relationship is represented by a horizontal line where $\frac{PV}{nT}$ equals the ideal gas constant.

Real Gas Behavior:

- Real gases deviate from ideal gas behavior, especially at high pressures and low temperatures. This deviation is depicted by the red curves in the graph.
- The red curves show how real gases approach the behavior of an ideal gas under different conditions.
- Kinetic theory of gases is based on the molecules picture of matter. According to which :-

(1) A given amount of gas is a mixture of very large number of identical molecules of the order of Avogadro's number.

(2) The molecules are moving randomly in all directions.

(3) At ordinary temperature and pressure, the size of the molecules is very small as compared to the distances between them. Thus, the interaction between them is negligible. Hence, they move according to Newton's laws of motion.

(4) The molecules do not exert any force of attraction or repulsion on each other, except during collisions.

(5) The collisions of molecules against each other or with the walls of the container are perfectly elastic. Such that the momentum and the kinetic energy of the molecules are conserved during collisions, though their velocities change.

Q.13. Select the appropriate property of an ideal gas

(a) Its molecules are infinitesimally small/

(b) There are no forces of interaction between i molecules/

(c) It strictly obeys the ideal gas laws/

(d) All of these

Sol. (d)

Q.14. The figure shows the plot of $\frac{PV}{nT}$ vs P, for oxygen gas at two different temperatures then

(a) $T_1 > T_2$

(b) $T_2 > T_1$

(c) $T_1 = T_2$

(d) May be (a) and (b)

Sol. (a)

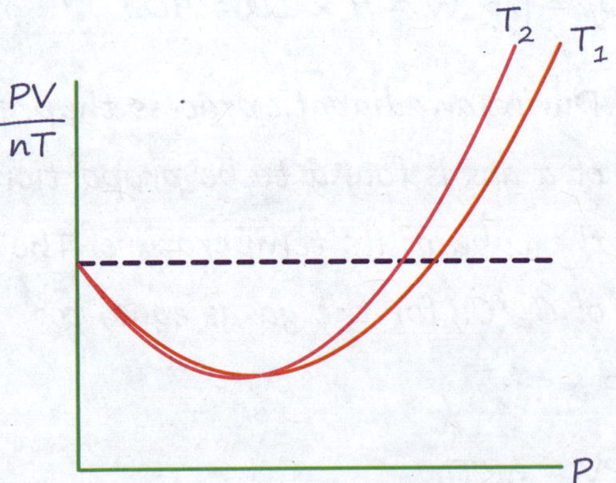

Mixing of gases

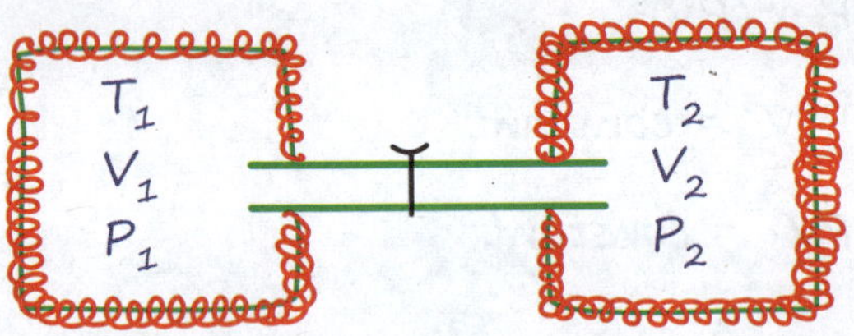

The valve joining the two vessels is opened :-

$T_{mix} = ?$

$$T_{mix} = \frac{n_1T_1 + n_2T_2}{n_1 + n_2} = \frac{\frac{P_1V_1}{R} + \frac{P_2V_2}{R}}{\frac{P_1V_1}{RT_1} + \frac{P_2V_2}{RT_2}}$$

$$T_{mix} = \frac{(P_1V_1 + P_2V_2)\,T_1T_2}{P_1V_1T_2 + P_2V_2T_1}$$

Pressure of ideal gas :-

$$P = \frac{1}{3}nmV_{rms}^2 \qquad n = \text{no. density} = \frac{N}{V}$$

$$V_{rms} = \sqrt{\frac{3P}{nm}} \qquad m = \text{mass of each molecule}$$

Kinetic interpretation of temperature :-

$$KE = \frac{3RT}{2}$$

Energy density

$$\frac{KE}{V} = \frac{3P}{2}$$

Rotational KE :-

$$KE = \frac{fRT}{2N_A} \qquad KE = \frac{fKT}{2}$$

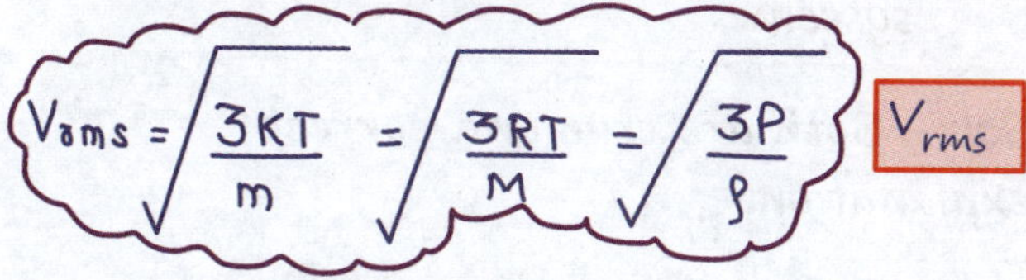

V_{rms}

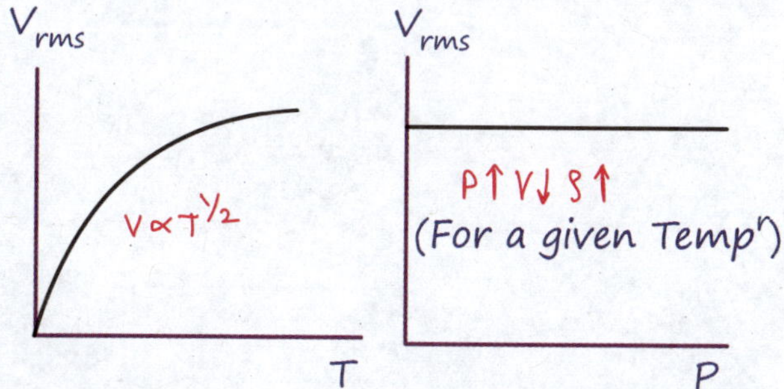

Averge speed of the molecule of gas:-

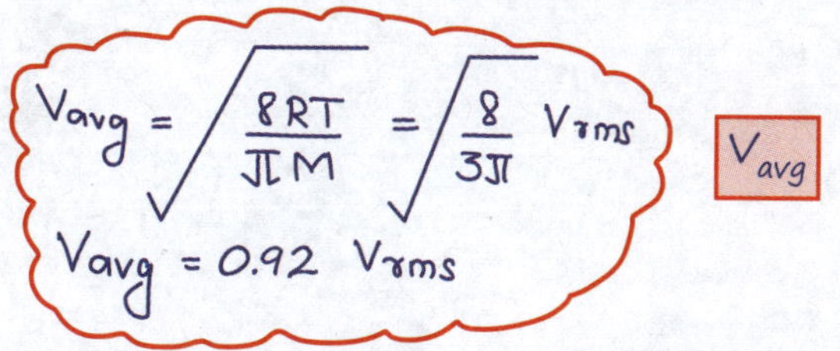

V_{avg}

Most proable speed of the molecule of gas:-

$$V_{MP} = \sqrt{\frac{2RT}{M}}$$

V_{MP}

Q.15. Four molecules of a gas have speeds 1, 2, 3 and 4 km/s. The value of the r.m.s. speed of the gas molecules is

(a) $\frac{1}{2}\sqrt{15}$ km/s (b) $\frac{1}{2}\sqrt{10}$ km/s

(c) 2.5 km/s (d) $\sqrt{\frac{15}{2}}$ km/s

Sol.

$$l_{r.m.s} = \sqrt{\frac{1^2 + 2^2 + 3^3 + 4^2}{4}} = \sqrt{\frac{1 + 4 + 9 + 16}{4}}$$

$$= \sqrt{\frac{30}{4}} = \sqrt{\frac{15}{2}}$$

Maxwell Distribution Curve :-

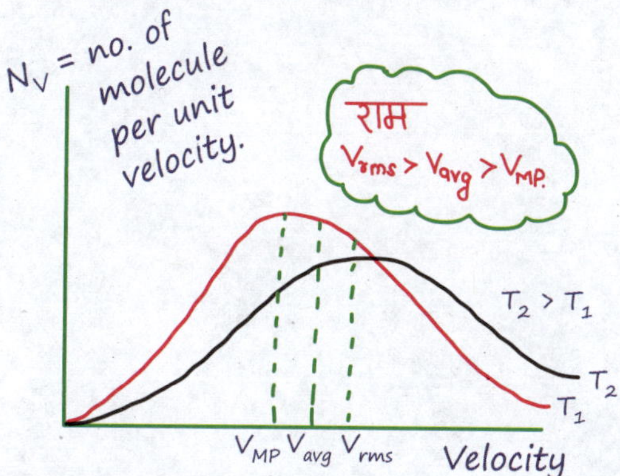

Area = $\int N_V\, dv$ = no. of molecules.

→ Temp^r independent.

Mean free path :-

$$\lambda = \frac{1}{\sqrt{2}\pi n d^2} \qquad n = \text{number density} = \frac{N}{V}$$

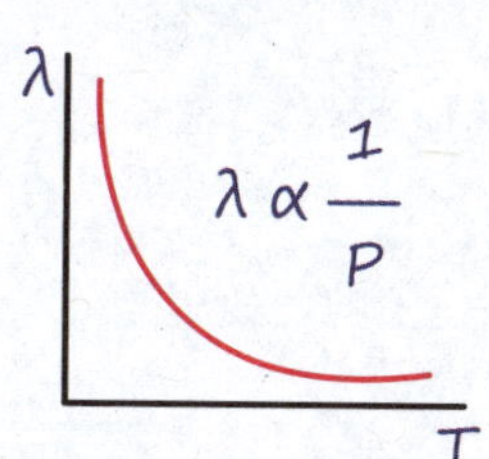

$$\lambda = \frac{RT}{\sqrt{2}\pi N_A P}$$ d = diameter = 2r

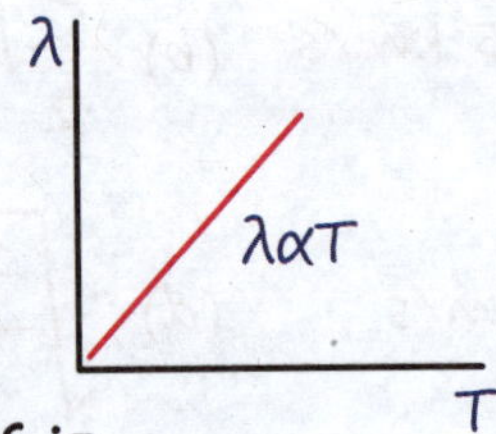

Density of gas :-

$$d = \frac{PM}{RT} = \frac{Pm}{K_B T}$$

Assertion and reason based questions:-

A → A liquid is filled in container which is moving with high speed does not have higher temperature.

R → Temp^r of liquid related to internal energy, not to K.E. of liquid.

Sol.:- Both are true and correct explanation.

A → A mixture of petrol and air when ignited is not in equlibrium state.

R → Its temperature and pressure not uniform.

Sol.:- Both are correct and correct explanation.

A → ΔQ is extensive.

R → It is propotional to total mass of system.

Sol.:- Both are true and correct explanation.

19 Simple Harmonic Motion

1 Periodic Motion

(A motion which Repeat itself after a regular interval of time)

(repeat about mean) (To & Fro or back and forth) → **Oscillation**

(repeat not about mean) → **Not Oscillation**

Oscillation:
- large Amplitude → **Not S.H.M**
- • Small amplitude • Energy conserved • $F \propto -x^n$ → **Simple harmonic Motion**

✦ Vibration → Oscillation with high frequency.

✦ Periodic but not oscillation
  → all uniform circular motion
  → earth around sun

✦ All SHM is oscillatory and periodic

✦ All oscillatory is periodic but need not S.H.M

✦ All periodic need not to be oscillatory & S.H.M.

2 S.H.M

(Amplitude is small)

✦ $\vec{F} \propto -\vec{x}$

✦ $\vec{a} = -\omega^2 x$

✦ $\frac{d^2\vec{x}}{dt^2} + \omega^2\vec{x} = 0$

$\vec{a} \propto -\vec{x}$

F vs x: $K = \tan\theta$, $F = -Kx$

a vs x: $a = -\omega^2 x$, $\tan\theta = \omega^2$

Equation of SHM:-

$$x = A\sin(\omega t + \phi)$$

$$v = \frac{dx}{dt} = A\omega\cos(\omega t + \phi)$$

$$a = \frac{dv}{dt} = -A\omega^2\sin(\omega t + \phi)$$

Where,

A = Amplitude

ω = Angular Frequency

$= \frac{2\pi}{T} = 2\pi f$

ϕ = initial phase, x = displacement from mean position.

Note:-

✦ $\phi = 0$ for mean से start

✦ $\phi = \pi/2$ for extreme से start

✦ $\phi = \pi/6$ for half of extreme से start

✦ $\phi = \pi/4$ for $x = \frac{A}{\sqrt{2}}$ से start

✦ $a \propto -x^n$ → n = even = Translational
  → n = odd but not 1. → "Oscillation but not SHM"

$a \propto -x^1 \Rightarrow$ SHM!

Fire Concept MR*

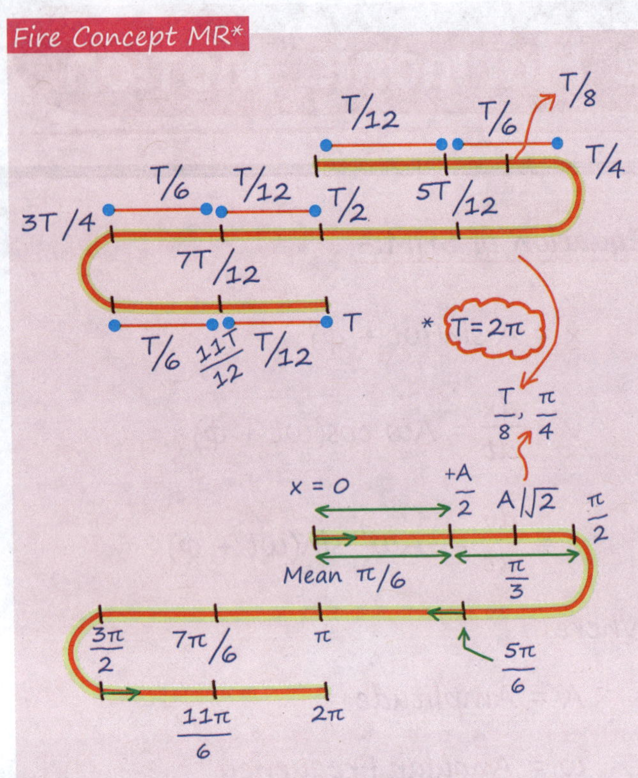

3 Motion From Mean position

Position,

+ $x = A\sin\omega t$,

velocity,

+ $v = A\omega\cos(\omega t)$
+ $v = A\omega\sqrt{1-\sin^2\omega t}$
+ $v = \omega\sqrt{A^2 - x^2}$ $\qquad \omega = \sqrt{\frac{k}{m}}$

Acceleration:-

$a = -A\omega^2\sin\omega t, \quad a = -\omega^2 x$

Note:-

Mean	Extreme
$x = 0$	$x = A$
$v_{max} = A\omega$	$v = 0$
$a = 0$	$a_{max} = -\omega^2 A$

MR*

Mai tere piche, wo mere piche hai re kismat mai tum se na mil pau, wo mere se na mil pay

Mai → Velocity $\qquad$ tum → acc^n

Wo → Position ☺

4 Graphs

x – t:-

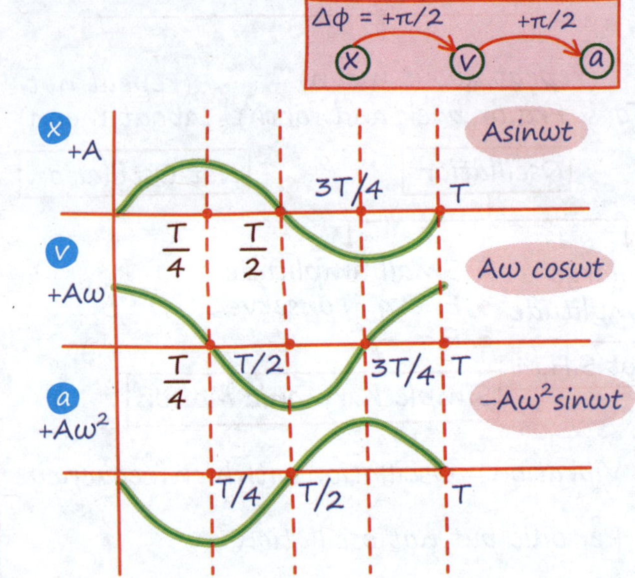

$a = -\omega^2 x$:-

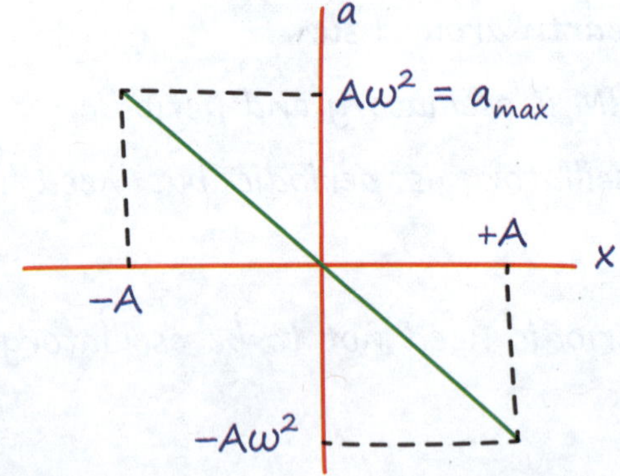

Note :-

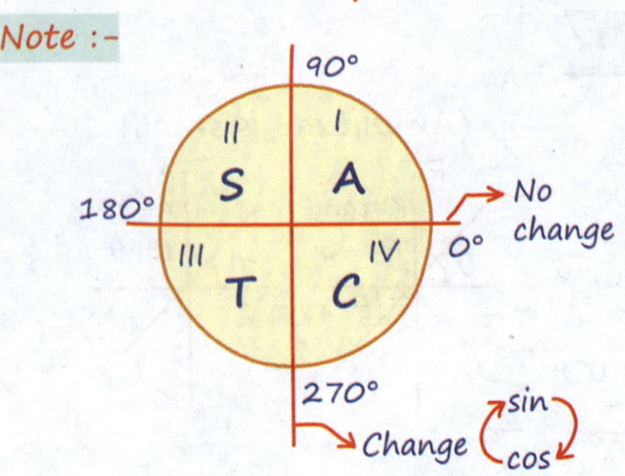

v – x :-

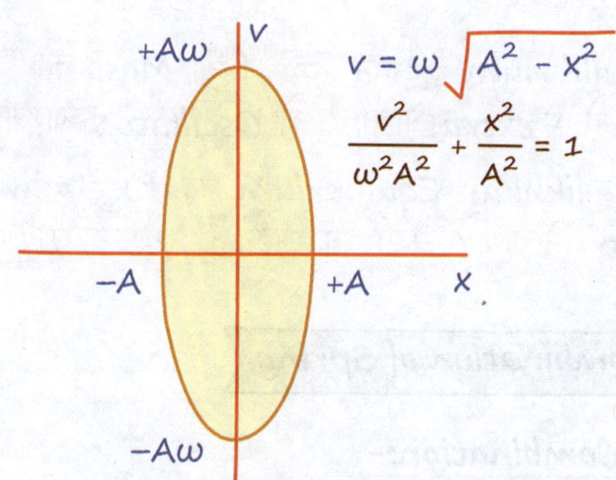

a – v :-

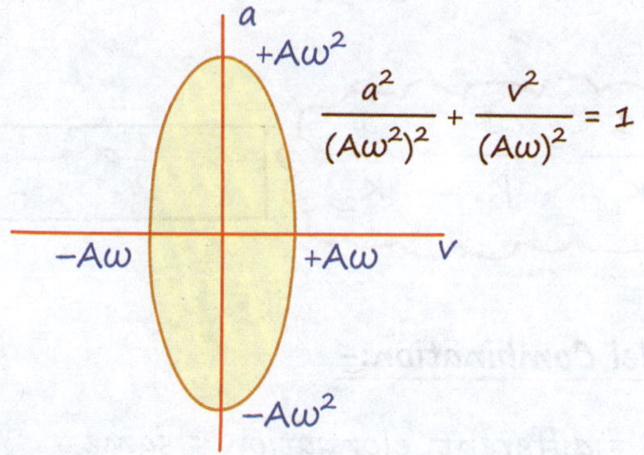

Important

Time period (TP) of a particle executing SHM along straight line with its velocity at position x_1 and x_2 from mean are V_1 and V_2 & Then TP:-

$$T = 2\pi\sqrt{\frac{x_2^2 - x_1^2}{v_1^2 - v_2^2}}$$

5 Energies

$$\text{Kinetic energy, } = \frac{1}{2}mv^2$$

$$\text{we know } m = \frac{K}{\omega^2}$$

Kinetic energy:-

$$KE = \frac{1}{2}K(A^2 - x^2) = \frac{1}{2}KA^2\cos^2(\omega t + \phi)$$

Potential energy:-

$$P.E = \frac{Kx^2}{2} + U_0 = \frac{1}{2}KA^2\sin^2\omega t + U_0$$

Total energy:-

$$TE = \frac{1}{2}KA^2 + U_0 = \text{constant through out the motion}$$

✦ $K = \text{Force Const}^n = m\omega^2$

Energy graphs:-

The graph visually represents how kinetic and potential energies vary with displacement in simple harmonic motion. The total energy of the system remains constant, and energy oscillates between kinetic and potential forms as the particle moves between the mean and extreme positions.

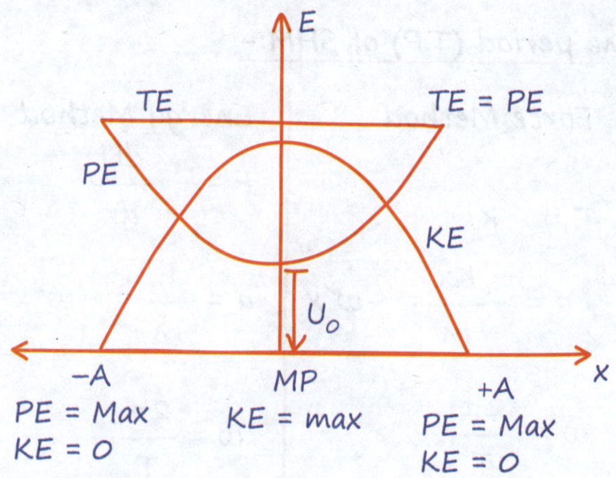

✦ At the mean position (MP):

- KE is at its maximum.
- PE is zero.
- Total Energy (TE) = Maximum Kinetic Energy.

- At the extreme positions (±A):
  - PE is at its maximum.
  - KE is zero.
  - Total Energy (TE) = Maximum Potential Energy.

MR*

Kitne time lagega Repeat Krne mein → Time Period

Kitne bar given time में अपने Pattern को Repeat करेगा. → Frequency

Physical Quantity	Time Period	Frequency
Position	T	f
Velocity	T	f
Speed	T/2	2f
Acceleration	T	f
KE	T/2	2f
PE	T/2	2f
KE-PE	T/2	2f
|KE-PE|	T/4	4f
ME = KE + PF	∞	0

Time period (T.P) of SHM:-

Force Method	Energy Method
$F = -Kx$	$F = -\frac{dU}{dr}$
$a = -\frac{Kx}{m} = -\omega^2 x$	$a = \frac{F}{m} = -\omega^2 x$
$\omega = \frac{2\pi}{T}$	$\omega = \frac{2\pi}{T}$

TP of Spring Mass System:-

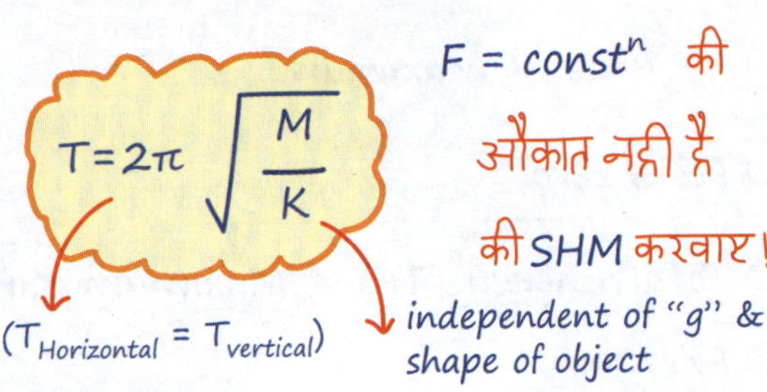

F = constⁿ की औकात नही है की SHM करवाए!

MR*

- Equilibrium से जब तूमने खिचा Mass को तो वो जो F extra आया है वो Oscillate करवाएगा!
- Eq^n likhkar Equilibrium का $Kx_0 = mg$ Put किया तो कटा था कटा है और कटेगा!

6 Combination of Spring

Series Combination:-

F = same, elongation different

$$\frac{1}{K_{eq}} = \frac{1}{K_1} + \frac{1}{K_2}$$

$$T^2 = T_1^2 + T_2^2$$

Parallel Combination:-

F = different, elongation = same

$$K_{eq} = K_1 + K_2$$

$$\frac{1}{T^2} = \frac{1}{T_1^2} + \frac{1}{T_2^2}$$

Cutting of Spring:-

Spring constant $K \propto \frac{1}{\text{Length of spring}}$

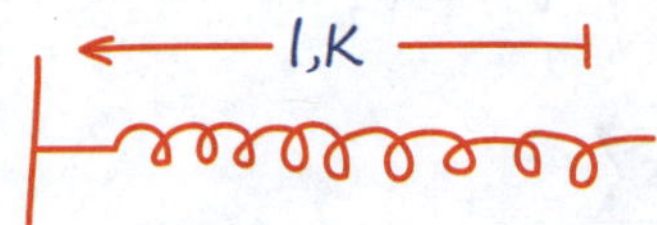

cut in two equal part then

Q.1. Cut into three part of ratio 1:2:3 then ratio of spring constant.

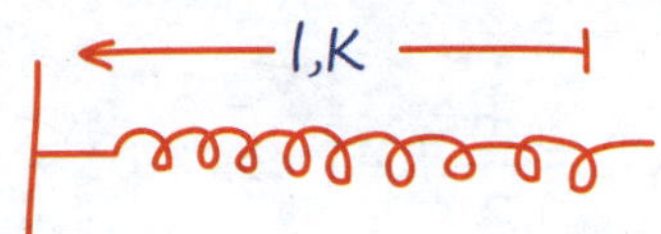

$l_1 : l_2 : l_3 = 1 : 2 : 3 = x : 2x : 3x$

$$K_1 : K_2 : K_3 = \frac{1}{x} : \frac{1}{2x} : \frac{1}{3x}$$

$K_1 : K_2 : K_3 = 6 : 3 : 2$

$K_1 = 6K \quad K_2 = 3K \quad K_3 = 2K$

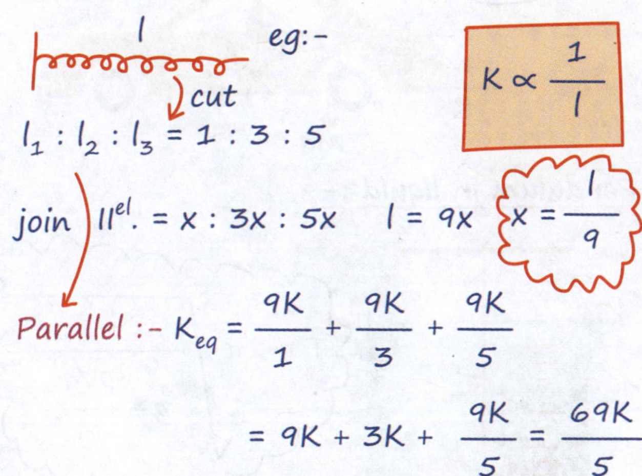

$K \propto \frac{1}{l}$

$l_1 : l_2 : l_3 = 1 : 3 : 5$

join $\|^{el}$. $= x : 3x : 5x \qquad l = 9x \qquad x = \frac{l}{9}$

Parallel :- $K_{eq} = \frac{9K}{1} + \frac{9K}{3} + \frac{9K}{5}$

$$= 9K + 3K + \frac{9K}{5} = \frac{69K}{5}$$

Reduced mass Concept :-

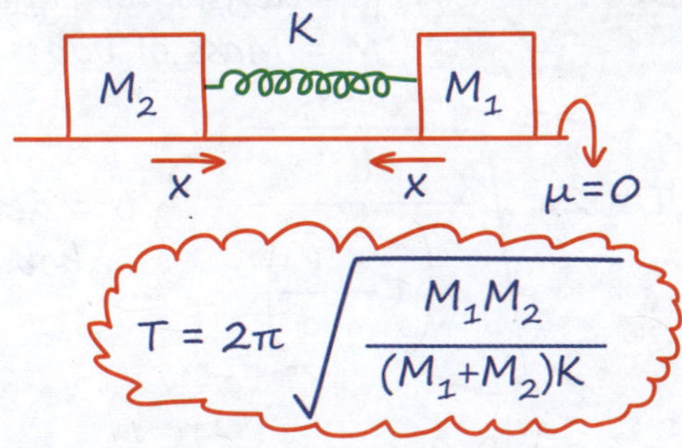

$$T = 2\pi\sqrt{\frac{M_1M_2}{(M_1+M_2)K}}$$

Rotation + Translation wale Que. :-

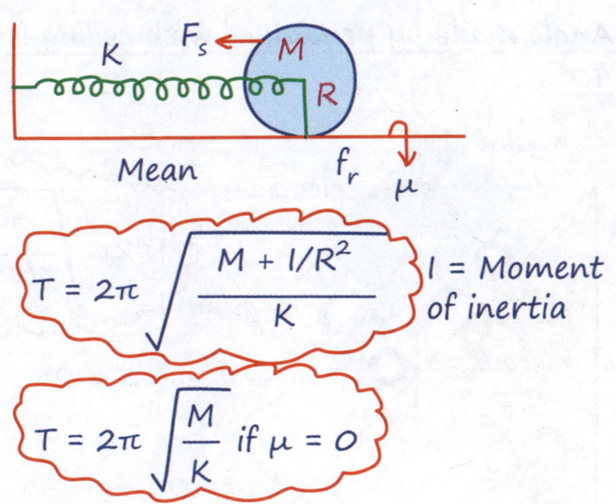

$$T = 2\pi\sqrt{\frac{M + I/R^2}{K}}$$ I = Moment of inertia

$$T = 2\pi\sqrt{\frac{M}{K}} \text{ if } \mu = 0$$

7 Constrain Motion

1>

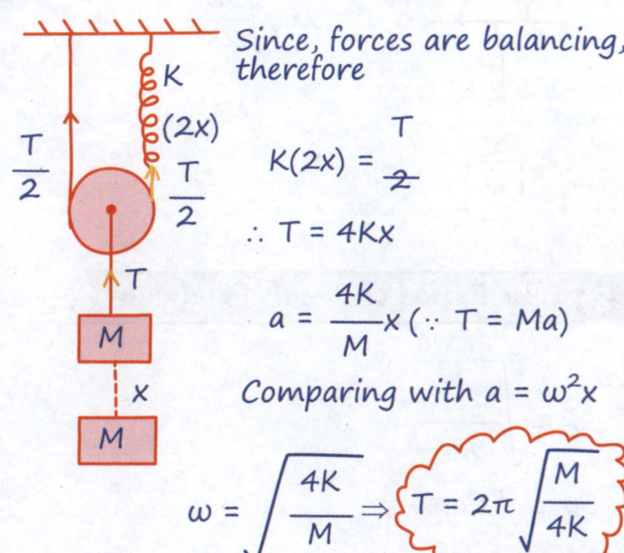

Since, forces are balancing, therefore

$K(2x) = \frac{T}{2}$

$\therefore T = 4Kx$

$a = \frac{4K}{M}x \; (\because T = Ma)$

Comparing with $a = \omega^2 x$

$$\omega = \sqrt{\frac{4K}{M}} \Rightarrow T = 2\pi\sqrt{\frac{M}{4K}}$$

2>

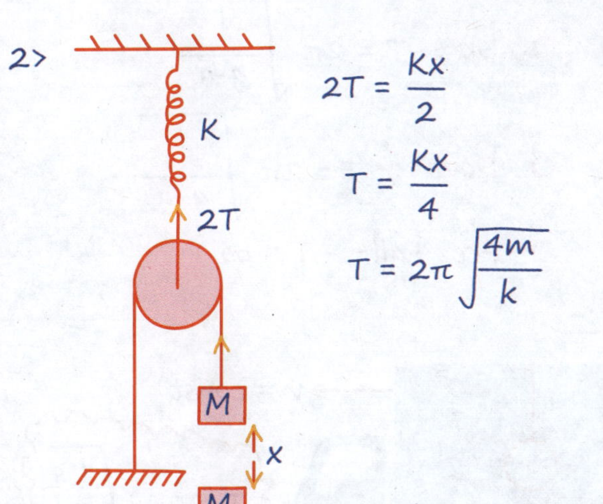

$2T = \frac{Kx}{2}$

$T = \frac{Kx}{4}$

$$T = 2\pi\sqrt{\frac{4m}{k}}$$

3>

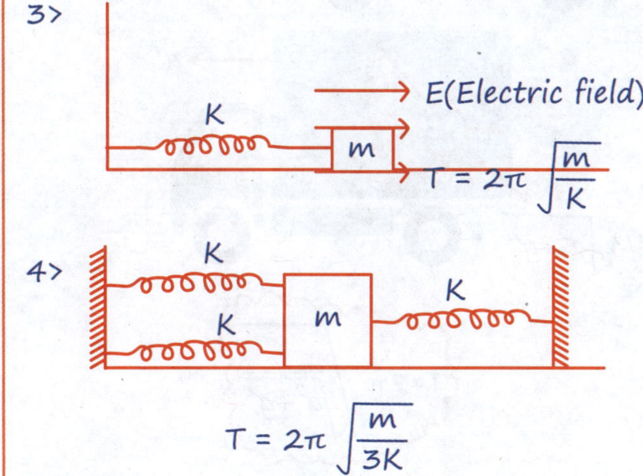

$$T = 2\pi\sqrt{\frac{m}{K}}$$

4>

$$T = 2\pi\sqrt{\frac{m}{3K}}$$

5>

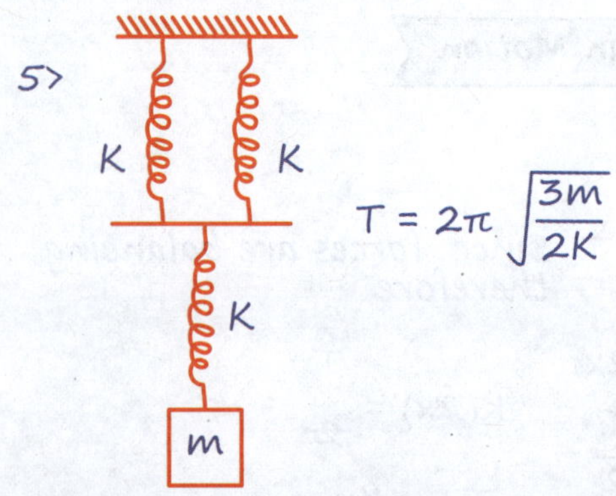

$$T = 2\pi\sqrt{\frac{3m}{2K}}$$

8 Time Period of Simple Pendulum

$$T = 2\pi\sqrt{\frac{l}{g_{eff.}}}$$

✦ Special Cases :

Lift :-

- Up :- $T = 2\pi\sqrt{\frac{l}{g+a}}$
- Down :- $T = 2\pi\sqrt{\frac{l}{g-a}}$
- Free Fall :- $T = \infty$

Car :-

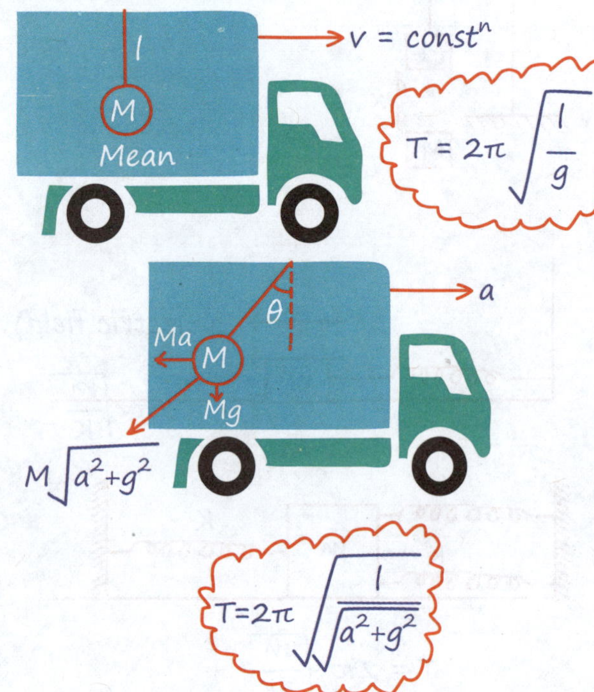

Car at Circular Track :-

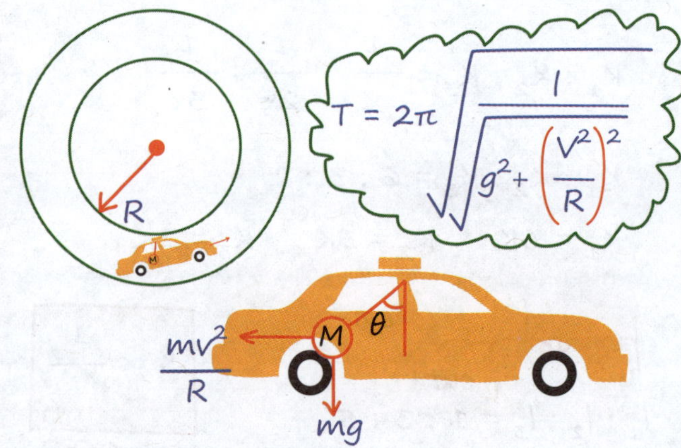

Pendulum in liquid :-

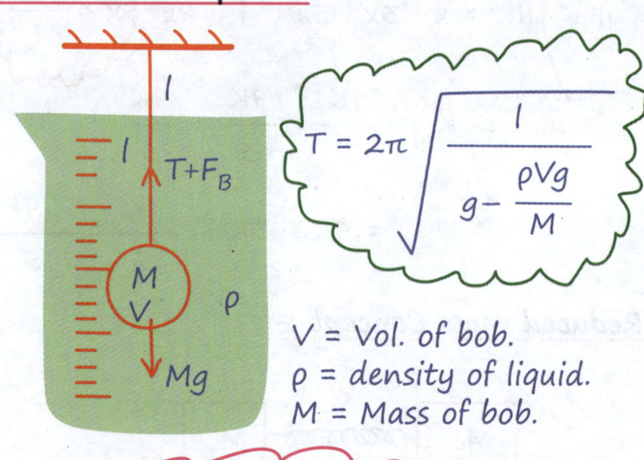

V = Vol. of bob.
ρ = density of liquid.
M = Mass of bob.

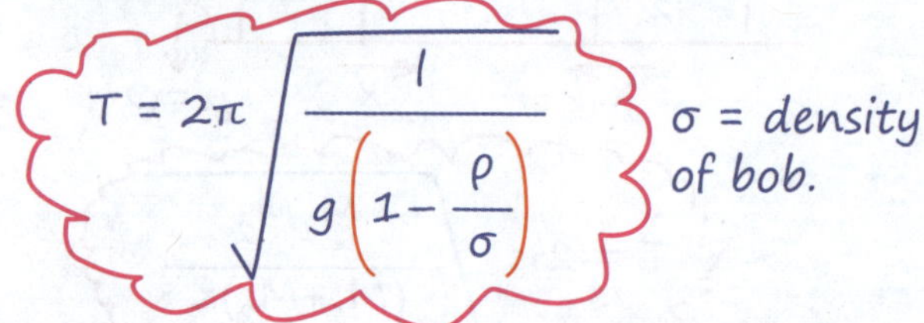

σ = density of bob.

Pendulum को किसी Fluid मे डाला तो उसका T.P. बढेगा!

Angle made by pendulum with ceiling is 90°.

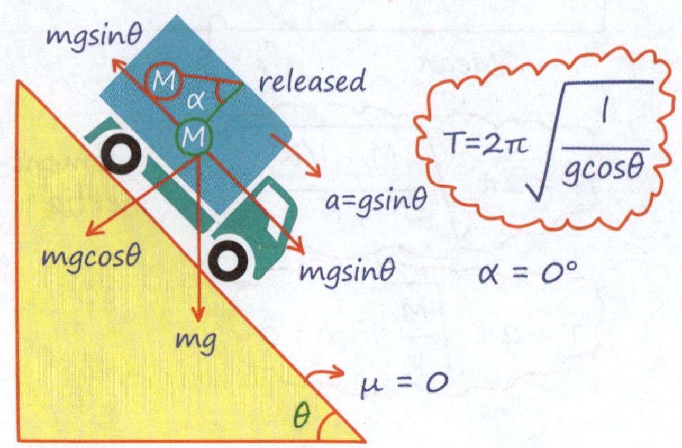

T.P. in Electric field:-

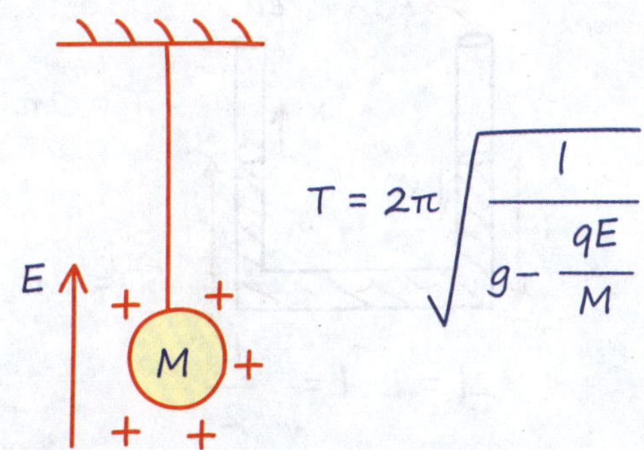

$$T = 2\pi\sqrt{\frac{l}{g - \frac{qE}{M}}}$$

- $$T = 2\pi\sqrt{\frac{l}{g + \frac{qE}{M}}}$$

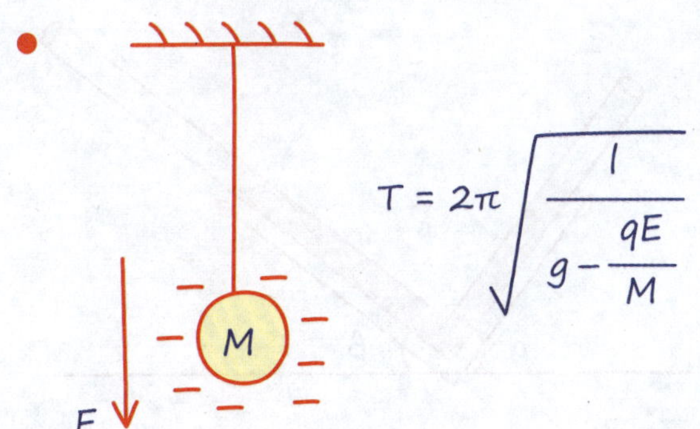

- $$T = 2\pi\sqrt{\frac{l}{g - \frac{qE}{M}}}$$

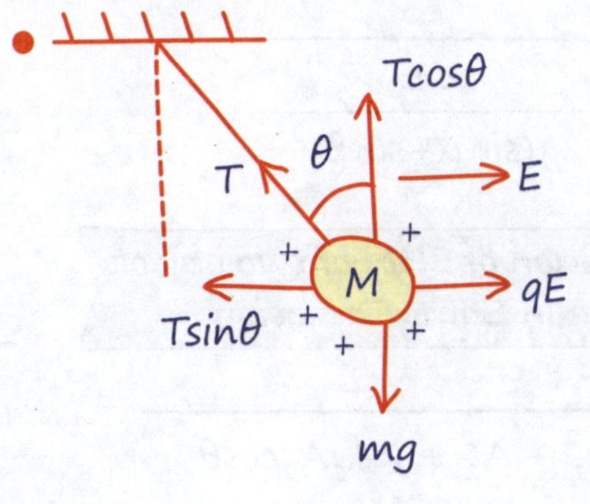

$$T = 2\pi\sqrt{\frac{l}{\sqrt{g^2 + \left(\frac{qE}{M}\right)^2}}}$$

T.P. when length of Simple Pendulum is very large :-

$*T = 2\pi\sqrt{\dfrac{1}{g\left(\frac{1}{l} + \frac{1}{R}\right)}}$	✦ $l >>> R$ $T = 2\pi\sqrt{\dfrac{R}{g}}$ $T = 84.6$ min.
$*l = R$ $T = 2\pi\sqrt{\dfrac{R}{2g}}$	✦ $l <<< R$ $T = 2\pi\sqrt{\dfrac{l}{g}}$

A mass released inside a tunnel through the Earth would oscillate back and forth through simple harmonic motion with the time period.

$$T = 2\pi\sqrt{\frac{R}{g}}$$

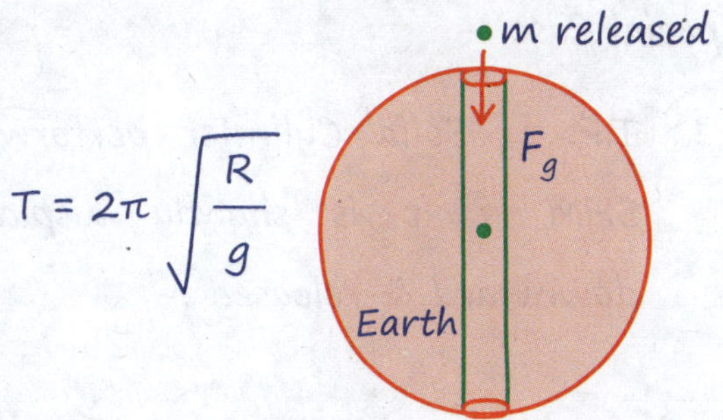

9> Second Pendulum :

length = 1m Time period = 2 sec

f = 0.5 Hz

TP of Physical Pendulum :-

$$T = 2\pi\sqrt{\frac{I}{mgd}}$$

I = M.O.I of object wrt point of suspension.

d = dist^n of O & CM !

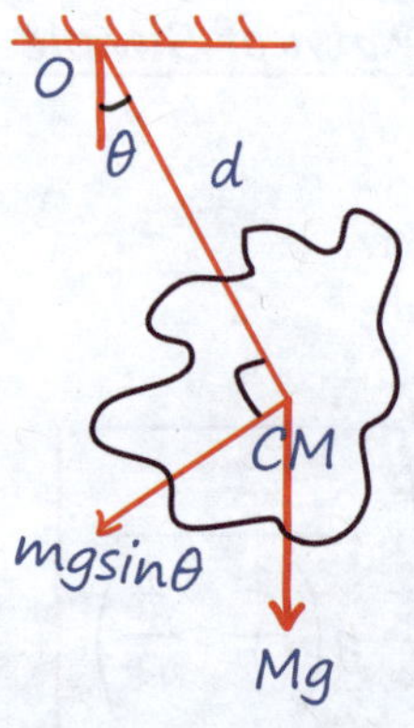

Vertical Rod hinged about one end

C.M

$$I = \frac{ML^2}{3}$$

$$T = 2\pi\sqrt{\frac{ML^2}{3mg\frac{L}{2}}}$$

$$T = 2\pi\sqrt{\frac{2L}{3g}}$$

✦ Special Case :-

1> T.P. of Solid Cylinder performing SHM if it is slightly displaced downward & released :-

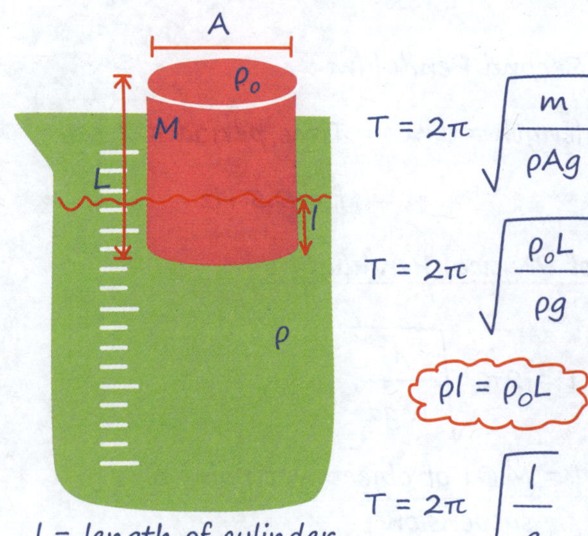

l = length of cylinder inside liquid.

$$T = 2\pi\sqrt{\frac{m}{\rho A g}}$$

$$T = 2\pi\sqrt{\frac{\rho_0 L}{\rho g}}$$

$$\rho l = \rho_0 L$$

$$T = 2\pi\sqrt{\frac{l}{g}}$$

2> Oscillation of liquid column (I) :-

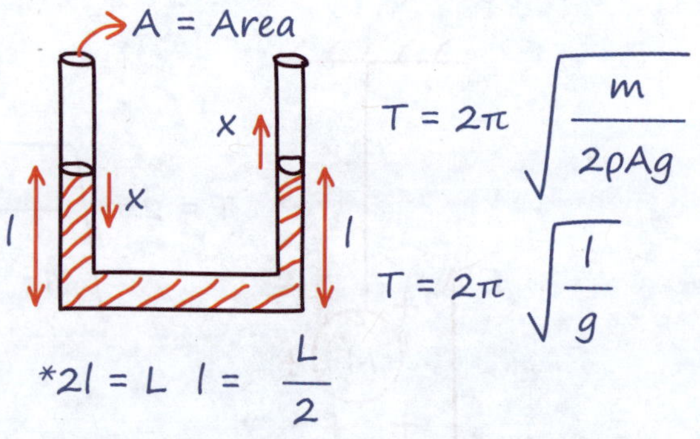

$$T = 2\pi\sqrt{\frac{m}{2\rho A g}}$$

$$T = 2\pi\sqrt{\frac{l}{g}}$$

$*2l = L \quad l = \frac{L}{2}$

m = mass of liquid

ρ = density of liquid

A = Area of liquid column.

L = Total length of liquid column.

Oscillation of liquid column (II) :-

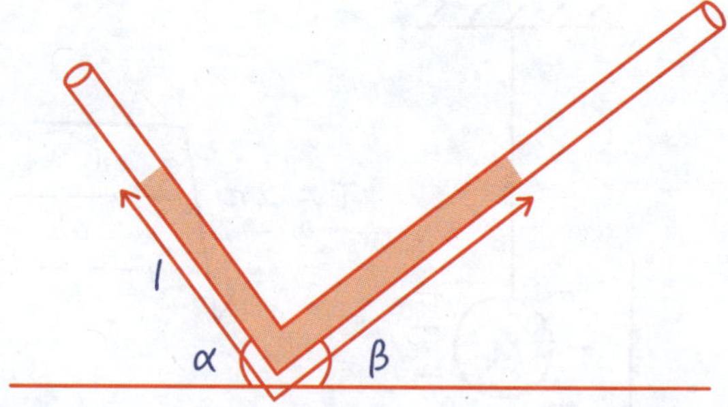

L = total length of water column

$$T = 2\pi\sqrt{\frac{2L}{g(\sin\alpha+\sin\beta)}}$$

9 Superposition of SHM of two object Oscillating in Same Direction

$$A_{net} = \sqrt{A_1^2 + A_2^2 + 2A_1A_2\cos\theta}$$

θ = initial phase difference.

$$\tan\alpha = \frac{A_2\sin\theta}{A_1 + A_2\cos\theta}$$

α = Angle b/n A_{net} & 1st SHM

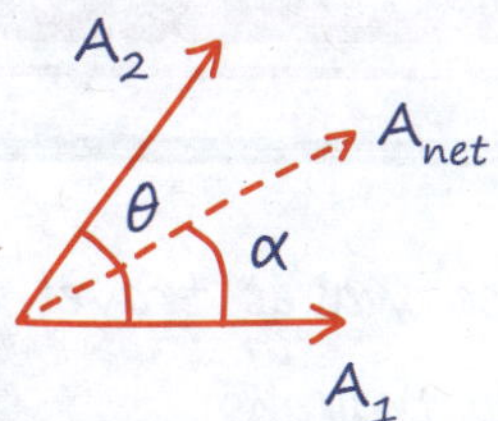

✦ Phasor diagram :-

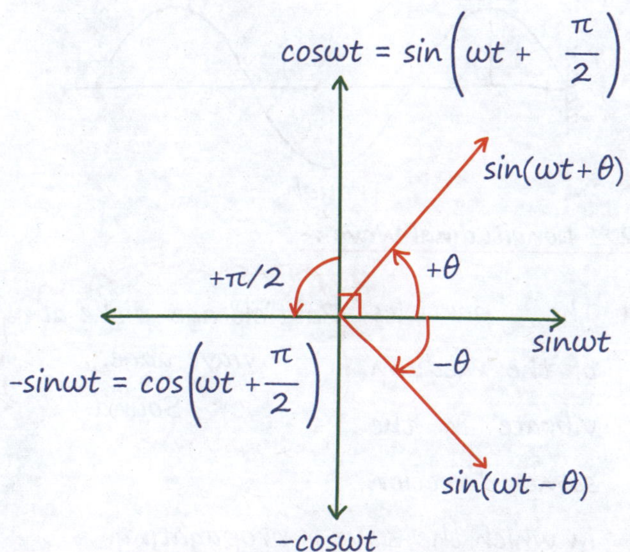

Equation of S.H.M of 1st	Equation of S.H.M of 2nd	Equation of Superimposed S.H.M
$x_1 = A\sin(\omega t)$	$x_2 = A\cos(\omega t)$	$x = \sqrt{2}A \sin(\omega t+\pi/4)$
$x_1 = A\sin(\omega t)$	$x_2 = A\cos(\omega t + \pi/6)$	$x = A \sin(\omega t+\pi/3)$
$x_1 = A\sin(\omega t - \pi/6)$	$x_2 = A\cos(\omega t - \pi/3)$	$x = \sqrt{3}A \sin(\omega t)$
$x_1 = A\sin(\omega t)$	$y_1 = A\cos(\omega t)$	**Circle** $x_1^2 + y_1^2 = A^2$
$x_1 = A\sin(\omega t)$	$y_2 = A\sin(\omega t)$	**Straight line** $x_1 = y_2$
$x_1 = A\sin(\omega t + \pi/3)$	$x_2 = A\cos(\omega t + \pi/6)$	$x = \sqrt{3}A \cos(\omega t)$

Q.2. Amplitude becomes half in 4-oscillation then find amplitude after 16 - oscillation.

Sol. $A = A_0 e^{-n}$

$$\frac{A_0}{2} = A_0 e^{-4}$$

$$e^{-4} = \frac{1}{2}$$

$$e^{-4} = \frac{1}{2}$$

$A = A_0 e^{-16}$

$$= A_0 e^{-4\times4}$$

$$= A_0 \left(\frac{1}{2}\right)^4$$

$$A = \frac{A_0}{16}$$

MR*

"Failure is the first step of success."

20 Wave and Sound Motion

Tum wave si ho main tumhara medium, Tum aati ho meri zindagi mein Mera use krti ho, main apne mean position ke about aage piche oscillate krta rehjata hu tumhare liye, aur tum Mera use krke aage nikal jaati ho...

WAVE

- **Mechanical Wave** → Medium ✓
  # Sound, Wave on string, wave on water.
- **Non-Mech. Wave** → Medium ✗
  # EM Wave i.e., Light Wave.

Electromagnetic Waves: Travel without needing any material medium to move through. They can move through a vacuum (empty space). Examples of electromagnetic waves include light waves and radio waves.

Matter Waves: Linked with tiny particles like electrons and protons. These waves are important in understanding the behavior of these fundamental particles.

1 Mechanical Waves

1> Transverse Wave:-

Particle upar niche wave aage.
Ex. :- String Wave

✦ The particles of the medium vibrate at right angles to the direction of propagation of wave. Transverse waves travel in the form of crests and troughs).

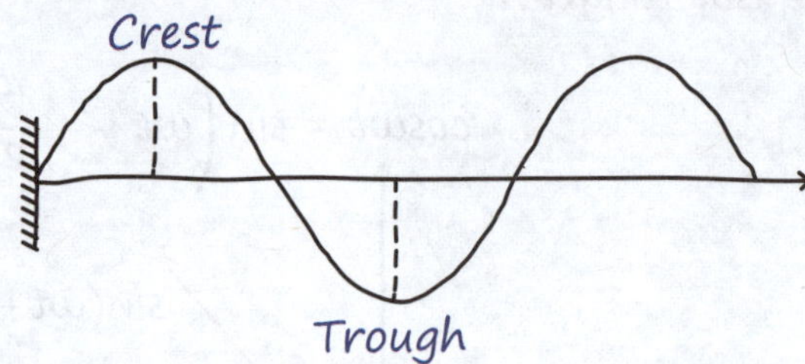

2> Longitudinal Wave :-

Particle age-piche aur wave aage.
Ex. :- Sound.

✦ The particles of the medium vibrate in the same direction in which the wave is propagating.

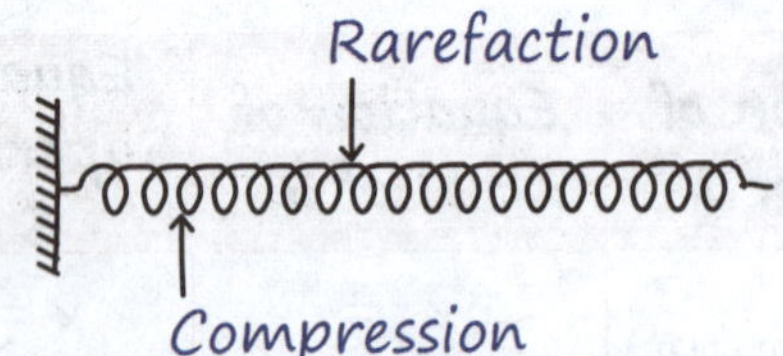

✦ Longitudinal waves travel in the form of compressions and rarefactions.

2 Equation of Propagating Harmonic Wave

SHM ek particle Ka oscillation hai wave hazaro lakho particle Ka oscillation hai !

$Y = A \sin(\omega t \pm Kx \pm \phi)$

Where, A = Amplitude of particle.

ω = Angular frequency

ϕ = initial phase $[t = 0 \quad x = 0]$

- Angular wave no., $K = \frac{2\pi}{\lambda}$
- wave no., $\bar{v} = \frac{1}{\lambda}$

Longitudinal Wave :-

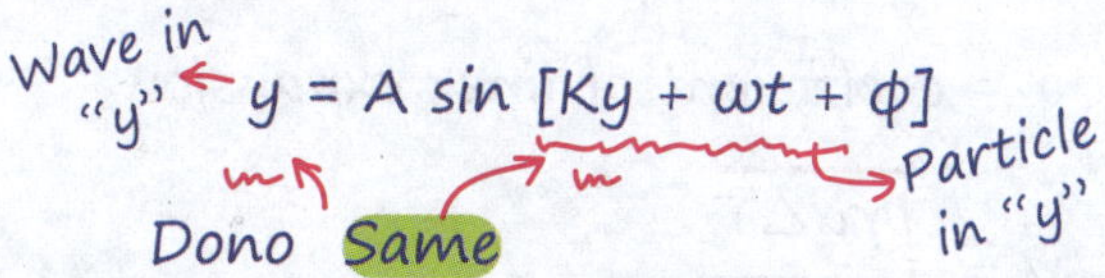

Transverse Wave :-

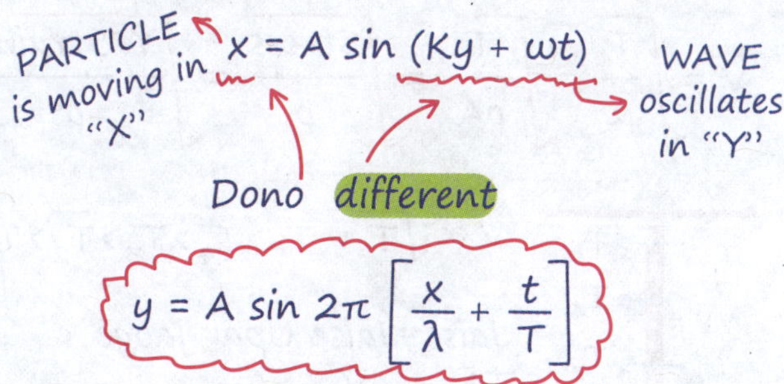

Velocity of Wave :-

$$V_{wave} = \frac{\omega}{K} = \frac{\lambda}{T} = \lambda f$$

Note:-

- $y = A \sin(Kx - \omega t)$ } ek ⊕ dusra ⊖ !
  Wave is moving in +x-axis.
- $y = A \sin(Kx + \omega t)$ } dono ⊕
  Wave is moving in −x-axis.
- $V_P = A\omega \cos(Kx + \omega t)$

  $(V_P)_{max} = A\omega$

Relation Between wave Velocity & maximum Particle Velocity:-

$$^*(V_P)_{max} = AKV_{wave}.$$

Estimation of Particle Whether IT'LL Go UP/ DOWN:-

$$V_{particle} = -V_{wave} \times \text{Slope of wave}$$

MR*

Na Tumhe Formula likhna na slope dhekhna direct answer bs wave jis direction mein travel Kr rahi hai uss direction mein thoda shift Krdo !

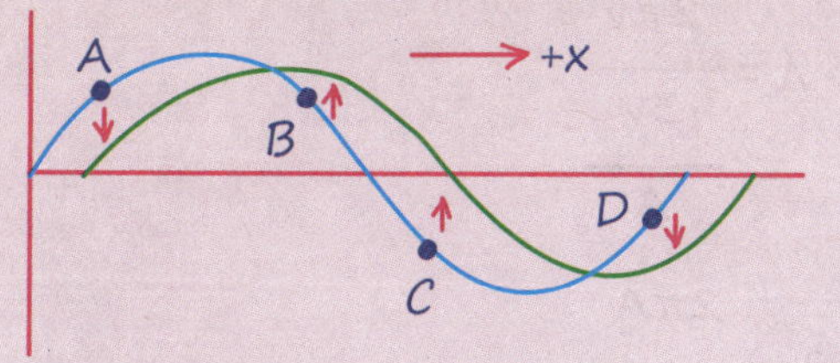

Condition of Wave equation.:-

$$\frac{d^2y}{dt^2} = \frac{\omega^2}{K^2} \cdot \frac{d^2y}{dx^2} \quad \Bigg| \quad \frac{d^2y}{dx^2} = \frac{d^2y}{dt^2} \cdot \frac{1}{V^2}$$

- "y" → Finite hona chahiye at all position of "x".

$$y = f(ax + bt) = \frac{f}{(ax + bt)}$$

Important Relation :-

$$\frac{2\pi}{\phi} = \frac{\lambda}{\Delta x} = \frac{T}{\Delta t}$$

Q.1. The maximum particle velocity is 3 times the wave velocity of a progressive wave. If A is the amplitude of oscillating particle, find phase difference between two particles of separation x.

Sol. The maximum particle velocity = 3 time wave velocity

$A\omega = 3v$

$\therefore \omega = \frac{3v}{A}$

As we know

$v = \lambda f$

$\therefore \lambda = \frac{v}{f}$

or $\lambda = \frac{2\pi v}{\frac{3v}{A}}$

$\therefore \lambda = \frac{2\pi A}{3}$

ϕ = phase difference = $\frac{2\pi}{\lambda} \times x$

$= \left(\frac{2\pi}{\frac{2\pi A}{3}}\right) x$

$= \frac{3x}{A}$

Q.2. The velocity of waves in a string fixed at both ends is 2 m/s. The string forms standing waves with nodes 5.0 cm apart. The frequency of vibration of the string in Hz is

Sol. Here the distance between the two nodes is half of the wavelength

$\frac{\lambda}{2} = 5.0\text{cm} \Rightarrow \lambda = 10\text{cm}$

Hence $n = \frac{v}{\lambda} = \frac{200}{10} = 20\text{Hz}$

3 Velocity of transverse Wave in A String

$\text{Strain} = \frac{\Delta l}{l} = \alpha \Delta T$

α = coefficient of liner expansion

$V = \sqrt{\frac{\gamma \alpha \Delta T}{\rho}}$

$\mu = M/L$ Y = Young Modulus

$V = \sqrt{\frac{T}{\mu}} = \sqrt{\frac{T}{\rho A}} = \sqrt{\frac{\text{Stress}}{\rho}} = \sqrt{\frac{Y.\text{Strain}}{\rho}}$

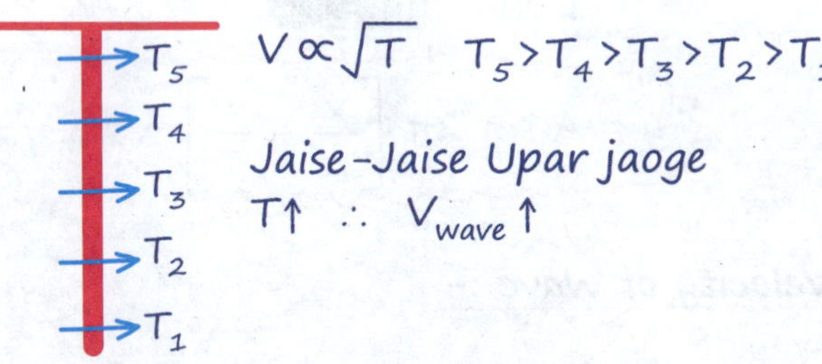

$V \propto \sqrt{T}$ $T_5 > T_4 > T_3 > T_2 > T_1$

Jaise-Jaise Upar jaoge

$T\uparrow \therefore V_{wave}\uparrow$

MR*

$m' = \mu x$ $T_x = m'g = \mu g x$

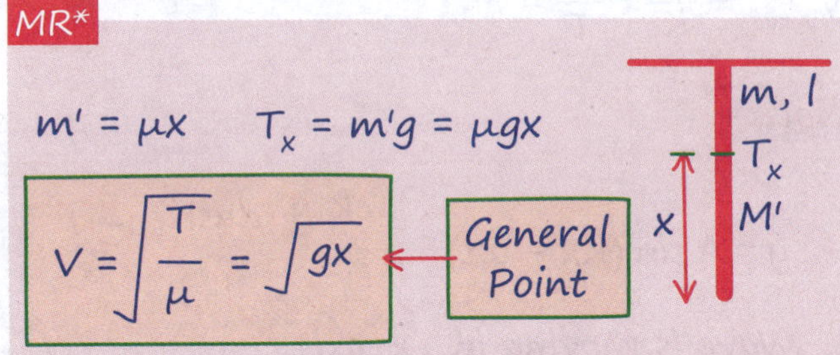

$V = \sqrt{\frac{T}{\mu}} = \sqrt{gx}$

$\therefore$ V & a of wave at L/2 :-

$V = \sqrt{\frac{gL}{2}}$ $a = \frac{g}{2} = \text{Const}^n$

✦ Time taken to reach top point

$T = 2\sqrt{\frac{L}{g}} = \sqrt{\frac{4L}{g}}$

Wave

MR*

- Ratio of wave length of transvers wave at bottom to top

$V = \lambda f \rightarrow$ frequency same

$\lambda \propto V \propto \sqrt{\text{tension}}$

$$\frac{\lambda_1}{\lambda_2} = \sqrt{\frac{m_2}{m_1 + m_2}}$$

λ_2, m_1, λ_1, m_2

4 Sound Wave

- Sound is a type of energy that makes it possible for us to hear things. It travels in waves, and these waves move in a pattern where particles of the medium (like air) vibrate back and forth in the same direction as the wave. This type of wave is called a longitudinal wave.

- Speed of :-

  1> Sound :- $V_s > V_{liq} > V_{gas}$

  2> Light :- $V_{vacuum} > V_{gas} > V_{liq} > V_s$

Sound waves can Reflect, Refract, Interfere and Diffract but cannot be polarized. Polarization is a property that only transverse waves (like light waves) can have.

- Range :-
  - Infrasonic Sound :- $f < 20$ Hz, → cannot hear
  - Audible Sound :- $20 \text{ Hz} \le f \le 20$ KHz, can hear
  - Ultrasonic Sound :- $f > 20$ KHz, can't hear

5 Speed of Sound

$$V_S = \sqrt{\frac{Y}{\rho}} \qquad V_G \text{ or } V_L = \sqrt{\frac{\beta}{\rho}}$$

ρ = Density of gas.

Y = Young modules β = Bulk modules

V_G or V_L = speed of sound in gas or liquid.

Newton's Formula :- (isothermal Process)

$$V_{G\,\&\,L} = \sqrt{\frac{P}{\rho}}, \beta = P$$

$V = 280$ m/s ↓ Typical speed of sound

Laplace's Correction :- (Adiabatic Process)

$$V_{G\,\&\,L} = \sqrt{\frac{\gamma P}{\rho}}, \gamma = 1 + \frac{2}{f}, \beta = \gamma P$$

$$V = \sqrt{\frac{\gamma RT}{M}}$$

$$V \propto \sqrt{T} \propto \frac{1}{\sqrt{M}}$$

Where,

γ : Adiabatic index.

R: Universal gas constant.

T: Absolute temperature.

M: Molar mass of the gas.

Note :-

- $V_{moist\ air} > V_{dry\ air}$
- $\rho_{moist\ air} < \rho_{dry\ air}$
- $V_{rms} > V_{sound\ in\ gas}$
- Speed of sound in gas varies with temperature. Let speed of sound is V_o at 0°C. Speed of sound is V_t at T°C. Then find the relation between them

$$V_t = V_0\left[1 + \frac{t}{546}\right]$$ } Sab °C mein Chalega

$$\Delta V = \frac{V_0 t}{546} = 0.61t$$

- Change in Velocity of Sound is **0.61 m/s**

- Per unit raise in temperature i.e. **0.18%**

Sound Wave Travel Due to Pressure & Density Variation:-

- In compressions, the pressure and density of the air molecules are at their highest. In rarefactions, the pressure and density are at their lowest. This alternating pattern of compressions and rarefactions enables sound waves to propagate through a medium.

	Compression	Rarefaction
Pressure (P)	Maximum (high pressure)	Minimum (low pressure)
Density (ρ)	Maximum (high density)	Minimum (low density)
Displacement (S)	Minimum	Maximum

Note:-

ϕ between disp^m (s) & Pressure wave in sound wave is $\pi/2$.

Relation Between Pressure Amplitude & Displacement Amplitude:-

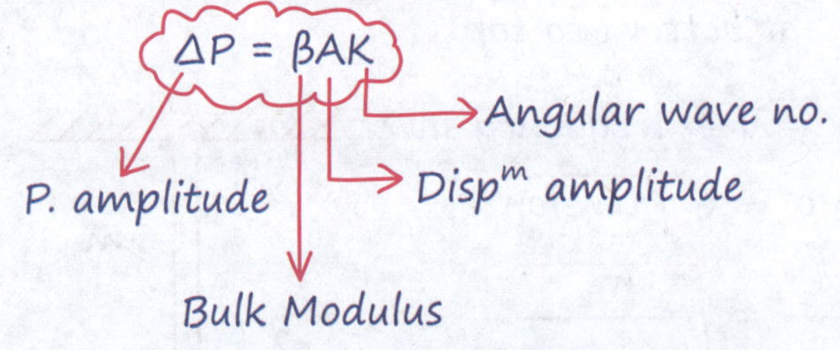

↓

Measure medium resistance to compression.

Intensity of Wave :-

- Intensity (I) is the amount of energy (E) passing through a unit area (A) per unit time (t).

$$I = \frac{E}{At} = \frac{1}{2}\rho V A^2 \omega^2$$

- Unit $= \frac{J}{m^2 sec} = \frac{watt}{m^2}$

Where,

I: Intensity of the wave

E: Energy

A: Area

t: Time

ρ: Density of the medium

V: Velocity of the wave

A: Amplitude of the wave

w: Angular frequency

Energy Density of Wave:-

- Energy density (u_e) is the amount of energy stored in a unit volume of the medium through which the wave is propagating.

$$u_e = \frac{Energy}{Volume} = \frac{I(Intensity)}{V(Velocity)} = \frac{1}{2}\rho A^2 \omega^2$$

$$\text{Where, } I = \frac{1}{2}\rho V A^2 \omega^2$$

MR* → direction se feel karo

Intensity Variation Due to Diff. Sources:-

Point source	Linear Source	Planar Source
$I \propto \frac{A^2}{r^2}$	$I \propto \frac{A^2}{r}$	$I \propto r^0$ $A \propto r^0$

6 Loudness of Sound Wave

- The reference intensity level, I_o, is the quietest sound that the average human ear can hear, typically:

$$I_o = 10^{-12}\ \frac{\text{Watt}}{m^2}$$

- The sound level or loudness (L) in decibels (dB) is calculated using the following formula:

$$L = 10 \log_{10}\left(\frac{I}{I_o}\right) dB$$

- 1Bel = 10 dB

MR*

Change in loudness:

$$\Delta L = 10 \log_{10}\left(\frac{I_f}{I_i}\right) = 10 \log_{10}\left(\frac{I_{max}}{I_{min}}\right)$$

7 Principle of Superposition of Waves

- Angle (α) Between Resultant and One of the Components: $\tan\alpha = \frac{A_2 \sin\phi}{A_1 + A_2 \cos\phi}$

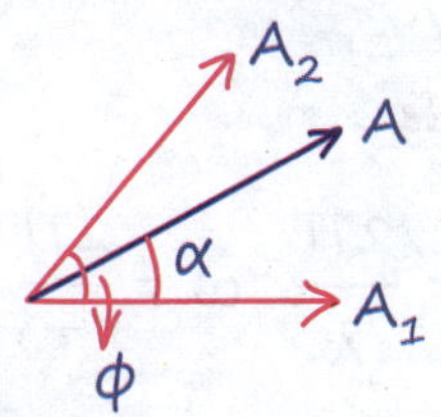

- Resultant amplitude

$$A^2 = A_1^2 + A_2^2 + 2A_1A_2 \cos\phi$$

- Intensity of resultant wave.

$$I = I_1 + I_2 + 2\sqrt{I_1}\sqrt{I_2}\cos\phi \quad \because I \propto A^2$$

- $$\frac{I_{max}}{I_{min}} = \frac{\left(\sqrt{I_1}+\sqrt{I_2}\right)^2}{\left(\sqrt{I_1}-\sqrt{I_2}\right)^2} = \frac{(A_1 + A_2)^2}{(A_1 - A_2)^2}$$

- $$\frac{I_{max}}{I_{min}} = \frac{\left(\sqrt{\frac{I_1}{I_2}}+1\right)^2}{\left(\sqrt{\frac{I_1}{I_2}}-1\right)^2} = \frac{\left(\frac{A_1}{A_2}+1\right)^2}{\left(\frac{A_1}{A_2}-1\right)^2}$$

8 Interference

Aaise doh wave Ka superposition jiska λ & f same hai !

Constructive	Destructive
$A_{max} = A_1 + A_2$	$A_{min} = A_1 - A_2$
$I_{max} = \left(\sqrt{I_1}+\sqrt{I_2}\right)^2$	$I_{min} = \left(\sqrt{I_1}-\sqrt{I_2}\right)^2$
$\cos\phi = 1$	$\cos\phi = -1$
$\phi = 0, 2\pi, 4\pi, 6\pi,$	$\phi = \pi, 3\pi, 5\pi, 7\pi,$
$\phi = 2n\pi$	$\phi = (2n+1)\pi$
$\Delta x = 0, \lambda, 2\lambda, 3\lambda,$	$x = \frac{\lambda}{2}, \frac{3\lambda}{2}, \frac{5\lambda}{2}, ...$
$\Delta x = n\lambda$	$\Delta x = (2n+1)\frac{\lambda}{2}$

Two wave same amplitude Ao :-

Constructive	Destructive
$A_{max} = 2A_0$	$A_{min} = 0$
$I_{max} = \left(2\sqrt{I}\right)^2 = 4I$	$I_{min} = 0$

Q.3. The intensity ratio of the two interfering beams of light is β. What is the value of $[(I_{max} - I_{min}) / (I_{max} + I_{min})]$?

(a) $\sqrt{\beta}$ (b) $2\sqrt{\beta} / (1 + \beta)$

(c) $2 / (1 + \beta)$ (d) $(1 + \beta) / 2\sqrt{\beta}$

Sol. ⇒ $\frac{2\sqrt{\beta}}{1 + \beta}$

Apply MR*

If. $\beta = 1$, then $\frac{I_{max} - I_{min}}{I_{max} + I_{min}} = 1$

9 Reflection of Transverse Waves in String

+ Transverse waves can reflect (or refract) when they encounter a boundary or a change in medium. This reflection can occur when a wave travels from a rarer (less dense) medium to a denser medium or vice versa.

(String और Sound का Same hai)

(a) Rarer ⟶ Denser :-

	Reflected	Transmited
Frequency, f	f	f
Speed (υ)	Same	Decrease υ↓
Wavelength (λ)	Same	Decrease λ↓
Phase diff. (ϕ)	π	0 (No phase change)

(b) Denser ⟶ Rarer :-

	Reflected	Transmited
Frequency, f	f	f
Speed (υ)	Same	Increase υ↑
Wavelength (λ)	Same	Increase λ↑
Phase diff. (ϕ)	0	0 (No phase change)

10 Stationary Wave

When two waves with similar properties (same frequency and amplitude) travel in opposite directions within a confined medium, their superposition creates a new type of wave. This resulting wave appears to stand still, rather than traveling through the medium. This phenomenon is known as a stationary wave or standing wave.

Aaise doh wave ka superposition jiska sab Kuch same hoga bs direction opposite

General equation:-

$$y = 2A \sin(Kx) \cos(\omega t)$$

$2A \sin(Kx)$ → Amplitude; $\cos(\omega t)$ → SHM.

Where, $K = \frac{2\pi}{\lambda}$, $\omega = \frac{2\pi}{T}$

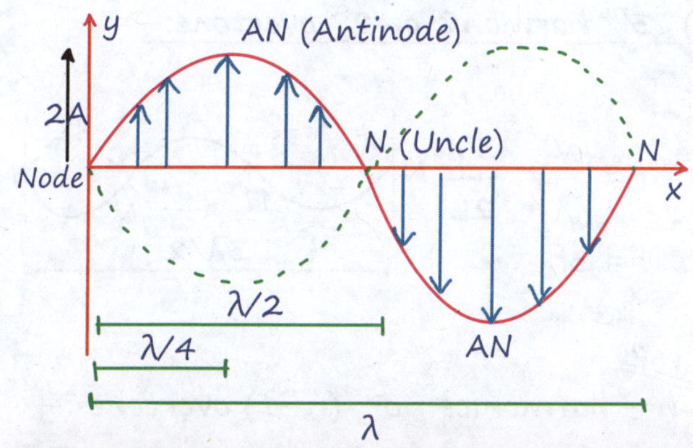

	Node	Antinode
Condition:	$\phi = Kx = n\pi$	$\phi = Kx = (2n+1)\frac{\pi}{2}$
Position:	$x = \frac{n\lambda}{2}$	$x = (2n+1)\frac{\lambda}{4}$
Occur at:	integral multiples of half wavelength	odd multiples of quarter wavelength

Time evolution of stationary wave:

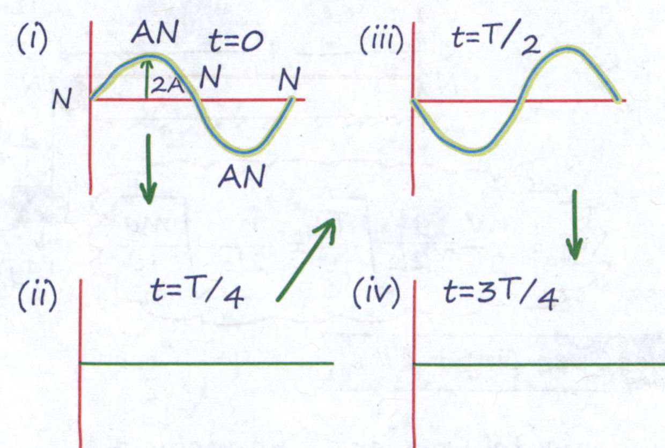

✦ A flat line, indicating that the wave's displacement is zero throughout the medium

Time	Description
$t = 0$	The wave has its initial form with nodes and antinodes positioned as per the general equation.
$t = \frac{T}{4}$ and $t = \frac{3T}{4}$	The wave passes through its equilibrium position, with maximum speed of particles at these points.
$t = \frac{T}{2}$	The wave returns to a form similar to its initial form but inverted, with nodes and antinodes remaining fixed in position.

✦ Different equation of Stationary Wave :-

$$Y = 2A \sin (Kx) \cdot \cos (\omega t)$$

$$Y = 2A \cos (Kx) \cdot \sin(\omega t)$$

$$Y = 2A \sin (Kx) \cdot \sin (\omega t)$$

$$Y = 2A \cos (Kx) \cdot \cos (\omega t)$$

Q.4. If node is formed at origin then amplitude which is at between node and antinode.

Sol. Midway between node and Anitnode

$$x = \frac{\lambda}{8}, \quad x = 2A \sin\left(\frac{2\pi}{\lambda} \times \frac{\lambda}{8}\right)$$

$$= 2A \sin \frac{\pi}{4} = \sqrt{2}\,A$$

Difference Between Stationary & Progressive Wave:-

Stationary wave	Progressive wave
1> Particle at Node → rest	1> No Particle at rest.
2> No Transfer of Energy and momentum.	2> Transfer of E & P occurs.
3> Sabka "A" diff.	3> "A" same.
4> All in same phase between N.	4> All are in diff. phase

Stationary wave	Progressive wave
5> All particles cross mean position with diff. speed at same time.	5> All particles cross MP with same speed at diff. time.

11 Formation of Stationary Wave in String Sonometre Wire

Frequency:-

$$f' = \frac{V}{2L} = \frac{1}{2L}\sqrt{\frac{T}{\mu}} = \frac{1}{2L}\sqrt{\frac{T}{\rho A}}$$

$$f' = \frac{1}{2L}\sqrt{\frac{Stress}{\rho}}$$

Where,

- V: Wave speed
- L: Length of the string
- T: Tension in the string
- μ: Linear mass density
- A is the cross-sectional area)

(a) Fundamental or 1st harmonic :-

$$f' = \frac{V}{\lambda} = \frac{V}{2L}$$

N AN N

(b) 2nd harmonic or 1st overtone :-

$$f = \frac{V}{\lambda} = \frac{V}{L} = 2f'$$

N A N A N

(c) 3rd harmonic or 2nd overtone:-

$$f = \frac{V}{\lambda} = \frac{3V}{2L}$$

$$f = 3f'$$

N A N A N A N

$3\lambda/2$

Vimp

n - harmonics or (n−1) overtone :-

$$f = n\left(\frac{V}{2L}\right)$$

If, No. of antinode = n, then No. of Node = n + 1

Difference between any two consecutive harmonics ;-

$$\Delta f = \frac{V}{2L} = f$$ NEET

- Sonometer Wire :-

μ = linear mass density of wire.

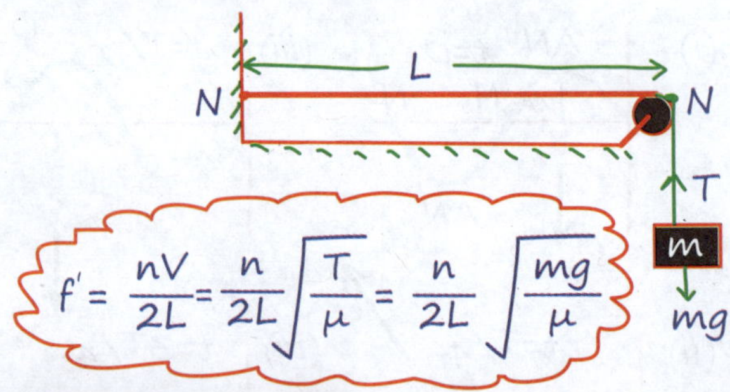

$$f' = \frac{nV}{2L} = \frac{n}{2L}\sqrt{\frac{T}{\mu}} = \frac{n}{2L}\sqrt{\frac{mg}{\mu}}$$

12 Open Organ Pipe

Fundamental Freq. or 1st harmonic :-

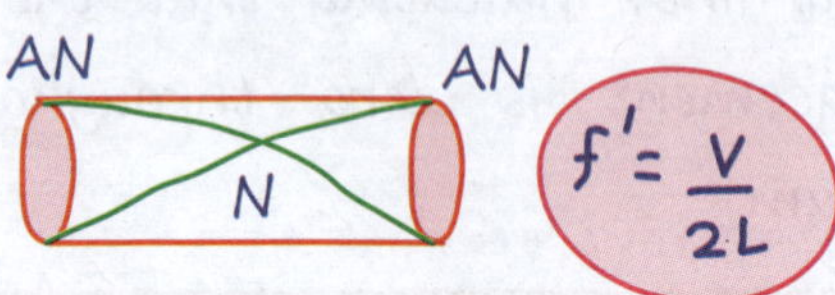

$$f' = \frac{V}{2L}$$

2nd harmonic or 1st overtone :-

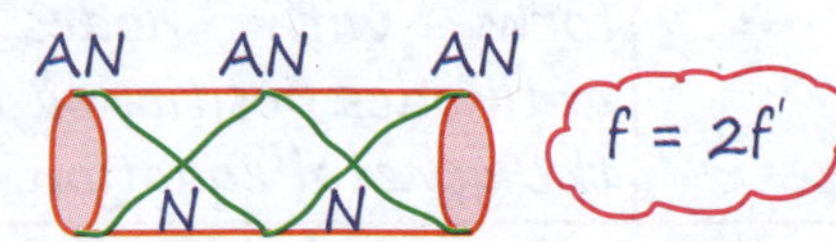

$$f = 2f'$$

3rd harmonic or 2nd overtone :-

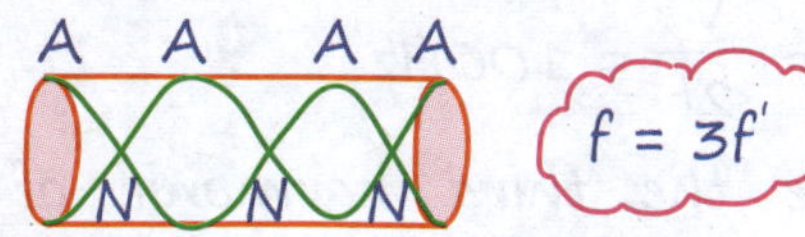

$$f = \frac{nV}{2L} = \frac{n}{2L}\sqrt{\frac{\gamma P}{\rho}} = \frac{n}{2L}\sqrt{\frac{\gamma RT}{M}}$$

- n = harmonic $\quad (n-1)$ = overtone

No. of Node = n $\quad$ No. of Antinode = n + 1

Vimp

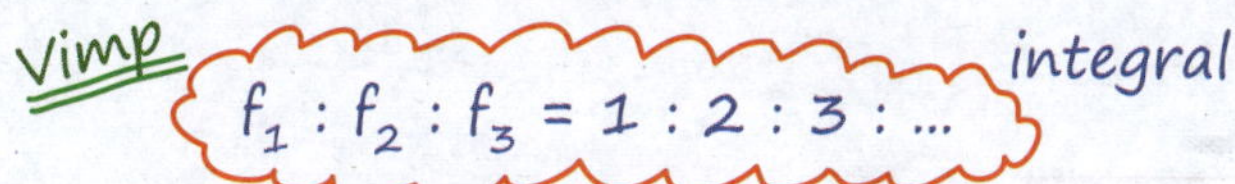

For Pressure wave :-

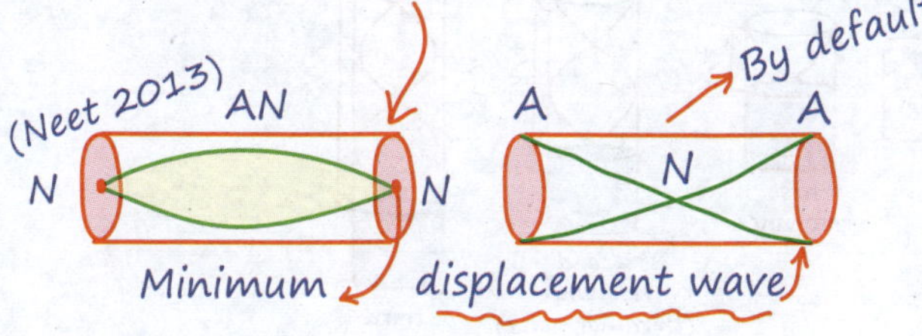

$\therefore$ Disp^m wave & Pressure wave have

$\phi = \pi/2$.

Note:-

- Antinode = Maximum pressure variation.
- Nodes = Minimum pressure variation.

13 Closed Organ Pipe

Fundamental or 1st harmonic :-

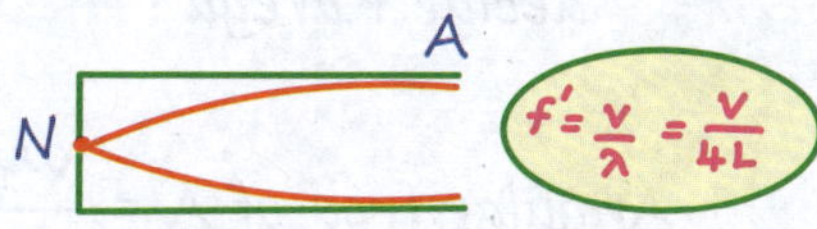

3rd harmonic or 1st overtone :-

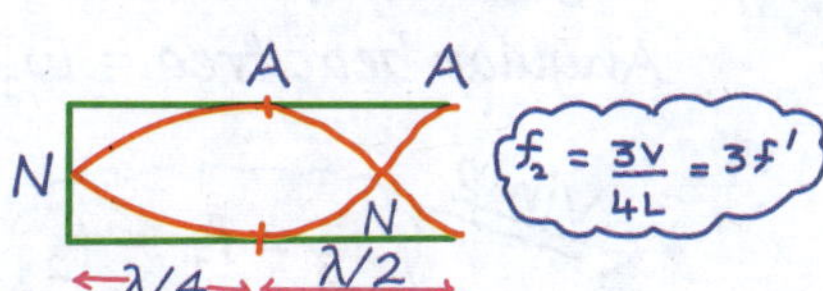

5th harmonic or 2nd overtone :-

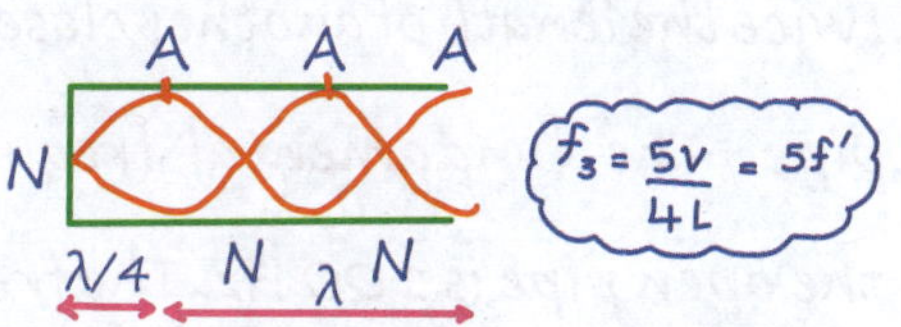

(2n + 1) = Hormonic $\quad$ n = Overtone

$$f = (2n+1)\frac{V}{4L} \qquad f_1 : f_2 : f_3 = 1 : 3 : 5$$

Difference between any two consecutive harmonics :-

$$\Delta f = 2f' = \frac{2V}{4L}$$ **NEET**

The MR*

Frequency Ka ratio likhdo aur uss ratio mein dhekho Konse wave form Ki baat chalrahi jiss wave form Ki baat chal rahi hogi wahi tumhara N & AN hoga.

$f_1 : f_3 : f_5 : f_7 = 1 : 3 : 5 : 7$

2nd overtone	3rd overtone
3 wave form	4 wave form
3N & 3AN	4N & 4AN

End Correction:-

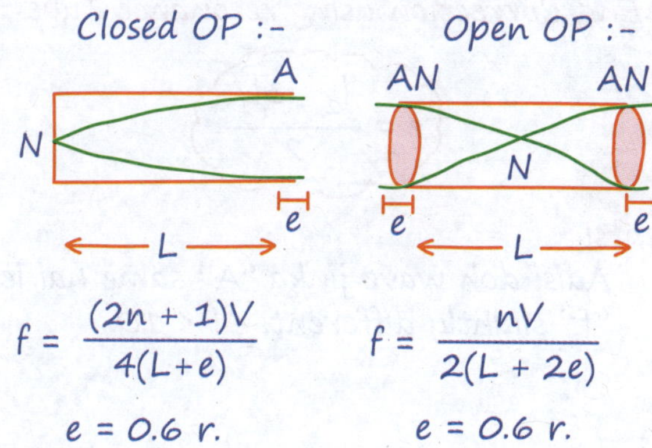

Closed OP: $f = \dfrac{(2n+1)V}{4(L+e)}$, $e = 0.6\ r$.

Open OP: $f = \dfrac{nV}{2(L+2e)}$, $e = 0.6\ r$.

Q.5. The length of an open organ pipe is twice the length of another closed organ pipe. The fundamental frequency of the open pipe is 100 Hz. The frequency of the third harmonic of the closed pipe is.

Sol. The fundamental frequency of open pipe

$$= f_o = \frac{V}{2l} = 100\text{Hz}$$

Hence the third harmonic of closed organ pipe is $\frac{3V}{4(l/2)}$

$$= 300\text{Hz}$$

14 Resonance Tube

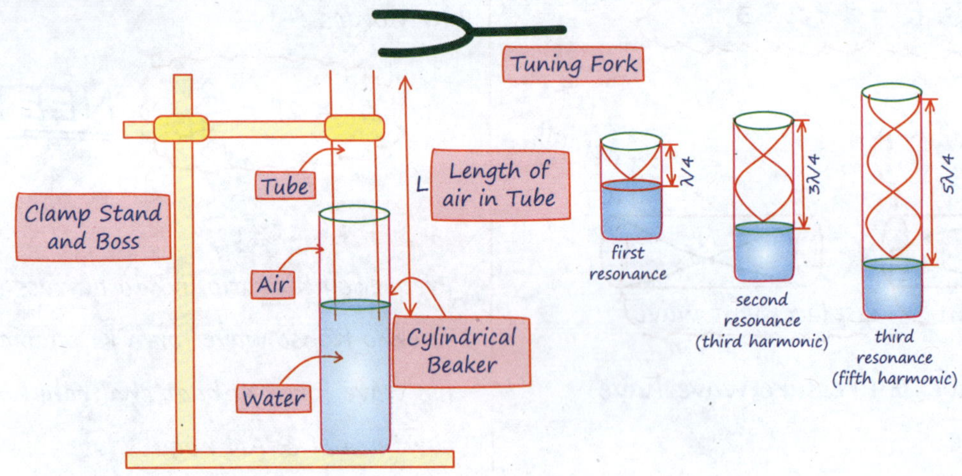

$f_1 = \frac{V}{4L_1} = f_T$ $\qquad$ $f_2 = \frac{3V}{4L_2} = f_T$

"Resonance"

NEET $\quad \frac{V}{4L_1} = \frac{3V}{4L_2}$ $\quad$ * $L_2 = 3L_1$ *

✦ End Correction using resonance Tube :-

$$e = \frac{l_2 - 3l_1}{2}$$

Beats!

Aaisi doh wave jinka "A" same hai lekin "f" slightly different. $\Delta f < 10$.

$$Y = 2A\left[\sin\left(\frac{\omega_1 + \omega_2}{2}t\right) \cdot \cos\left(\frac{\omega_1 - \omega_2}{2}t\right)\right]$$

2A → Amplitude; sin term → oscill^n; cos term → oscill^n

Ye Amplitude Ko decide Karega !

Angular freq of A $= \frac{\omega_1 - \omega_2}{2}$

Angular freq of I $= \omega_1 - \omega_2$

Angular beat freq $= \omega_1 - \omega_2$

Vimp $\quad f_B = f_1 - f_2$

✦ Tuning Fork :-

Sharp :- f↑
Waxing :- f↓

Waxing :-
Waning :- $\dfrac{I_{max}}{I_{min}} = \dfrac{(A_1 + A_2)^2}{(A_1 - A_2)^2}$

Q.6. Two tuning forks when sounded together produced 4 beats/sec. The frequency of one fork is 256. The number of beats heard increases when the fork of frequency 256 is loaded with wax. The frequency of the other fork is

Sol. $f_A - f_B = 4Hz$

$f_A = 256$ Hz

$f_B = ??$ → 260Hz, → 252Hz (Possible)

In waxing of A f_A will decrease but frequency difference is increasing hence 260 Hz will be answer.

Q.7. A tuning fork A produces 2 beats per second when sounded with a tuning fork of frequency 200Hz. When A is loaded with wax the beats stop. What is the frequency of fork A?

Sol. There are 2 beats per second

$\therefore f_1 - f_2 = 2$

$\Rightarrow 200 - f_2 = \pm 2$

$f_2 = 200 \pm 2$

$f_2 = 202$ or $f_2 = 198$

When loaded with wax the beats stop since the frequency decreases on loading.

$\therefore f_2 = 202$ Hz

21 Electrostatics

1 Charge

Charge is a fundamental property of matter that causes it to experience a force when placed in an electric or magnetic field. It is the property by which objects exert an electrostatic force of repulsion or attraction on each other.

- There are two Types of Charges:

Positive Charge (+): due to the deficiency of electrons

Negative Charge (–): due to the excess of electrons

Behaviour of charge under different condition:-

Charge	Electric field	Magnetic field	EM Wave
Rest	✓	✗	✗
V=Const^n	✓	✓	✗
Accelerated	✓	✓	✓

Properties of Charge:-

- Scalar, Conserved, Quantized
  $Q = ne$, n = integer
- **Invariant:** does not depends upon speed:-
- Charge can't exist without mass but mass can exist without charge.
- Same charge → repel (may attract)
- Opposite charge → must attract
- SI unit 1 C = 3×10^9 esu. IC = $\frac{1}{10}$ emu.
- Smallest unit frankline = 1 esu
- Largest unit faraday, 1 faraday = 96500 C
- One charge may attract other Neutral

Sureity check of a charge body is repulsion:-

Attraction can occur between charged and neutral objects due to polarization effects, but repulsion only occurs between like charges.

Quarks :- Does not exist in free state

up	$+\frac{2e}{3}$
down	$-\frac{1}{3}e$

Q.1 Which of the following charge is possible:

(a) $\frac{1C}{100}$ (b) $\frac{1e}{50}$

(c) $4.8\times10^{-21}C$ (d) $1.56\times10^{-18}C$

Sol. Use $Q = ne$ (a) is possible

Note: Charge को Sharp Point पसंद है:-

2 Method of Charging a body

Charging of Body by conduction:-

- Conduction involves transferring charge through direct contact between materials.
- For conductor only

  $Q \propto R$. [R Radius]

Ex When two conductors with charges Q_1 & Q_2 and Radius R_1 & R_2 are connected: If Q_1' is the new charge on first conductor then

$$\text{New Charge} = \frac{\text{Uska R}}{\text{Total R}}[Q_{Total}]$$

$$Q_1' = \frac{R_1}{R_1 + R_2}[Q_1 + Q_2]$$

Charging of body by friction :-

- Friction involves rubbing two different materials together, causing a transfer of electrons.
- For insulator only equal and opposite charge on two rubbing object

Charging by induction :-

- Induction involves bringing a charged object close to another object without direct contact, causing a redistribution of charges in the second object.
- For conductor and dielectric → equal or lesser charge of positive nature induced.

Charge density :-

Charge on Linear object	Charge on Areal object	Charge on Volumetric object
$dq = \lambda dl$	$dq = \sigma dA$	$dq = \rho dV$
$\lambda = \frac{dq}{dl}$ C/m	$\sigma = \frac{dq}{dA}$ C/m^2	$\rho = \frac{dq}{dv}$ C/m^3
↑ Linear charge density	↑ Areal charge density	↑ Charge density

3 Gold leaf experiment

- Device used to detect charge, not use to measure charge.
- Method involves conduction or induction.
- Diverge angle $\propto$ charge of leaf.

4 Coulomb's Law

- Valid for point or spherical charge symmetry
- The force of interaction between two point charges is directly proportional to the product of the charges and Inversely proportional to the square of the distance between them.

$$F = \frac{Kq_1q_2}{r^2} \qquad \varepsilon_0 = 8.85 \times 10^{-12} \frac{C^2}{Nm^2}$$

$$K = \frac{1}{4\pi\varepsilon_0} = 9\times10^9 \frac{Nm^2}{C^2}$$

Mag:- Formula से
Dir^n:- Buddhi से.

* Another Medium between Charges:-

$$K = \varepsilon_r = \frac{\varepsilon_m}{\varepsilon_0} \qquad F' = \frac{F_0}{K}$$

dec. to K times it's initial value.

- Conservative, long range, follow inverse square law $\left(F \propto \frac{1}{r^2}\right)$, central, mediated by photon.
- Electrostatic force on q_2 due to q_1 does not depends upon medium or presence of other charge → True.
- Net electrostatics force on q_2 depends upon medium or presence of other charge → True.

Coulomb's Law in Vector Form :-

$$\vec{F}_{21} = \frac{Kq_1q_2}{|\vec{r}_{21}|^2}\hat{r}_{21} = \frac{Kq_1q_2}{|\vec{r}_{21}|^3}\vec{r}_{21}$$

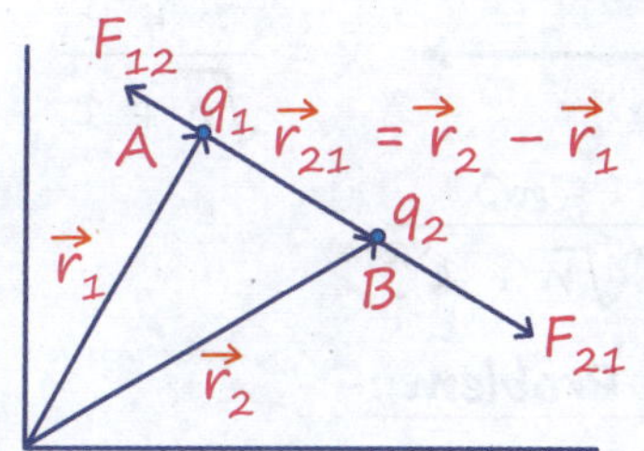

Q.2 Two identical charge 'q' repel each other with a force of 100 N, one of the charge is increased by 10% and decreased by 10% then new force of repulsion at the same distance?

Ans. $q'_1 = 110\%q,\ q'_2 = 90\%q$

$$F = \frac{Kq_1q_2}{r^2},\ F' = \frac{Kq'_1q'_2}{r^2} \Rightarrow F' = 99N$$

Neutral Point :-

- Find position of 3rd charge where force on that charge will be zero

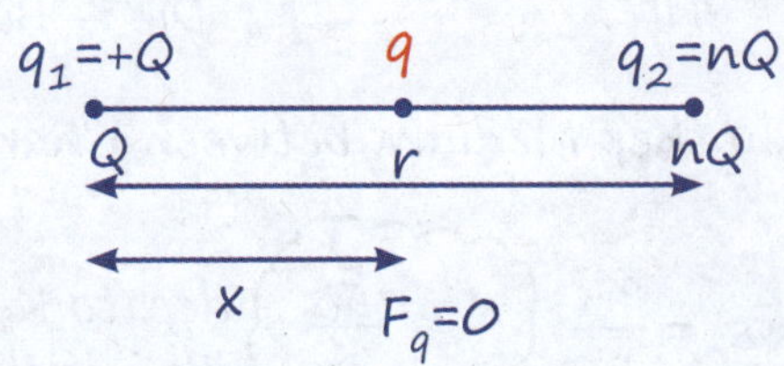

Like Charge

$$x = \frac{d}{\sqrt{n} + 1}$$

Unlike Charge

$$x = \frac{d}{\sqrt{n} - 1}$$

$$n = \frac{\text{Bada charge}}{\text{Chota charge}}$$

Note:-

If we Divide charge Equally, they repel each other with F_{max}.

Q.3 Find Position & Value of "q" so that System will be in Equilibrium:-

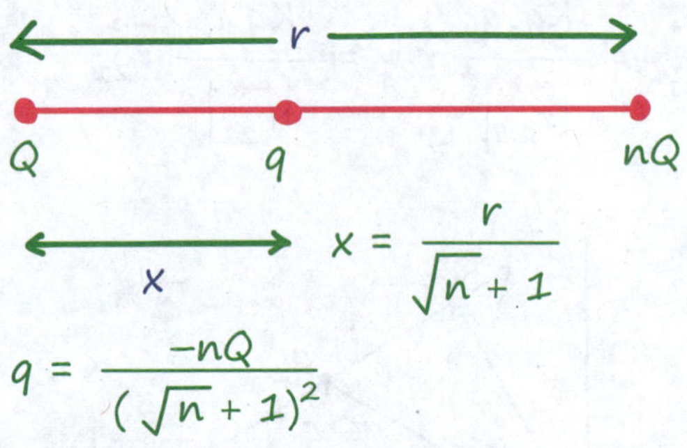

$$x = \frac{r}{\sqrt{n} + 1}$$

$$q = \frac{-nQ}{(\sqrt{n} + 1)^2}$$

Pendulum Problem :-

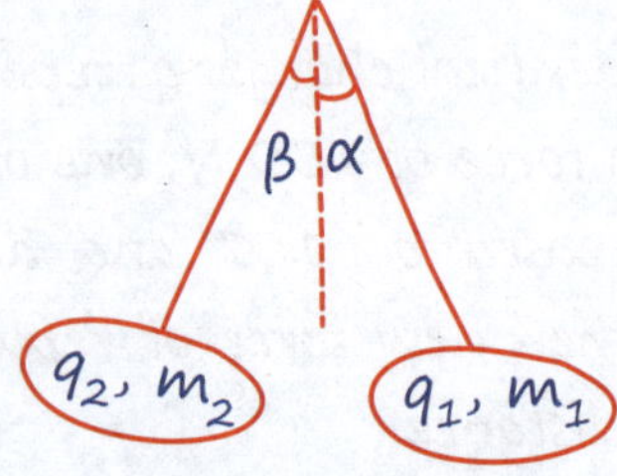

- Value of α and β depends on charges and masses but ratio of α and β does not depends on charges and depends only on masses

$$\frac{\tan\alpha}{\tan\beta} = \frac{m_2}{m_1}$$

- If $m_1 = m_2$ then $\alpha = \beta$
- If $m_1 > m_2$ then $\alpha < \beta$
- $\tan\theta = \frac{F_e}{mg}$

$$T = \sqrt{F_e^2 + (mg)^2}$$

Q.4 If $q_1 > q_2$ but $m_1 = m_2$ then which of the following is correct?

(a) $\alpha = \beta$ (b) $\alpha > \beta$ (c) $\alpha < \beta$

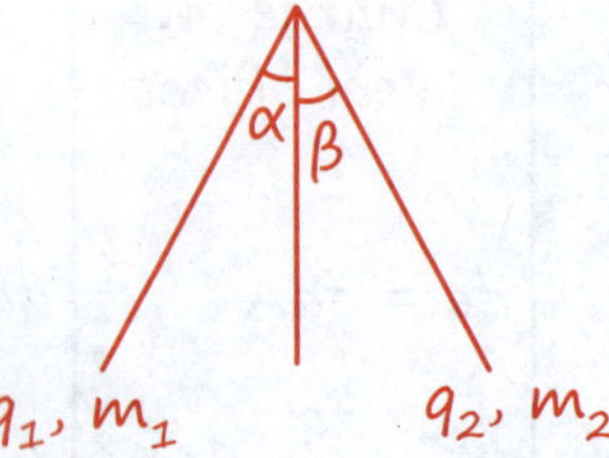

Ans. (a)

Force on Rod Due to Point Charge :-

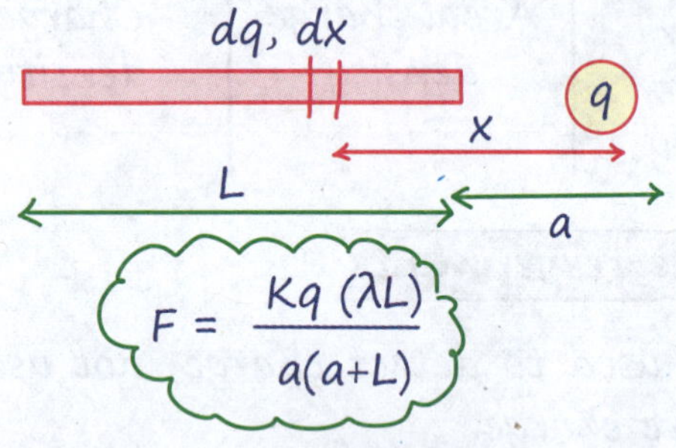

$$F = \frac{Kq(\lambda L)}{a(a+L)}$$

5 Electric Field Intensity

- Electrostatic force per unit positive charge.

$$\vec{E} = \frac{\vec{F}}{q} = N/C. \quad \text{Vector !}$$

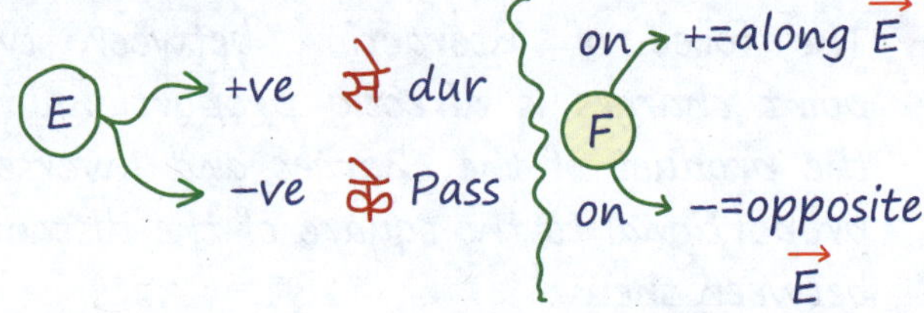

6 Electric field lines

- Electric field lines are imaginary lines or curves that represent the direction and strength of the electric field in a region of space. At any point on the line, the tangent to the line gives the direction of the electric field vector at that point.

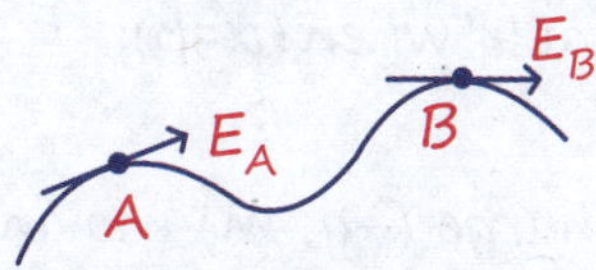

- The density of the lines at any point indicates the magnitude of the electric field: closer lines mean a stronger field.

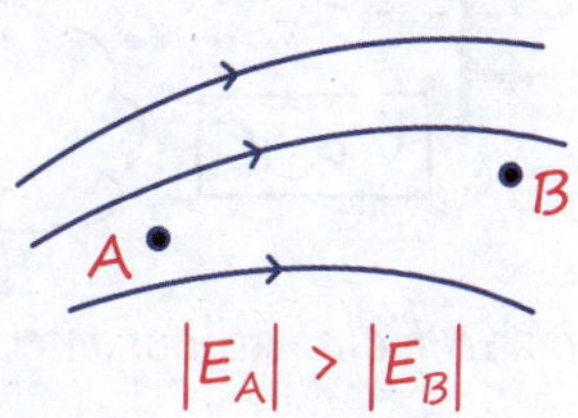

Properties of electric field lines:-

- Electric field lines originate from positive charges and terminate at negative charges.
- Electric field lines never cross each other. If they did, it would imply two different directions of the electric field at the same point, which is impossible.
- Electric field lines do not form closed loops. They begin at a positive charge and end at a negative charge, or extend to infinity in the case of isolated charges.
- In regions where there are no charges, electric field lines are smooth, continuous curves without any breaks.

Superposition Principle:

- When dealing with multiple charges, each pair of charges interacts independently.
- The total force on any charge is found by adding up all the individual forces exerted on it by every other charge.

Mathematically

Consider a system of n point charges; $q_1, q_2, q_3, ..., q_n$ distributed in space. Let the charges $q_2, q_3, ..., q_n$ exert forces $\vec{F}_{12}, \vec{F}_{13}, ..., \vec{F}_{1n}$ on charge q_1. The total force on charge q_1 $(\vec{F}_1)$ is given by:

$$F_1 = \frac{1}{4\pi\varepsilon_0}\left(\frac{q_1q_2}{r_{12}^2}\hat{r}_{12} + \frac{q_1q_3}{r_{13}^2}\hat{r}_{13} + ... + \frac{q_1q_n}{r_{1n}^2}\hat{r}_{1n}\right)$$

Electric field due to Point charge :-

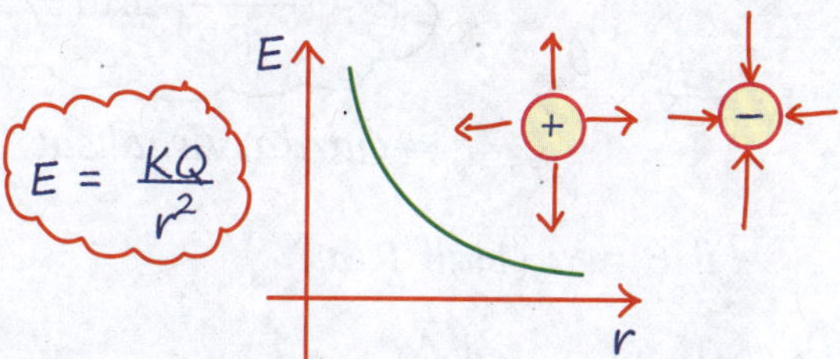

Electric field due to Ring :-

Q, R, X, P, $\sqrt{R^2+X^2}$

$$E_{max} = \frac{\text{Radius}}{\sqrt{2}}$$

$$E_P = \frac{KQx}{(R^2+X^2)^{3/2}}$$

✦ Graph B/w electric field and distance

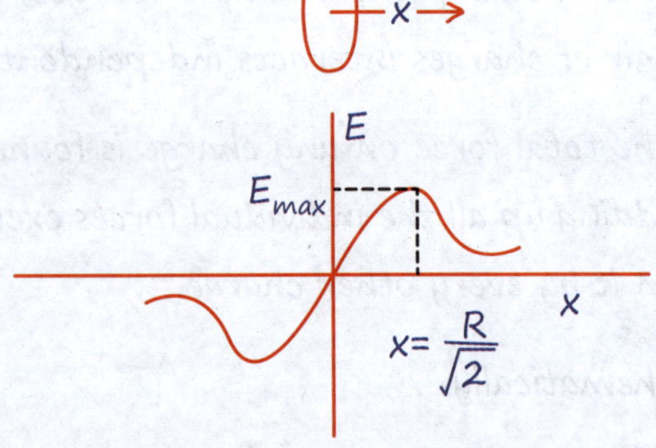

Electric field due to A Half ring :-

$E = \frac{2K\lambda}{R}$

$E = \frac{2KQ}{\pi R^2}$

MR*

Q

R

θ

$E = \frac{2K\lambda}{R} \sin(\theta/2)$

#Garda Visualisat^n.

if $\theta = \pi$ (Half Ring)

then, $E = \frac{2K\lambda}{r}$

if $\theta = 2\pi$ (Ring)

then $E = 0$

Electric Field Due to Line Charge :-

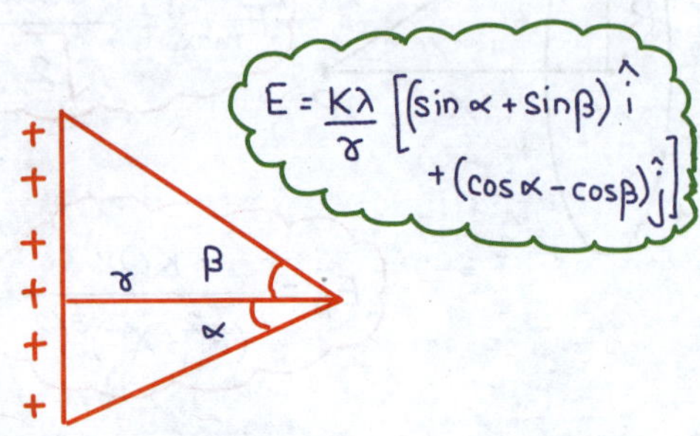

$E = \frac{K\lambda}{r}\left[(\sin\alpha + \sin\beta)\hat{i} + (\cos\alpha - \cos\beta)\hat{j}\right]$

✦ Infinite – line:- $E = \frac{2K\lambda}{r}$

✦ Semi- Infinite line:- $E = \frac{K\lambda}{r}(\hat{i} - \hat{j})$

✦ Finite line:- $E = \frac{K\lambda}{r}[(\sin\alpha + \sin\beta)\hat{i} + (\cos\alpha - \cos\beta)\hat{j}]$

✦ Finite line at 'r' when $(\alpha = \beta)$:- $E = \frac{2K\lambda}{r} \sin\alpha i$

IMP Case: Charge (–q, m) moving around infinite line charge with speed V.

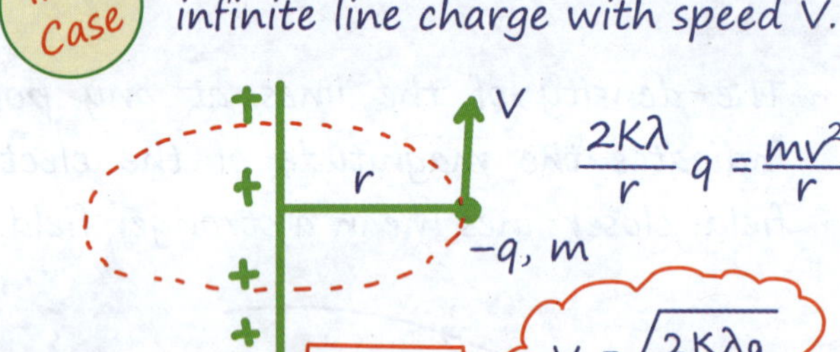

$\frac{2K\lambda}{r} q = \frac{mv^2}{r}$

$V \propto r^0$

$V = \sqrt{\frac{2K\lambda q}{m}}$

Q.5 Find electric field at centre ?

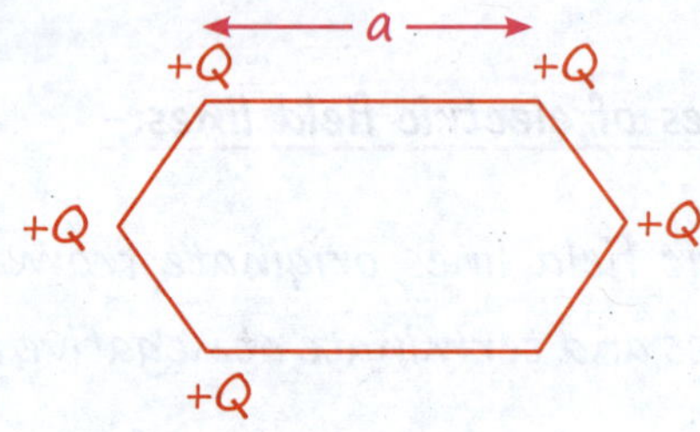

Sol. $E = \frac{KQ}{a^2}$

MR*

Put +Q and –Q charge on the corner, where no charge is present.

Motion of Charged Particle in Uniform E.F. :-

Charge is drop/released :-

Acceleration, $a = \frac{qE}{m}$, Velocity, $v = \frac{qEt}{m}$

Momentum, $p = qEt$

Distance covered, $s = \frac{1}{2}\frac{qEt^2}{m}$

Kinetic energy, $K.E = \frac{1}{2}\frac{q^2E^2t^2}{m}$

- Positron, P^+ = e,m. Deuteron = e, 2m.

  ∝-particle = 2e, 4m.

Charge is projected with "U_o" in dir^n of E.F. :-

- $a = \frac{qE}{m}$ = Const^n
- $V = U_o + \frac{qE}{m}t$
- $\Delta P = qEt$
- $S = U_o t + \frac{1}{2}\frac{qEt^2}{m}$
- $P_t = m\left[U_o + \frac{qEt}{m}\right]$

3> Charge is projected ⊥er to E.F. :-

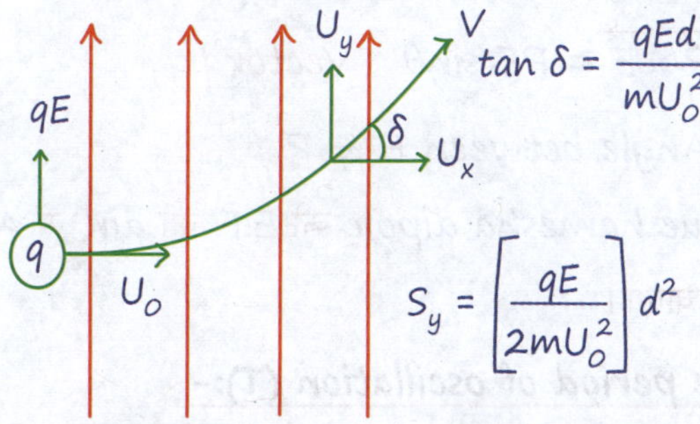

$\tan\delta = \frac{qEd}{mU_o^2}$

$S_y = \left[\frac{qE}{2mU_o^2}\right]d^2$

Ex Proton, Deutron and α-particle projected in electric field perpendicular to it, then find ratio of deviation if:

(i) They projected with same speed

$\delta_P : \delta_D : \delta_\alpha = 2 : 1 : 1$ $\boxed{\delta \propto q/m}$

(ii) They are projected with same K.E.

$\delta_P : \delta_D : \delta_\alpha = 1 : 1 : 2$ $\boxed{\delta \propto q}$

(iii) Projected with same momentum

$\delta_P : \delta_D : \delta_\alpha = 1 : 2 : 8$ $\boxed{\delta \propto qm}$

Graph of Electric field with R due to Combination of Point Charge :-

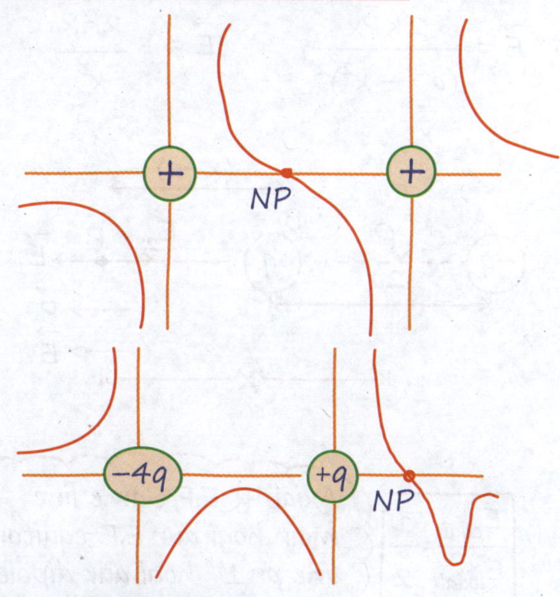

7 Electric Dipole

- Two equal and opposite charge placed at small distance.
- Unit = C.m, Net charge = 0, Ideal dipole = small dipole

$\vec{P} = \left[\begin{matrix}\text{Mag. of} \\ \text{Q.}\end{matrix}\right] \times \left[\begin{matrix}\text{Dist}^n\text{ between} \\ \text{two charge}\end{matrix}\right]$

- Direction of dipole moment -ve to +ve charge.

Equitorial line :-

$\vec{E} = \frac{-K\vec{P}}{(a^2 + x^2)^{3/2}}$ $\qquad \vec{E} = \frac{-K\vec{P}}{x^3}$

E
E sinθ
$\vec{P}$
$\vec{E}$
2E cosθ
θ
A
θ
E
E sinθ
$\sqrt{a^2+x^2}$
x
θ
a
a
θ
-q
+q

Axial line :-

$$\vec{E} = \frac{2KPx}{(a^2 - x^2)^2} \qquad \vec{E} = +\frac{2K\vec{P}}{x^3}$$

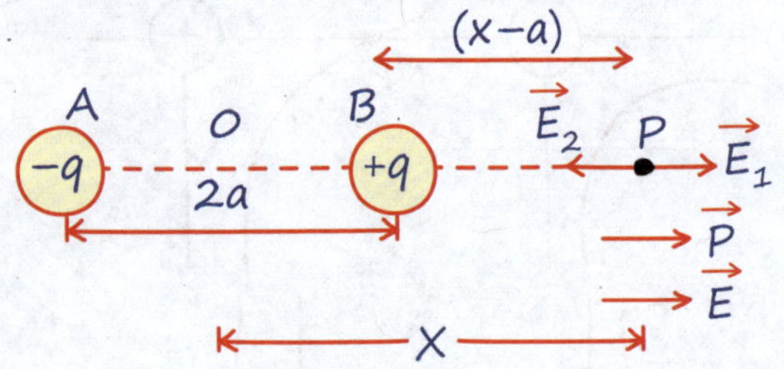

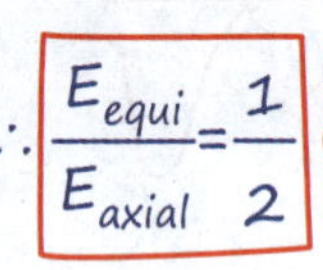

Axial पर E.F. Same line mein hogi aur E.F. equitorial line pr $\perp^{er}$ hogi aur dipole Ke opposite hogi.

At an angle θ from ideal dipole:-

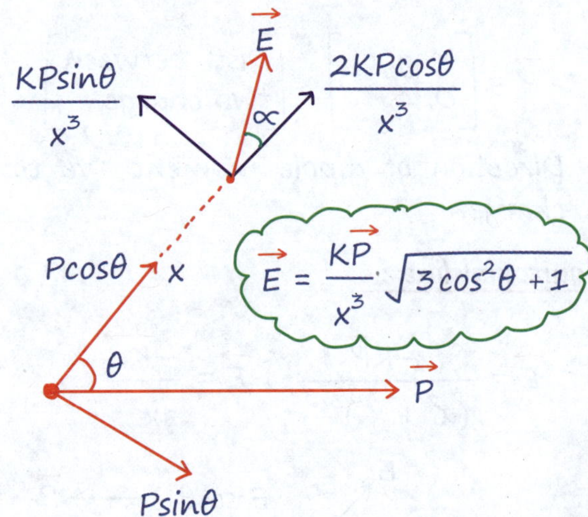

$\therefore$ Angle between $\vec{E}$ & $\vec{P}$ = $\theta + \propto$

$$\tan \propto = \frac{KP\sin\theta/x^3}{2KP\cos\theta/x^3} = \frac{\tan\theta}{2}$$

$\theta = \tan^{-1}(\sqrt{2})$ The Angle from dipole at which $\vec{E} \perp \vec{P}$

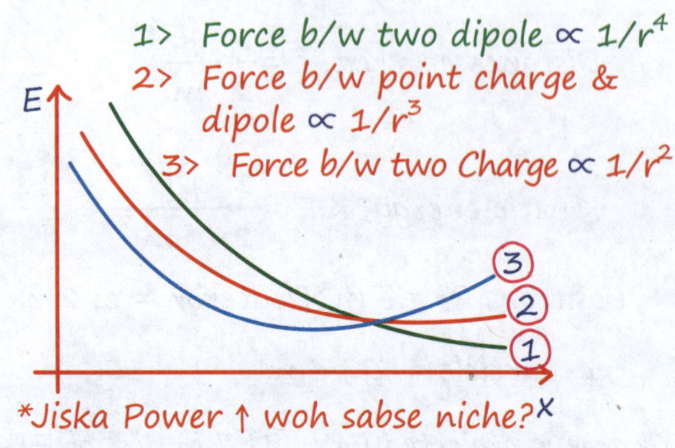

8 Electric oscillation in electric field

When an electric dipole is placed in a uniform electric field, it experiences a torque that tends to align the dipole with the field direction. This setup can lead to oscillatory motion if the dipole is displaced from its equilibrium position.

Torque on an Electric Dipole:-

$\vec{\tau} = \vec{P} \times \vec{E} = PE\sin\theta$ Vector !

θ = Angle between E & P.

Torque hamesha dipole को E.F. की dir^n में Align करना चाएगा!

Time period of oscillation (T):-

$$T = 2\pi\sqrt{\frac{I}{PE}}$$

I = Moment of Inertia

T = Time period of oscillation

P = Electric dipole moment

E = Electric field.

- Electric field on the axis of electric dipole is always parallel to electric dipole moment → False
- Electric field on the equitorial line of dipole is anti-parallel to dipole → True

Electrostatic P.E. Stored in Dipole in Uniform Electric field :-

Scalar.

$U_\theta = -\vec{P}\cdot\vec{E} = -PE\cos\theta.$

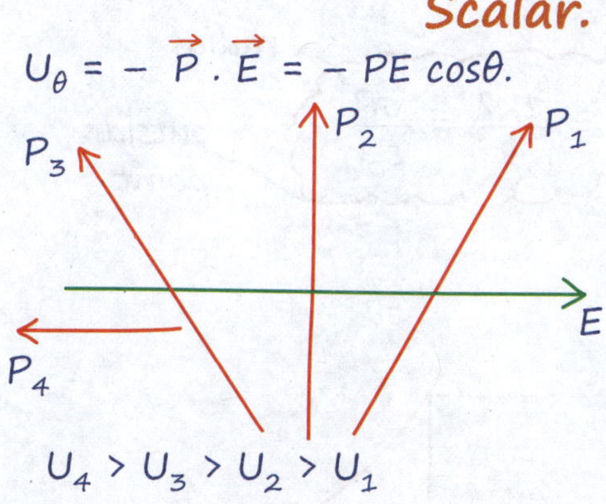

$U_4 > U_3 > U_2 > U_1$

$\theta = 0^0$	$\theta = 90^\circ$	$\theta = 180^\circ$
$\vec{P} \parallel^{el} \vec{E}$	$\vec{P} \perp \vec{E}$	$\vec{P}_{anti} \parallel^{el} \vec{E}$
$F = 0$	$F = 0$	$F = 0$
$\tau = 0$	$\tau = PE$ ⊗	$\tau = 0$
$U_{min} = -PE$	$U = 0$	$U_{max} = PE$

- ✦ Work Done to Rotate dipole $W = \Delta U = U_f - U_i$
- ✦ Work done by E.F. to rotate dipole $= w = -\Delta U$

*Special Case :-

- ✦ Work Done to Rotate from Stable to Unstable :- $U = 2PE$.

अगर जहासे Rotate किया वही पर वापस लाया तो W.D. = 0!

9 Electric Flux

$\phi = \text{flux} = \vec{E}\cdot\vec{A} = EA\cos\theta$

θ = Angle between $\vec{E}$ & $\vec{A}$

→ Gives the idea of electrostatic energy passing through given area.

→ Counting of field lines passing through given crosss-section area

→ Scalar

→ Unit (Volt - meter) & $(Nm^{+2}\ C^{-1})$

→ $d\phi = \oint \vec{E}\cdot d\vec{A}$ variable electric field.

Flux :- Aaisi line joh area Ko aarpar ched banakr jarahi hai !

3-D body :- $\phi_{in} = -ve\ \phi_{out} = +ve\ \phi_{Total} = \phi_{in} + \phi_{out}$

Uniform E.F.:- $(\phi_{Total})_{\text{close surface}} = 0.$

Q.6 A charge Q placed on the cornor of square plate of side 'L' then find flux through that square plate → zero.

10 Gauss Law

$\phi = \oint \vec{E}\cdot d\vec{A} = \dfrac{q_{enclosed}}{\varepsilon_0}$

ϕ = सिर्फ inside charge से आएगा

E = inside और outside दोनों से !

→ Flux from close surface does not depends shape, size of surface and location of charge inside surface and charge outside the surface.

→ Always valid

→ Only applicable to calculate electric field for symmetric charge distribution

Special Case :-

- ✦ Flux through any one of the faces not touching the corner charges :-

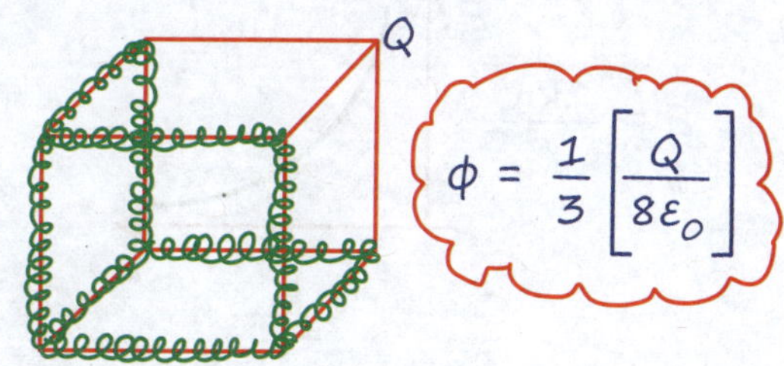

$\phi = \dfrac{1}{3}\left[\dfrac{Q}{8\varepsilon_0}\right]$

✦

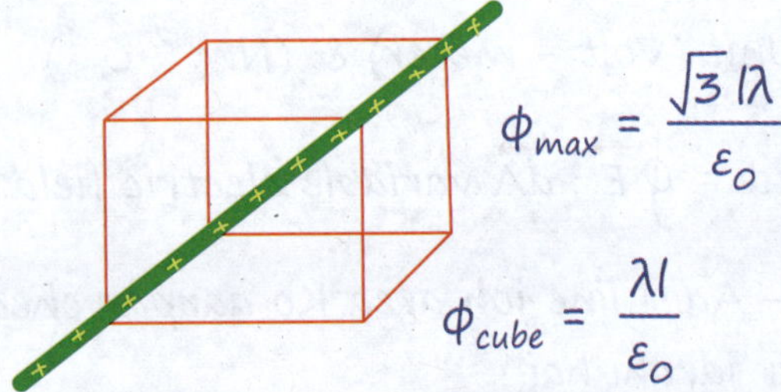

$$\phi_{max} = \frac{\sqrt{3}\,l\lambda}{\varepsilon_0}$$

$$\phi_{cube} = \frac{\lambda l}{\varepsilon_0}$$

Aaise Question mein pahile length nikalo, charge distribution dekho

$\phi = \frac{\lambda l}{\varepsilon_0}$ lagado !

✦ Relation between plane angle & solid angle :-

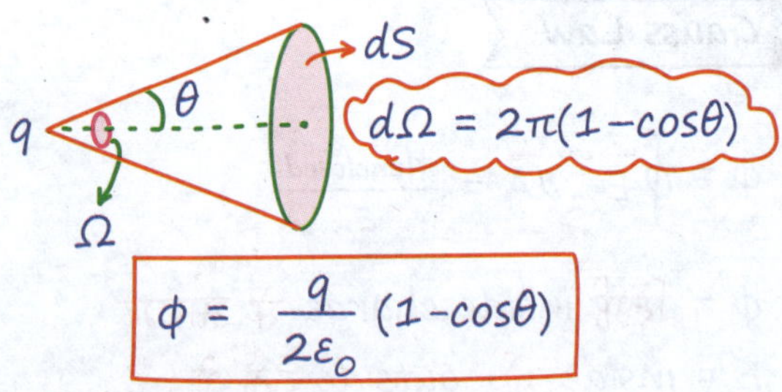

$$d\Omega = 2\pi(1-\cos\theta)$$

$$\phi = \frac{q}{2\varepsilon_0}(1-\cos\theta)$$

11 Application of gauss law

E.F. Due to point charge :-

$$E = \frac{KQ}{r^2}$$

E, r

Infinite line charge :-

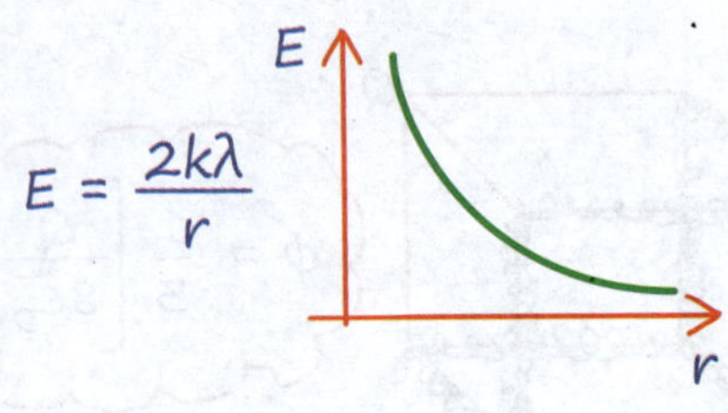

$$E = \frac{2k\lambda}{r}$$

Infinite hollow/solid conducting or hollow non-conducting cylinder:-

$$E = \frac{2K\lambda}{x} = \frac{\sigma R}{\varepsilon_0 r}$$

R → radius

r → outside point

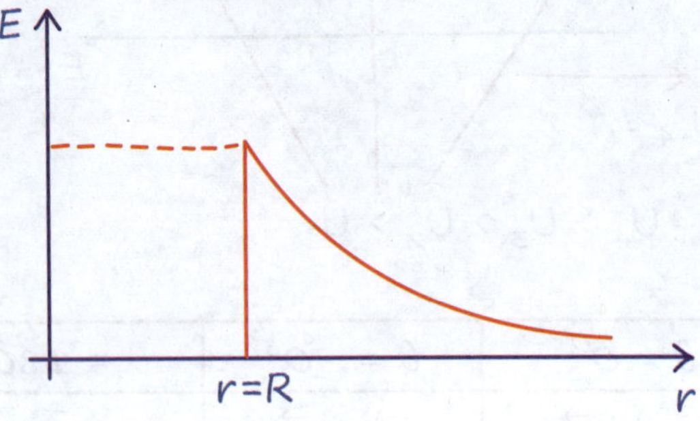

Solid non-conducting cylinder :-

$$E_{out} = \frac{\rho R^2}{2\varepsilon_0 r} \qquad E_{in} = \frac{\rho r}{2\varepsilon_0}$$

$$E_{surface} = \frac{\rho R}{2\varepsilon_0}$$

E, r = R, r

MR*

$$q_{in} = \lambda l = \sigma[2\pi R]l = \rho[\pi R^2]l$$

$$E = \frac{2K\lambda}{r}$$

E.F. Due to hollow conducting/N.C. Solid conducting sphere :-

$E_{in} = 0$

$E_{out} = \frac{KQ}{r^2}$

$E_{sur} = \frac{KQ}{R^2}$

E, r

E.F due to solid non-conducting sphere :-

$E_{in} = \frac{KQr}{R^3}$ $\quad E_{out} = \frac{KQ}{r^2}$

$\rho = \frac{Q}{\frac{4}{3}\pi R^3}$

$E_{sur} = \frac{KQ}{R^2}$ $\quad E_{in} = \frac{\rho r}{3\varepsilon_0}$

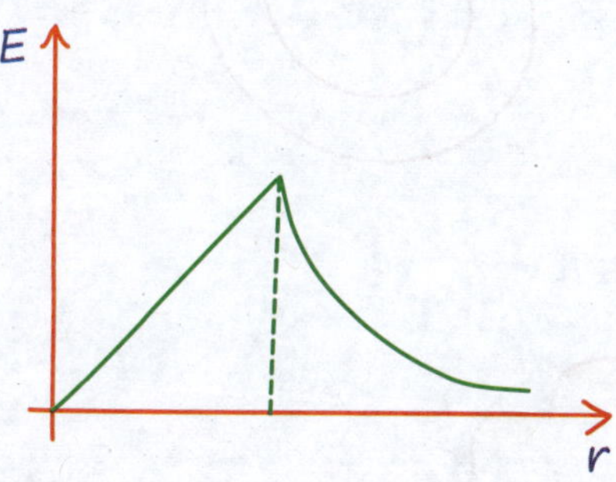

Spherical Shell:-

(like Hollow Sphere).

$E_{out} = \frac{KQ}{r^2}$ $\quad E_{sur} = \frac{KQ}{R^2}$ $\quad E_{in} = 0$

Conducting Shell :-

E.F. inside isolated Conductor is zero.

E.F. Inside cavity of non-conducting solid sphere :-

distn b/w centre of sphere to centre of cavity

$\vec{E}_P = \frac{\rho \vec{r}}{3\varepsilon_0}$ (uniform) $\vec{r} = \vec{r}_0 - \vec{r}_1$

$\vec{r}_0$ from Sphere, $\vec{r}_1$ from Cavity

E.F. Due to infinite large non conducting plate/sheet :-

$$E = \frac{\sigma}{2\varepsilon_0} = \frac{Q}{2A\varepsilon_0}$$

Kuch na bola jaye toh by default Non-conducting consider Karo!

E.F. Due to conducting plate :-

$$E = \frac{\sigma}{\varepsilon_0} = \frac{Q}{2A\varepsilon_0}$$

Charge Ke Terms मे दोनो sheet Ka Formula same है!

Electric field due to charge disc:-

$$E = \frac{\sigma}{2\varepsilon_0}\left(1 - \frac{x}{\sqrt{R^2 + x^2}}\right)$$

Charge Distribution :-

Charges on outer surface of plate = $\frac{\text{Total Charge}}{2}$

$$\begin{array}{c|c} 2Q & 3Q \end{array} \Rightarrow \begin{array}{c|c|c|c} \frac{5Q}{2} & \frac{-Q}{2} & \frac{+Q}{2} & \frac{5Q}{2} \end{array}$$

$$\frac{5Q}{2} - \frac{Q}{2} = \underline{\underline{2Q}}$$

$$\frac{5Q}{2} + \frac{Q}{2} = \underline{\underline{3Q}}$$

- Charge distribution on concentric sphere :-

$$q\left(\begin{array}{l}\text{final charge on outer}\\ \text{surface of outer sphere}\end{array}\right) = q_1 + q_2$$

- Charge on inner surface of outer sphere $= q_2 - q$

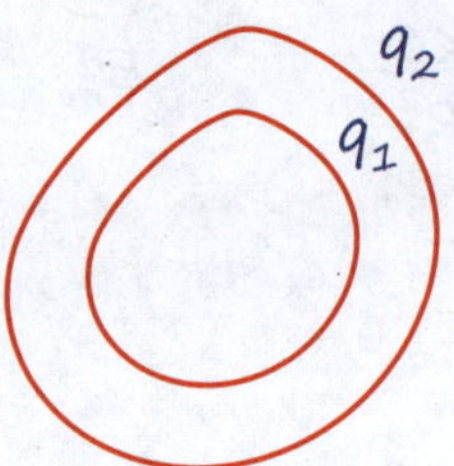

MR*

'मुश्किल नहीं है कुछ दुनिया में,
तू जरा हिम्मत तो कर।
ख्वाब बदलेंगे हकीकत में,
तू जरा कोशिश तो कर॥'

22

Electric Potential

1 Electric Potential

Work done in bringing unit positive charge from infinity to the point without change in K.E. is called electric potential at that point. (or) Negative work done by electric force in bringing unit positive charge from infinity to that point.

- depends upon refrence frame
- unit: volt, J/C, weber/sec, N-m/C

Potential :-

$$\Delta V = -\int \vec{E}\cdot\vec{dr} = \text{Area of } -E/r \text{ graph}$$

∫v dr Area ← नाली ← V → E → slope E/r नाली (diff)

$$E = -\frac{dV}{dr} \text{ (diff}^n\text{)}$$

$$E = -(\text{slope of } V/r \text{ graph})$$

$$\Delta V = -\int_{r_A}^{r_B} \vec{E}\cdot\vec{dr} = -\int_{r_A}^{r_B} \frac{F_{CF}}{q}\cdot dr = -\frac{W_{CF}}{q}$$

- Scalar
- Ref. at ∞, V = 0.
- Electric potential energy is defined only in a conservative field.

$$\# \quad E = -\frac{dV}{dr} = -\text{[Slope of } V-r \text{ graph]}$$

$$E = -\left[\frac{\partial v}{\partial x}\hat{i} + \frac{\partial v}{\partial y}\hat{j} + \frac{\partial v}{\partial z}\hat{k}\right]$$

- Potential decrease in the direction of Electric Field.
- Potential due to positive charge at some distance must be positive → False
- Due to positive charge potential may be +ve, −ve, zero depends on refrence → True

2 ← x_2 → (−Q) ← x_1 → 1 (nQ), separation r

\# At two point potential will be zero one, b/w the charges and 2nd left of smaller charge (V = 0 karne ke liye dono charge opposite nature ke hona Chahiye)

- $x_1 = \frac{r}{n+1}$ • $x_2 = \frac{r}{n-1}$

2 Potential due to

Point charge :-

$$\oplus = V_P = \frac{KQ}{r}$$

$$\ominus \; V_P = -\frac{KQ}{r}$$

(V vs r graph)

- For n point charges $q_1, q_2, \ldots, q_n$ located at distances $r_1, r_2, \ldots, r_n$ from a point P, the electric potential V at P is given by:

$$V = \frac{1}{4\pi\epsilon_0} \sum_{i=1}^{n} \frac{q_i}{r_i}$$

Ring :-

$$V_{axis} = \frac{KQ}{\sqrt{R^2 + X^2}},\ V_C = \frac{KQ}{R},\ V_{(X>>>R)} = \frac{KQ}{X}$$

V

r

Charges at corner of square :-

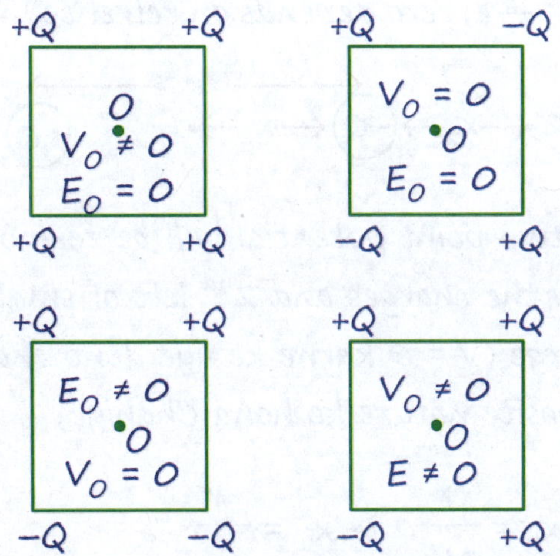

Axis of line charge rod :-

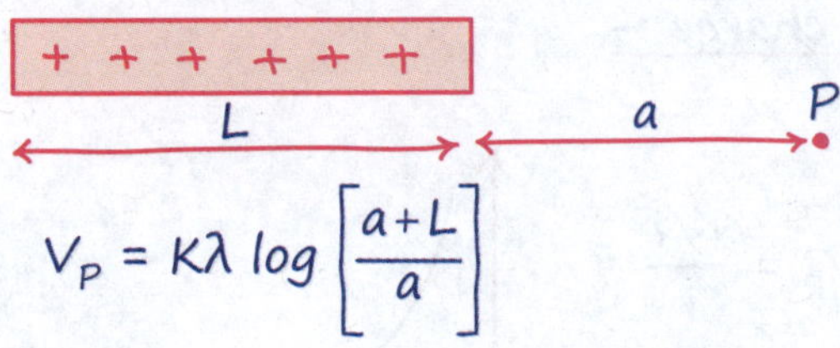

$$V_P = K\lambda \log\left[\frac{a+L}{a}\right]$$

Half ring :-

$$V_O = \frac{KQ}{R}$$

Q

O

R

Hollow conducting / non-conducting or solid conducting sphere :-

- $V_O = \frac{KQ}{r}$
- $V_S = \frac{KQ}{R}$
- $V_i = \frac{KQ}{R}$
- $V_C = \frac{KQ}{R}$

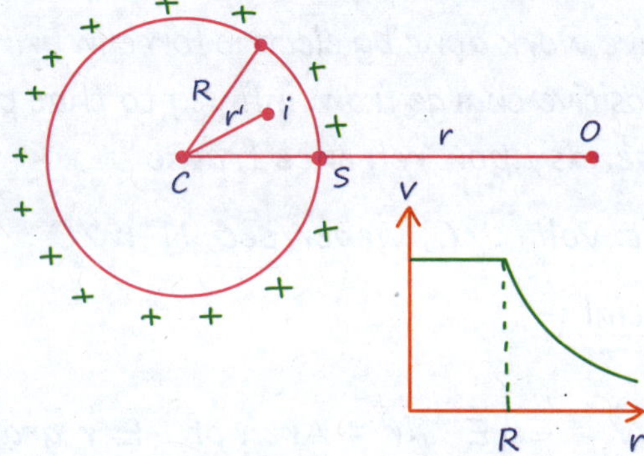

Solid non-conducting sphere :-

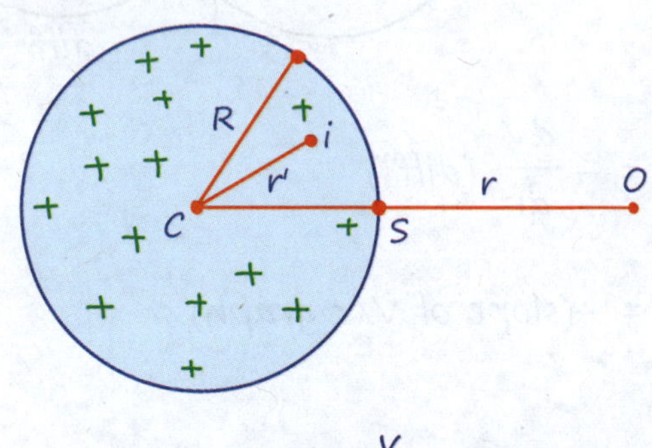

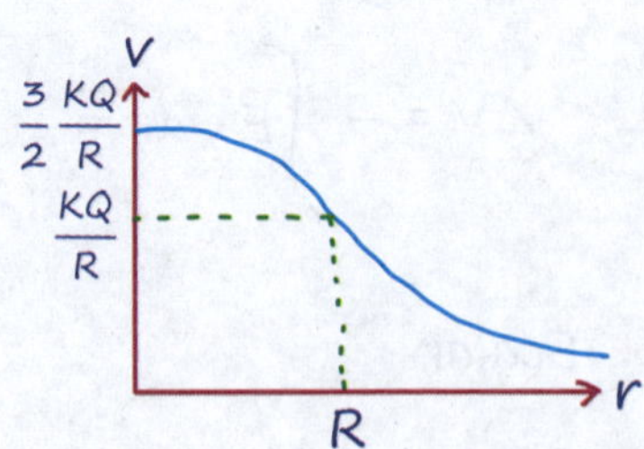

- $V_O = \frac{KQ}{r}$
- $V_S = \frac{KQ}{R}$
- $V_{in} = \frac{KQ}{2R^3}[3R^2 - r^2]$
- $V_C = \frac{3KQ}{2R}$

Electric Potential due to dipole at general point:-

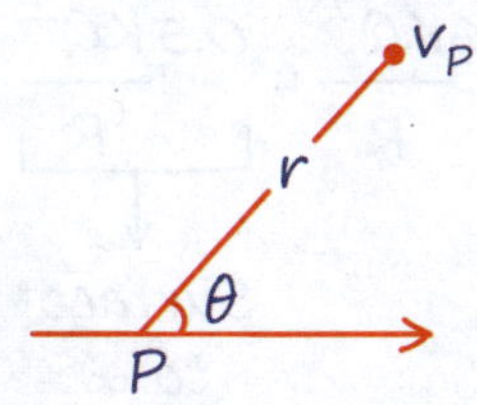

- $V_P = \dfrac{KP\cos\theta}{r^2}$
- $V_{axial} = \dfrac{KP}{(r^2 - a^2)}$
- $V_{equitorial} = 0$

- If potential zero then electric field must be zero – Flase
- If field is zero then potential must be zero – False
- If potential constant then field is zero – True
- If electric field constant then potential is zero – False
- If we move perpendicular to field potential remains constant. – True
- Potential increases if we move opposite to the direction of electric field.– True

Note:-

Potential difference does not depends upon reference point its always same between two points!

Potential Energy of System of Charged Particles :-

- For a charge q in a potential V: $U = qV$
- Using the electric field E: $U = q[-\int \vec{E} \cdot d\vec{r}]$

(i) Two charges system:

$+Q_1$ —— r —— $+Q_2$

$$U = \frac{KQ_1Q_2}{r}$$

(ii) Three charges system:

- For a system consisting of three point charges Q_1, Q_2, and Q_3, arranged in a triangular configuration with distances r_1, r_2, and r_3 between them, the total potential energy U of the system is given by the sum of the potential energy contributions from each pair of charges.

$$U = \frac{1}{4\pi\epsilon_0}\left(\frac{Q_1Q_2}{r_1} + \frac{Q_2Q_3}{r_2} + \frac{Q_3Q_1}{r_3}\right)$$

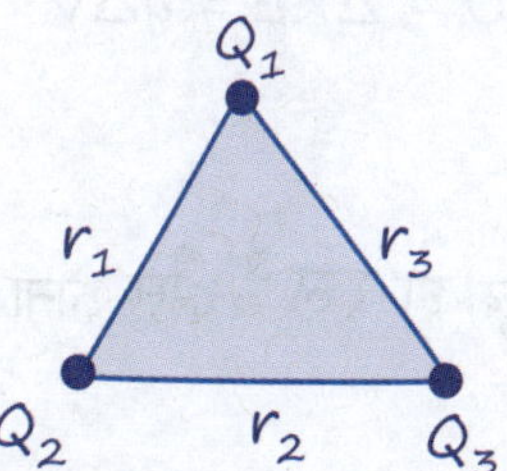

- If $Q_1 = Q_2 = Q_3 = Q$, then

$$U_{net} = \frac{3KQ^2}{L}$$

- For a system of four point charges Q_1, Q_2, Q_3, and Q_4, the electric potential energy U is

$$U = \frac{1}{4\pi\epsilon_0}\left(\frac{Q_4Q_3}{r_{43}} + \frac{Q_4Q_2}{r_{42}} + \frac{Q_4Q_1}{r_{41}} + \frac{Q_3Q_2}{r_{32}} + \frac{Q_3Q_1}{r_{31}} + \frac{Q_2Q_1}{r_{21}}\right)$$

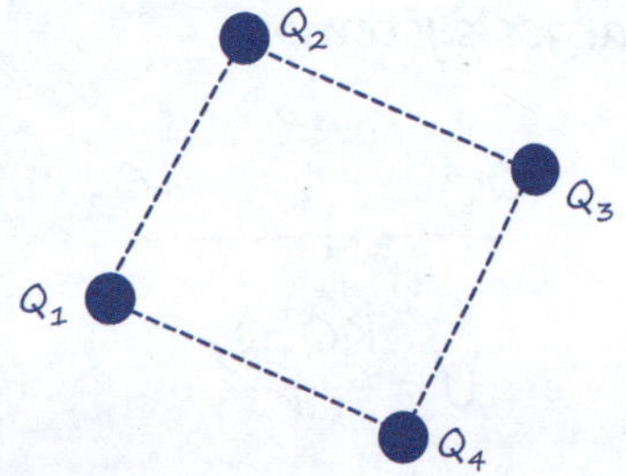

- If $Q_1 = Q_2 = Q_3 = Q_4 = Q$

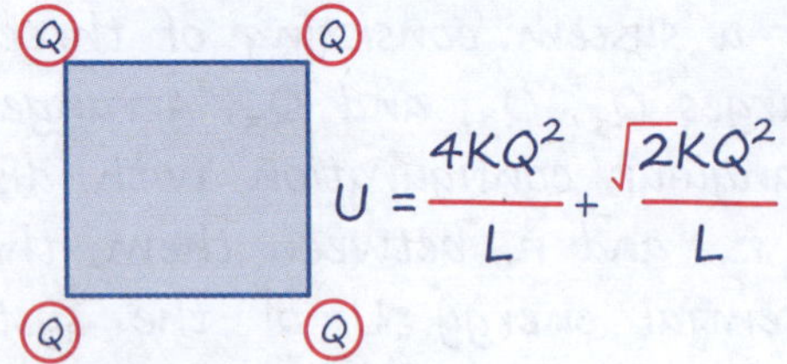

$$U = \frac{4KQ^2}{L} + \frac{\sqrt{2}KQ^2}{L}$$

Vimp.

Total no. of terms of PE :- $\frac{n(n-1)}{2}$

n = no. of Charges

- $W_{ext} = +\Delta U = \Delta KE = q\Delta V$

Note:-

जो काम खुद हो रहा है जैसा होना चाहिए वैसा हो रहा है तो :-

U↓ ⊕→←⊖ r↓

अगर जबरदस्ती किए :-

U↑ ⊕→←⊕ r↓

- Pressure :- $\frac{\sigma^2}{2\varepsilon_0}$ $\quad E = \frac{\sigma}{\varepsilon_0}$

3 Self Energy

Hollow conducting / non-conducting or solid conducting sphere :-

- $U = \frac{KQ^2}{2R}$

Solid non-conducting sphere :-

$$U_{Total} = \frac{0.6KQ^2}{R} = \underbrace{\frac{0.5KQ^2}{R}}_{\text{Surface to } \infty} + \underbrace{\frac{0.1KQ^2}{R}}_{\text{Centre to Surface}}$$

4 Equipotential Surface

- An equipotential surface is an imaginary surface where every point has the same electric potential. This means there is no potential difference between any two points on this surface.

Types of Equipotential Surfaces

Charge Distribution	Shape of Equipotential Surface
Point Charge	Spherical
Infinite Line Charge	Cylindrical
Infinite Charged Plate	Plane
Large Distance from Ring/Charge Distribution	Spherical

Characteristics:-

- Potential Difference (ΔV) is Zero.
- Work (W) Done is Zero.
- Equipotential surfaces are always at right angles (perpendicular) to electric field lines.

5 Electrostatics of Conductor

Electrostatic shielding

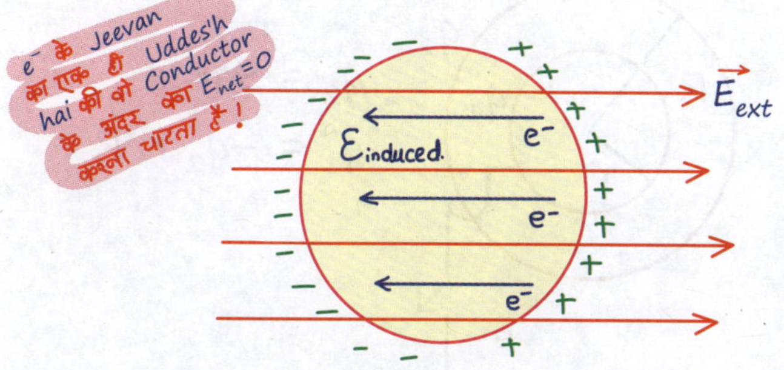

$\therefore\ \varepsilon_{net} = \varepsilon_{ext} + \varepsilon_{ind}$

$\varepsilon_{net} = 0.$

$\varepsilon_{ext} = -\varepsilon_0$

$\varepsilon_{induce} = \varepsilon_0$

\# Jitna External E. Field hoga Utna opposite में ε_{induce} hoga.

Q.1 Neutral conducting sphere, find electric field of center due to + q charge.

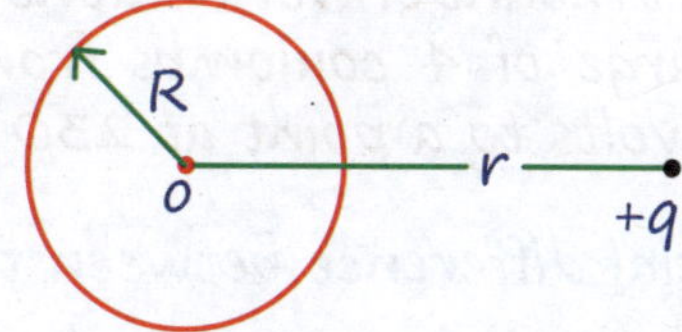

Ans. $\vec{E}_q = \frac{Kq}{r^2}$ $\quad E_{net}=0$

$\vec{E}_{induced} = -\vec{E}_q$

6 Important points

✦ 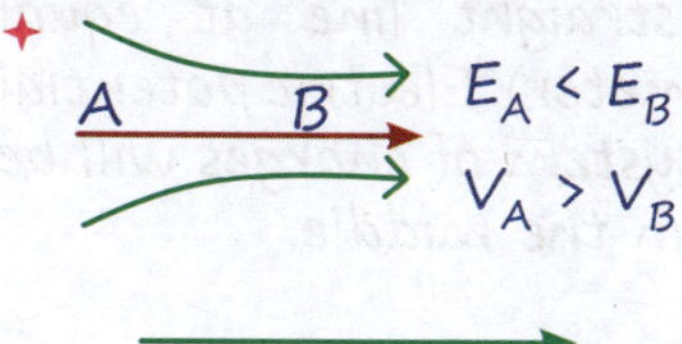

$E_A < E_B$

$V_A > V_B$

✦ 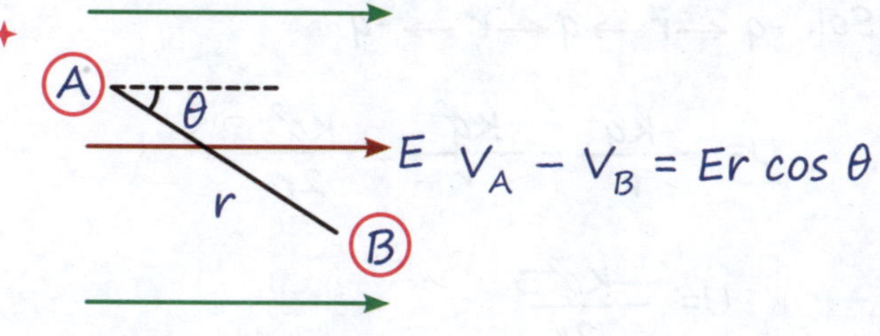

$V_A - V_B = Er\cos\theta$

✦ 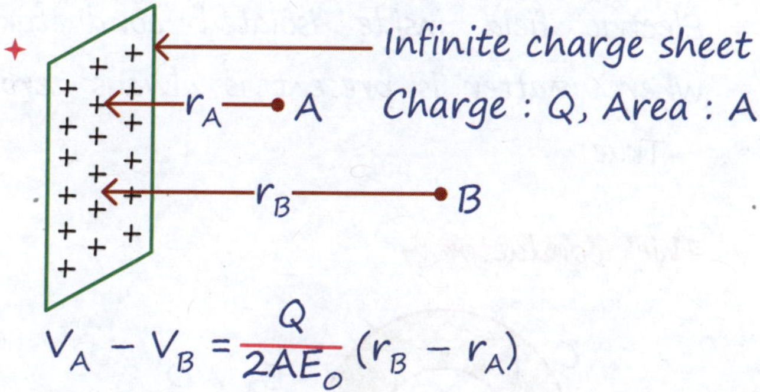

$$V_A - V_B = \frac{Q}{2A\varepsilon_0}(r_B - r_A)$$

✦ Find potential difference b/w two concentric sphere

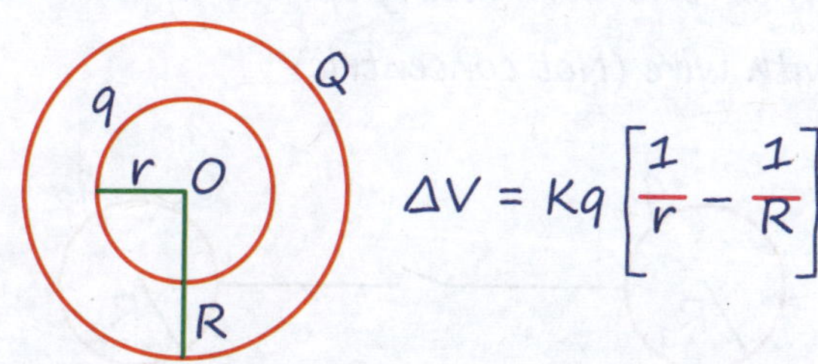

$$\Delta V = Kq\left[\frac{1}{r} - \frac{1}{R}\right]$$

✦ Potential difference does not depends on charge on outer sphere only on inner charge.

✦ When two concentric spherical charge connected with wire the potential difference becomes zero, all the charge of inner sphere will flow to outer sphere.

✦ $W_{ext} = \Delta U = q\Delta V$

✦ $W_{electric\ field} = -\Delta U = -q\Delta V$

✦ If external force is absent, we can apply (C.O.M.E.) conservation of mechanical energy or $\Delta U_{loss} = K.E._{gain}$

$$K.E. = q\Delta V$$

✦ Electric field inside conductor is zero – False

- Electric field inside isolated conductor where matter is present is always zero – True

- Thick conductor :-

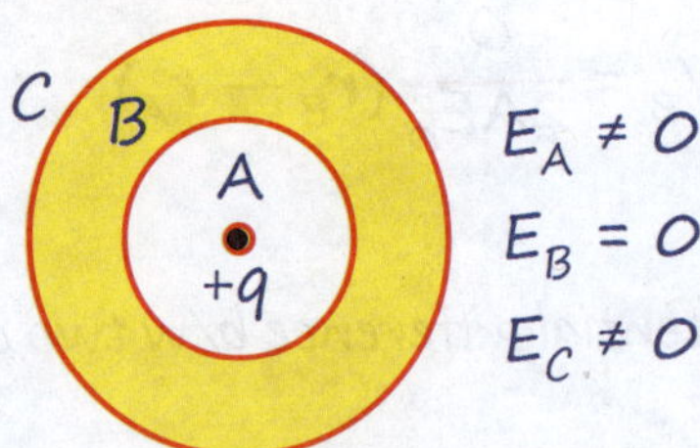

$E_A \neq 0$

$E_B = 0$

$E_C \neq 0$

- When two conducting sphere is connected with wire (Not concentric)

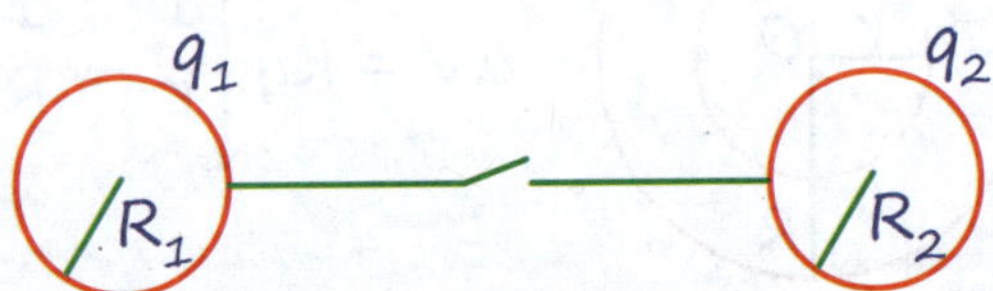

Final charge $q \propto$ Radius (R)

Potential V = Same

Surface charge density $\sigma \propto \frac{1}{\text{Radius (R)}}$

Electric field on surface $E \propto \frac{1}{\text{Radius (R)}}$

- After earthing potential of conductor must be zero, charge may or may not be zero.

- Final charge on each sphere when outer sphere is grounded.

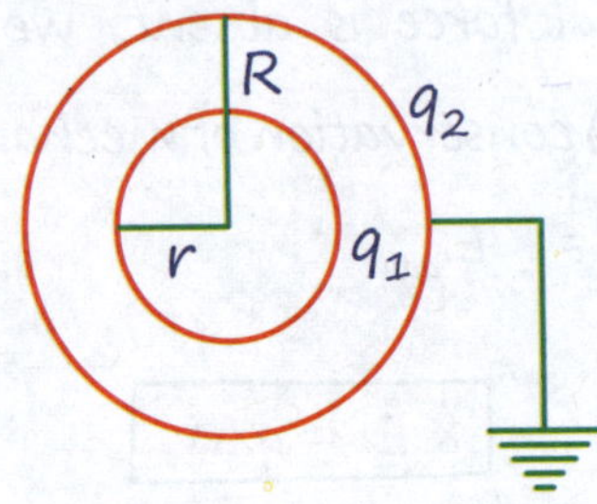

Jisko ground nahi kiya hai uska charge same rahega. Jisko ground kiya hai, uska potential zero hoga and final charge on outer sphere

$$q'_2 = -q_1$$

- 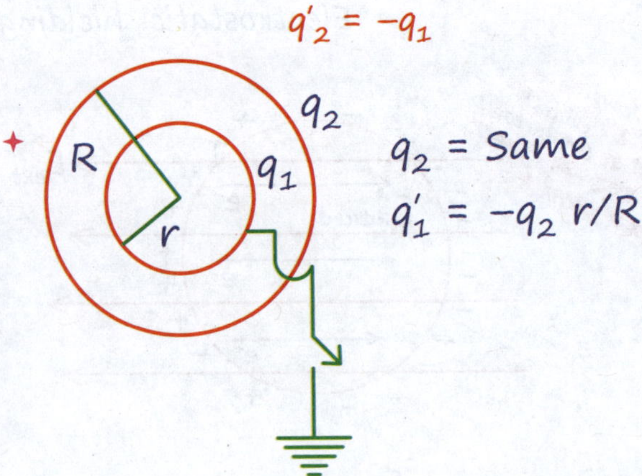

q_2 = Same

$q'_1 = -q_2\, r/R$

- From large distance then distance of closest approach

$q \rightarrow V_0$ fixed $+Q$

$$r = \frac{2KQq}{mV_0^2}$$

Q.2 What amount of work is done in moving a charge of 4 coulombs from a point 220 volts to a point at 230 volts?

Sol. Potential difference between the two points

$\Delta V = 230 - 220$ volts

$\therefore \Delta V = 10$ volts

Amount of charge moving q = 4 coulombs

Thus work done $W = q\Delta V$

$\therefore W = 4 \times 10 = 40$ joules

Q.3 Three point charge –q, +q and –q are placed along straight line at equal distance (say r meter). Electric potential energy of this system of charges will be if +q charge is in the middle.

Sol. $-q \leftarrow r \rightarrow q \leftarrow r \rightarrow -q$

$$U = \frac{-Kq^2}{r} - \frac{Kq^2}{r} + \frac{Kq^2}{2r}$$

$$U = \frac{-Kq^2 3}{2r}$$

Q.4 In an uniform electric field E = 10N/C as shown in figure, find $V_A - V_B$:

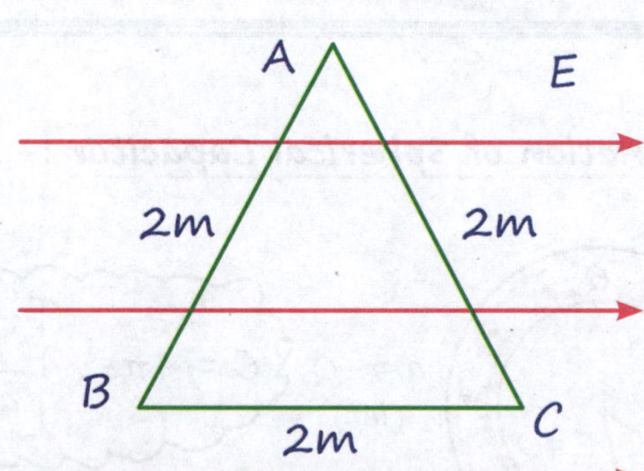

Sol. We know that Electric field, distance and voltage are related by formula,

E.d = V

Subsituting values,

V = 10 × 2 × cos 120°, V = – 10V

Q.5 The work done to move a charge on an equipotential surface is?

Sol. Workdone in equipotential surface is zero as $W=q(V_A-V_B)$ & $V_A=V_B$

Q.6 Equipotential surfaces are shown in fig. then the electric field strength will be

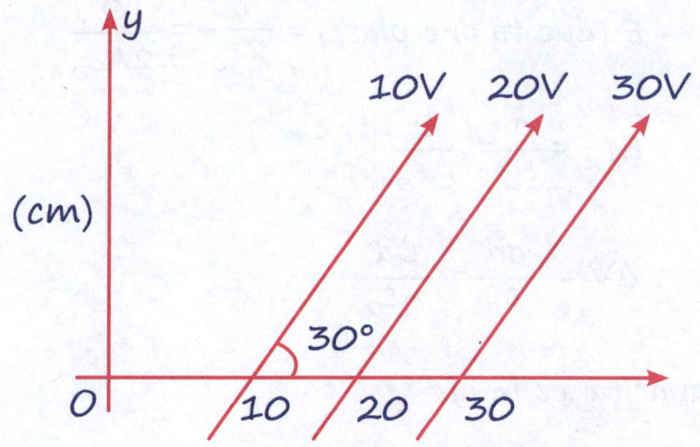

Sol. $\Delta V = -E.\Delta r\cos\theta$

$$E = \frac{-\Delta V}{\Delta r\cos\theta},\ E = \frac{-(20-10)}{10\times10^{-2}\cos120°}$$

$$= \frac{-10}{10\times10^{-2}(-\sin30°)}$$

$$= \frac{-10^2}{-1/2} = 200\ V/m$$

Q.7 If potential at centre of Non conducting sphere is zero then find potential at surface of sphere?

Sol. ΔV = Same, does not depends on reference

$$V_O - V_S = \frac{3}{2}\frac{KQ}{R} - \frac{KQ}{R}$$

$$O - V_S = \frac{KQ}{2R}$$

$$V_S = -\frac{KQ}{2R}$$

Q.8 Equipotential surface due to point charge then compair r_1 and r_2.

Sol. Hint $E = -\frac{dv}{dR}$

$r_1 < r_2$

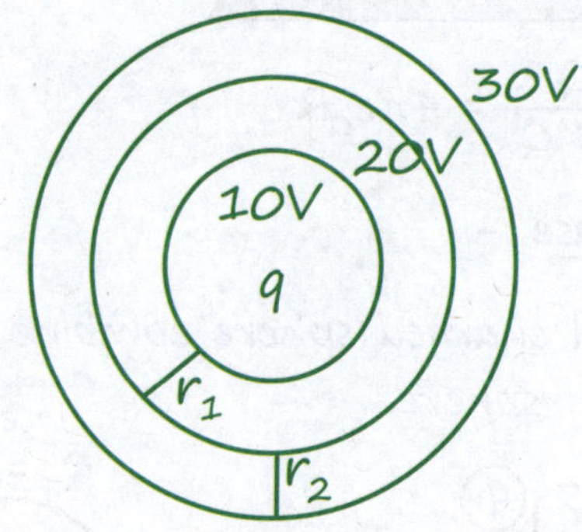

MR*

❛ Aasma bhi jhukega tere aage yu hi junun ke had se guzarte raho pura jeevan ek sangarsh hai ladte raho or aage badhte raho. ❜

23

Capacitor

1 Capacitor

- An electrical device which store electrostatic energy by storing charge.
- Generally it is combination of two conductor, having equal and opposite charge.

$Q = CV$ $\quad \frac{Q}{Volt} = Farad$

* Depends on size, shape & Medium between them.

$C = \frac{Q}{\Delta V} = \frac{Q}{V_1 - V_2}$ $\quad$ +Q V_1 $\quad$ −Q V_2

→ Slope of Q/v graph

→ Scalar

→ unit q/v = farad

→ Dim → $[M^{-1}L^{-2}T^{4}A^{2}]$

2 Spherical Capacitor

$C = \frac{Q.R}{KQ} = 4\pi\varepsilon_0 R$

Special Case :-

- n small charged sphere combine to form a bigger sphere.

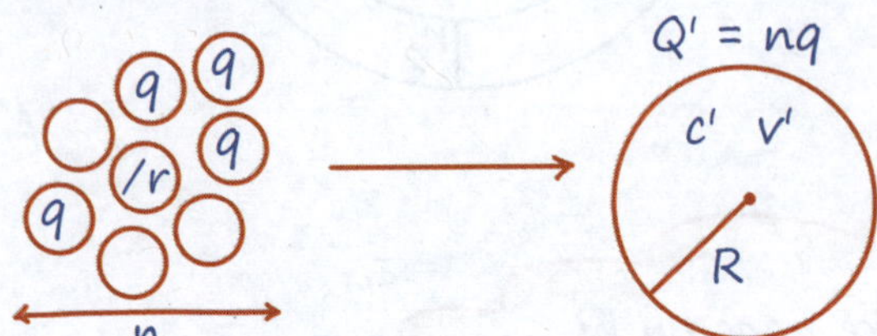

For bigger sphere,

$C' = n^{1/3}C$ $\quad V' = n^{2/3}V$

$U' = n^{5/3}U_0$ $\quad R = n^{1/3}r$

Combination of Spherical Capacitor :-

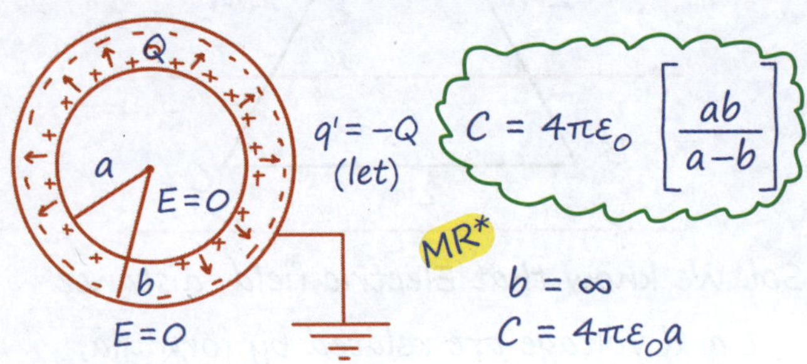

$C = 4\pi\varepsilon_0 \left[\frac{ab}{a-b}\right]$

MR*

$b = \infty$

$C = 4\pi\varepsilon_0 a$

Parallel Plate Capacitor :-

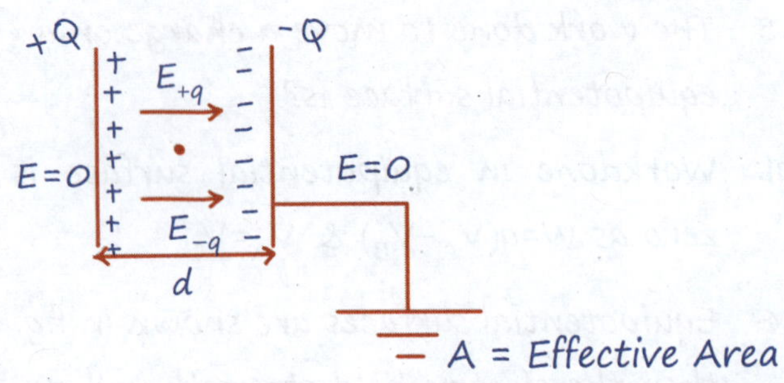

A = Effective Area

- $E \text{ (due to one plate)} = \frac{\sigma}{2\varepsilon_0} = \frac{Q}{2A\varepsilon_0}$
- $E_{net} = \frac{\sigma}{\varepsilon_0} = \frac{Q}{A\varepsilon_0}$
- $\Delta V = \frac{\sigma d}{\varepsilon_0} = \frac{Qd}{A\varepsilon_0}$

Capacitance in air

- $\text{(air) } C_0 = \frac{Q}{\Delta V} = \frac{A\varepsilon_0}{d}$
- $\text{(medium)} C' = \frac{A\varepsilon_m}{d}, \; C' = k\frac{A\varepsilon_0}{d}$

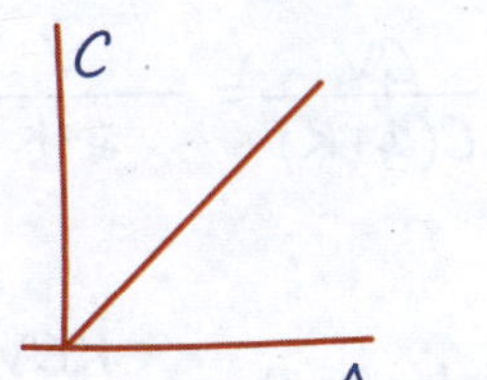

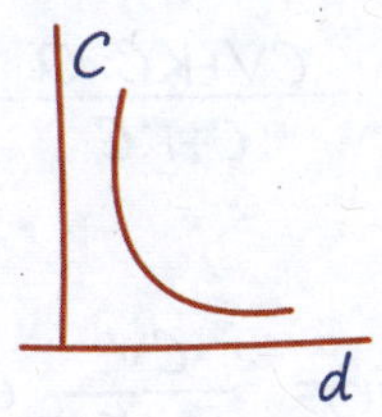

- Dielectric introduce between plates :-

$$C' = KC_o$$

Force between the Plates :-

$$F = \frac{Q^2}{2A\varepsilon_o} = \frac{\sigma^2 A}{2\varepsilon_o} \qquad (\because\ F = qE)$$

Pressure on the plate :-

$$P = \frac{F}{A} = \frac{Q^2}{2A^2\varepsilon_o} = \frac{\sigma^2}{2\varepsilon_o}$$

Energy density :-

$$U = \frac{\sigma^2}{2\varepsilon_o} = \frac{\text{Energy}}{\text{Volume}} = \frac{Q^2}{2CAd}$$

- Between the plate net electric field only due to charge on Inner surface of plate $E_{net} = \frac{Q}{A\varepsilon_o}$, Q = charge of Inner surface.

Note :-

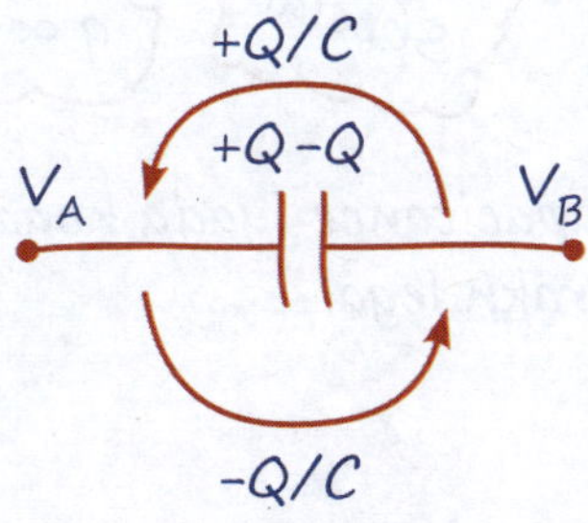

A to B	B to A
$V_A - Q/C = V_B$	$V_B + Q/C = V_A$
$V_A - V_B = Q/C$	$V_A - V_B = Q/C$

Energy on Capacitor :-

$$E = \frac{1}{2}CV^2 = \frac{1}{2}\frac{Q^2}{C} = \frac{1}{2}QV$$

Self Energy :-

For Spherical Capacitor :-

$$U = \frac{Q^2}{8\pi\varepsilon_o R}$$

Cylindrical Capacitor :-

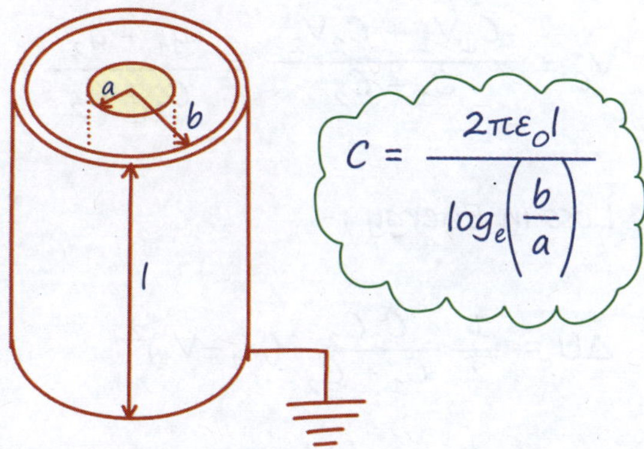

$$C = \frac{2\pi\varepsilon_o l}{\log_e\left(\frac{b}{a}\right)}$$

Charging of Capacitor :-

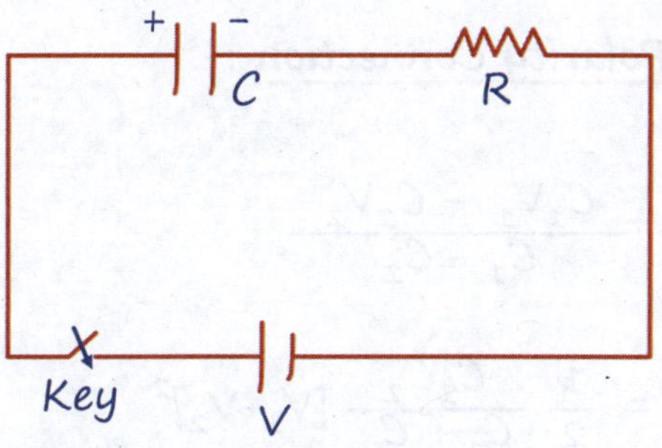

- WD by battery in Charging Capacitor $= CV^2$
- WD in Charging Capacitor $= \frac{1}{2}CV^2$
- $U_{loss} = \frac{1}{2}CV^2$ Heat

MR* for Calculating E_{loss}

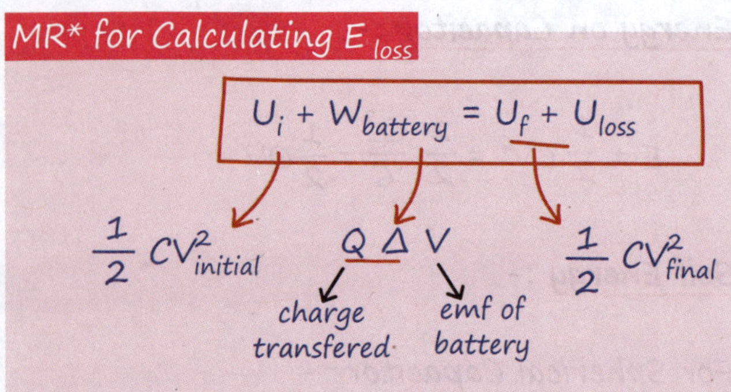

3 Connection of Two Capacitor

Same Terminal Connection :-

$$V_C = \frac{C_1V_1 + C_2V_2}{C_1 + C_2} = \frac{q_1 + q_2}{C_1 + C_2}$$

Loss in Energy :-

$$\Delta U = \frac{1}{2}\frac{C_1C_2}{C_1 + C_2}(V_1 - V_2)^2$$

Final Charge :- $q'_1 = C_1V_C$

$q'_2 = C_2V_C$ Transfer :- $[q_1 - q'_1]$

Reverse Polarity Connection :-

$$V_C = \frac{C_1V_1 - C_2V_2}{C_1 + C_2}$$

$$\Delta U = \frac{1}{2}\frac{C_1C_2}{C_1 + C_2}[V_1 + V_2]^2$$

- An air filled capacitor charged with potential V connected with identical uncharged capacitor filled with k-dielectric then common potential and final charge on each.

$$V_C = \frac{CV + KC\times O}{C + KC} = \frac{CV}{C(1+K)} = \frac{V}{1+K}$$

$$Q(air) = \frac{CV}{1+K} \quad Q(dielectric) = \frac{KCV}{1+K}$$

4 Combination of Capacitor

Series combination of capacitors :-

Q = Same

$V = V_1 + V_2 + \ldots$

$$\frac{1}{C_{eq}} = \frac{1}{C_1} + \frac{1}{C_2} + \frac{1}{C_3} + \ldots$$

Law of Potential drop :-

$Q = CV$ = same.

Jiska Capacitance Jyada hoga Usmein Voltage drop Kam hoga.

Parallel combination of capacitors :-

V = Same

$Q = q_1 + q_2 + \ldots$

$C_{eq} = C_1 + C_2 + C_3 + \ldots$

Law of Charge Distribution :-

$q = CV$ MR* Special $q \propto C$

Jiska Capacitance jyada hoga woh jyada charge rakh lega!

MR*

V_1 —| C_1 |— V_C —| C_2 |— V_2

If $V_1 > V_2$

$$V_C = \frac{C_1V_1 + C_2V_2}{C_1 + C_2}$$

$Q_1 = C_1(V_1 - V_C)$

$Q_2 = C_2(V_C - V_2)$

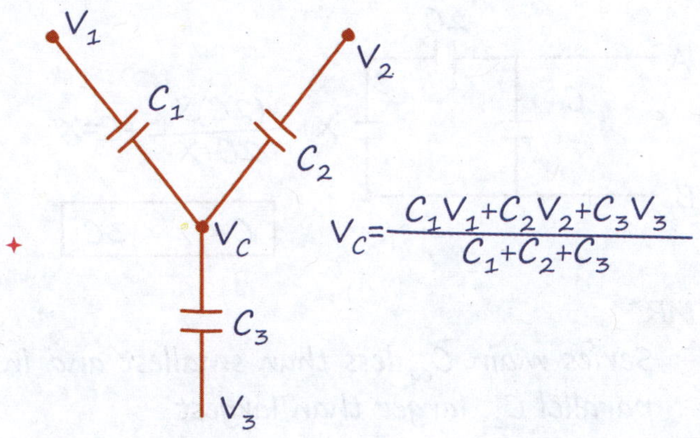

$$V_C = \frac{C_1V_1 + C_2V_2 + C_3V_3}{C_1 + C_2 + C_3}$$

Note:-

+ If n-indentical Connected in Series & Parallel :-

* Series :-

$$C_{eq} = \frac{C}{n}$$

C_{eq} is smaller than Smallest Capacitor

* Parallel :-

$$C_{eq} = nC.$$

C_{eq} is larger than largest Capacitor.

+ Breakdown Voltage :-

V_{max} that can be applied across a Capacitor.

* Series :-

$$V_B = V_1 + V_2 + V_3 + ...$$

* Parallel :-

$$V_B = V$$

+ Special Case :-

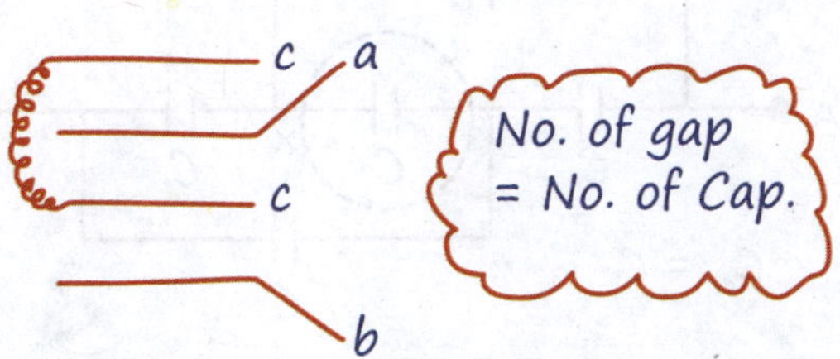

No. of gap = No. of Cap.

Q.1. Find charge and potential difference across each capacitor.

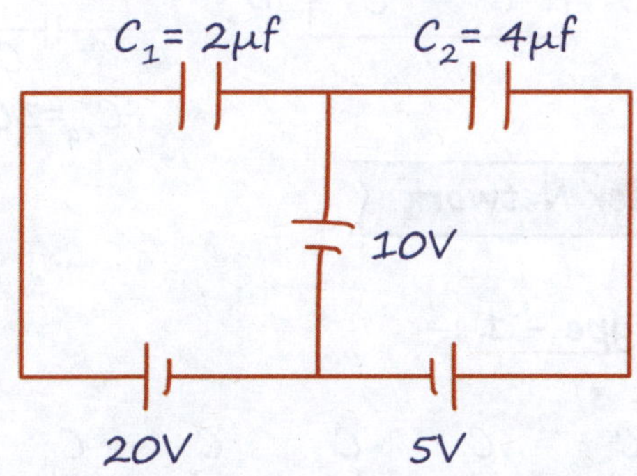

Sol.

MR* → You can assume potential ka refrence zero potential at any one point of circuit.

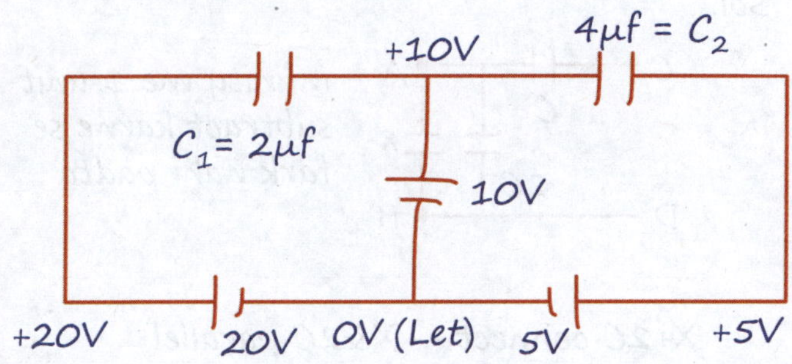

(Change on C_1) $Q_1 = C_1 (\Delta V) = 2\mu f \times (10V) = 20\mu c$

(Change on C_2) $Q_2 = C_2 (\Delta V) = 4\mu f \times (10-5) = 20\mu c$

+

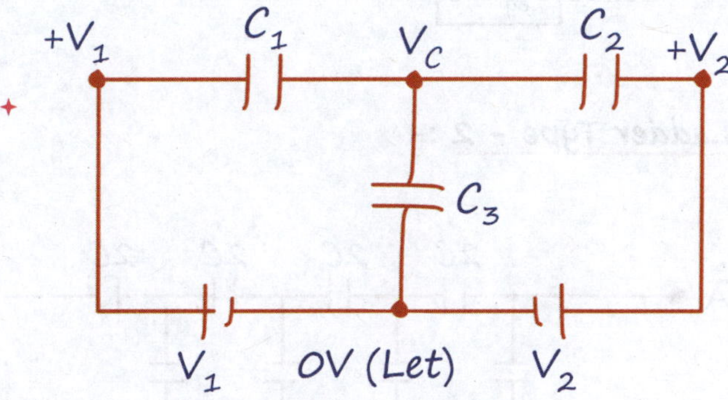

MR* Bade aaram से

$$V_C = \frac{C_1V_1 + C_2V_2 + C_3 \times 0}{C_1 + C_2 + C_3}$$

+ 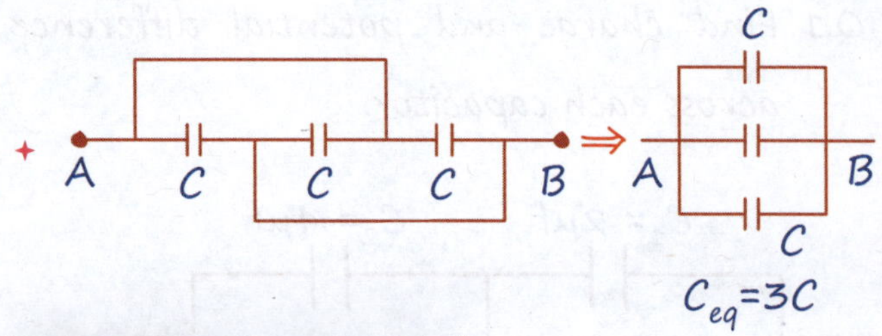

5 Ladder Network

Ladder Type – 1 :-

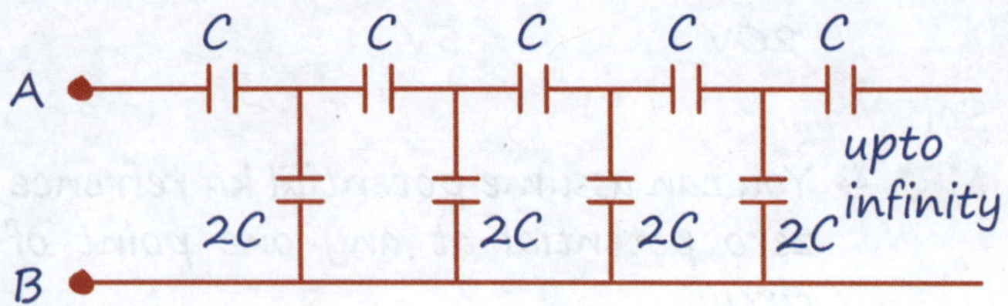

Sol.

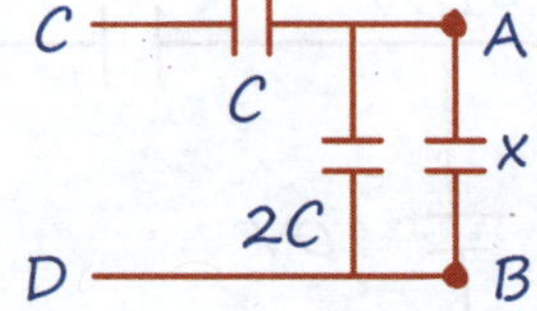

infinity me 1 unit subtract karne se fark nahi padta

X+2C connect ← X&2C parallel

$C_{eq} = \frac{(X+2C)C}{(X+2C)+C}$ ← then it with c in series to get C_{eq}

MR* In series C_{eq} must be less than smallest

hence $\boxed{C_{eq} < C}$

Ladder Type – 2 :-

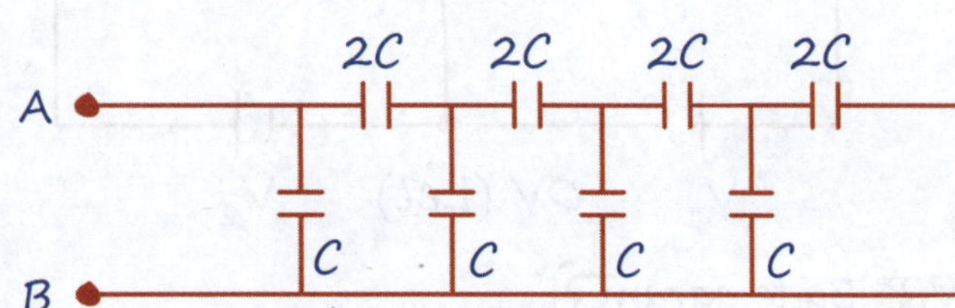

add 2C & X in series then in parallel with C.

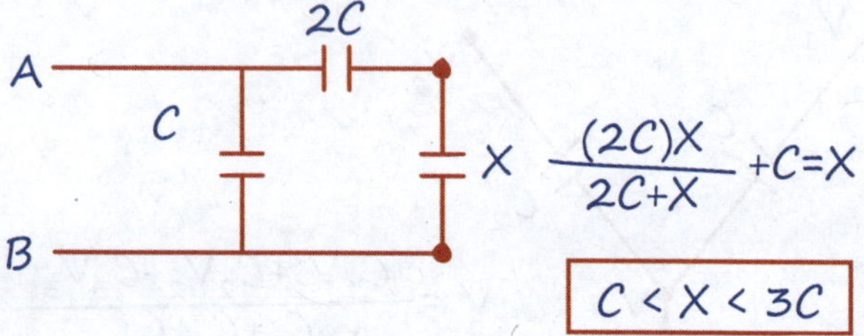

MR*

+ Series main C_{eq} less than smallest and in parallel C_{eq} larger than largest

6 Wheat stone Bridge

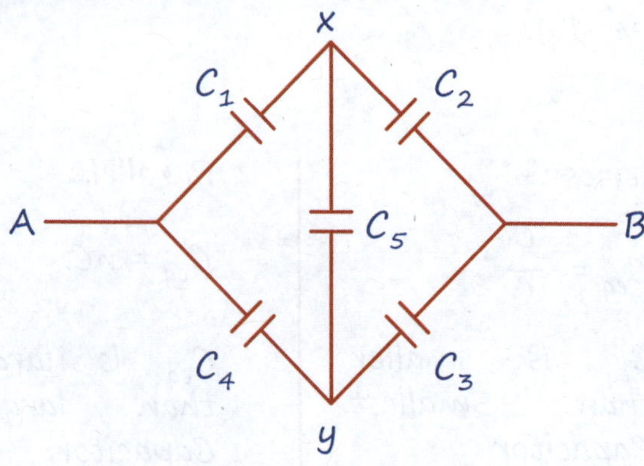

+ Balanced condition of a wheat stone bridge

$$\frac{C_1}{C_4} = \frac{C_2}{C_3}$$

+ $V_x = V_y$

(∵ Potential difference between x and y is zero)

+ Charge on $C_5 = 0$

Some example of Wheat Stone Bridge :-

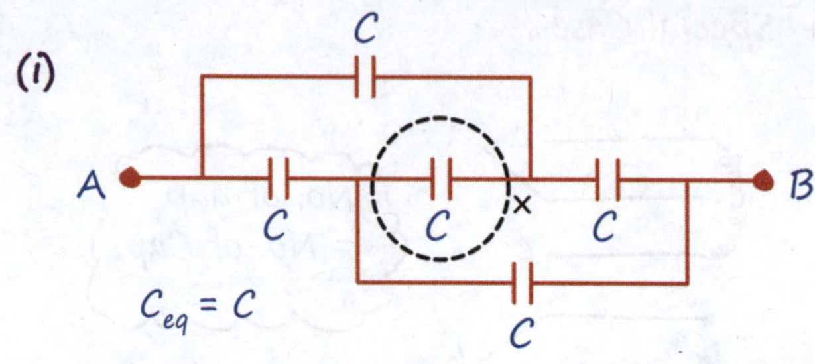

(ii)

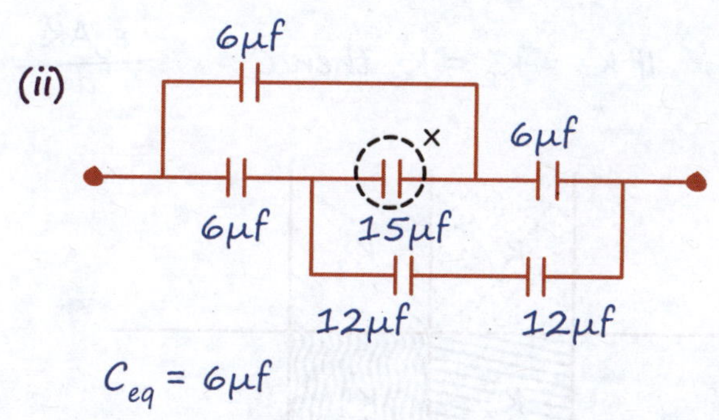

$C_{eq} = 6\mu f$

(iii)

$C_{eq} = \frac{\varepsilon_0 A}{d}$

Some symmetric circuit :-

(i)

A, B; q_1, q_2, $q=0$; C

$C_{eq} = 2C/3$

(ii)

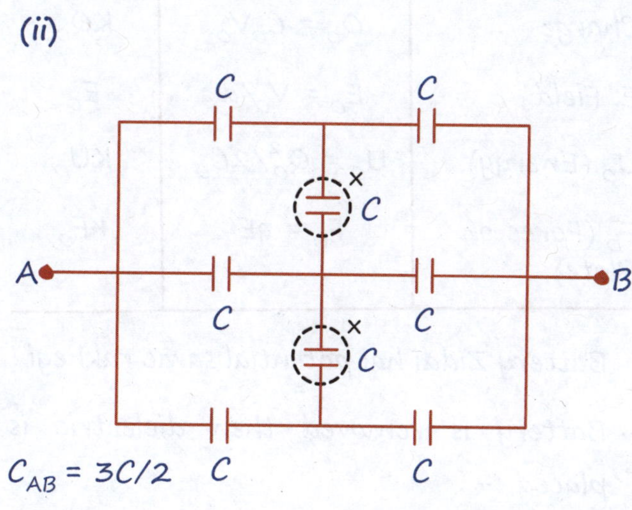

$C_{AB} = 3C/2$

(iii)

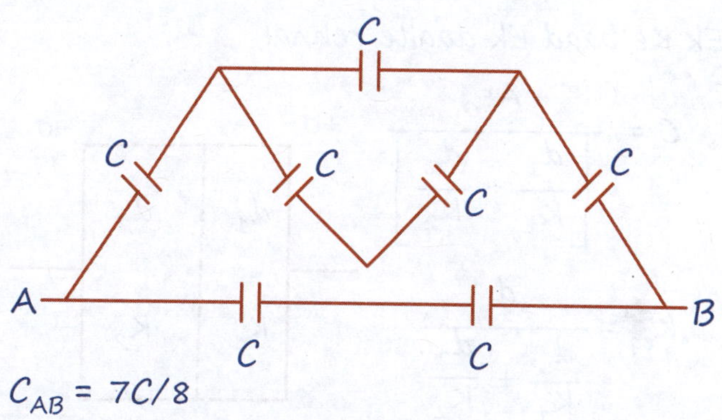

$C_{AB} = 7C/8$

+ 12-Identical capacitors or resistors in cubical form :-

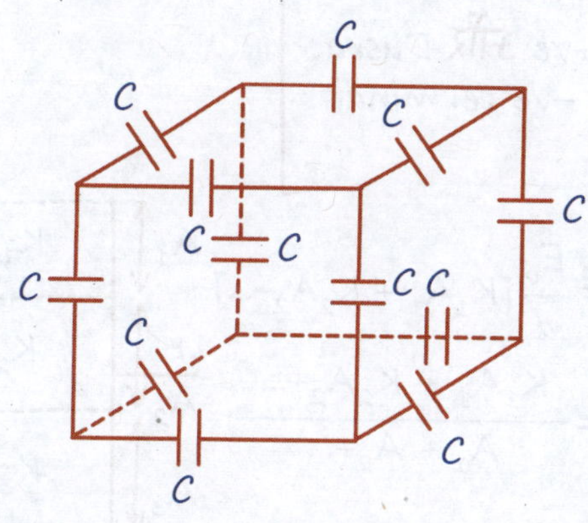

MR*

Along side	Along Face diag.	Along Body diag.
$C_{eq} = \frac{12}{7} C$	$C_{eq} = \frac{4}{3} C$	$C_{eq} = \frac{6}{5} C$
$R_{eq} = \frac{7R}{12} \Omega$	$R_{eq} = \frac{3R}{4} \Omega$	$R_{eq} = \frac{5R}{6} \Omega$

7 Dielectric Medium in an External E. Field

$E_{net} = \frac{E_o}{K}$ Jaise Coulomb F. dec. "K" time
Similarly E. Field dec. "K" times.

Series Combination of Dielectrics :-

Ek Ke baad Ek daalte rehna!

+ $C = \dfrac{A\varepsilon_0}{\left[\dfrac{d_1}{K_1} + \dfrac{d_2}{K_2}\right]}$

+σ −σ

d_1 d_2

K_1 K_2

+ $K_{eq} = \dfrac{d}{\dfrac{d_1}{K_1} + \dfrac{d_2}{K_2}}$

Parallel Combination of Dielectrics :-

[Saare dielectric ka ek terminal +ve और Dusra −ve terminal]

+ $C_{eq} = \dfrac{E_0}{d}[K_1A_1 + K_2A_2 + ...]$

+ $K_{eq} = \dfrac{K_1A_1 + K_2A_2 + ...}{A_1 + A_2 + ...}$

A_1 K_1

\+ A_2 K_2 −

A_3 K_3

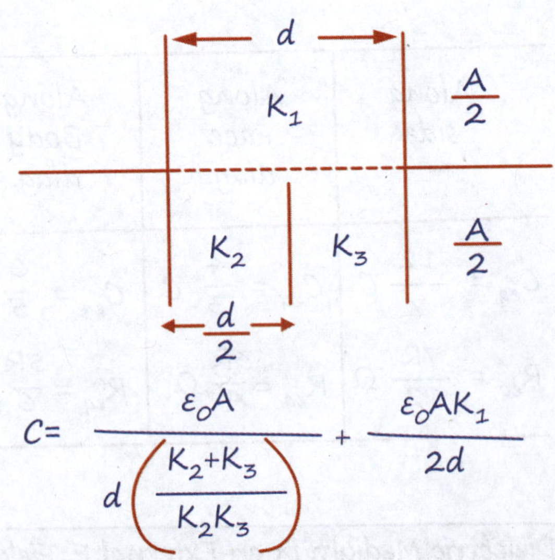

$$C = \frac{\varepsilon_0 A}{d\left(\dfrac{K_2+K_3}{K_2K_3}\right)} + \frac{\varepsilon_0 AK_1}{2d}$$

$$C = \frac{\varepsilon_0 A}{d}\left[\frac{K_2K_3}{K_2+K_3} + \frac{K_1}{2}\right]$$

MR* If $K_1 = K_2 = K_3$ then $C = \dfrac{\varepsilon_0 AK}{d}$

+

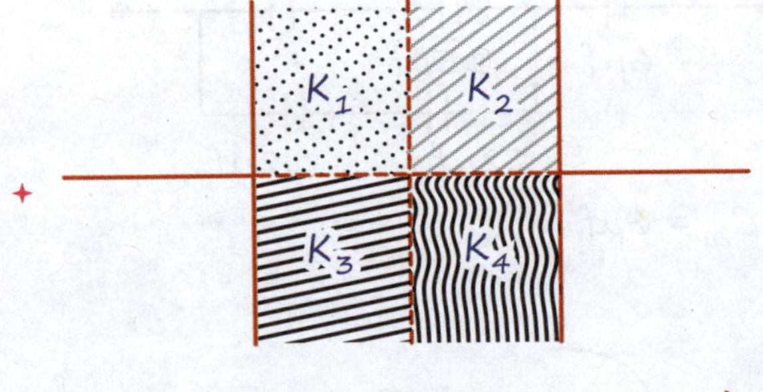

$$C_{eq} = \frac{\varepsilon_0 A}{d}\left(\frac{K_3K_4}{K_3+K_4} + \frac{K_1K_2}{K_1+K_2}\right)$$

MR* If $K_1 = K_2 = K_3 = K_4$ then $C_{eq} = \dfrac{\varepsilon_0 AK}{d}$

MR* If K=1, $C' = C$ | If $K=\infty$ Conductor $C' = \infty C$

Introduction of dielectric between plates of capacitor :-

+ Battery is Connected :-

	Initially	Finally
Capacitance	C_0	KC_0
Potential	V_0	$*V_0$
Charge	$Q_0 = C_0V_0$	KQ_0
E. Field	$E_0 = V_0/d$	E_0
U_0 (Energy)	$U_0 = Q_0^2/2C_0$	KU_0
F_0 (Force on Plate)	$F_0 = qE$	KF_0

+ Battery Ziddi hai potential same rakhegi !

+ Battery is removed then dielectric is placed :-

	Initially	Finally
Capacitance	C_0	KC_0
Potential	$V_0 = Q_0 \div C_0$	$*V_0/K$
Charge	$Q_0 = C_0V_0$	$*Q_0$ conserved
E. Field	$E_0 = V_0/d$	E_0/K
U_0 (Energy)	$U_0 = Q_0^2/2C_0$	U_0/K
F_0 (Force on Plate)	$F_0 = qE$	F_0/K

Plate isolated hai to charge conserved rahega.

Dielectric of width t placed between capacitor :-

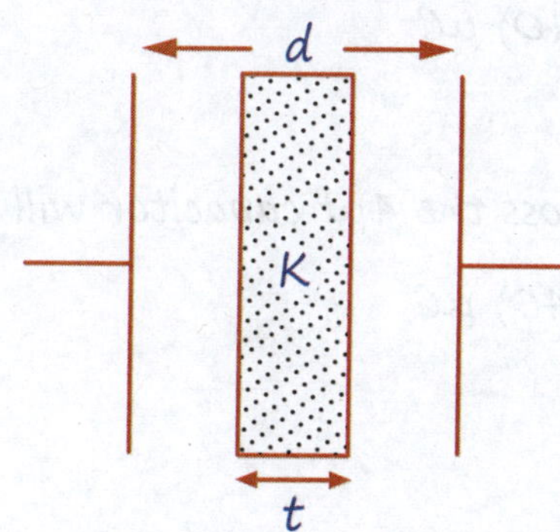

$$C = \frac{\varepsilon_0 A}{(d-t)+\frac{t}{K}}$$

if t=0

MR*

$$C = \frac{\varepsilon_0 A}{d} \quad \Big| \quad d=t \quad C = \frac{\varepsilon_0 KA}{d}$$

Q.2. A parallel plate air capacitor has capacitance C. Half of space between the plate is filled with dielectric K as shown to Fig. Then new Capacitance C' is

(a) $C' = C\left[\frac{K}{K+1}\right]$

(b) $C' = C\left[\frac{2K}{K+1}\right]$

(c) $C' = \frac{2C}{K+1}$

(d) $C' = C\left[1+\frac{K}{2}\right]$

Sol. (b) $C = \left[\frac{\epsilon_0 A}{d-t+\frac{t}{K}}\right] = \left[\frac{\epsilon_0 A}{d-\frac{d}{2}+\frac{d}{2K}}\right]$

$$= \left[\frac{2\epsilon_0 AK}{d(K+1)}\right] = 2C\left[\frac{k}{(K+1)}\right]$$

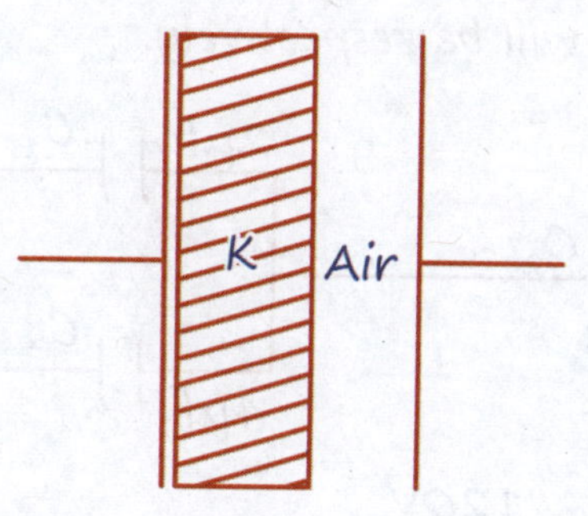

Q.3. Find charge and potential drop across each capacitor.

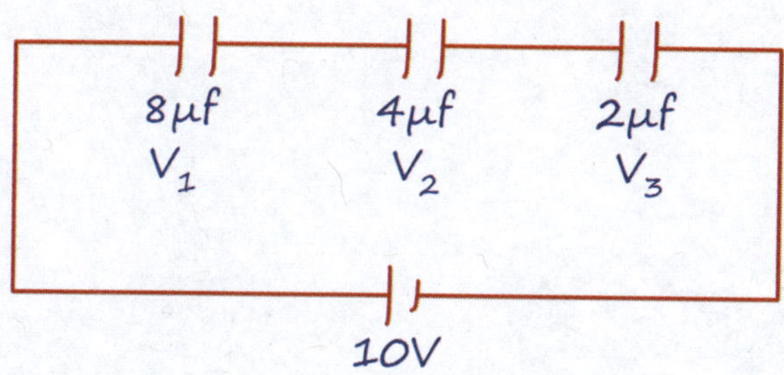

Sol. MR* $V = \frac{q}{C}$

$V_1 = x$(let) $V_2 = 2x$ $V_3 = 4x$

$x + 2x + 4x = 10V$

$x = \frac{10}{7}$ volt

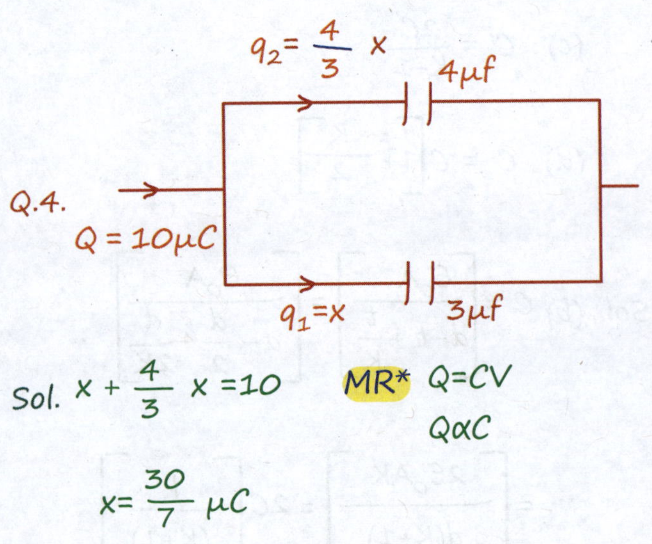

Sol. $x + \frac{4}{3}x = 10$ MR* $Q = CV$

$Q \alpha C$

$x = \frac{30}{7}\,\mu C$

Q.5. The charge on capacitors shown in the figure and the potential difference across each will be respectively.

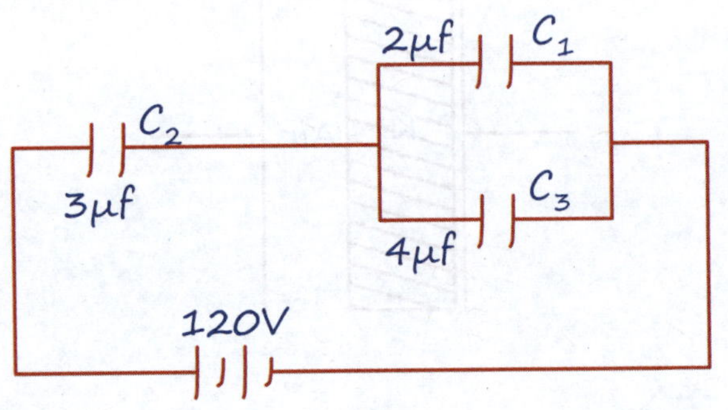

Sol. The total capacitance of the circuit is,

$C = 2\mu f$

$Q = CV$

So, $Q = 240\mu C$

Voltage across 3μf capacitor will be,

$V_1 = \frac{240}{3}V = 80V$

Voltage across 2μf and 4μf capacitor will be,

$V_2 = (120 - \frac{240}{3})\,V$

$= 40V$

Charge across the 2μf capacitor will be,

$Q_1 = (2 \times 40)\,\mu C$

$= 80\mu C$

Charge across the 4μf capacitor will be,

$Q_2 = (4 \times 40)\,\mu C$

$= 160\mu C$

24 Current Electricity

1 Electric Current

The rate of directional flow of electric charge is called electric current.

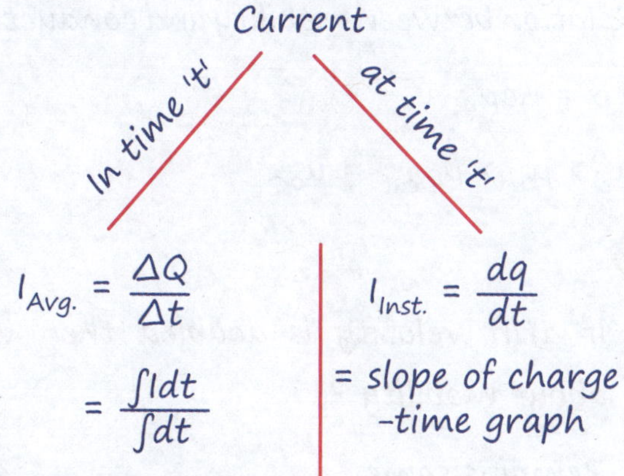

$I_{Avg.} = \frac{\Delta Q}{\Delta t}$

$= \frac{\int I dt}{\int dt}$

$I_{Inst.} = \frac{dq}{dt}$

= slope of charge -time graph

*** Charge on circular path**

$I_{Avg.} = \frac{Q}{T} = \frac{ne}{T}$

$= nef = \frac{nev}{2\pi r}$

$\int dq = \int I dt$

ΔQ = Area of current time graph

- scalar,
- unit (Ampere),
- dimension $[A^1]$

Q.1 Through a given cross section n_1 electron per-sec are passing from left to right and n_2 proton are passing from right to left simultaneously then the electric current through this cross section.

Ans. $I = (n_1 + n_2)e$

Q.2 Find ratio of current.

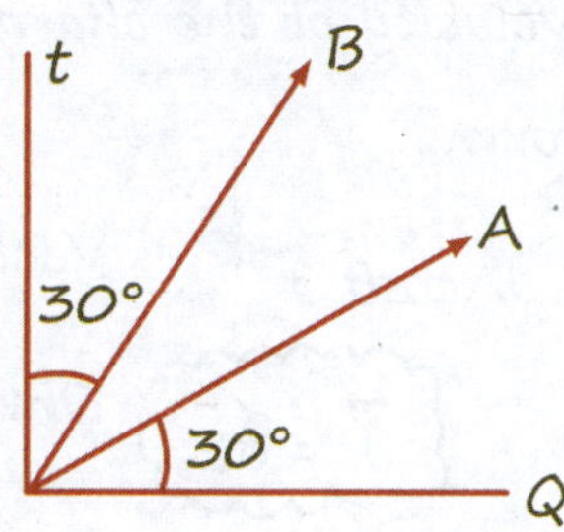

Ans. $\frac{I_A}{I_B} = \frac{\tan 60°}{\tan 30°} = \frac{3}{1}$

Isolated Conductor:-

- Number Density of Electrons (n)

  $ne^- = 10^{28} e^-/m^3$

- The average velocity of electrons in a conductor is

  $V = 10^3$ m/s

- In an isolated conductor, the electrons move in such a way that they cancel out any internal electric potential, resulting in a net average potential of zero.

  $V_{avg} = 0 \quad E_{in} = 0 \quad (\text{speed } V \propto \sqrt{T})$

- Kinetic Energy and Temperature

$\frac{1}{2} mv^2 = \frac{3}{2} K_B T \qquad K_B = 1.38 \times 10^{-23} \frac{Js}{K}$

2 Battery Connected to Conductor

$F_{e^-} = qE, \quad a = \frac{qE}{m_e} \quad E = \frac{V}{l}$

(F_{e^-} → Force on electron; a → acceleration)

Drift velocity

$V_D = \cancel{u}^{\,0} + AT \qquad I = neAV_D$

$V_D = a_e T = \frac{eE\tau}{m_e}$

Where:

E = electric field

V_d = drift velocity

l = length of conductor

n = no. of electron per unit volume

τ = relaxation time

m_e = mass of electron

V = emf of battery

e = charge of electron

- The electric field in the conductor is due to the potential difference applied by the battery, distributed along the length of the conductor.

- **Electrons gain velocity under the influence of the electric field, known as drift velocity. This is the average velocity attained by electrons in a conductor due to the applied electric field.**

Conductivity(σ) and Resistivity(ρ):

- **Conductivity is a measure of a material's ability to conduct electric current, while resistivity is the measure of how strongly a material opposes the flow of electric current. They are inversely related.**

MR* feel

$$\sigma = \frac{ne^2\tau}{m} \xrightarrow{\rho = \frac{1}{\sigma}} \rho = \frac{m}{ne^2\tau}$$

$$R = \frac{\rho l}{A}$$

$$R = \frac{ml}{ne^2\tau A} \xrightarrow{V=iR} V = \frac{mil}{ne^2\tau A}$$

Micro scopic form of Ohm's law!

Mobility :-

- Property of charge carrier.
- Does not depends on drift velocity and electric field.

$$\mu = \frac{V_d}{E} = \frac{e\tau}{m},$$

- Relation between mobility and conductivity

$$\sigma = ne\mu$$

- $\mu_e > \mu_p > \mu_{Deut.} = \mu_\alpha$.

MR*

Q.3 If drift velocity is doubled then what about mobility ?

Ans. Remains same

Current Density :- (Vector)

Current density is the amount of electric current flowing per unit area of the cross-section of a conductor.

$$J = \frac{I}{A} = neV_d$$

Where

I : Electric current

A : Cross sectional area of the conductor

n : Number of charge carriers per unit volume

e : Charge of an electron

V_d : Drift velocity of the charge carriers

In vector form

$$i = \vec{J}\cdot\vec{A} = JA\cos\theta$$

$$\vec{J} = ne\vec{V_d}$$

$$\vec{J} = \sigma\vec{E}$$

Vector form of Ohm's law !

- The direction of the current density vector is the same as the direction of the flow of positive charge (or the direction of the conventional current).

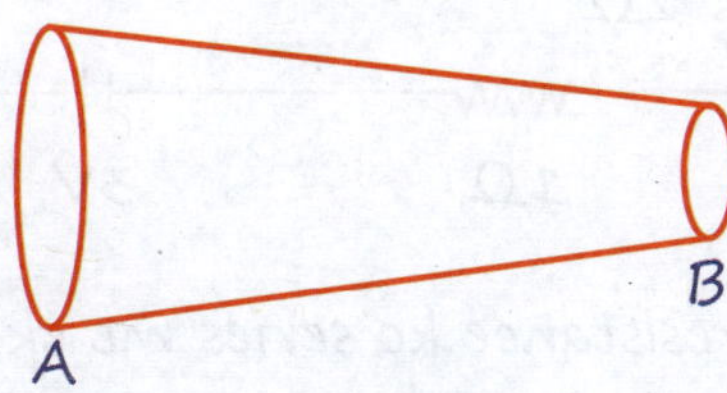

Current : $I_A = I_B$

Current density : $J_A < J_B$

Drfit velocity : $V_A < V_B$

Variation of Resistance :-

Material prop.

$$R = \frac{\rho L}{A}$$

where R is the resistance, ρ is the resistivity of the material, L is the length of the conductor, and A is the cross-sectional area of the conductor.

(a) $l \rightarrow$ Change $A \rightarrow$ Const^n $R \propto l$	(b) $l \rightarrow$ Change $V \rightarrow$ Const^n $R \propto l^2$
(c) $A \rightarrow$ Change $l \rightarrow$ Const^n $R \propto 1/A$	(d) $A \rightarrow$ Change $V \rightarrow$ Const^n $R \propto 1/A^2$
(e) $R = \frac{\rho l^2}{M} \times \text{density}$ $R \propto \frac{l^2}{M}$	(f) $R = \frac{\rho M}{\text{density } A^2}$ $R \propto \frac{M}{A^2}$

If a wire of Resistance R stretched to double of its length the new resistance becomes 4R.

Q.4 The masses of the wire in the ratio of 1 : 3 : 5 and their length are in ratio of 5 : 3 : 1. The ratio of their resistance.

Ans. $R \propto \frac{l^2}{m}$ $R_1 : R_2 : R_3 = 125 : 15 : 1$

Ohm;s Law :-

$$\vec{J} = \sigma\vec{E}, \quad V = \frac{mil}{ne^2\tau A} \quad V = iR$$

*Note :-

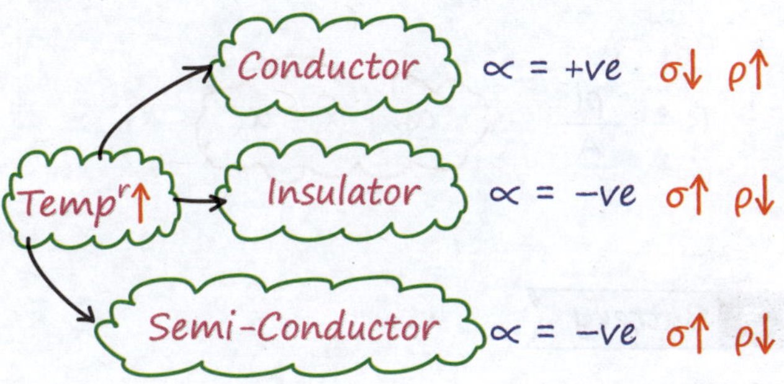

Conductor $\propto$ = +ve σ↓ ρ↑

Insulator $\propto$ = −ve σ↑ ρ↓

Semi-Conductor $\propto$ = −ve σ↑ ρ↓

Temperature Dependence of Resistance & Resistivity :-

$$\rho_T = \rho_0[1 + \alpha\,\Delta T], \qquad \alpha = \frac{\Delta\rho}{\rho_0\,\Delta T}$$

unit K^{-1}

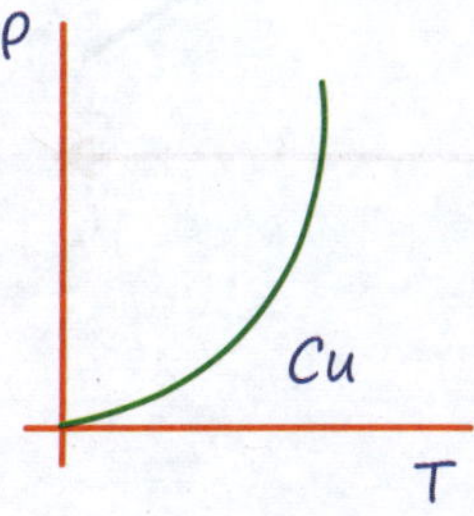

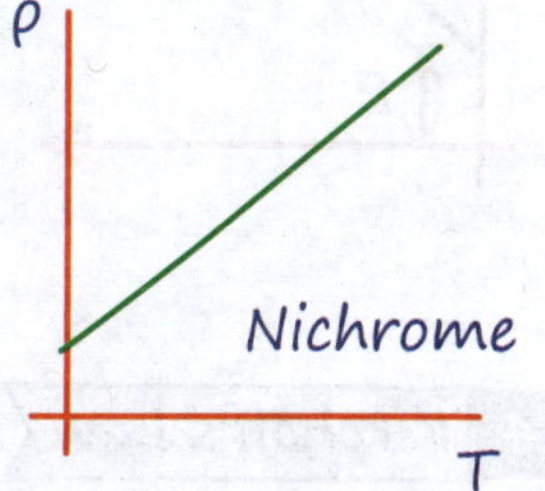

- Two resistance R_1 and R_2 connected in series and their R_{eq} does not depends upon temperature then $R_1\alpha_1 = -R_2\alpha_2$

✦ $R_{t°C} = R_0 (1 + \alpha T)$ always valid

$R_{t_2} = R_{t_1} [1 + \alpha (T_2 - T_1)]$

Valid for small change in temperature

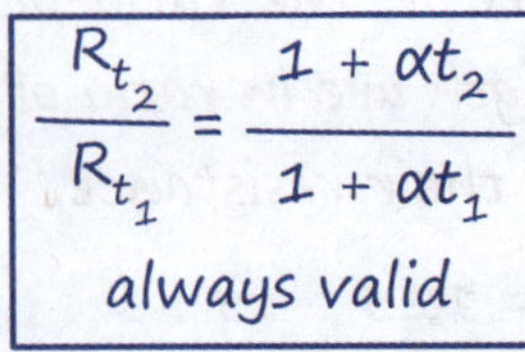

$$\frac{R_{t_2}}{R_{t_1}} = \frac{1 + \alpha t_2}{1 + \alpha t_1}$$

always valid

R_0 = resistance at 0°C

R_{t_1} = Resistance at t_1

R_{t_2} = Resistance at t_2

Relation between Coefficient of :-

(a) linear expansion (α)

(b) Resistance (α_R)

(c) Resistivity (α_ρ)

$R = \frac{\rho l}{A}$

$\alpha_R + \alpha = \alpha_\rho$

3 Battery

(a) Charging

$\Delta V = E + ir$

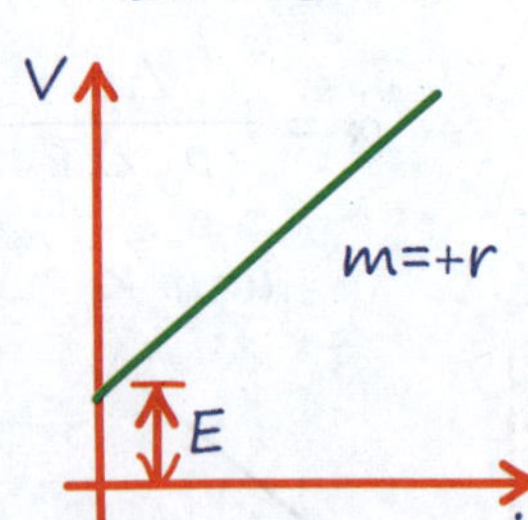

(b) Discharging

$\Delta V = E - ir$

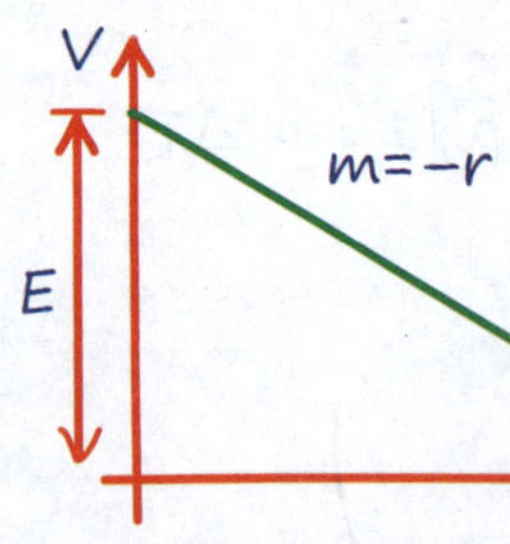

4 Kirchoff's Law

Law (i) $\Sigma i = 0$ [Charge Conserv^n]

Law (ii) Energy Conserv^n.

Note :- Household circuit maltab parallel circuit.

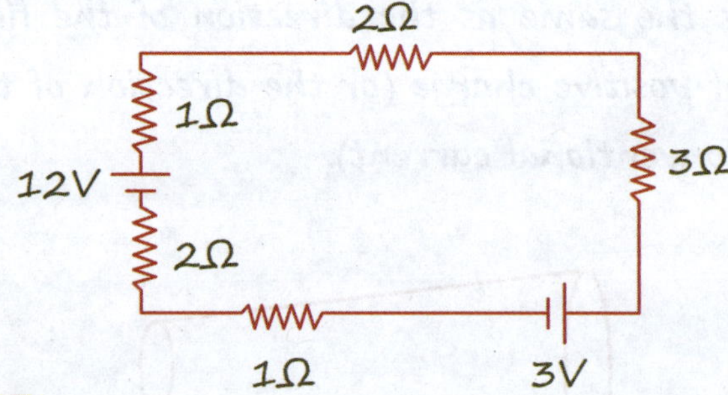

MR*

Sare resistance ko series me ek sath add kar ke battery ko ek sath polarity ke sath add kare

$I = \frac{9}{9} = 1A$

9Ω

9V

MR*

Point potential → current depends on potential difference not on potential and potential difference does not depends on reference, hence you can assume zero potential at any one point of circuit. (Sirf ek point pe hi zero man sakte hai)

Kishi bhi point ka Potential Apan zero Mann Sakte hai!

Q.5 Find current in 15Ω and 30Ω.

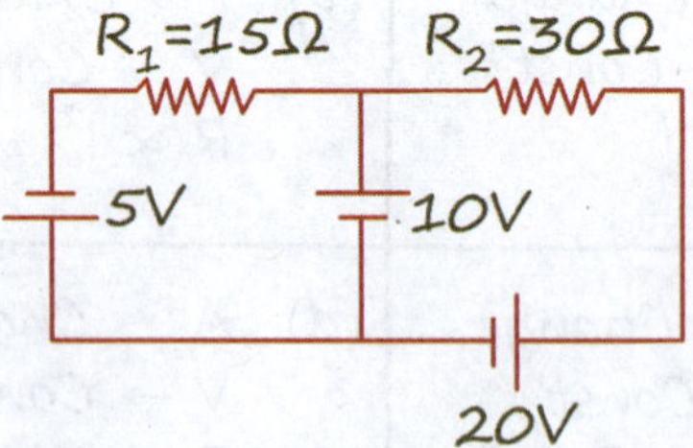

MR*

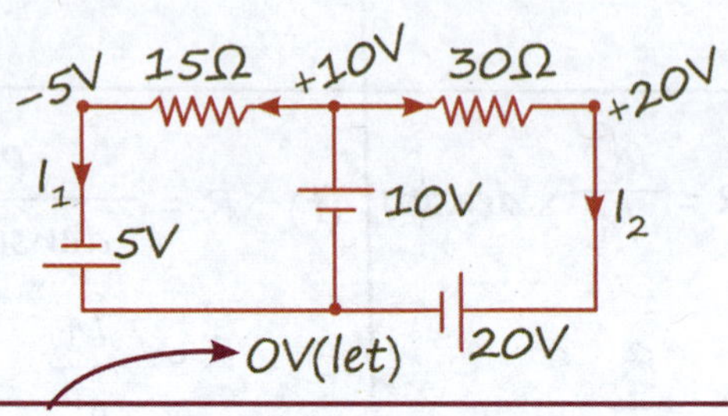

yaha ish liye kyki yha se sare point potential nikalna easy hai.

$$I_1 = \frac{10-(-5)V}{15} = \frac{15}{15} = 1A$$

$$I_2 = \frac{(20-10)}{30} = \frac{10}{30} = \frac{1}{3}A$$

5 Combination of Resistance

Series :- i = Same potential different

- $V_{total} = V_1+V_2+V_3+...$
- $R_{eq} = R_1+R_2+R_3+...$
- If n equal resistance, $R_{eq} = nR$
- R_{eq} will be larger than the largest Resistance

  $V \propto R$
- Jitna Jyada "R" Utna Jyada V_{drop}.

Parallel :- [v = same] current different

- $I_{total} = I_1+I_2+I_3+...$
- $\frac{1}{R_{eq}} = \frac{1}{R_1} + \frac{1}{R_2} + \frac{1}{R_3} + ...$
- If n equal resistance, $R_{eq} = R/n$
- $i \propto 1/R$
- Jitna Kam "R" Utna jyada "i"

*Household circuit matlab parallel circuit.

To Calculate Potential at Midpoint :-

$$V_{mid} = \frac{\frac{V_1}{R_1} + \frac{V_2}{R_2}}{\frac{1}{R_1} + \frac{1}{R_2}}$$

V_1 —R_1— V —R_2— V_2

Q.6 Find current through each resistance?

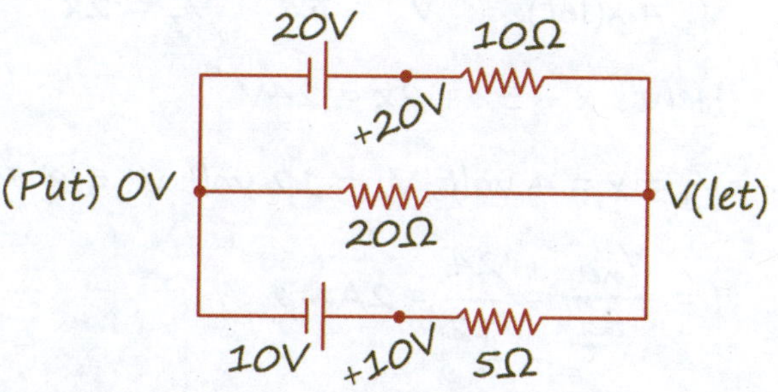

Ans.

MR*

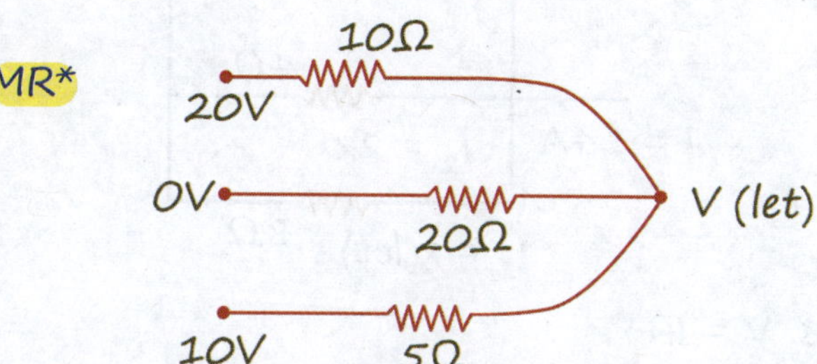

$$V = \frac{\frac{V_1}{R_1} + \frac{V_2}{R_2} + \frac{V_3}{R_3}}{\frac{1}{R_1} + \frac{1}{R_2} + \frac{1}{R_3}}$$

$$V = \frac{\frac{20}{10} + \frac{0}{20} + \frac{10}{5}}{\frac{1}{10} + \frac{1}{20} + \frac{1}{5}} = \frac{4}{\frac{2+1+4}{20}} = \frac{80}{7}$$

Now, we can calculate each current because we have $V = \frac{80}{7}$ hence we have potential difference.

Q.7 Find current and potential drop across each resistance

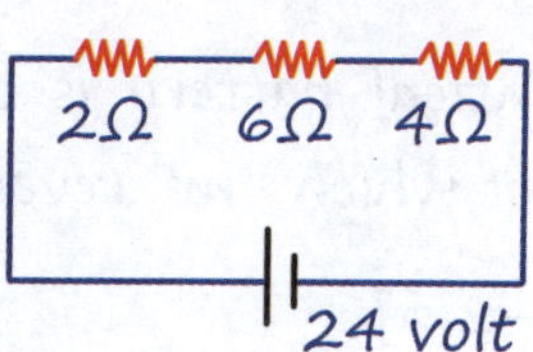

Ans. $V = IR$, $V \propto R$

$V_1 = x(\text{let})$ $V_2 = 3x$ $V_3 = 2x$

Hence $x + 3x + 2x = 24V$

$V_1 = x = 4$ volt, $V_2 = 12$ volt $V_3 = 8$ volt

$$I = \frac{V_{net}}{R_2} = \frac{24}{12} = 2\text{Amp}$$

Q.8 Find current through each resistance

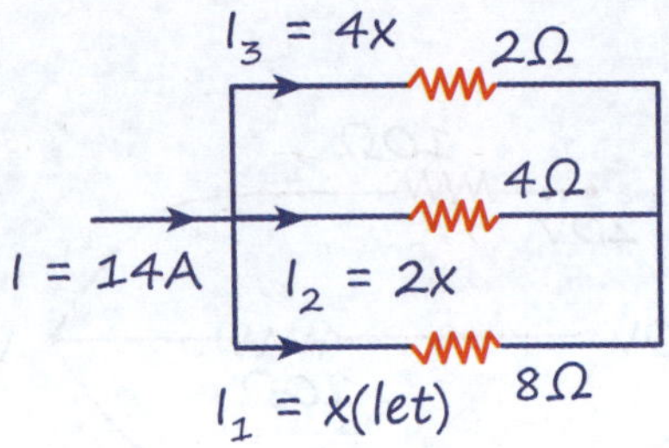

Ans. $V = IR$

$$I \propto \frac{1}{R}$$

$x + 2x + 4x = 14$

$X = 2$ Amp

6 Combination of Battery

Series :-

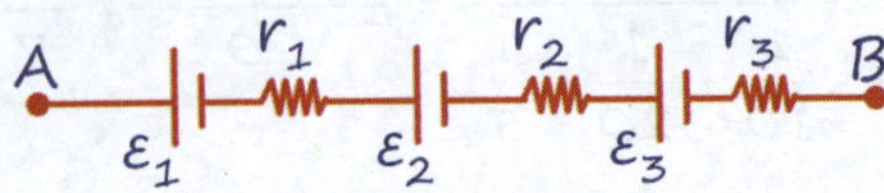

$\varepsilon_{net} = \varepsilon_1 + \varepsilon_1 + \varepsilon_3$

$r_{net} = r_1 + r_2 + r_3$

- If there are "n" identical battery is connected in series :

$\varepsilon_{net} = n\varepsilon$

$r_{net} = nr$

- If "n" identical battery is connected in series out of which "m" reversed :

$\varepsilon_{net} = [n-2m]E$

$r_{net} = nr$ [Maximum]

Parallel :-

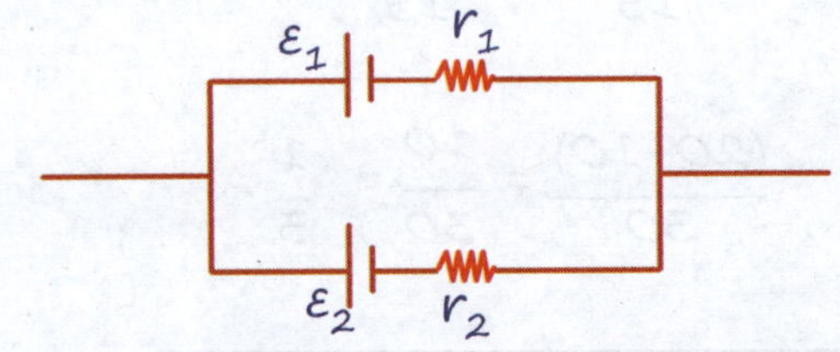

$$V = \frac{\frac{\varepsilon_1}{r_1} + \frac{\varepsilon_2}{r_2}}{\frac{1}{r_1} + \frac{1}{r_2}}, \ \& \ \frac{1}{r_{eq}} = \frac{1}{r_1} + \frac{1}{r_2}$$

- If there are n-identical cell in parallel then,
  - $\Sigma_{net} = \epsilon(\text{emf})$
  - $r_{eq} = r/n$ [Minimum]

Mixed Grouping :-

- n-Identical battery (E,r) connected in series then this series combination connected m-times parallel with external resistance R.

n – series $\Rightarrow nE, nr$

then m-times parallel $(nE)\left(\frac{nr}{m}\right)$

$$I = \frac{nE}{R + \frac{nr}{m}}$$

This I will be maximum when

$$R = \left(\frac{nr}{m}\right) \text{(Internal resistance)}$$

- Circuit Mai chalna Important Hai :

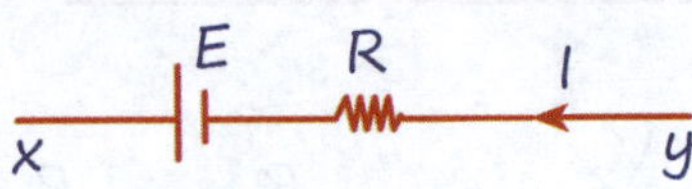

→ $V_y - IR + E = V_x$

→ $V_x - E + IR = V_y$

- Current ki direction me resistance ko cross karne par potential drop hoga (–IR).
- Current ke opposite potential increase hoga (+IR).
- Battery ko lower to higher cross karne pe potential increase hoga (+E).
- Higher to lower cross karne pe potential decrease (–E).
- Current ki direction se fark nahi pedega.

Q.9 Find V_P?

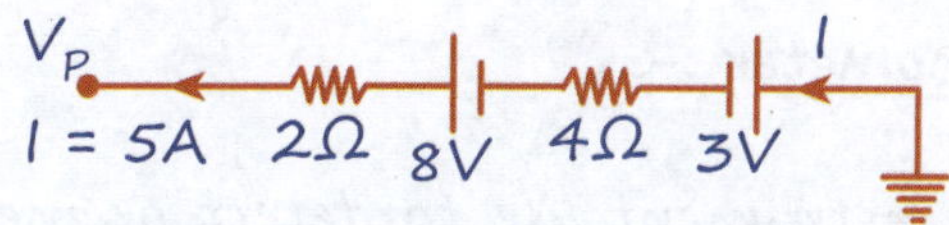

Ans. Move from 'P' to ground

$$V_P + 5 \times 2 - 8 + 4 \times 5 + 3 = 0$$

$$V_P = -33 + 8 = -25 \text{ volt}$$

7 Power

$$P = iV, \quad P = i^2R, \quad P = \frac{V^2}{R}$$

Series:– $P \propto R$

Parallel:– $P \propto 1/R$

8 Joule's Law of heating

Heat produced (H) in a resistor when current flows through it;

$$H = i^2Rt = ivt = ms\Delta\theta$$

If, i = Variable:– $H = R\int i^2 . dt$.

Special Case $r = \sqrt{R_1R_2}$

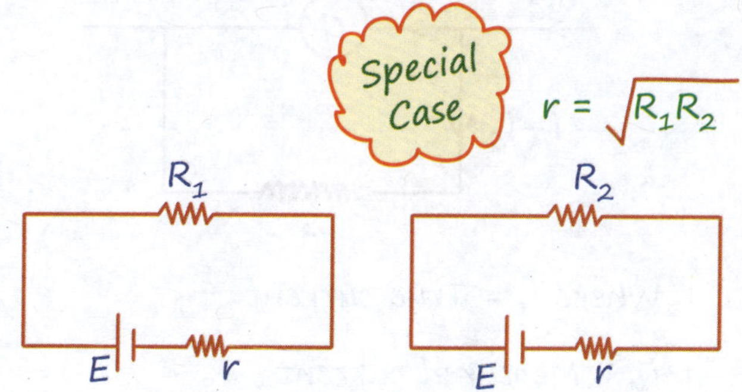

agar ye internal resistance between two circuit iss pattern mein raha toh P_{drop} Same hoga woh doh circuit mein!

Power drop in ext. Circuit with maximum power theorem :-

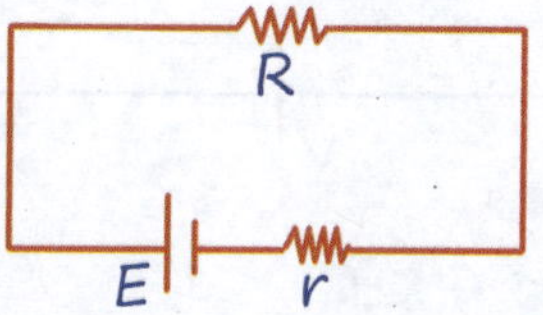

Power drop across R, $P = I^2R = \left[\frac{E}{R+r}\right]^2 R$

Power drop will be maximum when r=R

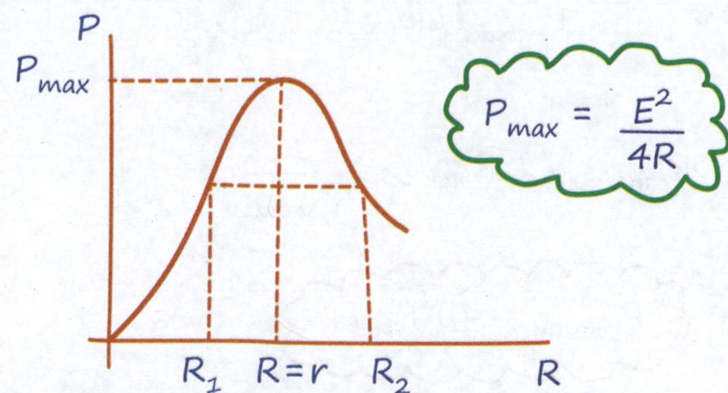

9 Bulb (Pure Resistance)

- Rated power and rated voltage given to calculate resistance of bulb.

$$R_{bulb} = \frac{V_R^2}{P_R}$$

$$R_{Bulb} \propto \frac{1}{P_{Rated}}$$

$$P_{Cons.} = \left[\frac{V_S}{V_R}\right]^2 P_{Rated}$$

Q.10 If two bulb of power (60W, 110V) and (100W, 110V) are connected in series with supply of 220V then?

Ans. Potential drop across 60W bulb will be greater than 110V hence it will fuse.

Series Combination :-

Bakwas Combination

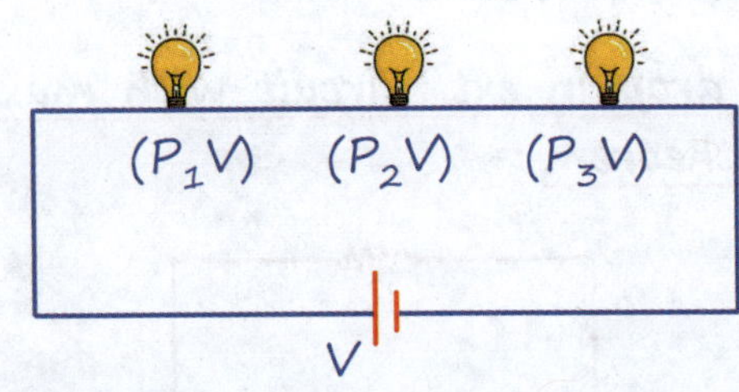

$$\frac{1}{P_{cons.}} = \frac{1}{P_1} + \frac{1}{P_2} + \frac{1}{P_3}$$

If all are identical bulb then,

$P_{cons.} = \frac{P}{n}$ $P_{cons.}$ is smaller than the smallest P_{rated} bulb.

$$P_{consumed.} \propto R_{bulb} \propto \frac{1}{P_{rated}}$$

$$P_{consumed} = i^2 R_{bulb}$$

Joh Kam P_{rated} Ka hoga woh Jyada Chamkega!

Parallel Combination :-

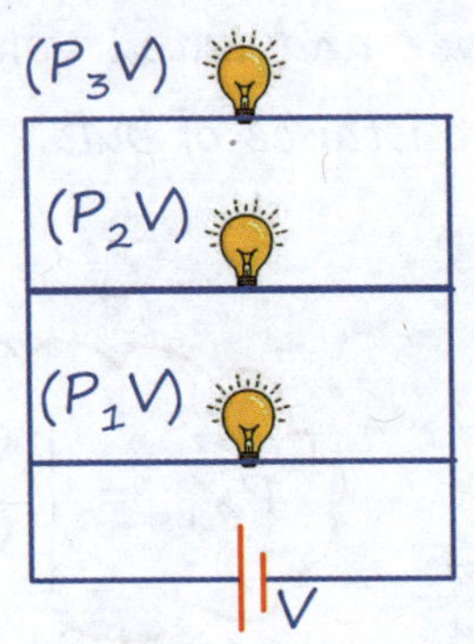

$P_{cons.} = P_1 + P_2 + P_3$

If all are identical bulb then,

$P_{cons.} = nP$ $\quad P_{consumed} \propto \frac{1}{R_{bulb}} \propto P_{rated}$

$$P_{consumed} = \frac{V^2}{R_{bulb}}$$

Joh Jyada "P_{rated}" Ka hoga woh Jyada Chamkega!

[1 kWh = 36 × 10^5J]

Time Taken by Heater Coils :-

- Series :- $t = t_1 + t_2$
- Parallel :- $t = \frac{t_1 t_2}{t_1 + t_2}$

10 Electrical Instruments

Galvanometer :-

- An instrument use to detect or measure small current.
- Very sensitive, produce large error.
- I_G = Maximum current that can flow.
- G = resistance of galvanometer
- As Ammeter :- Connected in series in circuit.
  - Small resistance shunt connected in parallel with galvanometer.
  - Ideal R = 0; Behave as simple wire.
  - % Error = $\frac{i_T - i_M}{i_T} \times 100$

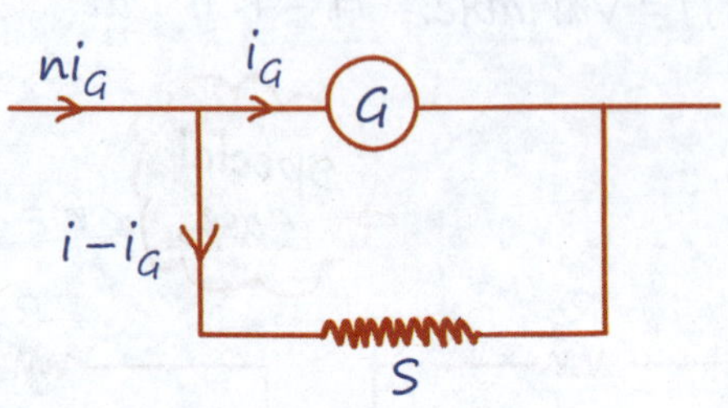

Where i_T = True current

i_M = Measured current

$$R_{ammeter} = \frac{GS}{G+S}$$

$$\boxed{S = \frac{G}{n-1}}$$ R ↓ % Error ↓

$$n = \frac{\text{i we want to Measure}}{\text{i jitna Galvano-meter se jayega.}}$$

$$= \frac{\text{Total current}}{\text{Current through galvanometer}}$$

+ **As Voltmeter :-** Connected parallel in circuit
   - large resistance connected in series with galvanometer
   - Ideal R = ∞ infinite (Behave as open wire)

i_G → G — S

$$\boxed{S = G(n-1)}$$

$$\% \text{ Error} = \frac{V_T - V_M}{V_T} \times 100$$

Meter Bridge :-

+ Use to find value of Resistance.
+ Working based on wheat-stone bridge.

$$\frac{P}{l_1} = \frac{Q}{100 - l_1}$$

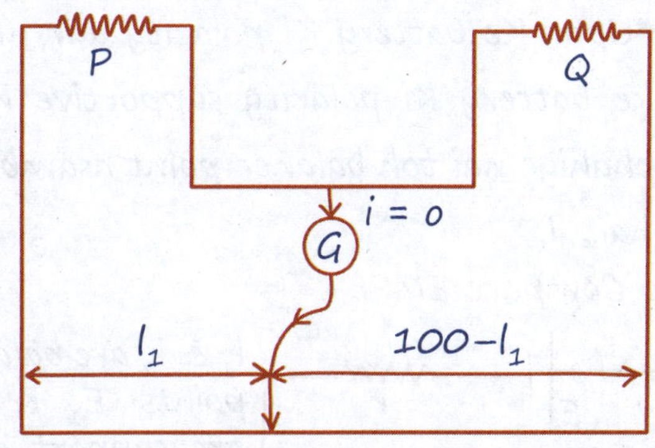

Potentiometer wire :-

Working based on potential gradient.

+ To Find EMF :-

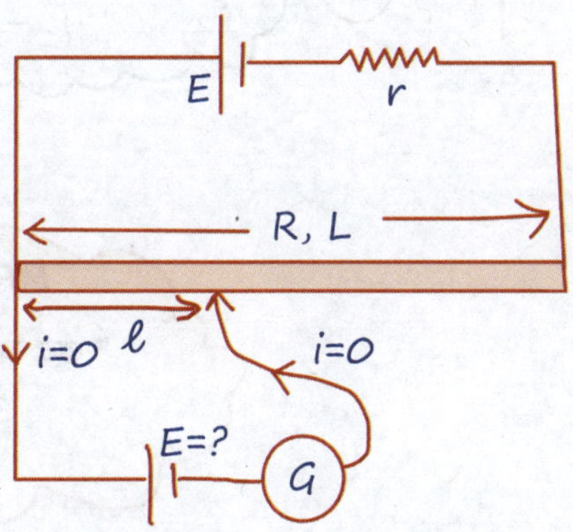

*Step-i :- $i = \frac{E}{R+r}$

*Step-ii :- $v = iR$

*Step-iii :- $K = \frac{V}{L}$

→ Potential gradient

Potential drop per unit length in wire.

$E = kl$

l = Balancing length where current through galvanometer is zero.

Caution !

*EMF of the battery must be less or equal to the Potential drop in wire.

*Upar Ke battery Ki polarity aur niche ke battery Ki polarity supportive honi chahiye nai toh balance point asambhav hai... !

+ To Compare EMF :-

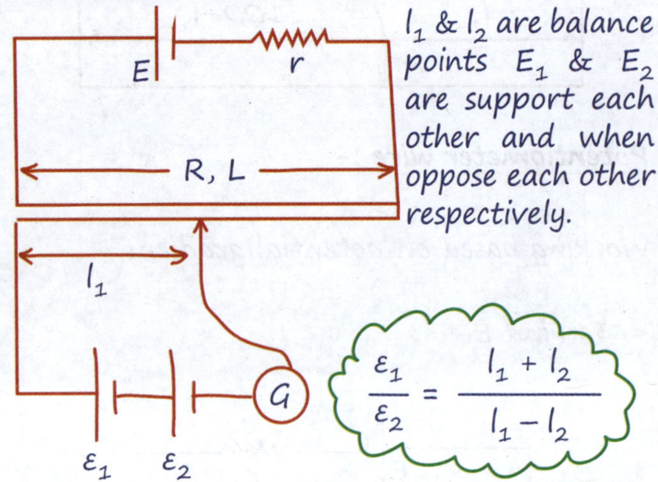

l_1 & l_2 are balance points E_1 & E_2 are support each other and when oppose each other respectively.

$$\frac{\varepsilon_1}{\varepsilon_2} = \frac{l_1 + l_2}{l_1 - l_2}$$

+ To Find internal Resistance :-

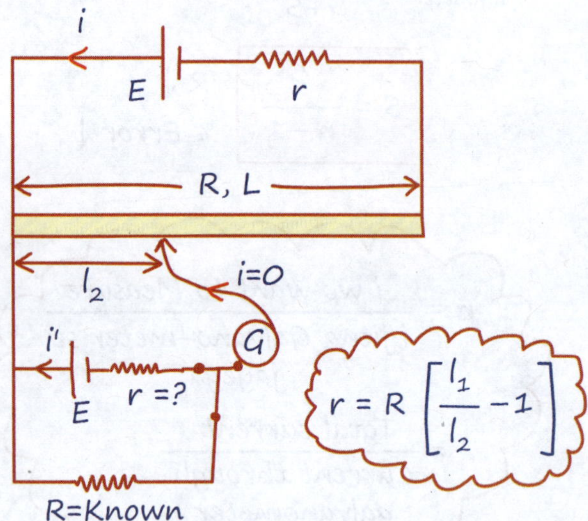

$$r = R\left[\frac{l_1}{l_2} - 1\right]$$

l_1 = Balancing length when key is open

l_2 = Balancing length when key is closed

$(l_1 > l_2)$

MR*

"खुद की समझदारी भी अहमियत रखती है,
वरना दुर्योधन और अर्जुन दोनों
के गुरु एक ही थे।"

25 Magnetic Effect of Electric Current

1 Oersted Experiment

- Electric field outside current carying wire is zero
- Electric field inside current carrying wire may or may not be zero.
- But moving charge near to current carrying wire experience force hence there must be a field that is magnetic field.

$$F_{rest\ Q} = 0 \rightarrow F_{moving\ Q} \neq 0$$

Note:- *Current carring wire produces MF around it.

2 BIOT-SAVART'S LAW

The magnetic field (dB) due to a small element of current carrying wire (idL) at a point is given by:

$$dB = \frac{\mu_0}{4\pi}\frac{idl\sin\theta}{r^2}$$

(Scalar form)

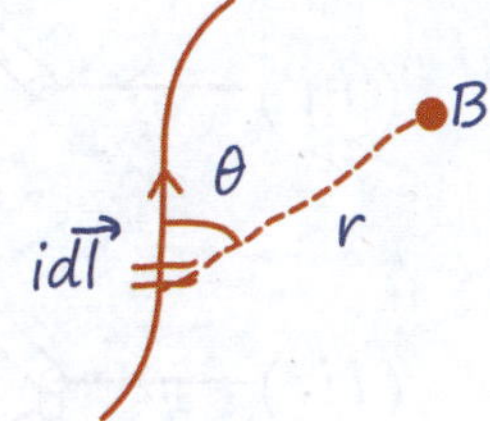

Where:

μ_0 = permeability of free space ($4\pi \times 10^{-7}$ Tm/A)

i = current through the wire.

dl = length of the small element of the wire.

θ = angle between the element dl and the line joining the element to the point where the field is being calculated.

r = distance from the element to the point.

Unit conversion:

$$*1T = 10^4\ G = 1\ \frac{wb}{m^2}$$

Vector Form :-

$$\vec{dB} = \frac{\mu_0}{4\pi}\frac{i\vec{dl} \times \hat{r}}{r^2} = \frac{\mu_0}{4\pi}\frac{(i\vec{dl} \times \vec{r})}{r^3}$$

MR*

To find direction of magnetic field

$\vec{dB} = i\vec{dl} \times \vec{r}$

(Result) (1st vector) (2nd vector)

Place your four-finger (palm) of right hand along 1st vector ($i\vec{dl}$) slap 2nd vector($\vec{r}$), thumb will represent $\vec{B}$.

3 Magnetic Field due to Straight Wire

- The magnetic field (B) at a point due to a straight current carrying wire is given by:

idl, i, r, α, β

$$B = \frac{\mu_0 i}{4\pi r}[\sin\alpha + \sin\beta]$$

α, β :- Hamesha point से lena hai !

Where

- r is the perpendiclar distance from the wire to the point.
- α and β are the angle subtended by the wire at the point of interest.

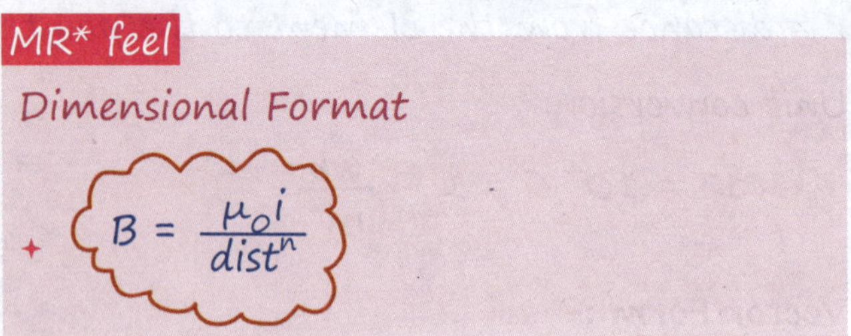

For infinitely long wire :-

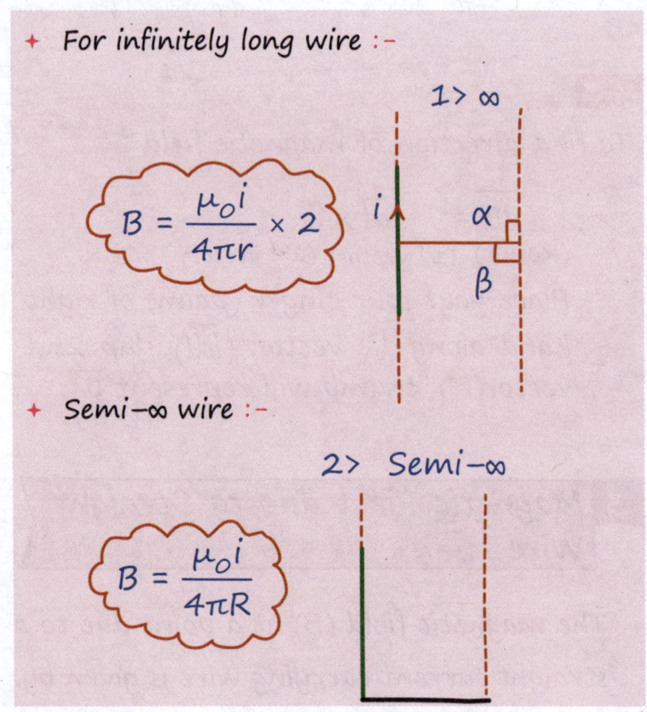

Direction of Magnetic Field :-

Right hand Rule:-

- Place thumb in the direction of current then curling finger will represent direction of magnetic field

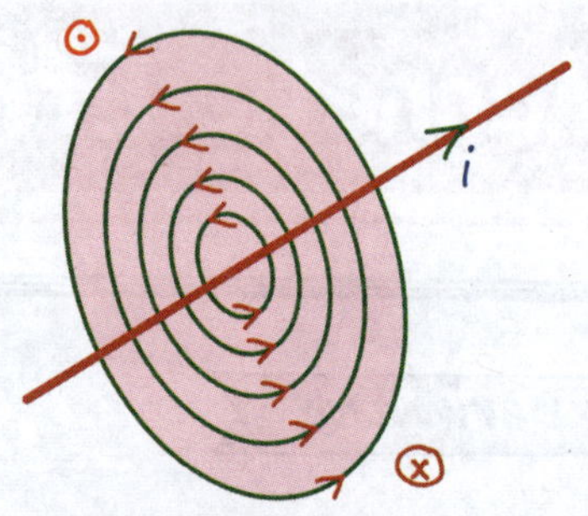

Symmetrical Object :-

Symmetrical objects with uniform current distribution often result in zero magnetic fields inside the object.

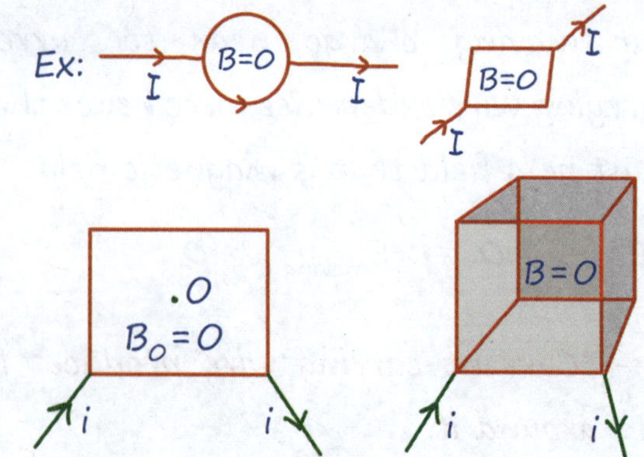

Combination of two thick and two thin wire :-

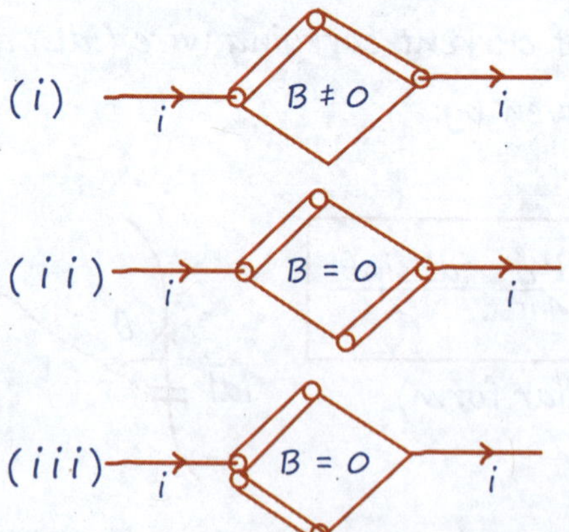

Parallel wires with different currents:

- Consider two parallel wires separated by a distance r, with one wire carrying current $I_1 = I$ and the other carrying current $nI = I_2$

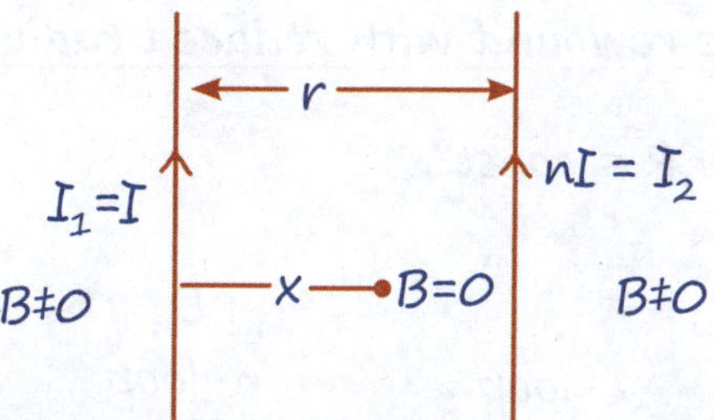

- At a point x from the wire carrying I and $r - x$ from the wire carrying current nI, the magnetic field due to these currents:

$$B_1+B_2=0$$

$$\frac{\mu_0 I}{2\pi x} = \frac{\mu_0 In}{2\pi(r-x)}$$

$$r-x = nx$$

$$\boxed{x = \frac{r}{n+1}}$$

Similarities and difference between Biot Savart law and coloumbs law :-

(i) Electric field is produced by scalar source "charge"	(i) Magnetic field produced by vector source "current element $i\vec{dl}$"
(ii) Electric field along the position vector from source	(ii) Magnetic field perpendicular to the position vector from source
(iii) follow inverse square law	(iii) also follow same
(iv) linear inside source	(iv) linear inside source

MR wala sawal

The magnetic field at the centre O of a rectangular loop with sides a and b carrying a current i is given by:

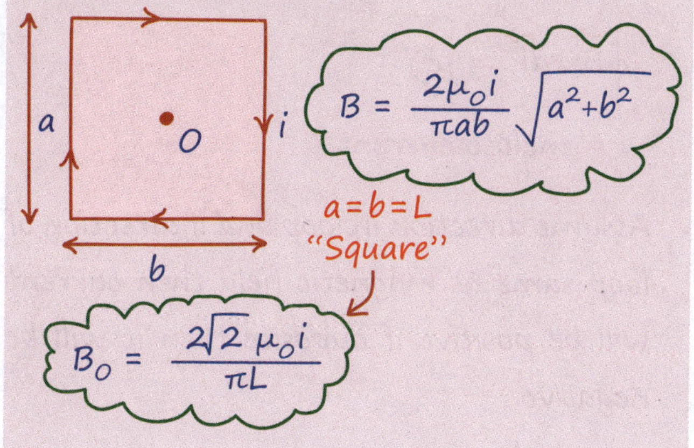

$$B = \frac{2\mu_0 i}{\pi ab}\sqrt{a^2+b^2}$$

$$B_O = \frac{2\sqrt{2}\,\mu_0 i}{\pi L}$$

For a regular hexagonal loop with the side length L carrying a current i, the magnetic field at the centre O is:

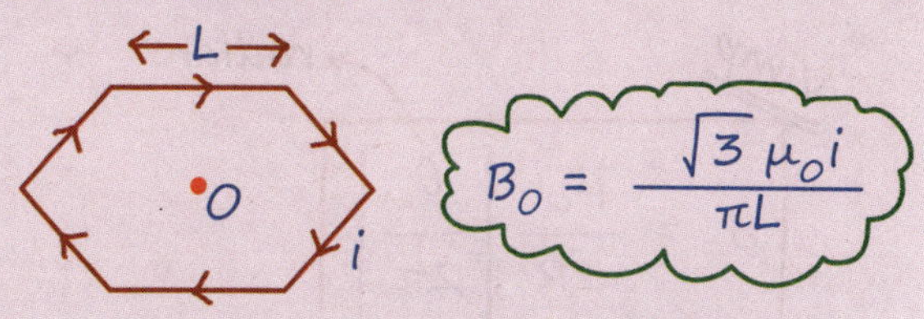

$$B_O = \frac{\sqrt{3}\,\mu_0 i}{\pi L}$$

4 Magnetic Field on the axis of a current carrying loop

For a circular loop of radius R carrying a current i the magnetic field at a distance x along the axis

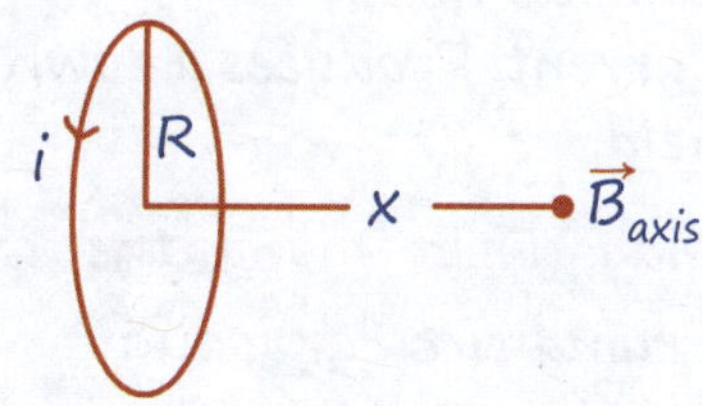

$$B_{axis} = \frac{\mu_0 i R^2}{2(R^2+x^2)^{3/2}}$$

Special cases :-

B at the centre of circular loop:-

$$B_{centre} = \frac{\mu_0 i}{2R}$$

B at the centre of semi-circular loop:-

$$B_O = \frac{\mu_0 i}{4R}$$

B at the centre of quarter circle:-

$$B_O = \frac{\mu_0 i}{8R}$$

Generalised Formula for Circular Arc :-

For an arc with a central angle θ

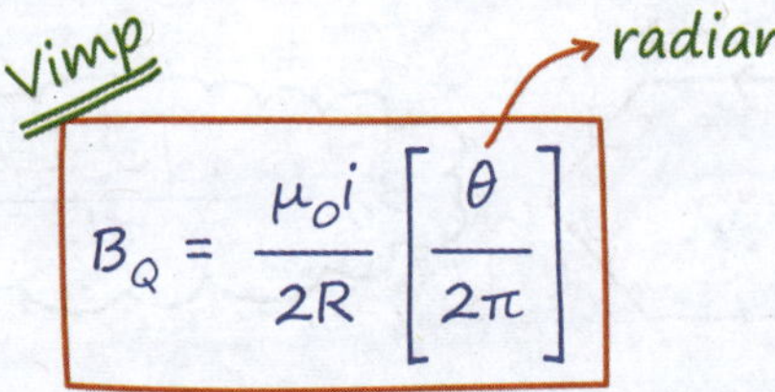

$$B_Q = \frac{\mu_0 i}{2R}\left[\frac{\theta}{2\pi}\right]$$

Direction of magnetic field

I (Anti-clock) I (clock-wise)

B ⊙ B ⊗

MR* **Anti-clockwise Current:** Produces an upward magnetic field.

Clockwise Current: Produces a downward magnetic field.

+ Graph of Mag. Field along the axis of Current Carrying circular loop :-

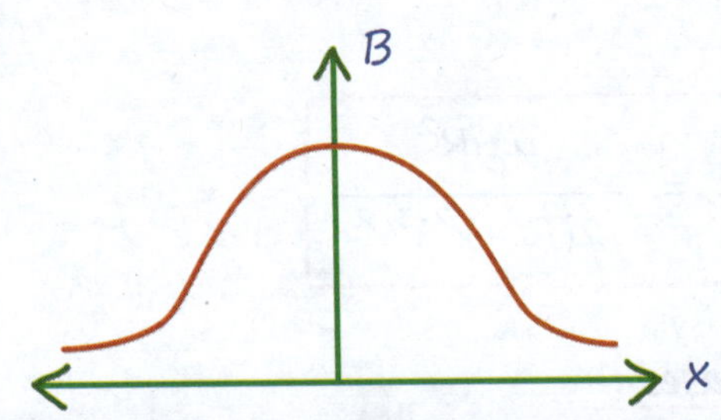

5 Magnetic Field due to looping of wire

"R = Constant." :-

1-loop

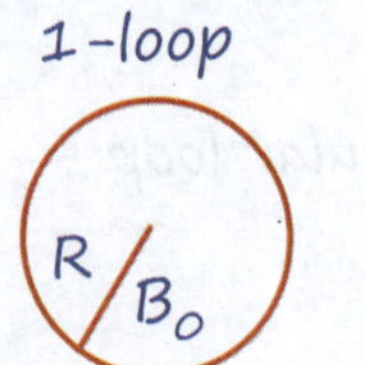

n-loop

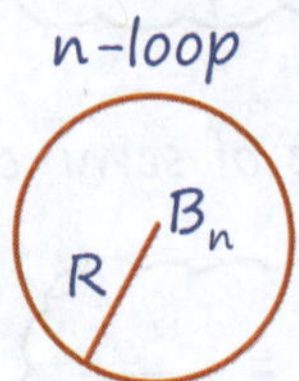

If the wire is wound into n loops, each with radius R, the total magnetic field at the center of the loops is:

$$B_n = n\,B_{1\ loop} = \frac{n\mu_0 I}{2R}$$

Same wire rewound with reduced radius:-

$L = 2\pi R = const^n.$

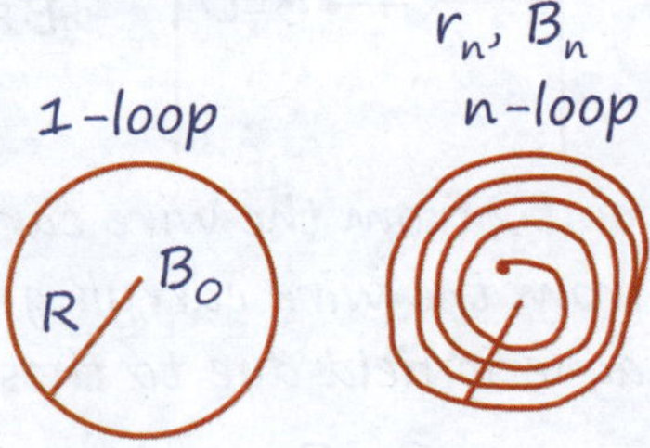

If the same wire is rewound into n loops, each with a reduced radius r_n:

$$r_n = \frac{R}{n}$$

The magnetic field

$$B_n = n^2 B_{1\text{-}loop} = \frac{n^2\mu_0 I}{2R}$$

6 Amperes Circuital Law

+ $\vec{B}$ = due to all inside or outside current.
+ Always valid for all type of current.
+ only applicable when current distribution is symmetric.
+ Not a magnetic flux, because here is close line integral.

$$\oint \vec{B}\cdot\vec{dl} = \mu_0 i.$$

i = enclose current.

+ Assume direction in loop and if direction of loop same as magnetic field then current will be positive if opposite then it will be negative.

B (current out of page), dl

$\oint \vec{B}.\vec{dl} = +\mu_o I$

because B also Anti-Clock wise and loop

B (current out of page), dl

$\oint \vec{B}.\vec{dl} = -\mu_o I$

$\vec{B}$ is Anti-Clock but loop is clock

Steps to apply Ampere's circuital law

1. Draw close symmetric Amperian loop, that must be passes through the point where field have to calculate.

Ex circular, square loop etc

2. Angle between loop and magnetic field must be 0°, 90° or 180°

3. Value of B must be constant at all point of loop.

Magnetic Field Due to a straight infinite current carrying wire :-

$$B = \frac{\mu_o i}{2\pi r}$$

Magnetic Field Due to a infinite Current hollow Cylinder :-

Outside the cylinder, $B_o = \frac{\mu_o i}{2\pi r}$

Inside the cylinder, $B_{in} = 0$

$B_o = \frac{\mu_o i}{2\pi r}$, $B_{in} = 0$

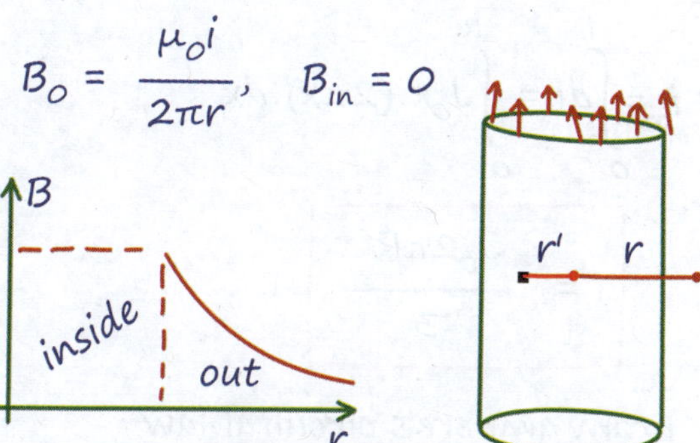

Magnetic Field Due to infinite Current Carrying solid Cylinder :-

Outside $B_o = \frac{\mu_o i}{2\pi r}$

$$\text{Inside } B_{in} = \frac{\mu_o i r}{2\pi R^2}$$

$$B_{in} = \frac{\mu_o J r}{2}$$

$$\vec{B}_{in} = \frac{\mu_o \vec{J} \times \vec{r}}{2}$$

$* \; J = \frac{I}{\pi R^2} \; *$

On surface $B_{surface} = \frac{\mu_o i}{2\pi R}$

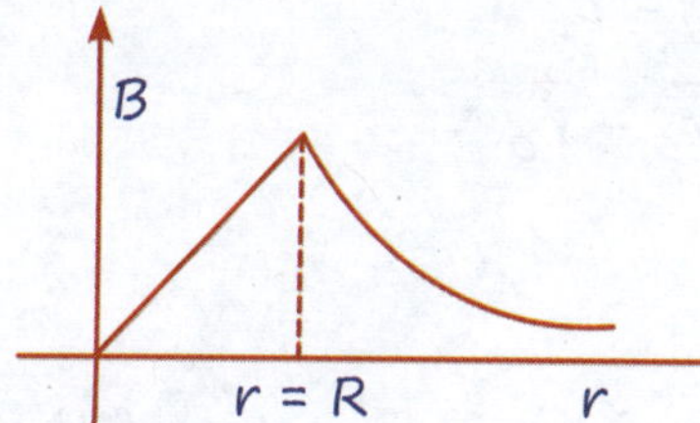

Q.1. Magnetic field Due to infinite current carrying solid cylinder where $J = J_o x$ find magnetic field outside the cylinder

Sol. Given $\hat{J} = j_o x$

Total current i_n enclosed within the cylinder up to radius R

$$i_{in} = \int_0^i di = \int_0^R J_0 x.(2\pi x).dx$$

$$i = \frac{J_0 2\pi R^3}{3}$$

From ampere's circuital law

$$B(2\pi r) = \mu_0 i_{in}$$

$$B = \frac{\mu_0 J_0 R^3}{3r}$$

Magnetic Field Inside cavity of solid cylinder :-

A cylindrical cavity is located inside the cylinder. The center of this cavity is at a distance r_0 from the center of the solid cylinder.

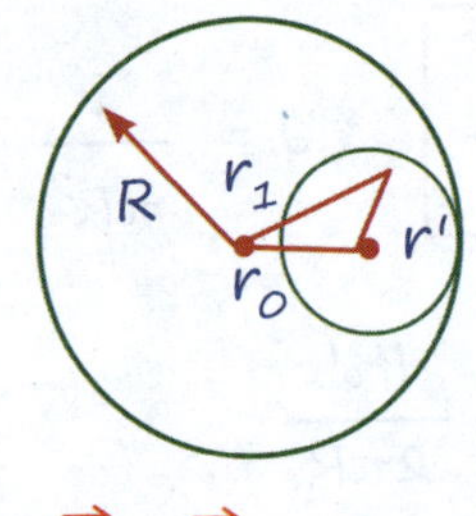

$$\vec{r_0} + \vec{r'} = \vec{r_1}$$

$$\vec{r'} = \vec{r_1} - \vec{r_0} \; *$$

Where

$\vec{r_0}$ = vector from the center of the solid cylinder to the center of the cavity.

$\vec{r_1}$ = vector from the center of the solid cylinder to the point of interest with the cavity.

$\vec{r'}$ = Distance from the center of the cavity to the point of interst.

$$^*B_{net} = \underbrace{\frac{\mu_0 J r_1}{2}}_{\text{Complete}} - \underbrace{\frac{\mu_0 J r'}{2}}_{\text{Cavity}} \; *$$

$$B = \frac{\mu_0 J r_0}{2}$$

+ Magnetic field wil be uniform inside cavity.

7 Solenoid

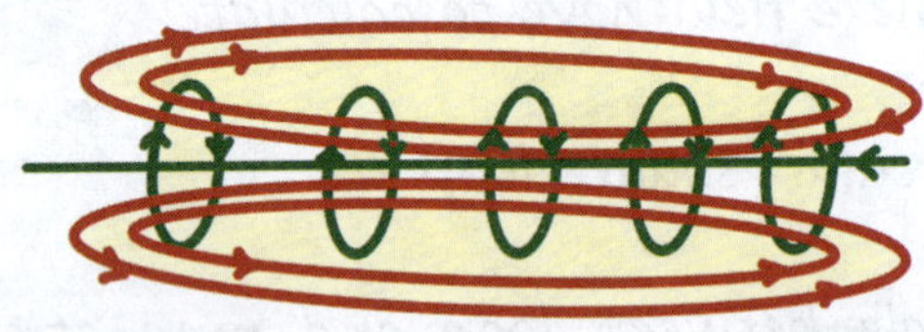

Finite Solenoid :-

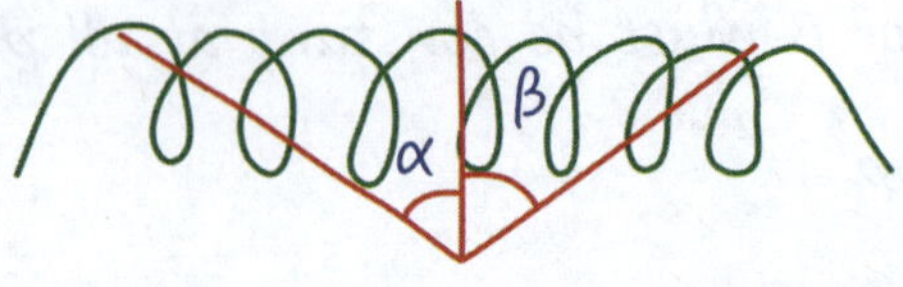

Magnetic field at a point inside a solenoid:

$$B = \frac{\mu_0 ni}{2}[\sin\alpha + \sin\beta]$$

At centre: $B_C = \mu_0 ni$, At ends: $B_{end} = \frac{\mu_0 ni}{2}$

Infinite Solenoid :-

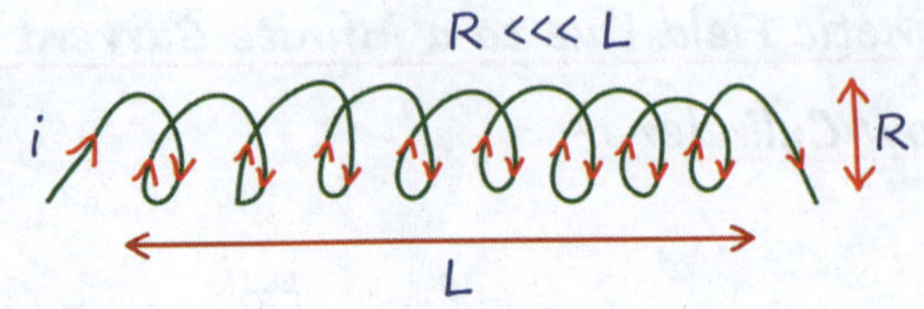

$$B = \mu_0 ni = \frac{\mu_0 Ni}{L}$$

N = Total turn

L = Total length

$n \Rightarrow$ Turns per unit length.

$$n = \frac{N}{l} = \frac{\text{Total turns}}{\text{length of solenoid}}$$

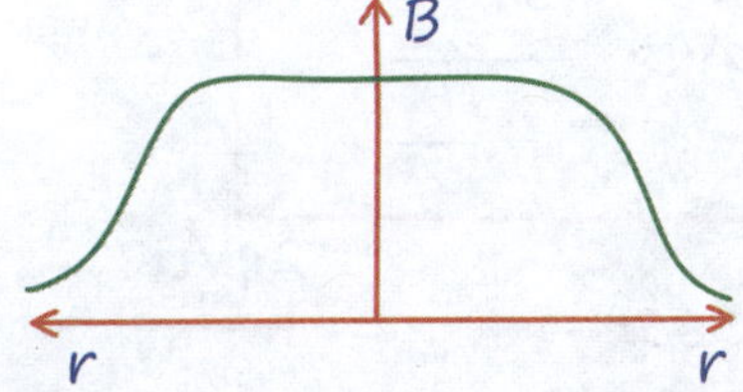

Toroid:-

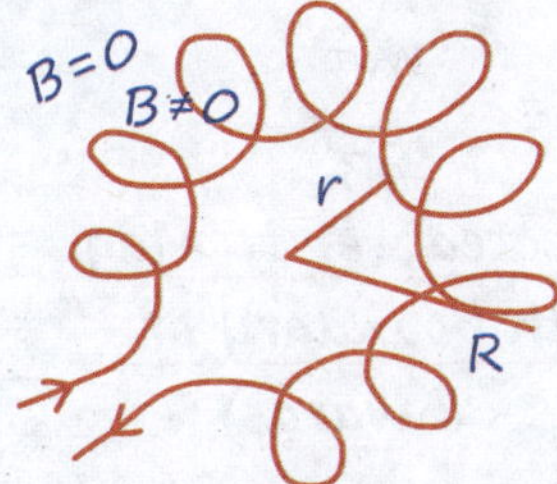

$$B = \frac{\mu_0 N i}{2\pi R_{avg}}$$

$$R_{avg} = \frac{R + r}{2}.$$

8 Magnetic Force

The magnetic force on a charge q moving with velocity $\vec{v}$ in a magnetic field $\vec{B}$ is

$$\vec{F} = qBv\sin\theta = q(\vec{V} \times \vec{B})$$

Magnetic field rest charge per force nahi lagata

* θ = Angle between $\vec{V}$ & $\vec{B}$.
* $\vec{F} \perp^{er} \vec{V}$ and $\vec{F} \perp^{er} \vec{B}$.
* $\vec{a} \perp^{er} \vec{v}$
* only direction will change, speed will remains constant.
* K.E = constant
* Work done = 0
* Power = $\vec{F} \cdot \vec{v} = 0$

$\therefore$ M.F. की औकात नही है "q" की speed change करे! # Garda visualize.

Lekin accl^n $\neq 0$

Kyuki dir^n change hokr $\vec{V}$ & $\vec{P}$. change hoga!

M.F. is like Centripetal Force ($F \perp V$)

9 Lorentz Force

$$\vec{F}_m = q[\vec{E} + \vec{V} \times \vec{B}]$$

$$\vec{F}_m = \vec{F}_{elec.} + \vec{F}_{Mag.}$$

MR* Law for direction

Place your four finger of right hand along velocity and then slap magnetic field by palm then thumb will represent direction of force.

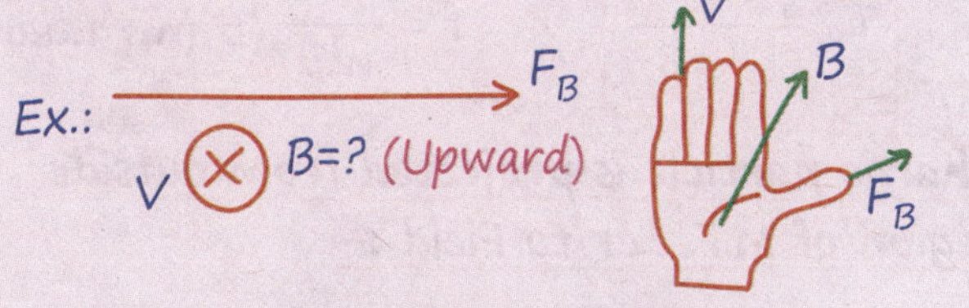

10 Motion of Charge Particle in Magnetic Field

Charge is projected IIel to Mag. Field :-

$\vec{F}_m = 0 \quad a = 0 \quad V = \text{Const}^n$

$$\text{dist}^n = vt$$

Charge is projected $\perp$er to Mag. Field :-

$\vec{V}$ = Variable

$F_m = qVB$

$$a = \frac{qVB}{m}$$

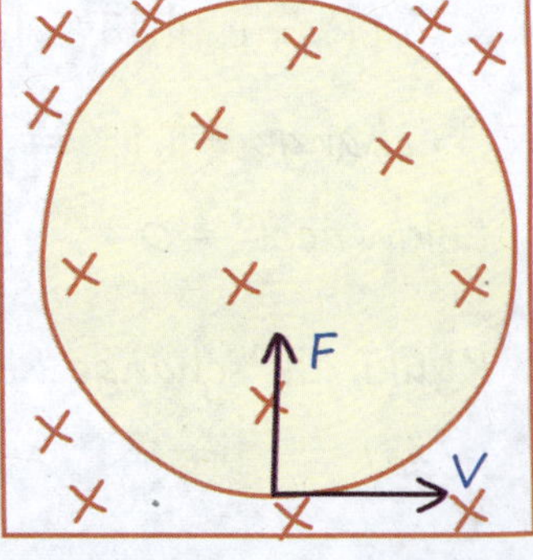

Radius of Circular path :-

$$R = \frac{mv}{qB} = \frac{P}{qB} = \frac{\sqrt{2m(K.E)}}{qB} = \frac{\sqrt{2mqV}}{qB}$$

Time Period & Frequency:-

Time taken to complete one rotation.

T does not depends upon speed

$$T = \frac{2\pi m}{qB} \quad f = \frac{qB}{2\pi m}$$

time taken to Rotate through an θ angle

$$T_\theta = \frac{m\theta}{qB} \quad f = \frac{qB}{m\theta} \ (\theta \text{ in radian})$$

Charge particle is projected from outside region of M.F. ⊥er to Field :-

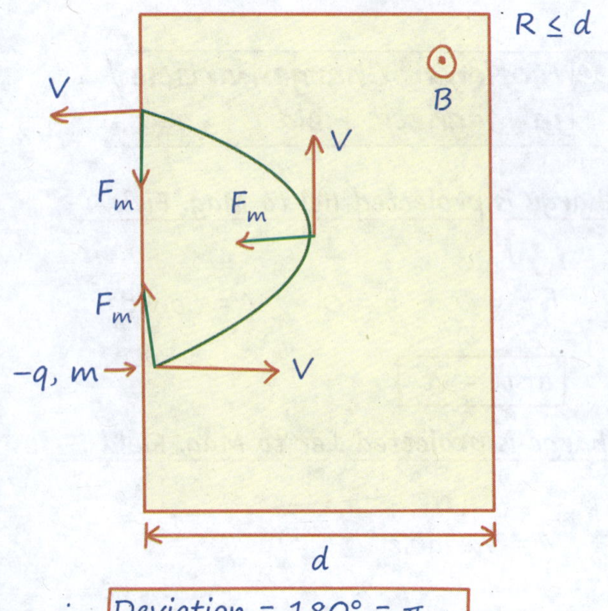

∴ Deviation = 180° = π

MR**

+ Average Force

$$F_{avg} = qVB \frac{\sin\frac{\theta}{2}}{\frac{\theta}{2}}$$

$$\text{for } \theta = 180^\circ \rightarrow F_{avg} = \frac{2qVB}{\pi}$$

$$\text{general formula for Time period } t_\theta = \frac{m\theta}{qB}$$

$$\text{for } \theta = 180^\circ \rightarrow T = \frac{m\pi}{qB}$$

Charge particle is projected ⊥er to Mag. Field at some angle with boundary of Magnetic Field :- (M.F. :- inwards)

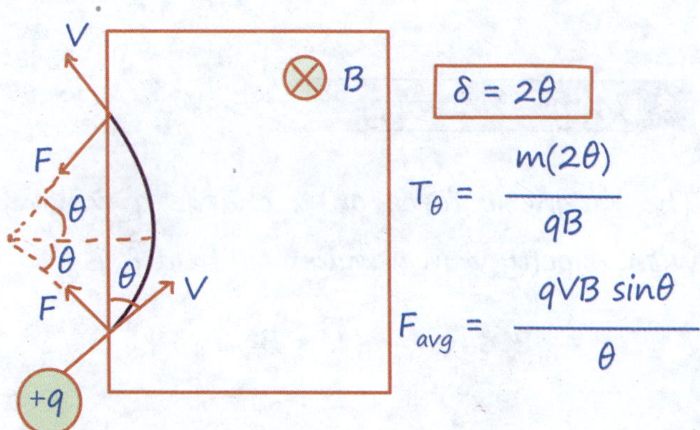

$$\delta = 2\theta$$

$$T_\theta = \frac{m(2\theta)}{qB}$$

$$F_{avg} = \frac{qVB\sin\theta}{\theta}$$

Charge particle is projected ⊥er to Mag. Field at some angle with boundary of outward magnetic field :-

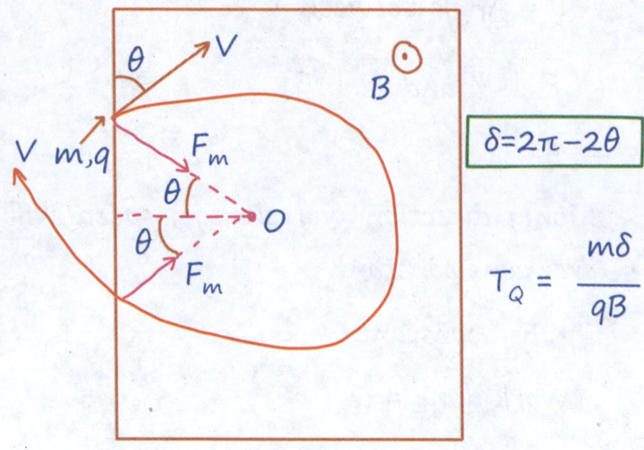

$$\delta = 2\pi - 2\theta$$

$$T_Q = \frac{m\delta}{qB}$$

+ θ = Angle between boundary & velocity

11 Important Questions

Q.2. Time required when "q" loose contact on smooth inclined plane

Magnetic field is outside the plane : B⊙

Sol. $N + qV_tB = mg\cos\theta$

$\because N = 0$ for losing contact

& $V_t = g\sin\theta . t$

$$t = \frac{m\cot\theta}{qB}$$

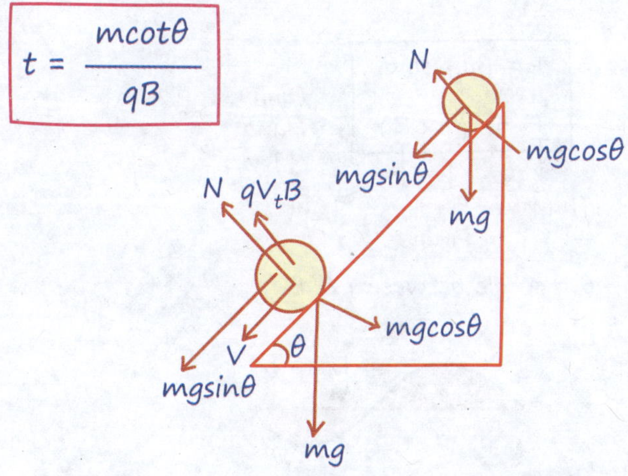

Q.3. Charge particle is projected ⊥er to Mag. Field where (d < R) then find δ.

Sol.

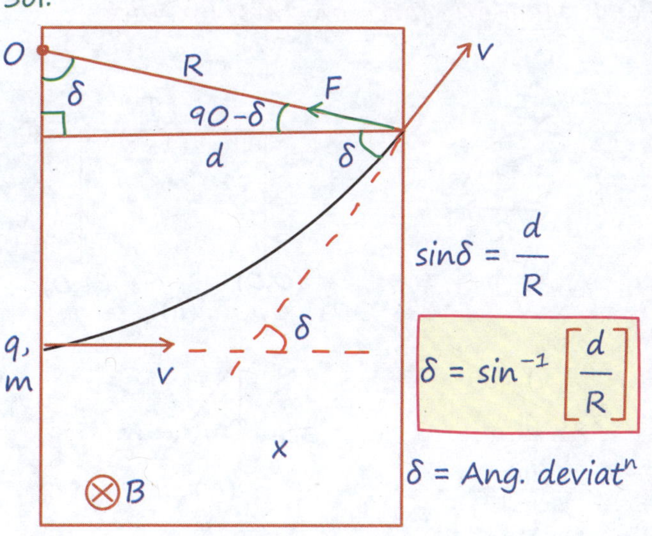

$$\sin\delta = \frac{d}{R}$$

$$\delta = \sin^{-1}\left[\frac{d}{R}\right]$$

δ = Ang. deviat^n

Important Cases :-

- Object is projected with speed "v" at an angle "θ" from Mag. Field :-

  Path of the particle will be helical.

  $\theta \neq 0°, 90°$ & $180°$

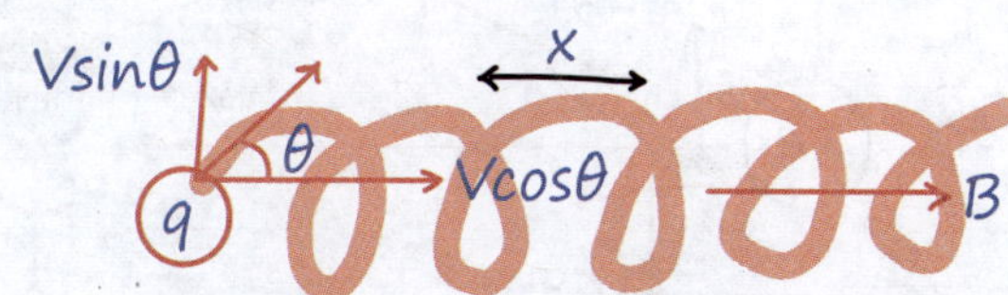

- The perpendicular component vsinθ causes the particle to move in a circular path with radius R and time period T:

$$R = \frac{mv\sin\theta}{qB} \qquad T = \frac{2\pi m}{qB}$$

- The parallel component vcosθ causes the particle to move along the magnetic field, resulting in helical motion with pitch:

$$\text{Pitch }(x) = u_x T = v\cos\theta \cdot \frac{2\pi m}{qB}$$

12 Velocity Selector

- A velocity selector is a device used to filter particles so that only those with a specific velocity pass through undeflected. This is achieved using
- Magnetic field, Electric field and velocity all are perpendicular to each other.

Condition for No Deviation:

- For a particle to pass through the velocity selector without deviation, the net force must be zero, i.e.,

$\vec{F}_E + \vec{F}_B = 0$

$qE_0 = qV_0B_0$

$$V_0 = \frac{E_0}{B_0}$$

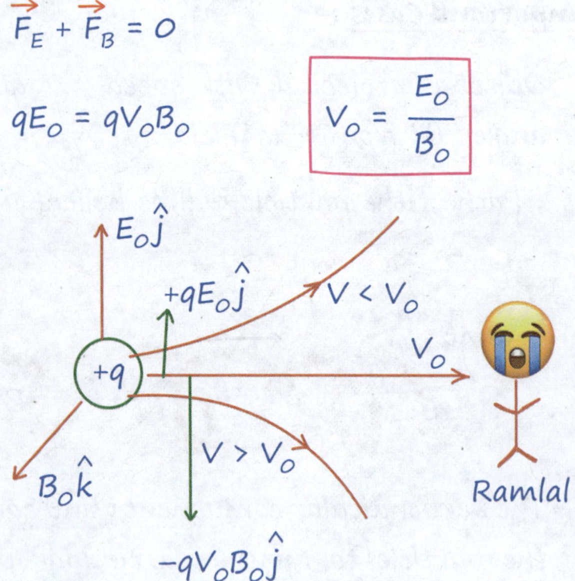

- Particle which have velocity $V > V_0$ will experience large magnetic force and deviates downward
- Particle which have velocity $V < V_0$ will experience small magnetic force and deviates upward.

13 Cyclotron

- Device use to acceleration charge particle like proton deuteron, α -particle but not for electron, we use betatron for electron.

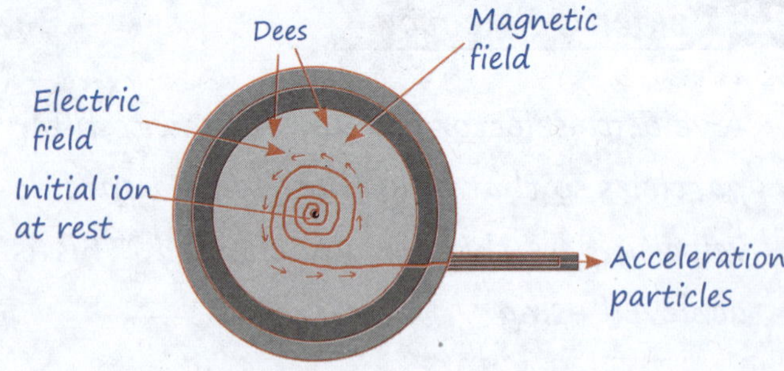

- Electric field used to acceleration and shift provide k.E. to the charge particle.
- Magnetic field used to keep the charge particle inside magnetic field.
- Freq. of oscillator = Freq. of charge particle

$$KE_{1rot^n} = 2q\Delta V$$

$$f_{req} = \frac{qB}{2\pi m}$$

$$KE = \frac{1}{2}\frac{q^2B^2R^2}{m} \quad \because \quad \frac{mv^2}{r} = qVB$$

14 Magnetic Force on Current Carrying Wire

- Magnetic force always perpendicular to the plane of $I\vec{dl}$ and $\vec{B}$.
- Force on close loop of any shape is zero ($l_{eff} = 0$)

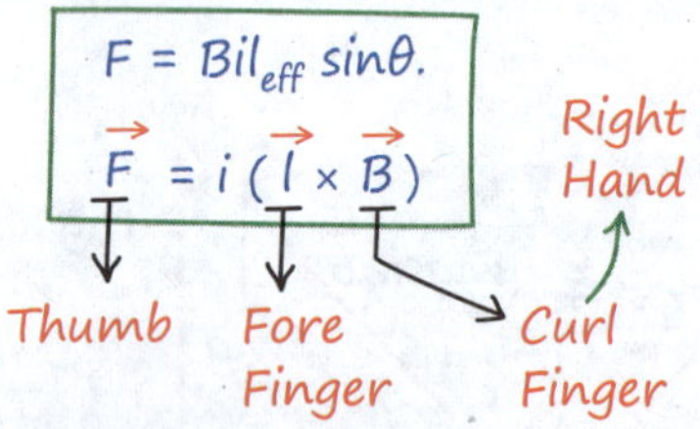

θ = Angle between i & B.

L_{eff} :-

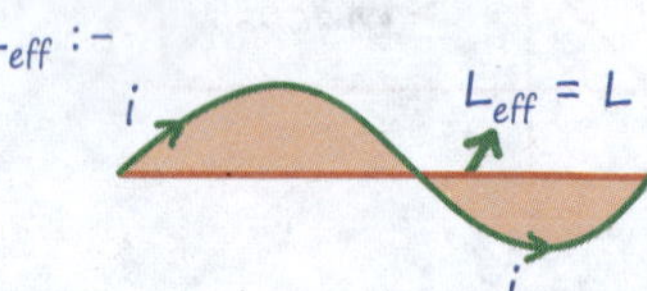

MR*for direction:- Place your four finger of right hand along $I\vec{l}$ and slap magnetic field thumb will represent force.

Q.4. Magnetic force on different wire.

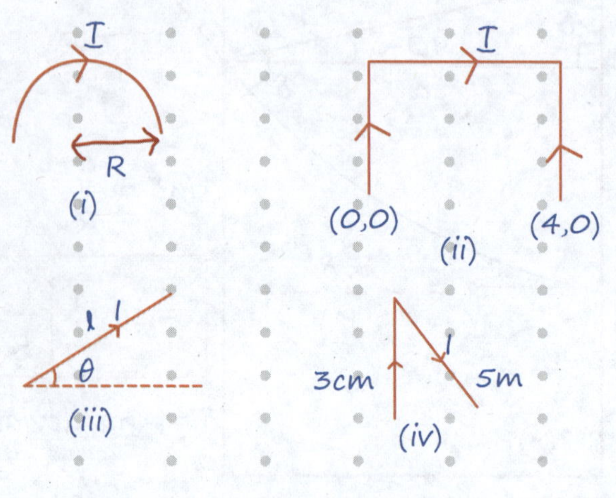

Sol. (i) $F = 2BIR$ (ii) $F = 4IB$

(iii) $F = BIl$ (iv) $F = 4BI$

I, O, R, B — $F = 0$

I, O, A, R, B — $F = BIR$ effective length perpendicular to B is R.

Force on small current carrying wire due to infinite large current carrying wire :-

Case 1 :-

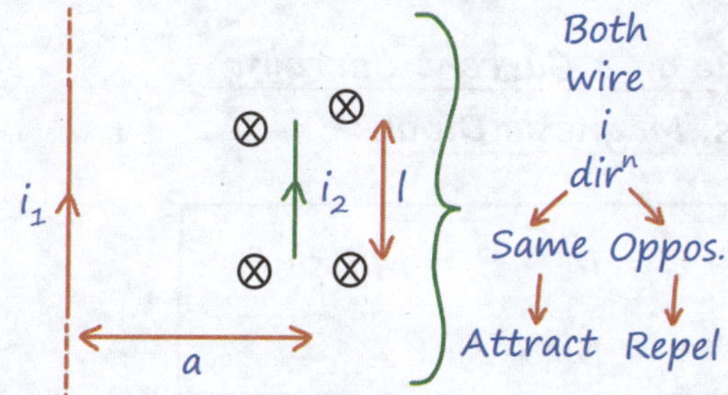

Magnetic field due to infinite wire:

$$B_1 = \frac{\mu_0 i_1}{2\pi a}$$

Force on small wire $\boxed{F = B_1 i_2 l = \frac{\mu_0 i_1}{2\pi a} i_2 l}$

Case 2 :-

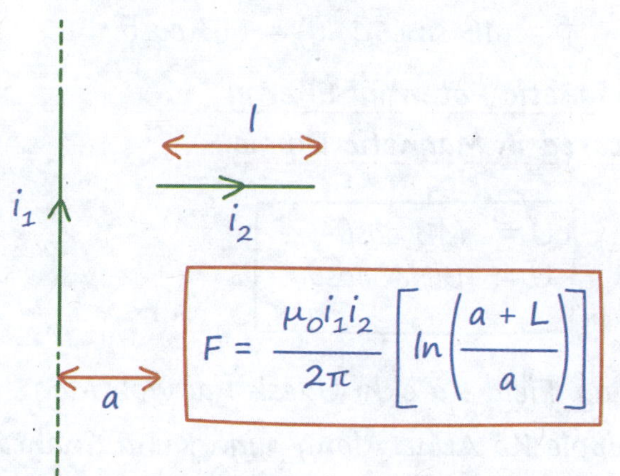

$$F = \frac{\mu_0 i_1 i_2}{2\pi}\left[\ln\left(\frac{a+L}{a}\right)\right]$$

#ye sab Mutual Force hai matlab jitna badi-choti wire pr Force lagayega utnahi Force choti-badi pr lagayegi.

Long prallel wires :-

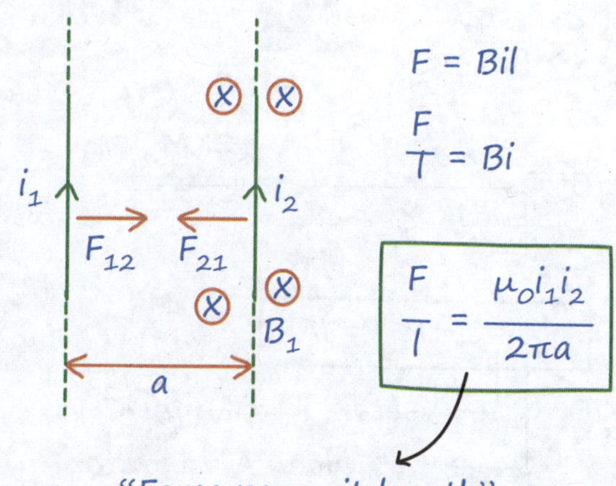

$F = Bil$

$\frac{F}{l} = Bi$

$$\boxed{\frac{F}{l} = \frac{\mu_0 i_1 i_2}{2\pi a}}$$

"Force per unit length"

Biot-savart Law in Terms of Velocity of Particle :-

$$\boxed{B = \frac{\mu_0}{4\pi}\frac{qV\sin\theta}{r^2}}$$

θ = Angle between I and r.

Note :- For electron, $q = e^-$

Ratio of magnetic force to electric force

$$\frac{F_{MF}}{F_{EF}} = \frac{\mu_0 e^2 V^2}{4\pi d^2} \times \frac{4\pi\varepsilon_0 d^2}{e^2}$$

$$\boxed{\frac{F_{MF}}{F_{EF}} = \frac{V^2}{C^2}} \qquad C^2 = \frac{1}{\mu_0 \varepsilon_0}$$

$F_{EF} >>>>>> F_{MF}$.

15 Circular Current Loop as a Magnetic Dipole

A current-carrying circular loop generates a magnetic field similar to that of a bar magnet, acting as a magnetic dipole.

Magnetic Field on the Axis of the Loop:

$$B = \frac{\mu_0 i R^2}{2(R^2+X^2)^{3/2}}, \text{ where, } K = \frac{\mu_0}{4\pi}$$

M = magnetic moment

At x >> R

$$B = \frac{2\mu_0}{4\pi}\,\frac{R^2\pi i}{X^3} = \frac{2KM}{X^3}$$

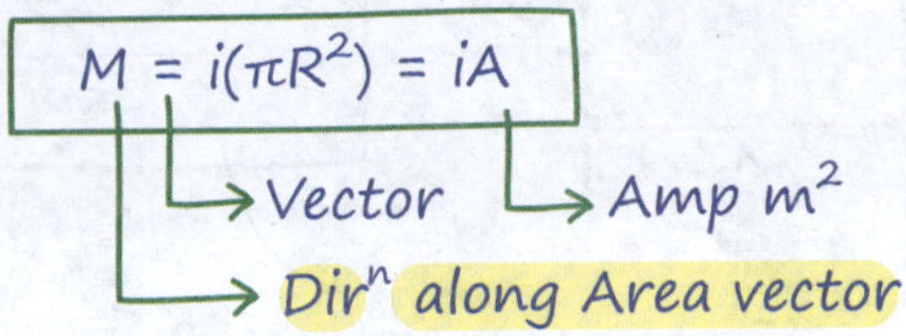

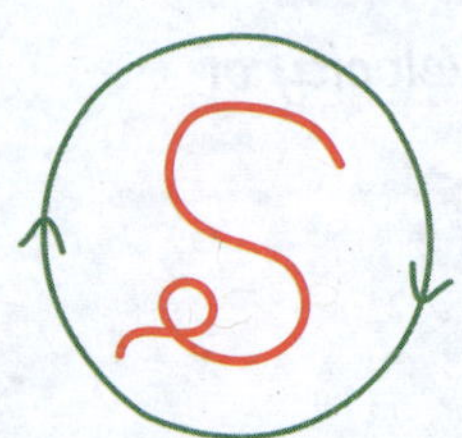
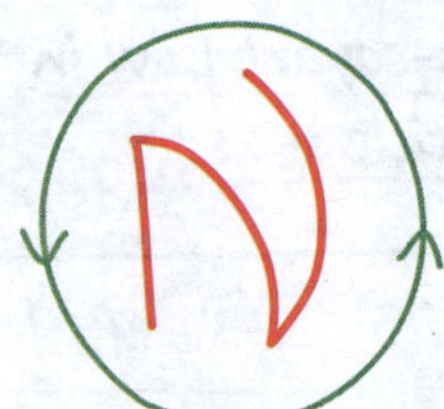

Clockwise
South-pole
M:- ⊗

Anticlockwise
North-pole
M:- ⊙

Directⁿ of $\vec{M}$ is along $\vec{B}$.

M = NAI → N = no. of turns.

Magnetic Moment of revolving electron :-

Frequency $f = \frac{V}{2\pi R}$,

Current $i = \frac{e}{T} = ef$

$$M = IA = ef\pi R^2 = \frac{eVR}{2}$$

Gyromagnetic Ratio:-

- Ratio of Magnetic Moment and Angular Momentum:-

$$\frac{M}{L} = \frac{e}{2m_e} = \frac{8.8\times10^{10}C}{Kg}$$

- Bohr Magneton:-

$$M_{e^-} = \frac{e}{2m_e}\left[\frac{h}{2\pi}\right]$$

$$\therefore\ M = \frac{e}{2m_{e^-}}[I\omega]$$

I = moment of inertia

Torque on a Current Carrying Loops, Magnetic Dipole:-

$$\vec{\tau} = \vec{M}\times\vec{B} = MB\sin\theta$$
$$\vec{\tau} = BiNA\sin\theta$$

θ = Angle between M & B.

Torque perpendicular to magnetic moment and magnetic field

Net force = zero

- If angle given from plane of loop

$$\tau = MB\sin(90-\theta) = MB\cos\theta$$

Magnetic Potential Energy Stored in Magnetic Dipole:-

$$U = -MB\cos\theta$$
$$U = -BiNA\cos\theta$$

Mag. Field Ka ekhi Udesh hai woh Mag. Dipole Ko Apne Along align Krna Chahta hai.

- Time Period :- $T = 2\pi\sqrt{\frac{I}{MB}}$

I = Moment of Inertia

M = Magnetic moment

- Work Done to Rotate dipole $W = \Delta U$.
- Work done by M.F. to rotate dipole = $W = -\Delta U$

16 Moving Coil Galvanometer

Working Principle: Torque on current carrying loop

- If No. of turns in moving coil-galvanometer is increase then current sensitivity will increases but voltage sensitivity remains same because resistance will also increase.

$V_s \propto \frac{N}{R}$

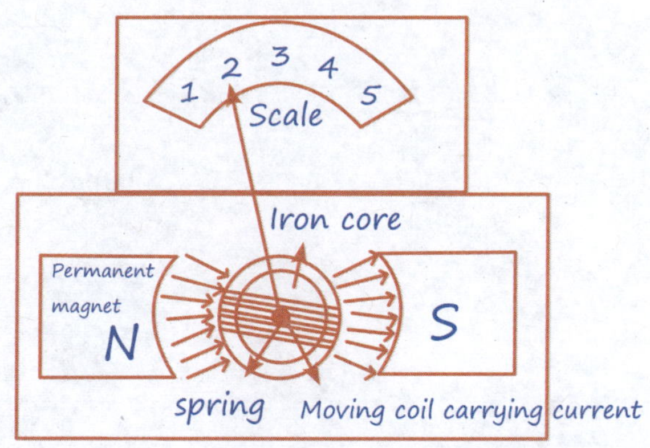

Torque on the coil:

$$\tau = MB = NIAB = C\theta$$

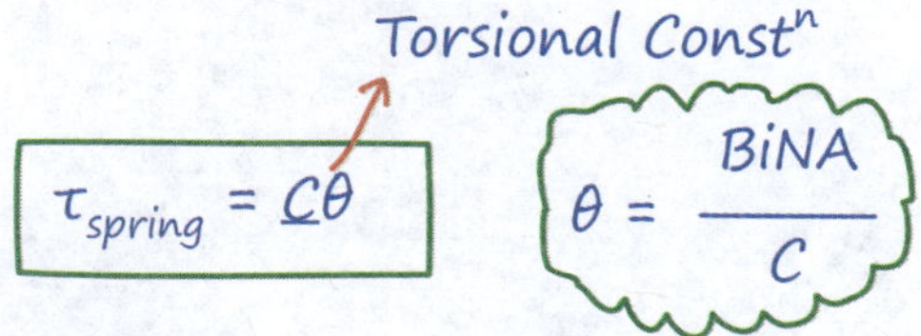

$\tau_{spring} = C\theta$ (C: Torsional Const^n) $\theta = \frac{BiNA}{C}$

$$I_s = \frac{\theta}{I} = \frac{BAN}{C}$$ (Current Sensitivity; A: Area of Loop)

Divided by R both sides

$$V_s = \frac{I_s}{R} = \frac{BAN}{CR} = \frac{\theta}{V}$$ (Voltage Sensitivity)

Visualisation 1.

If an electron is not deflected in passing through a certain region of space, can we be sure that there is no magnetic field in that region?

No, electron would not be deflected if $\vec{V}$ and $\vec{B}$ are in the same direction.

Visualisation 2.

If a moving electron is deflected sideways on passing through a certain region of space, can we be sure that a magnetic field exists in that region?

No, the sideways deflection may be due to Electric field as well. In the absence of electric field, the sideways deflection shows the presence of magnetic field in the region.

Visualisation 3.

If a charged particle at rest experiences no electromagnetic force, then the electric field must be zero or

The magnetic field may or may not be zero

Visualisation 4.

If a charged particle kept at rest experiences an electromagnetic force, then The electric field must not be zero

or

The magnetic field may or may not be zero.

Visualisation 5.

If a charged particle projected in a gravity-free room deflects, then Both fields cannot be zero

or

Both fields can be non-zero

Visualisation 6.

A charged particle moves in a gravity-free space without change in velocity. Possible cases are

E = 0 B = 0 or E = 0 B ≠ 0 or E ≠ 0 B ≠ 0

Visualisation 7.

A charged particle moves along a circle under the action of possible constant electric and magnetic fields. Possible case is

E = 0, B ≠ 0

Visualisation 8.

A charged particle goes undeflected in a region containing electric and magnetic field. It is possible that

$\vec{v} \parallel \vec{E}$, $\vec{E} \parallel \vec{B}$ or $\vec{E}$ is not parallel to $\vec{B}$

Visualisation 9.

If a charged particle goes unaccelerated in a region containing electric and magnetic fields, then

$\vec{E}$ must be perperdicular to $\vec{B}$ and $\vec{v}$ must be perpendicular to $\vec{E}$.

Magnetism & Matter

1 Bar Magnet

A bar magnet creates a magnetic field, which can be visualized using magnetic field lines.

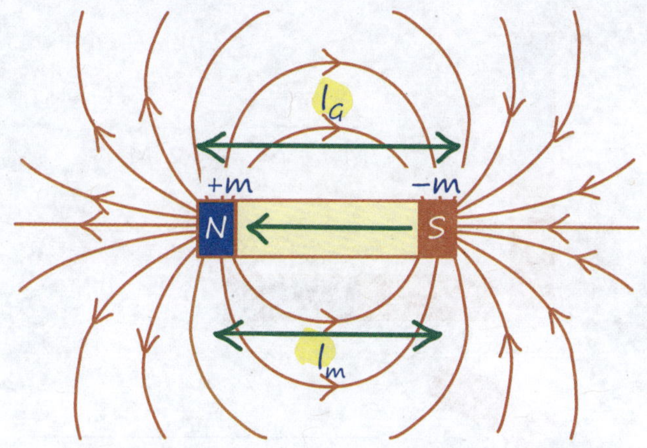

$$\frac{l_{Mag}}{l_{Geo}} = 0.84$$

- Magnetic field lines are also called magnetic force line → false because force acts perpendicular to magnetic field.
- Magnetic field lines always from N to S → false, Inside Magnet it is S to N

Properties of Magnetic field lines :-

- They form closed loop
- They never intersect each other
- Magnetic field lines are crowded near the Pole where magnetic field is strong. and spread apart from each other where field is weak.
- They flow from the South Pole to the north Pole within a magnet and north pole to South Pole outside
- They Comes out and go in at any angle from magnet.

2 Magnetic Dipole Moment

The magnetic dipole moment represents the strength and orientation of a magnetic source, such as a bar magnet or a current loop.

$\vec{M} = m\vec{l} = NiA$ m:- Pole Strength

$m\vec{l}$ → For bar magnet

NiA → For current loop

Direction:

- The direction of $\vec{M}$ is from the south pole to the North pole for a bar magnet.
- For a current loop, the direction of $\vec{M}$ is determined using the right hand rule: curl your fingers in the direction of the currrent, and your thumb points in the direction of $\vec{M}$

Cutting of bar Magnet :-

*m ∝ Area, M ∝ Volume

Cute type	Resulting Length (l′)	Pole Strength (m′)	Magnetic Moment (M′)
Cut along Length	$l' = l$	$m' = m/2$	$M' = m.l' = (m/2)l = M/2$
Cut Perpendicular to length	$l' = l/2$	$m' = m$	$M' = m.l' = m\,(l/2) = M/2$

Bending of bar Magnet :-

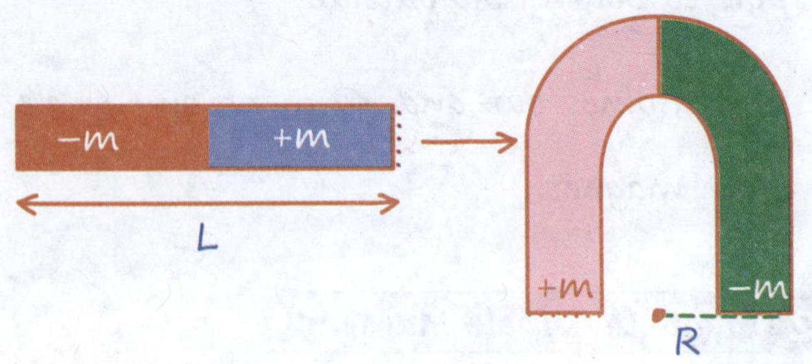

MR* **

Jab bhi mein Koi Circle dhekhu mera dil dewaana bole....

Magnetic moment circular Segment

$$M' = \frac{M \sin(\theta/2)}{\theta/2}$$

Complete Circle	Semi-Circle.
$\theta = 2\pi$	$\theta = \pi$
$M' = 0$	$M' = \frac{2M}{\pi}$

- Magnetic moment of 60° segment

$$M' = \frac{M \sin\left(\frac{60^\circ}{2}\right)}{\frac{\pi}{3\times 2}} = \frac{3M}{\pi} \quad \left(\theta = 60^\circ = \frac{\pi}{3}\text{ radian}\right)$$

60°

R

- Two Identical bar Magnet of magnetic Moment M

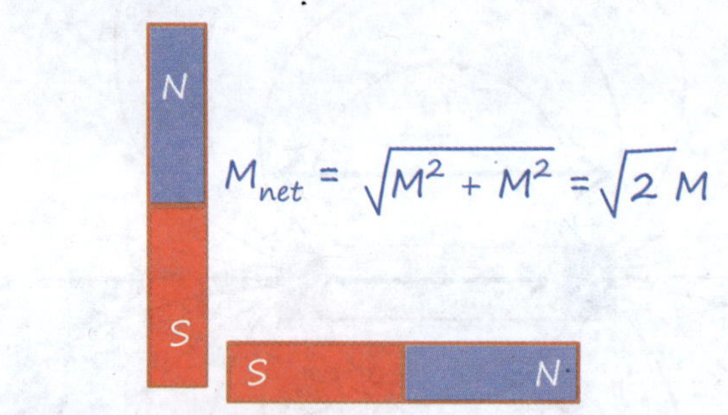

$$M_{net} = \sqrt{M^2 + M^2} = \sqrt{2}\,M$$

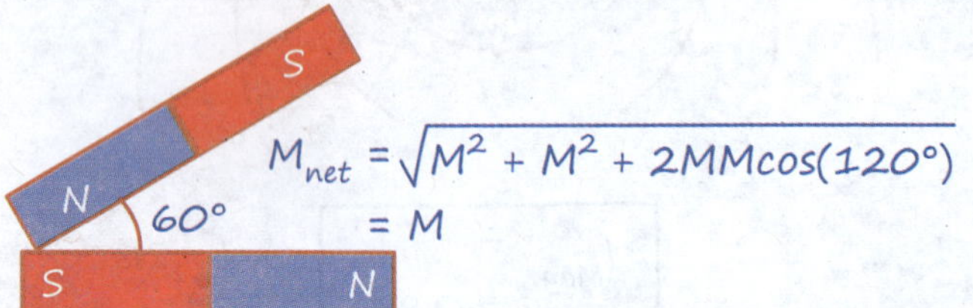

$$M_{net} = \sqrt{M^2 + M^2 + 2MM\cos(120^\circ)} = M$$

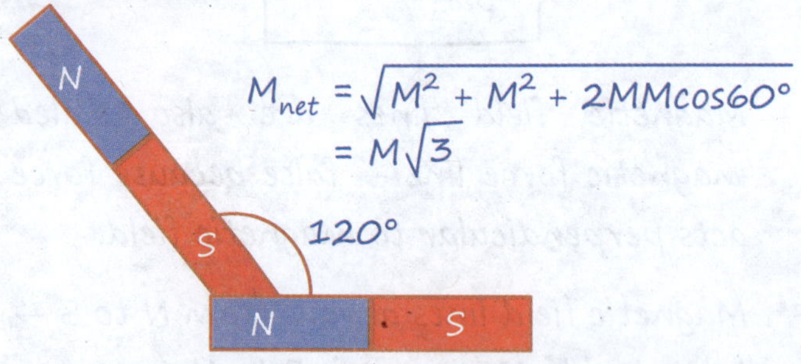

$$M_{net} = \sqrt{M^2 + M^2 + 2MM\cos 60^\circ} = M\sqrt{3}$$

3 Magnetic Field Due to Bar Magnet

On axial point :-

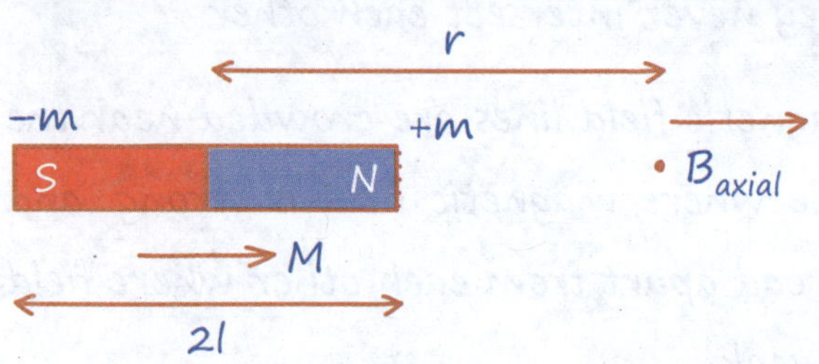

$$B_{axial} = \left(\frac{\mu_0}{4\pi}\right)\frac{2Mr}{(r^2-l^2)^2}$$

$$\text{For } r >> l,\ B_{axial} = \left(\frac{\mu_0}{4\pi}\right)\frac{2M}{r^3}$$

- Magnetic Field along dipole

On Normal Bisector :-

$$B_{eq} = \left(\frac{\mu_0}{4\pi}\right)\frac{2ml}{(r^2+l^2)^{3/2}}$$

For r >> l

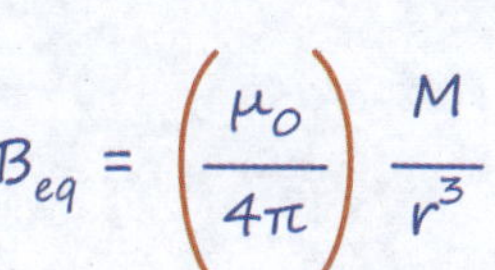

$$B_{eq} = \left(\frac{\mu_0}{4\pi}\right)\frac{M}{r^3}$$

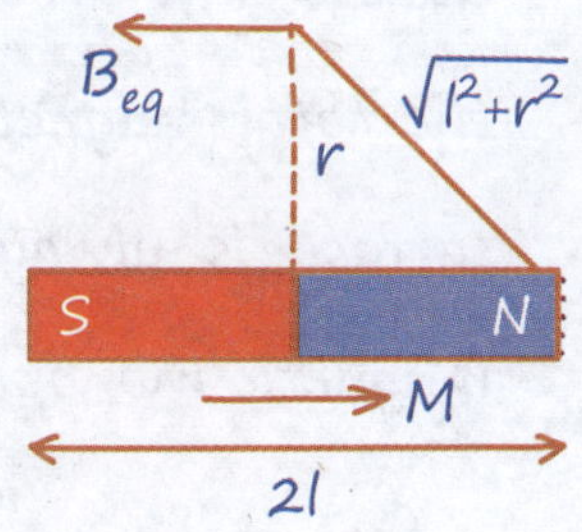

- Magnetic Field opposite to dipole.

Magnetic field due to Dipole at General Point :-

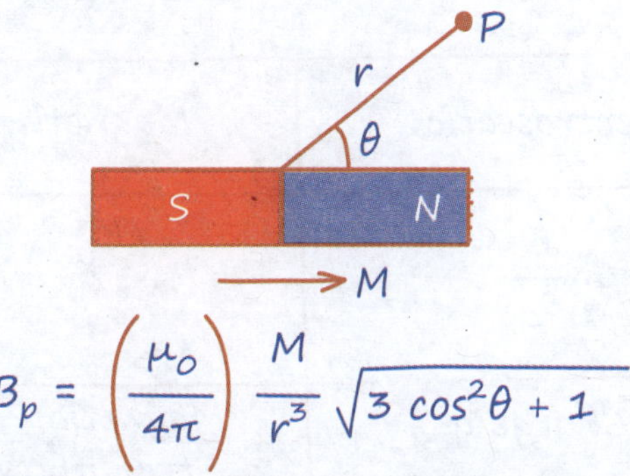

$$B_p = \left(\frac{\mu_0}{4\pi}\right)\frac{M}{r^3}\sqrt{3\cos^2\theta + 1}$$

Dipole in uniform Magnetic Field :-

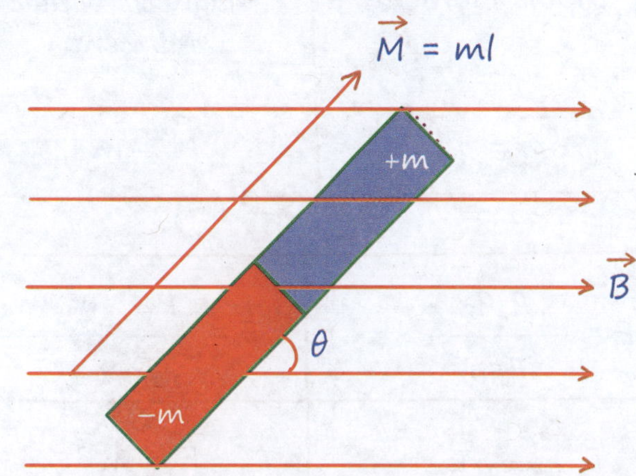

Condition	Torque (t)	Potential Energy(U)	Explanation
General	$\tau = \vec{M}\times\vec{B} = MB\sin\theta$	$U = -\vec{M}.\vec{B} = -MB\cos\theta$	
$\theta = 0°$	$\tau = 0$	$U = -MB$ (Minimum energy)	Dipole aligned with the magnetic field
$\theta = 90°$	$\tau = MB$ (Maximum Torque)	$U = 0$	Dipole perpendicular to the magnetic field
$\theta = 180°$	$\tau = 0$	$U = MB$ (Maximum energy)	Dipole opposite to the magnetic field

- $W_{ext} = \Delta U$ $\quad$ $W_B = -\Delta U$
- Stable equilibrium of $\theta = 0°$ (Diople opposite to field)
- Unstable equilibrium at $\theta = 180°$
- BAR magnet will oscillate in uniform magnetic field about stable equilibrium

$$T = 2\pi\sqrt{\frac{I}{MB}}$$

MR* feel

Magnetic Field Ka ekhi Udessh hai Ki woh Mag. dipole Ko apni taraf Khich Ke rakhna Chahta hai !

4 Analogy

Electrostatics	Magnetism
$\frac{1}{\varepsilon_0}$	μ_0
Charge q	Magnetic Pole Strength (m)
Dipole Moment $\vec{p} = q\vec{l}$	Magnetic Dipole Moment $\vec{M} = m\vec{l}$
$F = \frac{q_1 q_2}{4\pi\varepsilon_0 r^2}$	$F = \frac{\mu_0}{4\pi}\frac{m_1 m_2}{r^2}$
$\vec{F} = q\vec{E}$	$\vec{F} = m\vec{B}$
Axial Field $\vec{E} = \frac{2\vec{p}}{4\pi\varepsilon_0 r^3}$	$\vec{B} = \frac{\mu_0}{4\pi}\frac{2\vec{M}}{r^3}$
Equatorial field $\vec{E} = \frac{-\vec{p}}{4\pi\varepsilon_0 r^3}$	$\vec{B} = -\frac{\mu_0}{4\pi}\frac{\vec{M}}{r^3}$
Torque $\vec{\tau} = \vec{p} \times \vec{E}$	$\vec{\tau} = \vec{M} \times \vec{B}$
Potential Energy U $= -\vec{p}\cdot\vec{E}$	$U = -\vec{M}\cdot\vec{B}$
Work, W $= pE(\cos\theta_1 - \cos\theta_2)$	Work, W $= MB(\cos\theta_1 - \cos\theta_2)$

MR Speical***

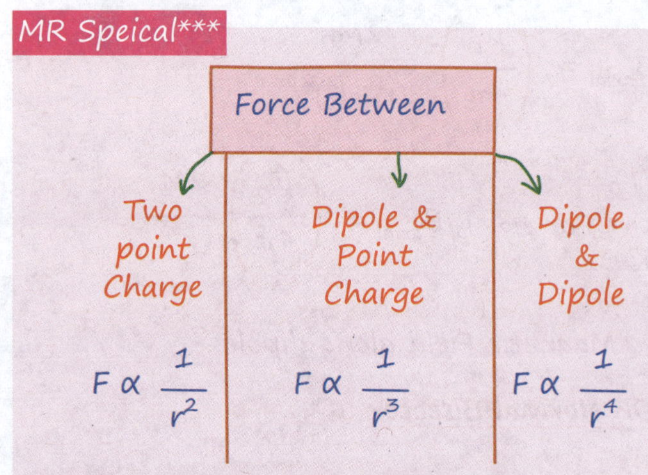

5 Gauss Law in Magnetism

- Gauss's law in magnetism states that the total magnetic flux through a closed surface is always zero, indicating that magnetic monopoles do not exist.

$$\int \vec{B}\cdot\vec{ds} = \phi_B$$

$$\oint \vec{B}\cdot\vec{ds} = 0 \text{ (Always)}$$

This means that the magnetic field lines are continuous and do not begin or end at any point, unlike electric field lines.

MR Speical***

1> $T = \text{Same}$ } When bar magnet cut along length.

2> $T' = \frac{T}{n}$ } When bar magnet cut ⊥er to length.

n = no. of equal cutted part.

$T = \frac{T}{n}$ $T' = T$

6 Neutral Point

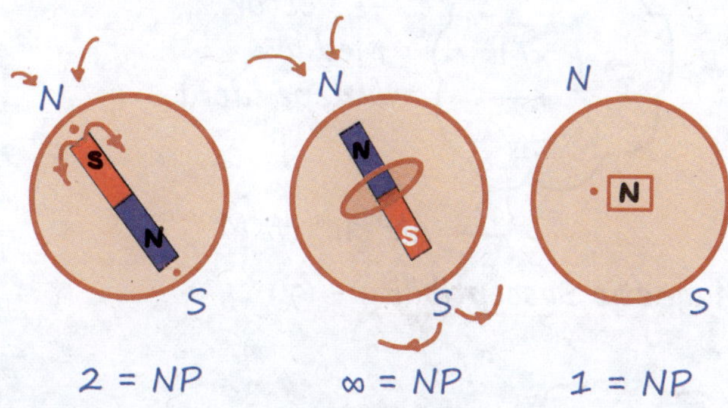

$2 = NP$ $\infty = NP$ $1 = NP$

7 Vibrational Magnetometer

(Oscillational Magnetometer)

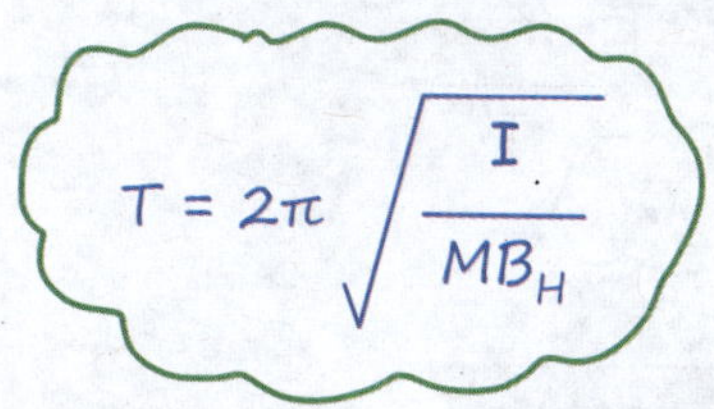

Application -1 :-

To Find Magnetic dipole Moment (M):-

$$M = \frac{4\pi^2 I}{T^2 B_H}$$

Application -2 :-

To Find ratio of Mag. dipole moment of two magnet of same size.

$$\frac{M_2}{M_1} = \left(\frac{T_1}{T_2}\right)^2$$

Application -3 :-

To Compare Mag. dipole moment of two magnet of diff. size.

$$\therefore \frac{M_1}{M_2} = \frac{T_2^2 + T_1^2}{T_2^2 - T_1^2} = \frac{f_1^2 + f_2^2}{f_1^2 - f_2^2}$$

$$T_1 = 2\pi\sqrt{\frac{I_1 + I_2}{(M_1 + M_2)B_H}}$$

$$T_2 = 2\pi\sqrt{\frac{I_1 + I_2}{(M_1 - M_2)B_H}}$$

8 Tangent Law

The tangent law in magnetism describes the relationship between two perpendicular magnetic fields acting on a magnetic needle, such as in a tangent galvanometer. When a magnetic needle is subjected to two perpendicular magnetic fields, B_1 and B_2, it aligns at an angle θ such that:

$$\tan\theta = \frac{B_2}{B_1}$$

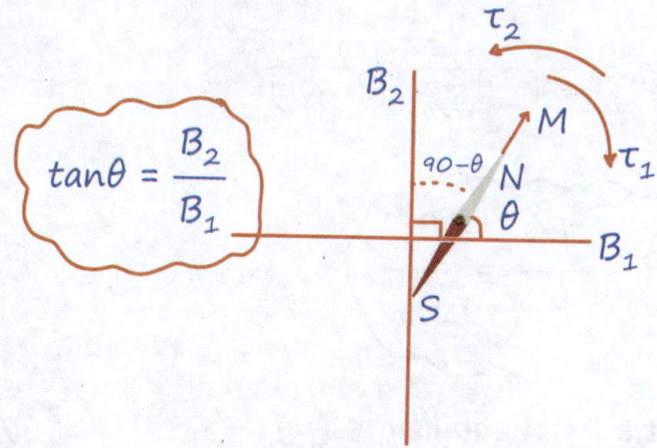

Tangent Galvanometer :-

A tangent galvanometer is an instrument used to measure the horizontal component of the Earth's magnetic field.

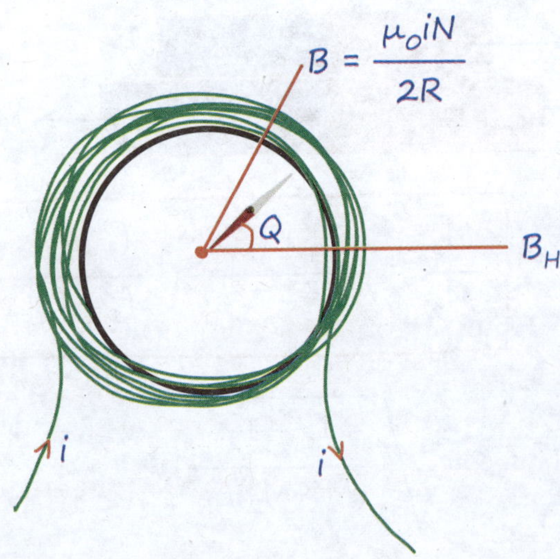

When placed in the Earth's magnetic field.

$$M \cancel{B_H} \sin\theta = \cancel{M}\left(\frac{\mu_0 iN}{2R}\right)\cos\theta.$$

$$\boxed{\tan\theta = \frac{\mu_0 iN}{2R\,B_H}}$$

"K = Reduction Factor."

$$K\tan\theta = i \quad \left(K = \frac{2R\,B_H}{\mu_0 N}\right)$$

Magnetic Field Intensity:- (H)

Source की असली औकात

$$H = \frac{B_0}{\mu_0} = \frac{B_m}{\mu_m}$$

Magnetic Permeability :- (μ)

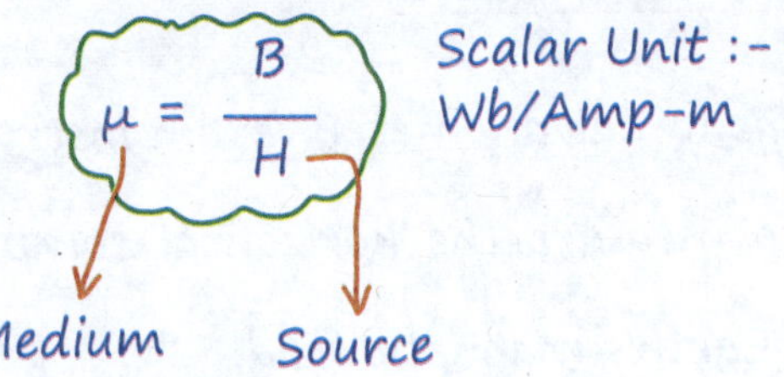

Scalar Unit :- Wb/Amp-m

Magnetisation & Magnetic Intensity:- (I)

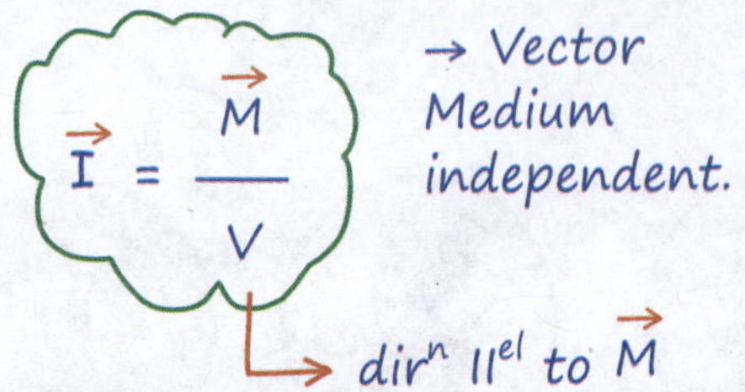

→ Vector
Medium independent.

Magnetic Susceptibility:- (χ)

$$\chi_m = \frac{I}{H}$$

→ Scalar
→ Unit & Dimensionless

$\chi_m \uparrow$ = Easily magnetised.

Relation Between μ_r & χ_m:-

$$\mu_r = 1 + \chi_m$$

9 Cause of Magnetism

Atom → (Nucleus + electron in rotational motion)

#revolving electron Produce magnetic field, (magnetic moment) along axis of Rotation.

#In Paired electron atom, two electrons are in Opposite spin.

$\vec{M}_{atom} = 0$

In unpaired electron atom.

$\vec{M}_{atom} \neq 0$ and $\vec{M}_{crystal} = 0$ due to random orientation of atoms.

10 Materials

Diamagnetic:-

+ Diamagnetic have tendency to Move from region of Stronger to weaker part of external Magnetic Field

∵ They are Magnetised in opposite direction.

- Magnetic field lines are expelled by these substances.

Imp → Diamagnetism is universal property

MR***

Magnetise hota hai lenz law (law of inertia) se.

$\chi_m = -ve$
$B < B_0$
$\mu < \mu_0$
$\mu_r < 1$

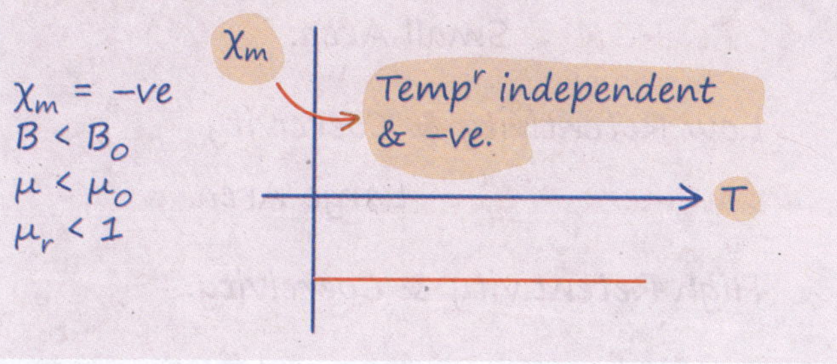

Paramagnetic:-

Paramagnetic substance

$M_{atom} \neq 0$

$M_{material} = 0$ (in absence of external mag. field)

$M_{material} \neq 0$ (⊕ presence of B_{ext})

Tendency to move from Weak Magnetic field Region to strong magnetic field.

MR***

#Magnetise hota hai τ_{ext} se.

χ_m = +ve (Small +ve)

$B > B_0$
$\mu > \mu_0$
$\mu_r > 1$

$\chi_m \propto \frac{1}{Temp^r}$

Ferromagnetic :-

Ferromagnetic Material

strongly attracted to Magnetic Substance

$\vec{M}_{domain} \neq 0$

$M_{material} = 0$ (∵ Random arrangement of domains)

$M_{material} \neq 0$ (⊕ magnetic field external)

Domain Formation.

$\chi_m >>> 1$
$B >>> B_0$
$\mu >>> \mu_0$
$\mu_r >>> 1$

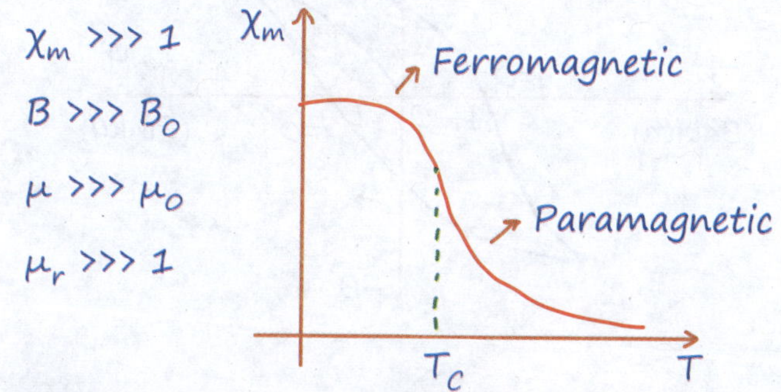

MR Speical***

Pinky — Babu Jalela Paramagnetic — Kalua Diamag. — Ramlal Ferromag.

Curie's Law:-

$$I = \frac{CB_0}{T}$$ Paramagnetic Sub^s.

$$\chi_m = \frac{C\mu_0}{T}$$

Curie Weiss Law :-

$$\chi_m = \frac{C}{T - T_C}$$

11 Hysteresis Loop or B/H Curve

MR Speical***

"Life of Ramlal"

Area under loop ∝ कितना कटा Ramlal का

Area under loop ∝ Energy loss

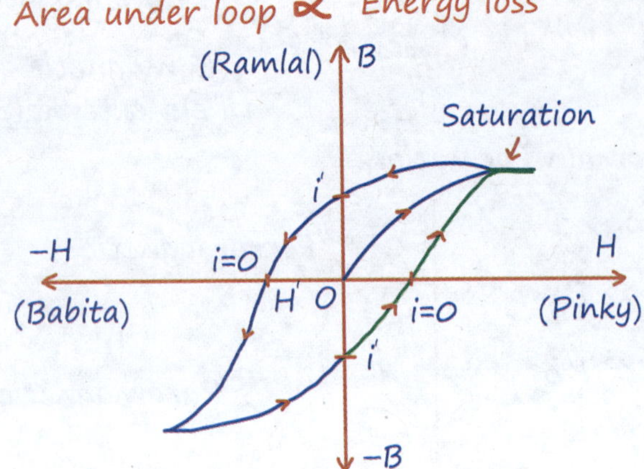

OI' = Retentivity (R)	OH' = Coercivity. (C)
Ramlal Pinky Ke Pyaar mein Kitne yaadein bachake rakh paya tha	Babita Ko Jitna Pyaar dikhana pada pinky Ki yaadein Ko mitane Keliye.

- SOFT IRON :- Small Area.

  Low Retentivity & Coercivity.

- Permanent Mag. :- Large Area.

  High Retentivity & Coercivity.

MR*

Everyone including your society, family, friends sirf success ko he salute karte hai

27 Electromagnetic Induction

1 Faradays Experiment in Uniform Magnetic Field

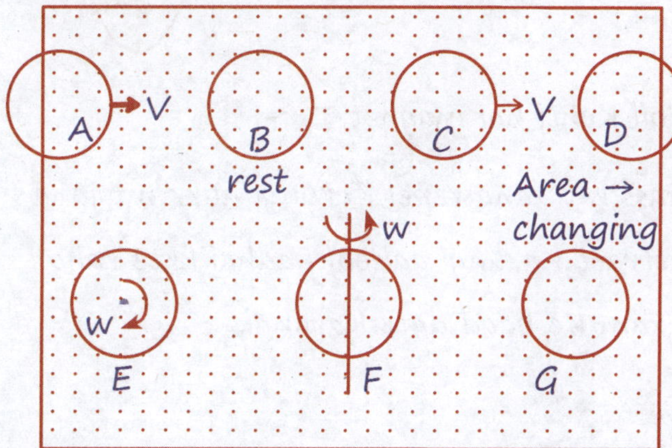

In loop A, D, F, G :-

- Current induced in loop because of change in magnetic flux.
- This can be due to either the movement of the loop or a change in the area of the loop within the magnetic field.

In loop B, C, E :-

- No current induced, because flux is constant.
- These loops could be stationary, or their area within the magnetic field might not be changing.

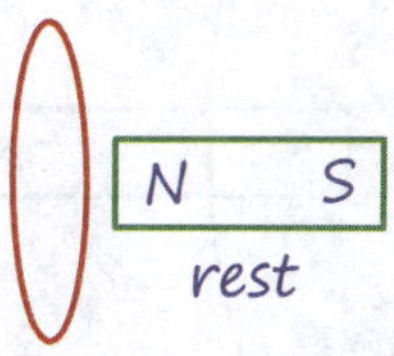

No current

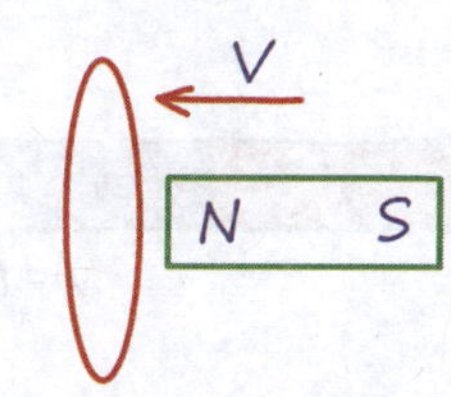

Current induced

2 Magnetic Flux

Counting of magnetic field lines passing through given cross-section area.

- gives the idea of magnetic energy.
- scalar S.I. unit → Tesla-m^2 = Weber C.G.S. unit → Gauss cm^2 = Maxwell

$$\phi = \vec{B}.\vec{A} = BA\cos\theta$$

θ = Angle between Mag. Field & Area

Unit :- 1 Wb = 10^8 Maxwell

- If $\vec{B}$ = Variable.

$$\phi = \int \vec{B}.d\vec{A}$$

- If magnetic field $\vec{B} = B_x\hat{i} + B_y\hat{j} + B_z\hat{k}$ and $\vec{A} = A_x\hat{i} + A_y\hat{j} + A_z\hat{k}$ then (flux), $\phi = B_xA_x + B_yA_y + B_zA_z$
- Flux through circular loop ($\ell >> R$)

$$\phi = \frac{2\sqrt{2}\mu_0 I\pi R^2}{\pi\ell}$$

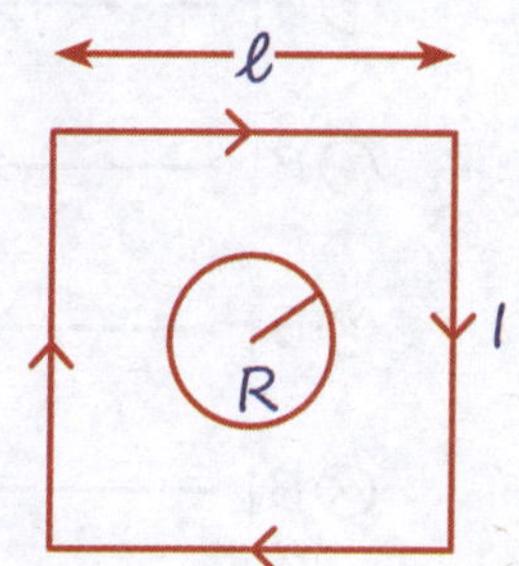

MR***

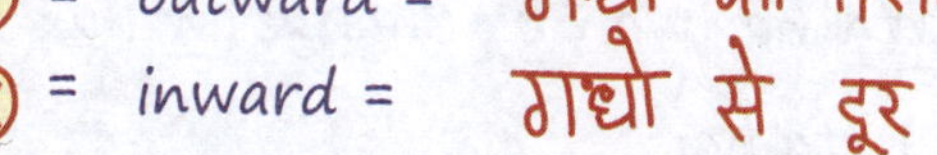

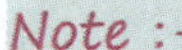

Note :-

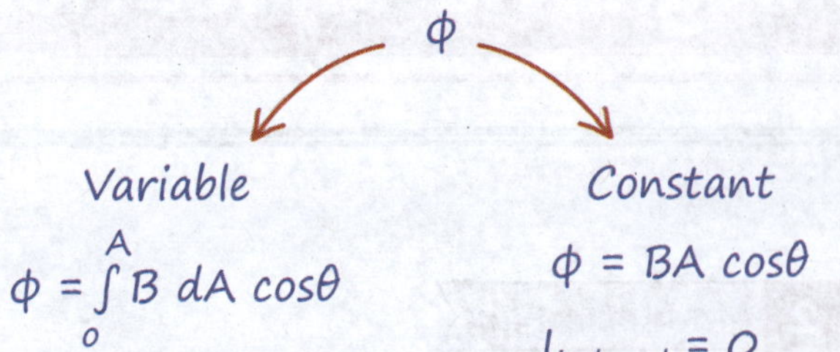

Variable: $\phi = \int_0^A B\, dA \cos\theta$

$I_{induced} \neq 0$

Constant: $\phi = BA \cos\theta$

$I_{induced} = 0$

Variable due to :-

B → time varying

A → Variable

θ = Changing

Induced current

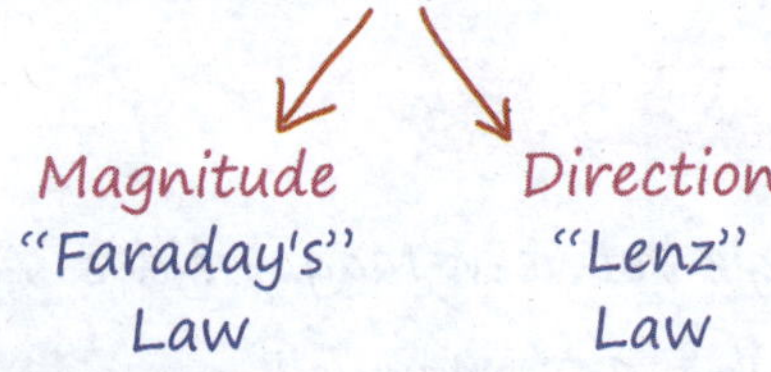

3 Lenz Law

*Joh "i" Ko Paida Krta hh "i" Ushika Oppose Krta hh.

MR* law :-

- Coil is in inertia (Pyaar) of Flux.

External Field	Induced Field B
⊙B↑ ⟶	⊗(Then I_{in} = CW)
⊙B↓ ⟶	⊙(Then I_{in} = ACW)
⊗B↑ ⟶	⊙(Then I_{in} = ACW)
⊗B↓ ⟶	⊗(Then I_{in} = CW)

"Up ⊙ Anti, down C"

- Coil Na Flux Ko pyaar Krta hai na Usko hate Krta hai woh bs Change in flux ko oppose Krta hai.

MR* law :-

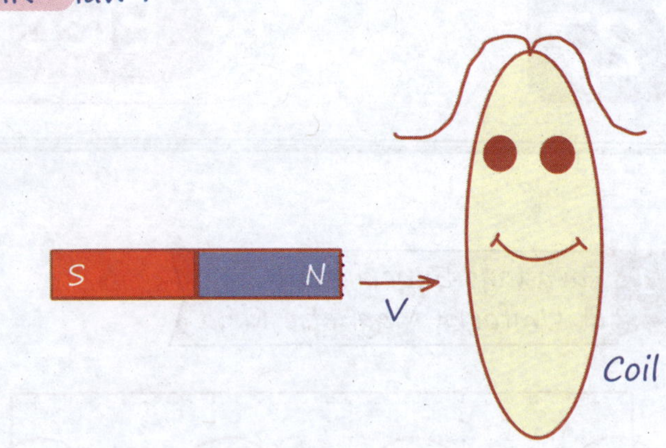

Coil Kehti hai magnet से :-

Pass mt aana repel Karungi durr mt jana attract Karungi aaisehi padhai krte raho Exam Ke baad date Karungi.

Case :- I

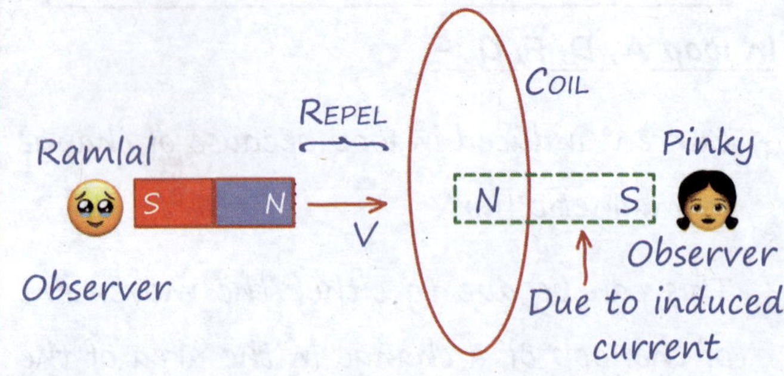

Current induced in loop

- w.r.t. Ramlal → ACW
- w.r.t. Pinky → CW

MR* → Aage se Anti then pichhe se clock

Case :- II

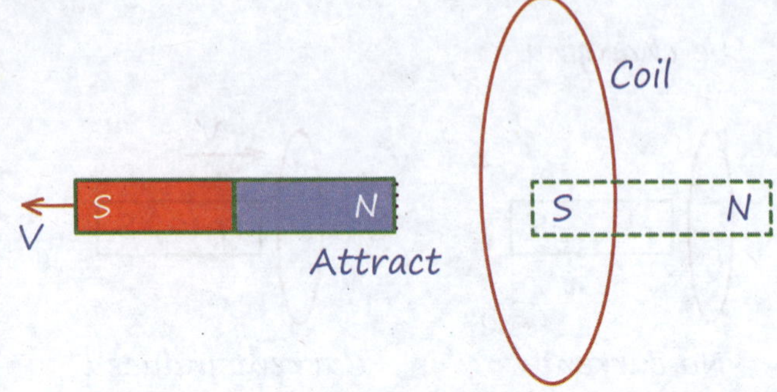

4 NOTE (AIIMS)

Q.1 Find time for which current will induced in ractangular loop?

Sol.

I> $a > b$

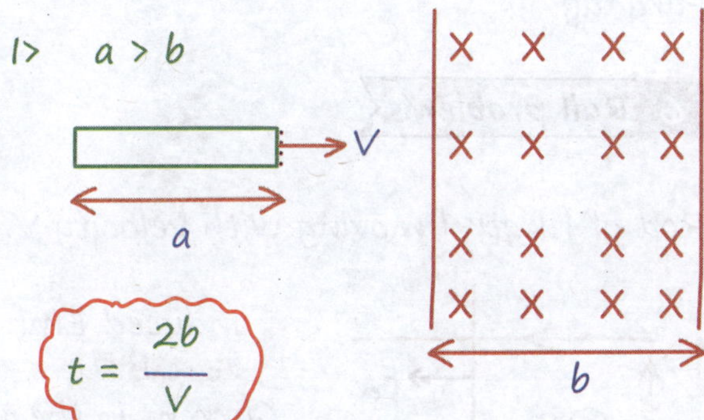

$$t = \frac{2b}{V}$$

II> $a < b$

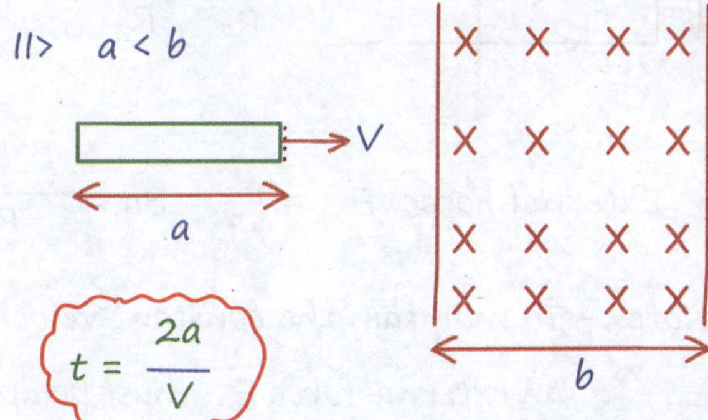

$$t = \frac{2a}{V}$$

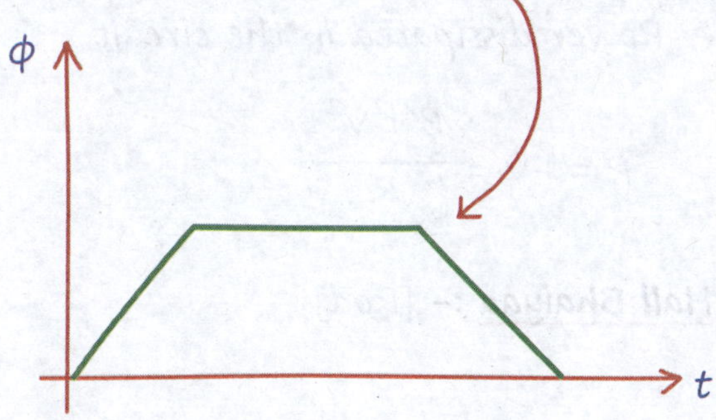

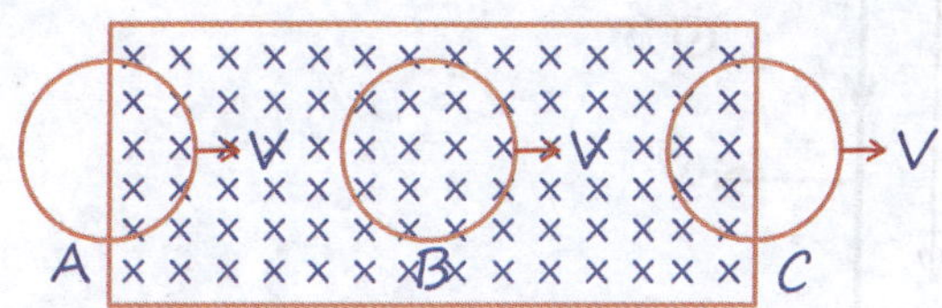

Current induced in loop

(A) → ACW (B) → 0 (C) → CW

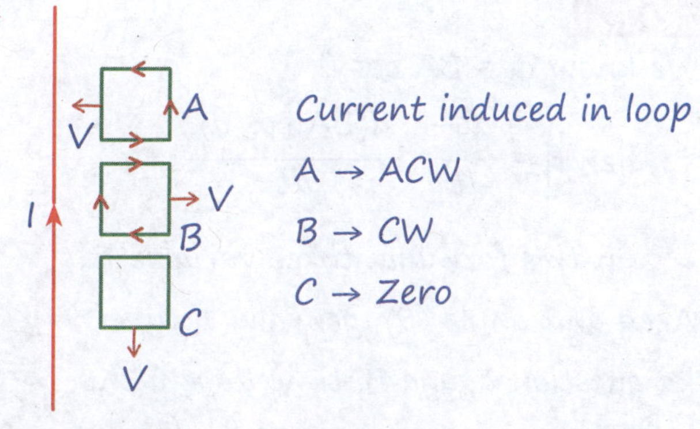

Current induced in loop

A → ACW

B → CW

C → Zero

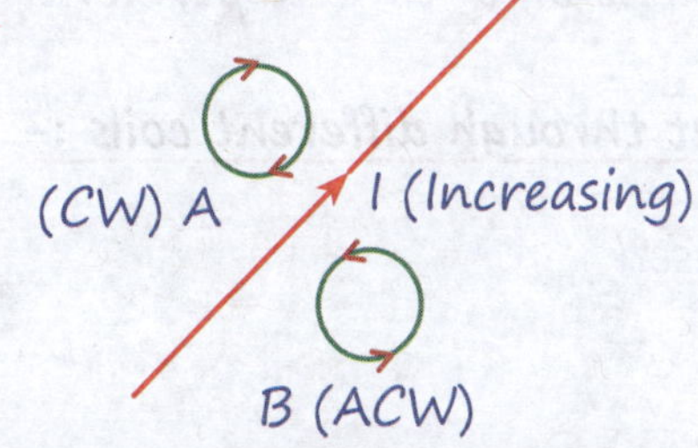

5 Faraday's Law of EMI

Based on Energy Conservation. [Loop]

$\varepsilon_{inst.}$	$\varepsilon_{avg.}$
$\varepsilon_{inst.} = \frac{-d\phi}{dt}$	$\varepsilon_{avg.} = -\frac{\Delta\phi}{\Delta t}$
$i_{inst.} = \frac{-d\phi}{R.dt}$	$i_{avg} = -\frac{\Delta\phi}{R\Delta t}$
$\lvert\varepsilon_{inst.}\rvert = \frac{d(BA\cos\theta)}{dt}$	$\Delta Q = -\frac{\Delta\phi}{R}$

Q.2 Radius of circular loop placed perpendicular to magnetic field increasing at rate r_o m/s then find induced emf in loop when radius is r.

Sol. emf

$$= -\frac{d(BA\cos 0°)}{dt} = -B\frac{d\,\pi r^2}{dr} \times dr/dt$$

$$\text{emf} = -B\pi 2 r_o r \text{ Volt } \left(\text{Given, } \frac{dr}{dt} = r_o\right)$$

MR* feel

We know $\phi = BA \cos\theta$

$$|emf| = \frac{d\phi}{dt} = \frac{d(BA\cos\theta)}{dt}$$

In this formula, three variable B, Area and angle (θ), genrally 3 type ka question aayga B-time dependent, A-time dependent or θ-time dependent.

Falling a magnet through different coils :-

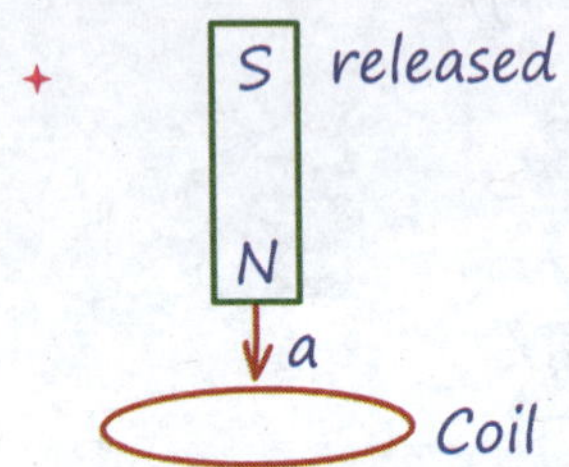

A non-conducting or a coil with a small cut:

- No current flows, hence no opposing magnetic field is created. The magnet falls with acceleration $a = g$.

Conducting coil:

- Induced current generates a magnetic field opposing the fall of the magnet, reducing the net acceleration i.e., $a < g$.

Q.3 In which loop induced emf will be uniform:

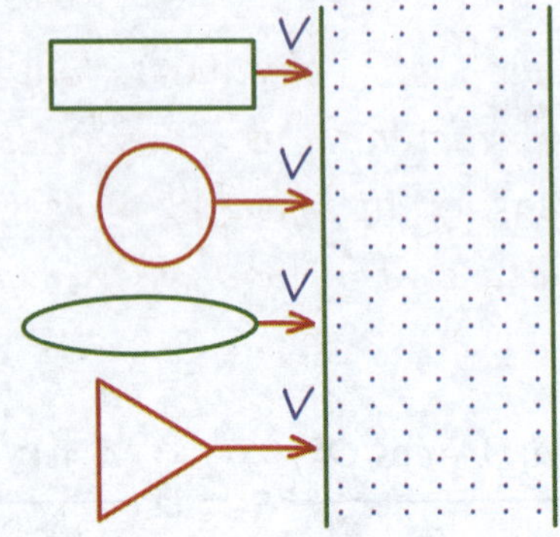

Sol. In rectangular loop because rate of change in flux is constant.

MR*

law :-

लोहा हो लकडी हो Plastic हो टूटा हुआ तार हो Kuch bhi ho sab mein EMF induced करवाएगा Tera Faraday.

6 Rail problems

Rod of length l moving with velocity V.

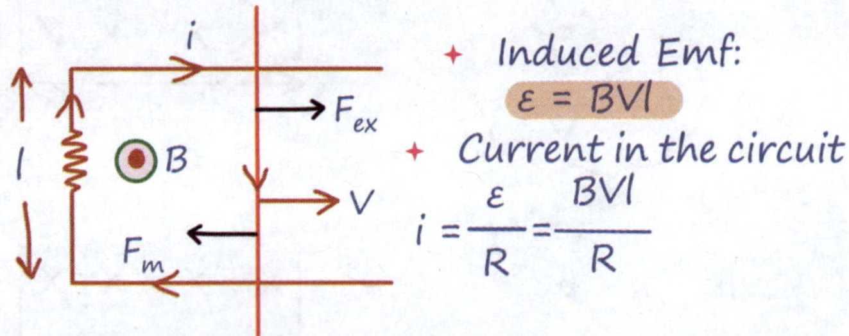

- Induced Emf: $\varepsilon = BVl$
- Current in the circuit $i = \frac{\varepsilon}{R} = \frac{BVl}{R}$

- External Force: $F_{ext} = F_m = Bil = \frac{B^2l^2V}{R}$

Note :- To maintain the constant velocity V, an external force F_{ext} must be applied to countereract the magnetic force.

- Power dissipated in the circuit

$$P = FV = \frac{B^2l^2V^2}{R}$$

Hall Bhaiyaa :- [Rod]

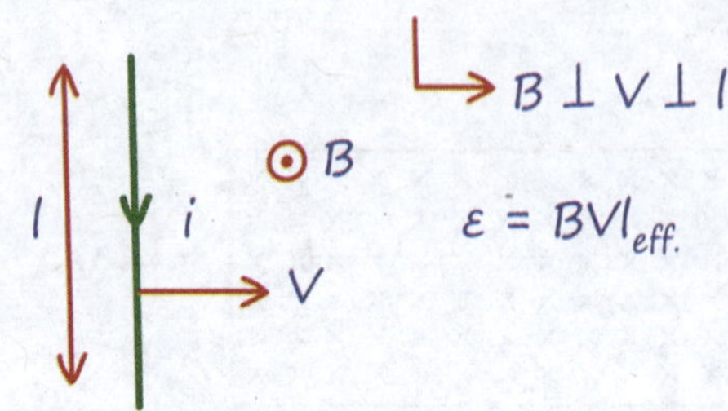

Scalar Triple Product :-

$\varepsilon = (\vec{V} \times \vec{B}) \cdot \vec{l}$ $\quad\quad \because (F_M = F_E)$

Direction of EMF :-

Plam :- Velocity
Curl Finger :- Magnetic Field
Thumb :- EMF.

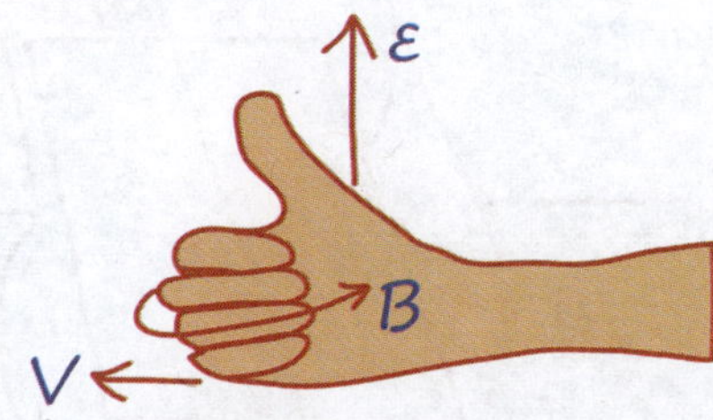

MR* for Direction of higher potential place four finger along velocity and slap magnetic field thumb will represent higher potential.

Induced emf in given Rod :-

The rod is divided into segments with specific lengths and velocities. The induced EMF in each segment depends on the length of the segment and the magnetic field.

$$emf = 0$$

$$emf = \frac{\mu_0 i V}{2\pi} \log\left(\frac{a+l}{a}\right)$$

$$emf = \frac{\mu_0 i l V}{2\pi a}$$

$$emf = 0$$

Rotational Motion in a Magnetic Field :-

When a conductor rotates in a magnetic field, an emf is induced across its length due to the motion of the conductor through the magnetic field lines.

$$(emf)_{OB} = \frac{1}{2} B\omega l^2$$

$$(emf)_{OA} = \frac{1}{2} \frac{B\omega l^2}{4}$$

$$(emf)_{AB} = \frac{3}{8} B\omega l^2$$

MR* :- *VVimp*

🔥Fire Concept🔥

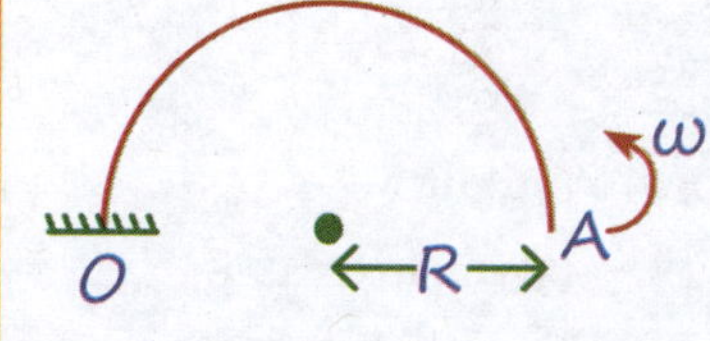

$$|emf|_{OA} = \frac{1}{2} B\omega(2R)^2$$

$$l_{eff} = 2R$$

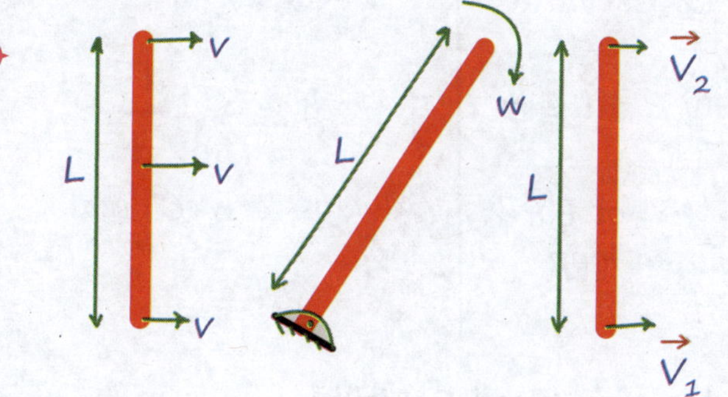

Translational Motion (T)	Rotational Motion (R)	T + R Motion
$E_{motional} = BVl$	$E_R = \frac{1}{2} B\omega l^2$	

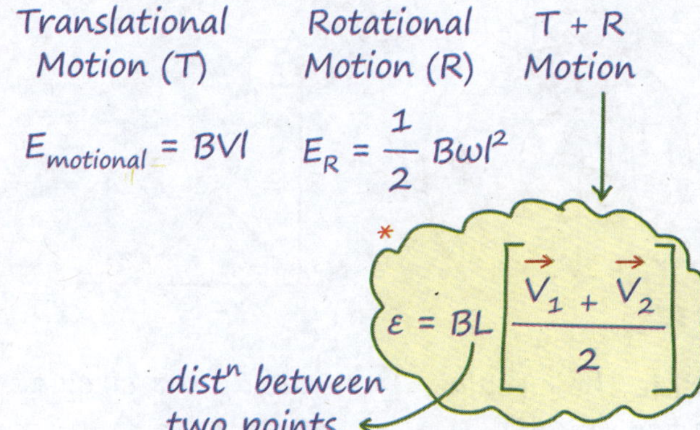

$$\varepsilon = BL\left[\frac{\vec{V}_1 + \vec{V}_2}{2}\right]$$

distⁿ between two points.

Note :-

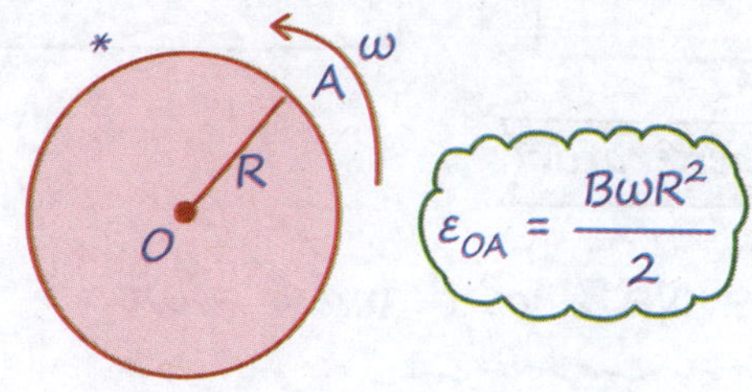

$$\varepsilon_{OA} = \frac{B\omega R^2}{2}$$

Disc. ⟶ Collection of ∞ Rods.

7 Induced Electric Field

MR*

Electric Field

असली	नकली
+ Due to Rest Charge.	+ Due to time varying M. Field
+ Electrostatic Field.	+ Induced E. Field

असली	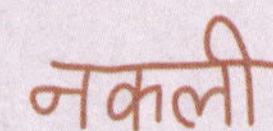
Conservative	Non-conservative
$\oint \vec{E} \cdot \vec{dl} = 0$	$\oint \vec{E} \cdot \vec{dl} = A \cdot \frac{d\vec{B}}{dt}$
Does not Form Closed loop	Always form Closed loop. (Concentric Circle)

Value of induced E. Field :-

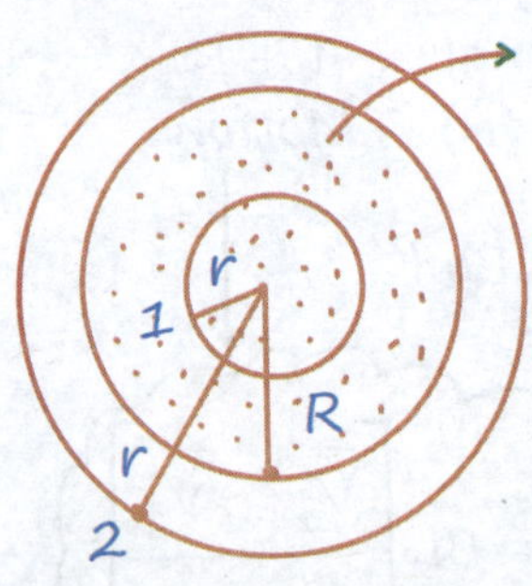

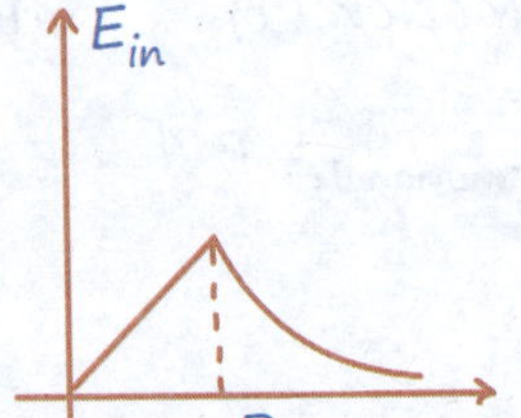

Inside the circular region (r < R)

$$E_1 = \frac{r\,dB}{2 \cdot dt}$$

Outside the circular region (r > R)

$$E_2 = \frac{R^2}{2r}\frac{dB}{dt}$$

8 Ac-generator

ϕ (flux) = NAB cos θ and $\theta = \omega t$

$$\varepsilon = BAN\omega \sin\theta$$

*Convert Mech. Energy into Electric. Energy.

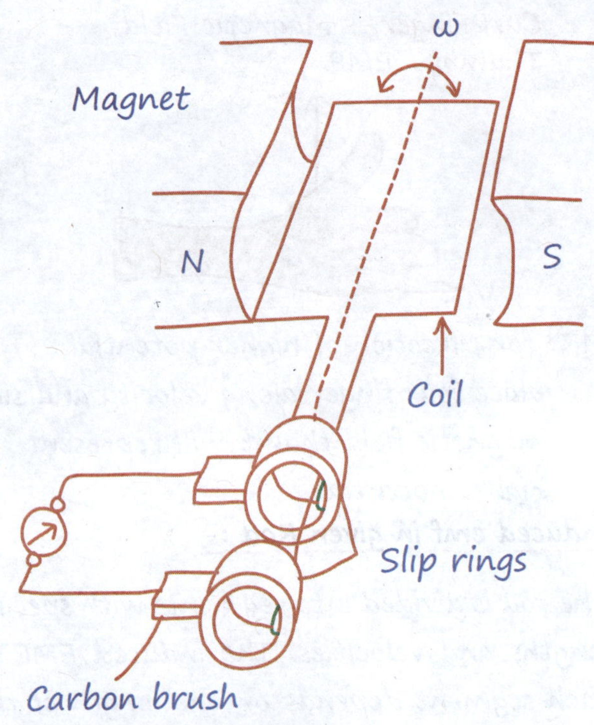

9 Self-inductance(L)

Aaj Kal Ke Rishte Kaha hai itne Ache isliye hm Akele hi hai Ache.

+ $L = \mu_0 n^2 Al = \frac{\mu_0 N^2 A}{l}$
+ $e = -L\frac{di}{dt}$
+ $\phi = Li$

L (self-inductance) $\propto l$ (n = cos t^n)
$\propto 1/l$ (N = cos t^n)

Where n = no of turns per unit length

N = total no of turns

Unit :- Henry (H)

+ ϕ = constant ;_ $L_1 i_1 = L_2 i_2$
+ Direction buddhi से, Magnitude Formula से

$i \uparrow$

$\varepsilon = L \dfrac{di}{dt}$

$I = cost^n$

$\varepsilon = 0$

$i \downarrow$

$\varepsilon = L \dfrac{di}{dt}$

$i \downarrow$

$\varepsilon = L \dfrac{di}{dt}$

Energy Store in Inductor :-

$$E = \frac{1}{2} Li^2 = \frac{B_0^2 Al}{2\mu_0}$$

- Energy Stored per unit Volume

$$= \frac{B_0^2}{2\mu_0}$$

Charging of Inductor :-

t = 0	$t = \infty$ [Steady State]
L = Open Wire C = Simple Wire	L = Simple Wire C = Open Wire
+ $V_L = L\left(\dfrac{di}{dt}\right) = E$ + $V_R = 0$	+ $V_L = 0$ + $V_R = E$ + $i = E/R$ $E_{max} = \dfrac{1}{2} L\left(\dfrac{E}{R}\right)^2$

Capacitor	Inductor
$Q = Q_0[1 - e^{-t/RC}]$	$i = i_0[1 - e^{-Rt/L}]$
$\tau = RC$	$\tau = L/R$
+ $t = \infty$ $Q = Q_0$	+ $t = \infty$ $i = i_0$
Q; Q_0; 63% Q_0; $\tau = RC$; t	i; i_0; 63% i_0; $\tau = \dfrac{L}{R}$; t

Graph Between di/dt and Time :-

Rate of Change in Current :-

$$\frac{di}{dt} = \left(\frac{E}{L}\right) e^{-t/\tau}$$

At $t = 0$, $\dfrac{di}{dt} = \left(\dfrac{E}{L}\right)$

At $t = \infty$ $\dfrac{di}{dt} = 0$

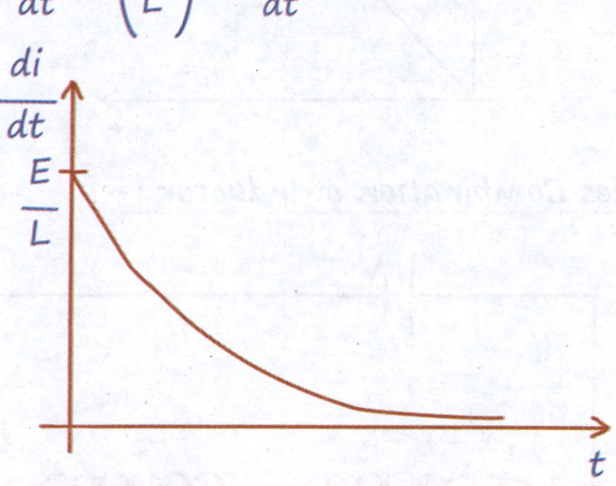

Graph Between V_R and Time :-

$$V_R = i_0 (1 - e^{-t/\tau}).R$$

V_R

t

$$V_L = L\left(\frac{di}{dt}\right) = Ee^{-t/\tau}$$

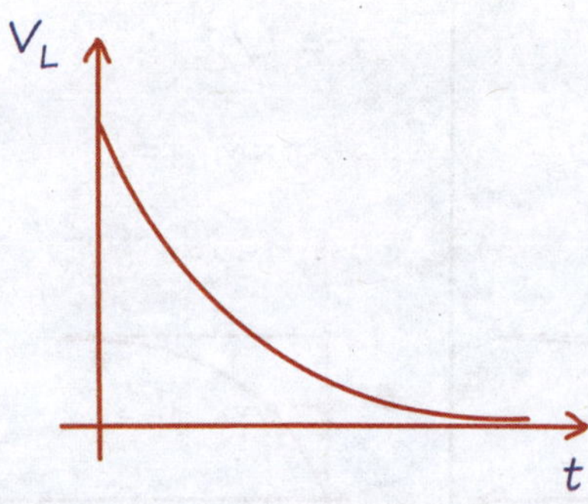

Current in Inductor as Function of Time :-

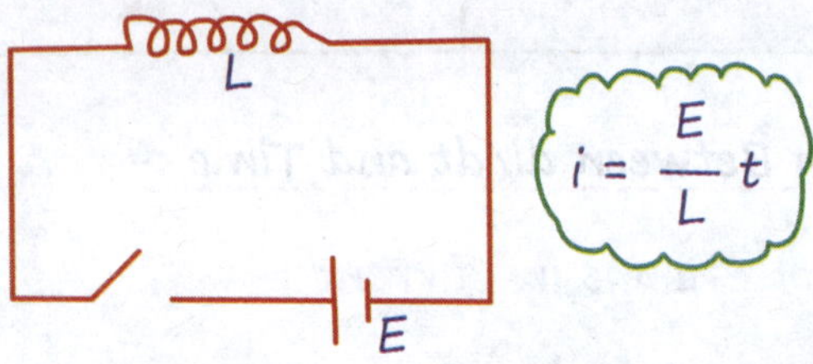

$$i = \frac{E}{L}t$$

Battery be like :- Raziya Gundo mein Fasgayi.

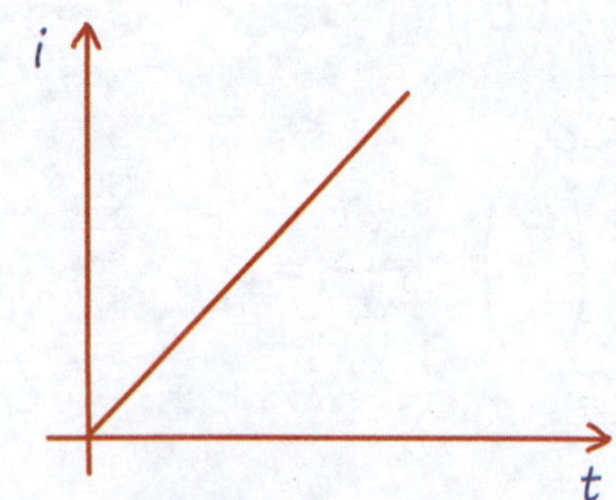

Series Combination of Inductor :-

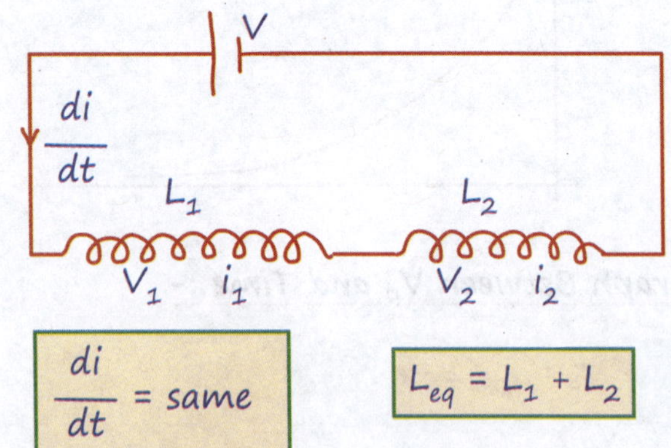

$$\frac{di}{dt} = \text{same}$$

$$L_{eq} = L_1 + L_2$$

Parallel Combination of Inductor :-

$$\frac{1}{Leg} = \frac{1}{L_1} + \frac{1}{L_2}$$

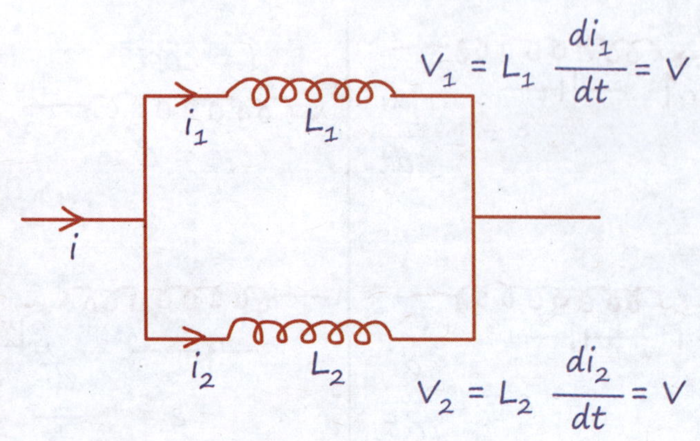

$$\frac{1}{L_{eq}} = \frac{1}{L_1} + \frac{1}{L_2}$$

10 Mutual Inductance

Mutual inductance is the property of two coils (or inductors) where a change in current in one coil induces an emf in the other coil.

MR*

Jitne mein tumhe तंग Karunga utna tum bhi mujhe तंग Krna Jyda kam mt Krna...

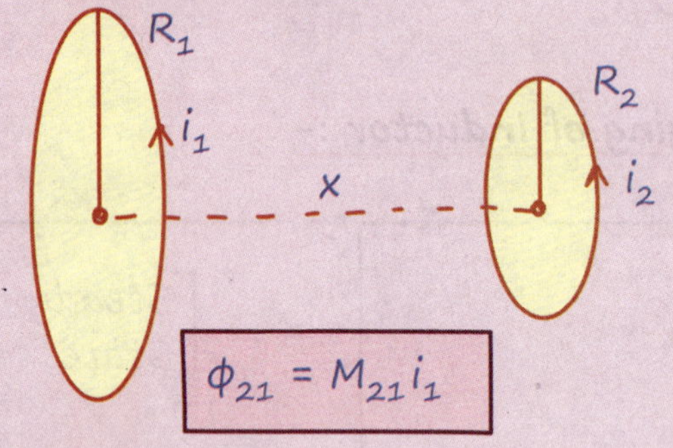

Mutual inductance of 2 due to 1 :-

$$\Rightarrow \quad M_{21} = \frac{\mu_0 N_1 N_2 R_1^2 R_2^2 \pi}{2x^3}$$

Reciprocity Theorem :-

$$M_{12} = M_{21}$$

Magnetic flux	$\phi_{12} = M_{12} i_2$	$\phi_{21} = M_{21} i_1$
Induced e.m.f.	$\varepsilon_{12} = M_{12} \frac{di_2}{dt}$	$\varepsilon_{21} = M_{21} \frac{di_1}{dt}$

+ Note :- Coupling Factor

$0 \le K \le 1$

$$M = K\sqrt{L_1 L_2}$$

+ L_1 & L_2 very close → K = 1.

Series Combination of Inductor (Considering M) :-

$L_{eq} = L_1 + L_2 \pm 2M$

$$L_{eq} = L_1 + L_2 \pm 2M$$

L_1 & L_2 are in same order	L_1 & L_2 are in Opposite order
$L_{eq} = L_1+L_2+2M$	$L_{eq} = L_1+L_2-2M$
$L_{eq}=L_1+L_2+2\sqrt{L_1L_2}$	$L_{eq}=L_1+L_2-2\sqrt{L_1L_2}$

11 Discharging of Inductor

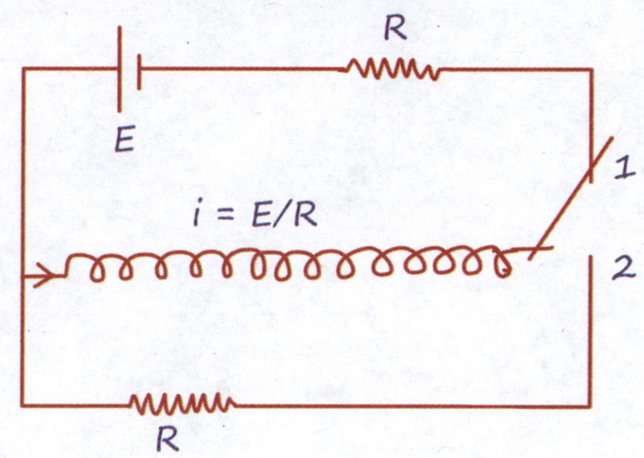

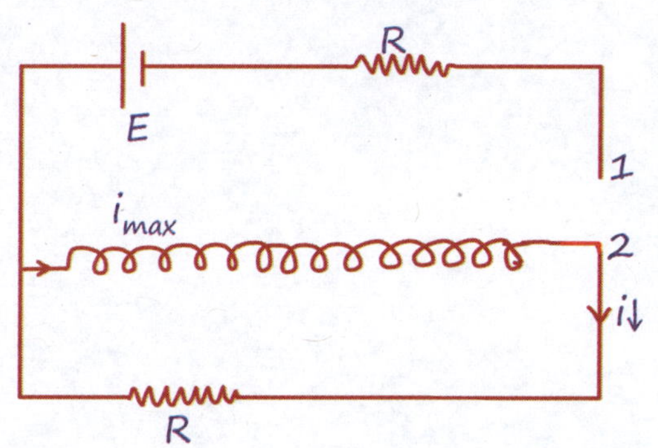

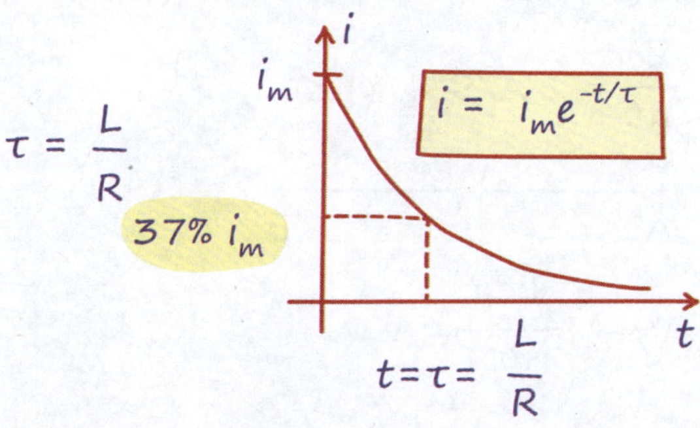

$i = i_m e^{-t/\tau}$ $\quad V_L = L\dfrac{di}{dt} = iR$

$\tau = \dfrac{L}{R}$

37% i_m

$t=\tau=\dfrac{L}{R}$

+ Find total heat loss across R_2 when Key is shifted from (2) to (1) :-

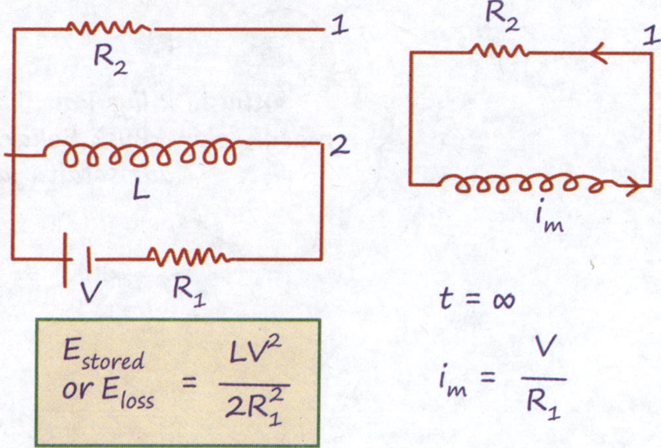

$t = \infty$

$i_m = \dfrac{V}{R_1}$

$$E_{stored} \text{ or } E_{loss} = \frac{LV^2}{2R_1^2}$$

12 Eddy Current

Agar Kishi तार मे Current Flow करेगा तो उसे Kehte है Current और Plate मे Current Flow करेगा तो Eddy Current.

Application :-

1> Magnetic braking in train

2> Electromagnetic damping.

3> Induction Furnance

4> Electric Power meters.

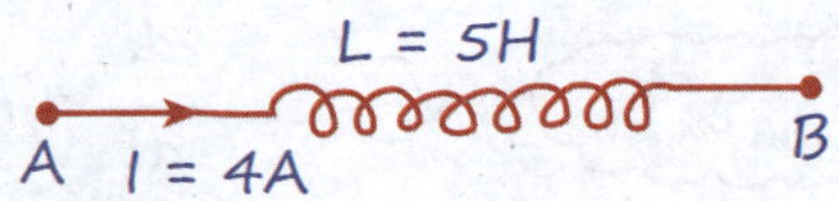

$$\frac{di}{dt} = 2\ A/S \rightarrow i(\uparrow es)$$

A B

$$V_A - \frac{Ldi}{dt} = V_B$$

$$V_A - V_B = \frac{Ldi}{dt} = 5 \times 2 = 10\ V$$

Current is increasing hence B is at lower potential.

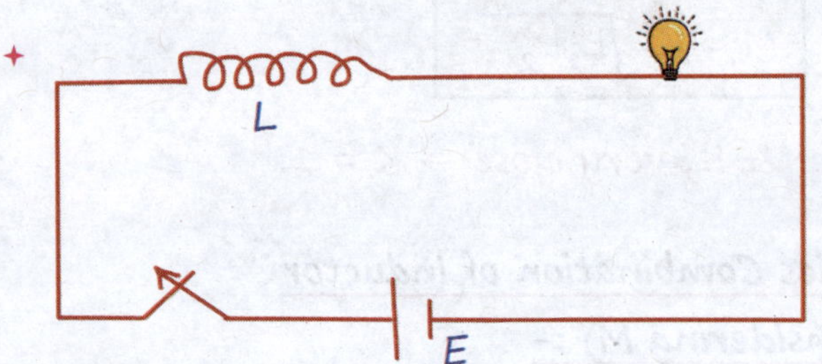

Brightness of bull will suddenly increase when key is just opened.

MR*

"Khushi k liye kam karoge to khushi nahi milegi, lekin khush hokar kam karoge to khushi aur safalta jarur milegi."

Alternating Current

1 Alternating Current

+ Alternating Current (A.C.) is an electric current that reverses its direction periodically. It is bidirectional, meaning it flows in both directions.

  Example: The household electricity supply is A.C., typically with a sinusoidal wave.

Direct Current (D.C.) :-

+ Direct Current (D.C.) is an electric current that flows in one direction only. It is unidirectional and usually has a constant magnitude.

  Example: The current from a battery is D.C., providing a steady flow of electricity.

+ A current of constant magnitude can be A/c current → Yes. It is square wave A/c.

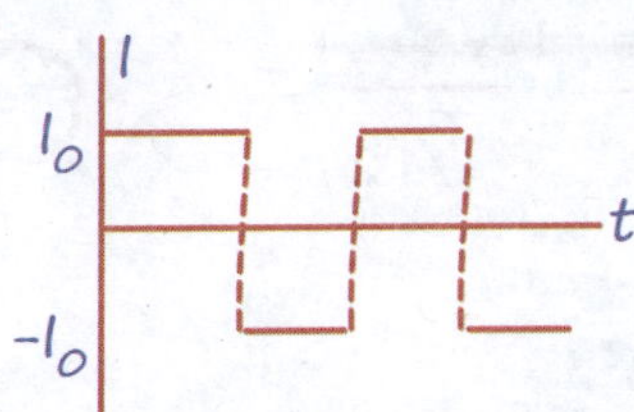

+ A Variable current in magnitude only is D/c

  # A current is varying from +5A to +15A sinusoidally then this current is D/c or (AC+DC) mixture, not A/c because it is not a bidirectional.

+ A/C → Bidirectional

+ D/C → Unidirectional (Steady Current)

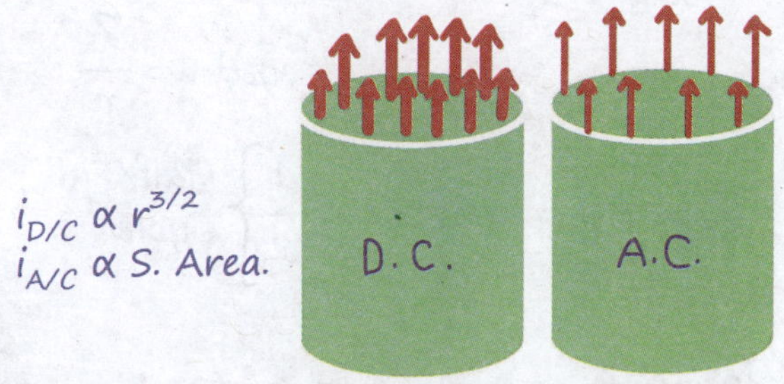

2 Measuring of Current

Moving coil galvanometer	Hot wire ammeter
+ "T" on Current carrying coil.	+ Heat loss
+ Only D/C	+ Both D/C & A/C
+ θ (angle) $\propto i$	+ θ (angle) $\propto H \propto i^2$
+ Linear Scale	+ Non-linear Scale

Average Value :-

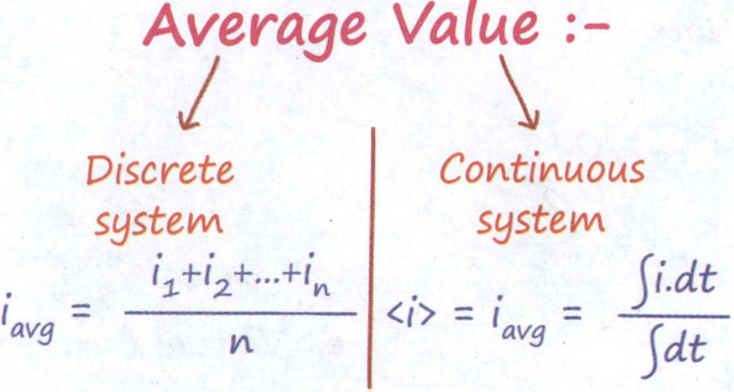

RMS Current :-

$$i_{rms} = \sqrt{\langle i^2 \rangle} = \sqrt{\frac{\int i^2.dt}{\int dt}}$$

RMS Voltage :-

$$V_{rms} = \sqrt{\langle V^2 \rangle} = \sqrt{\frac{\int V^2.dt}{\int dt}}$$

MR* Ratta

+ Full Cycle (FC)

$\langle \sin\theta \rangle = 0$
$\langle \cos\theta \rangle = 0$

+ Half Cycle (HC)

$\langle \sin\theta \rangle = \frac{2}{\pi}$

$\langle \cos\theta \rangle = \frac{2}{\pi}$

$\langle \sin^2\theta \rangle = \langle \cos^2\theta \rangle = \frac{1}{2}$ } Half/Full Cycle

+ $\langle 2\sin(\omega t).\cos(\omega t)\rangle_{F.cycle} = \langle \sin(2\omega t)\rangle_{F.cycle} = 0$
+ Avg. value of $I_0 \sin(\omega t)$ in half cycle may be zero or $\frac{2i_0}{\pi}$ because it will depends half cycle ki limit kaha se liya hai.

Mathematical representation Alternating Current :-

$$i = i_0 \sin[\omega t + \phi]$$

where i_0 = peak current

ω = Angular frequency

ϕ = Phase angle

+ Note :-

$\langle i \rangle_{FC} = 0$

$\langle i \rangle_{HC} = \frac{2i_0}{\pi}$

$\sqrt{\langle i^2 \rangle} = i_{rms} = \frac{i_0}{\sqrt{2}}$ } • Virtual Current • Effective Current

Reading of Ammeter.

$H = i^2_{rms} Rt.$

Voltage relation

+ $V = V_0 \sin(\omega t) = V_0 \sin(2\pi ft)$
+ $V_{rms} = \frac{V_0}{\sqrt{2}}$

MR* Table

	Sinusoidal A/c	Square (A/c)	Trianglular	Full wave rectifier	Half wave rectifier
Avg. Full cycle	0	0	0	$\frac{2i_0}{\pi}$	$\frac{i_0}{\pi}$
Avg. Half cycle	$\frac{2i_0}{\pi}$	i_0	$\frac{i_0}{2}$	$\frac{2i_0}{\pi}$	$\frac{2i_0}{\pi}$, 0
R.M.S. value	$\frac{i_0}{\sqrt{2}}$	i_0	$\frac{i_0}{\sqrt{3}}$	$\frac{i_0}{\sqrt{2}}$	$\frac{i_0}{2}$

Fire Concept MR*

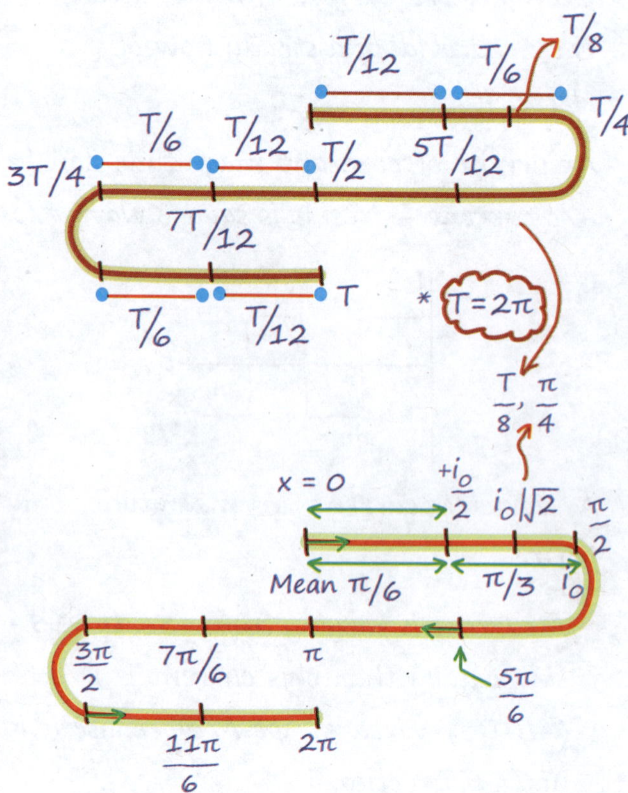

Alternating Current

MR* (i) $i = a + b\sin(\omega t)$

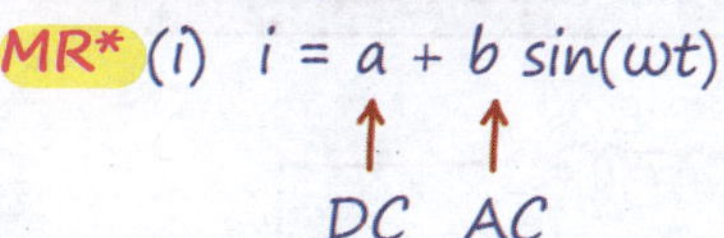

$$i_{rms} = \sqrt{a^2 + \frac{b^2}{2}} = \sqrt{\langle i^2 \rangle}$$

(ii) $i = a\sin\omega t + b\cos\omega t$

AC AC

$$i_{rms} = \sqrt{\frac{a^2}{2} + \frac{b^2}{2}} = \sqrt{\langle i^2 \rangle}$$

Alternating Current :-

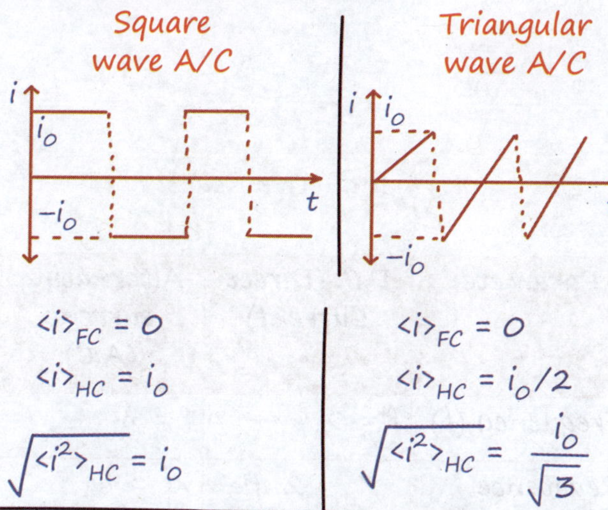

Square wave A/C:

$\langle i \rangle_{FC} = 0$

$\langle i \rangle_{HC} = i_0$

$\sqrt{\langle i^2 \rangle_{HC}} = i_0$

Triangular wave A/C:

$\langle i \rangle_{FC} = 0$

$\langle i \rangle_{HC} = i_0/2$

$\sqrt{\langle i^2 \rangle_{HC}} = \frac{i_0}{\sqrt{3}}$

Representation of A/c Current & Voltage by Phase Diagram :-

$i_1 = i_0\sin\omega t$

$i_2 = i_0\sin(\omega t + \phi)$

$i_3 = i_0\cos(\omega t - \delta)$

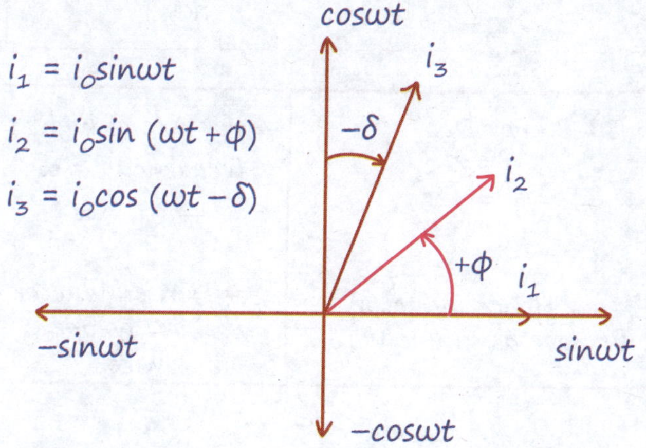

MR*

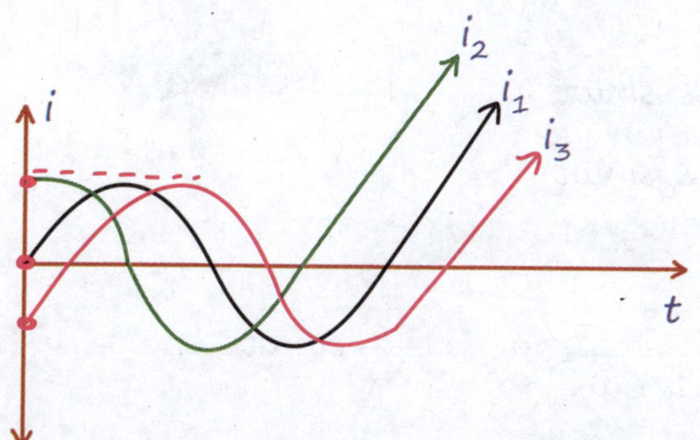

Joh jitna niche y-axis pe wo utna piche.

$*i_3 < i_1 < i_2*$

3 A/C Source Across Pure Resistance

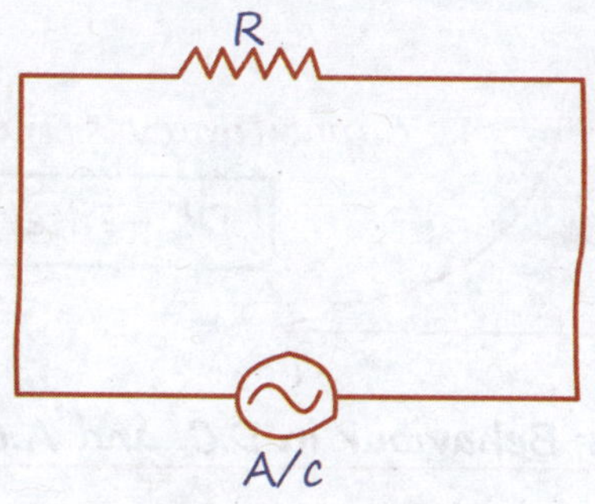

Same phase

$\varepsilon = \varepsilon_0\sin\omega t$ $\qquad i = i_0\sin\omega t$

$i_0 = \frac{\varepsilon_0}{R}$

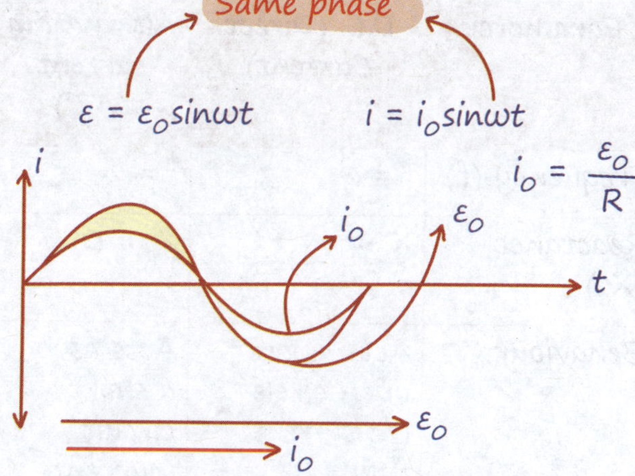

- Power factor, $\cos\phi = 1$

Power drop across R :-

$$\langle P \rangle = \varepsilon_{rms} i_{rms} = \frac{\varepsilon_{rms}^2}{R} = i_{rms}^2 R$$

4 A/C Source Across Pure Capacitor

+ $\varepsilon = \varepsilon_0 \sin\omega t$

"i" leads "V"

+ $\varepsilon = \varepsilon_0 \sin\omega t$

by $\pi/2$ i.e., T/4

+ $i = \dfrac{\varepsilon_0}{1/C\omega}\cos\omega t = \dfrac{\varepsilon_0}{X_C}\cos\omega t$

+ $i_0 = \dfrac{\varepsilon_0}{X_C}$

+ $V_C = iX_C$

$$X_C = \frac{1}{C\omega} = \frac{1}{2\pi fC}$$ } Capacitive Reactance

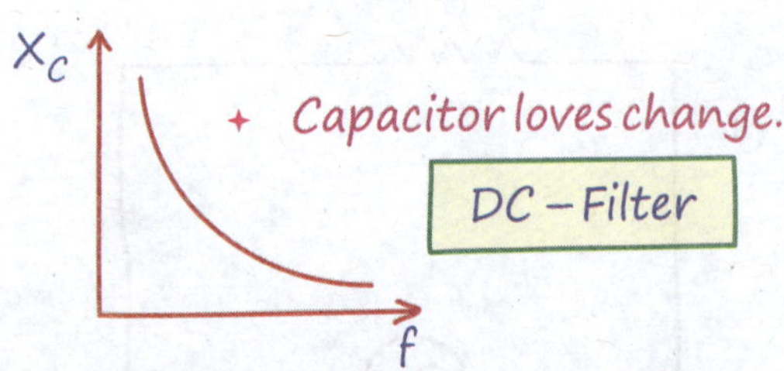

+ Capacitor loves change.

DC – Filter

Capacitor Behaviour in D.C. and A.C. Circuit :-

Parameter	D.C. (Direct Current)	Alternating current (A.C)
Frequency (f)	f = 0	$f = \infty$
Reactance (X_C)	$X_C = \infty$	$X_c = 0$
Behaviour	Acts as an open circle (no current flow)	Acts as a short circuit (current flow easily)

+ Power drop across pure Capacitor :-

$\cos\phi = 0$

$\langle P \rangle = 0$

Wattless Circuit

5 A/C Source Across Pure Inductor

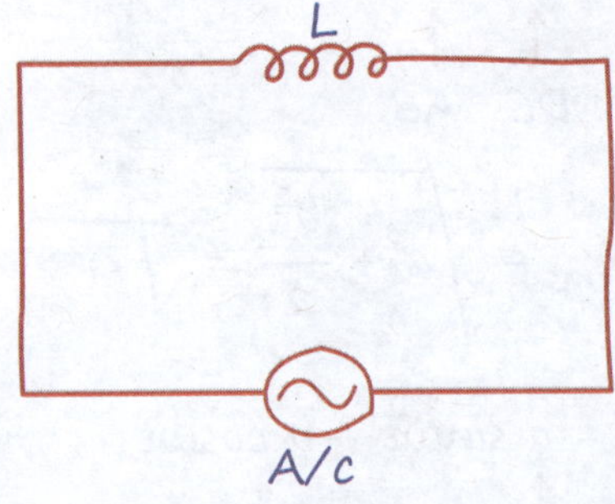

$V_L = \varepsilon_0 \sin\omega t$

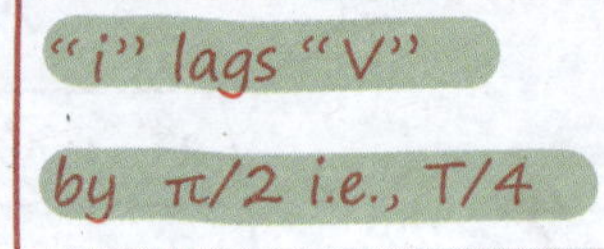

$$i = \frac{-\varepsilon_0}{\omega L}\cos\omega t = \frac{-\varepsilon_0}{X_L}\cos\omega t$$

+ $i_0 = \dfrac{\varepsilon_0}{\omega L}$ + $V_L = iX_L$

$X_L = \omega L$ } Inductive Reactance

Parameter	D.C. (Direct Current)	Alternating current (A.C)
Frequency (f)	f = 0	$f = \infty$
Reactance (X_L)	$X_L = 2\pi fL = 0$	$X_L = \infty$
Behaviour	Acts as a short circuit (simple wire)	Acts as a open circle (no current flow)

For D/c f = 0 ⇒	For A/c of high frequency $f = \infty$ ⇒ $X_L = \infty$
⇒ Hence inductor be have as simple wire.	⇒ Hence inductor behave as open wire.

✦ Power loss across pure inductor

$\langle P\rangle = 0$

$\cos\phi = 0$

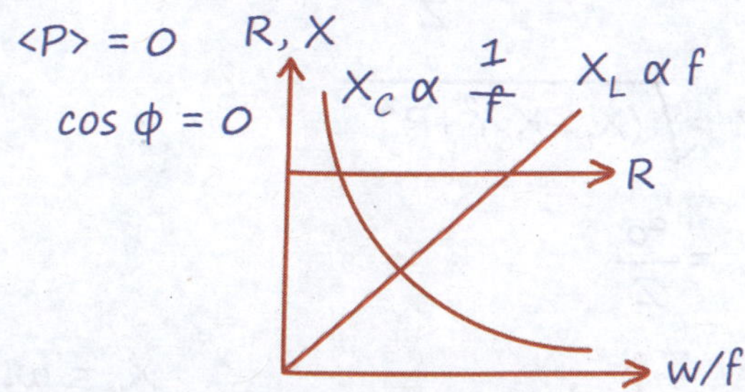

6 Series R-L Circuit

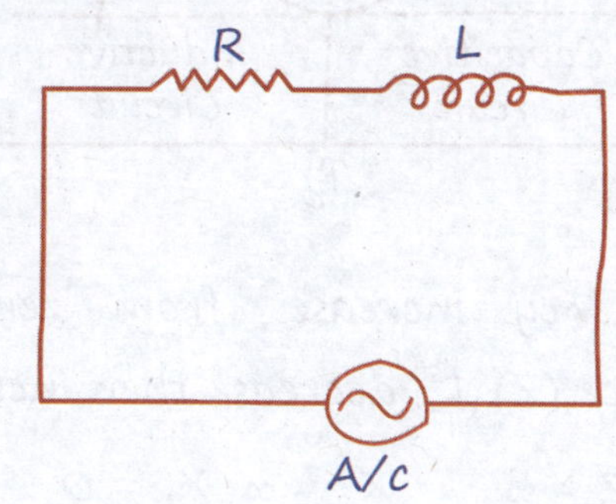

$V_{net} = V_B = \varepsilon_0 \sin\omega t$

$i = i_0 \sin(\omega t - \phi)$

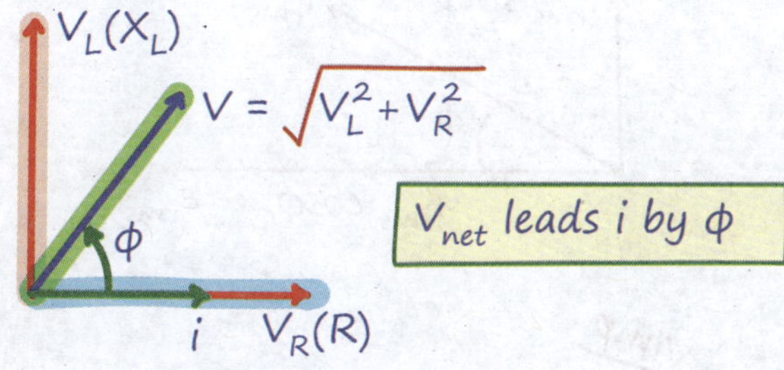

V_{net} leads i by ϕ

- Impedence $Z = \sqrt{R^2 + X_L^2}$
- $\tan\phi = \frac{V_L}{V_R} = \frac{X_L}{R}$
- $\cos\phi = \frac{V_R}{V_{net}} = \frac{R}{Z}$ } Power Factor

ϕ = Phase difference.

7 Series R-C Circuit

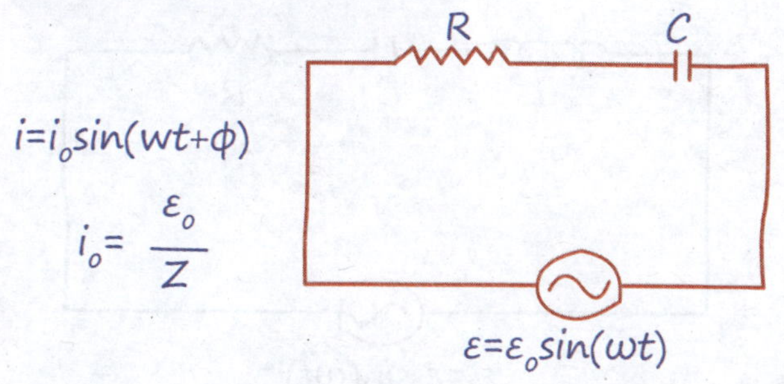

$i = i_0 \sin(\omega t + \phi)$

$i_0 = \frac{\varepsilon_0}{Z}$

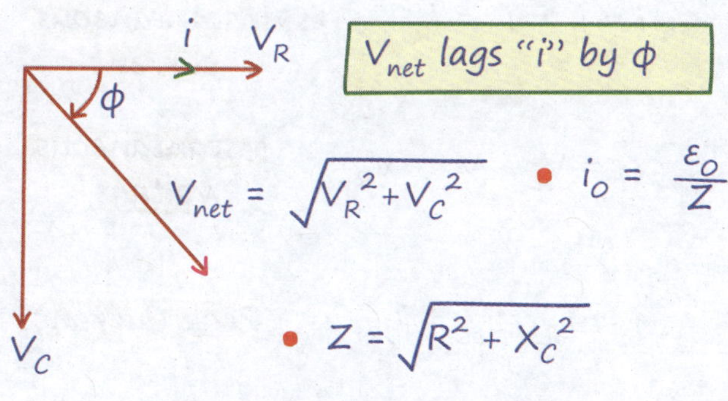

V_{net} lags "i" by ϕ

$V_{net} = \sqrt{V_R^2 + V_C^2}$

- $i_0 = \frac{\varepsilon_0}{Z}$
- $Z = \sqrt{R^2 + X_C^2}$

$\tan\phi = \frac{V_C}{V_R} = \frac{X_C}{R}$

$\cos\phi = \frac{V_R}{V_{net}} = \frac{R}{Z}$

8 Series L-C Circuit

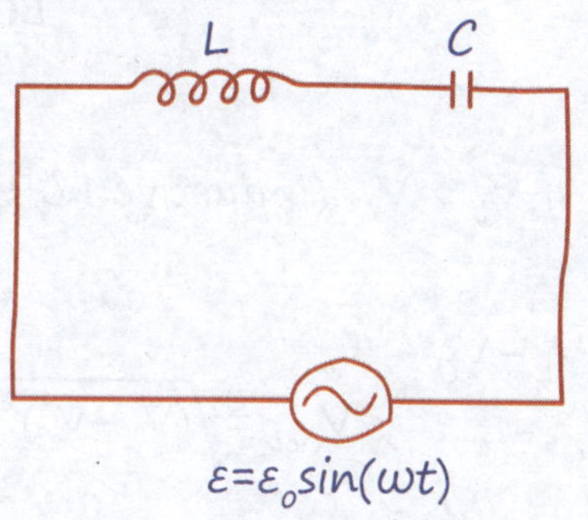

$V_L - V_C \quad (V_L > V_C)$

ϕ

i

V_{net} leads i by ϕ

Net impedence :-

$Z = X_L - X_C \qquad i_0 = \varepsilon_0 / Z$

9 Series LCR Circuit

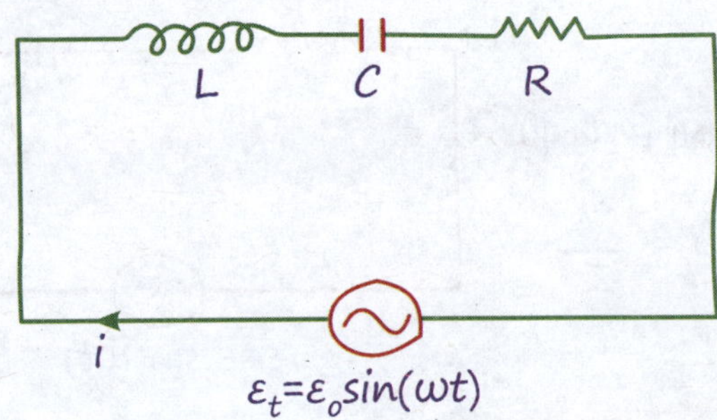

$i = i_L = i_C = i_R$ Instantaneous current

$\varepsilon_t = V_L + V_C + V_R$ Instantaneous voltage

$i_o \neq i_{oL} \neq i_{oC} \neq i_{oR}$ Peak Current

$\varepsilon_o = V_{oL} + V_{oC} + V_{oR}$ Peak voltage

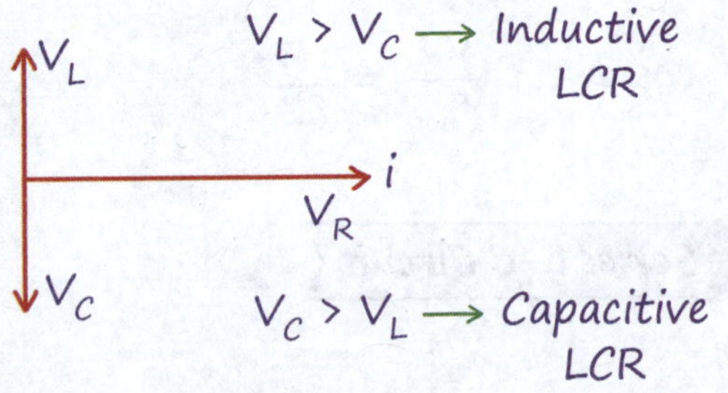

+ If $V_L > V_C$. (Inductive LCR)

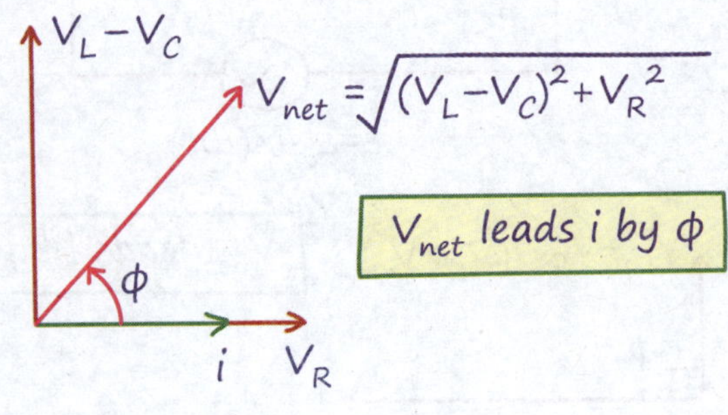

V_{net} leads i by ϕ

+ $\tan\phi = \dfrac{V_L - V_C}{V_R} = \dfrac{X_L - X_C}{R}$

+ $\cos\phi = \dfrac{V_R}{V_{net}} = \dfrac{R}{Z}$

+ $Z = \sqrt{(X_L-X_C)^2+R^2}$

+ $i_o = \dfrac{\varepsilon_o}{Z}$

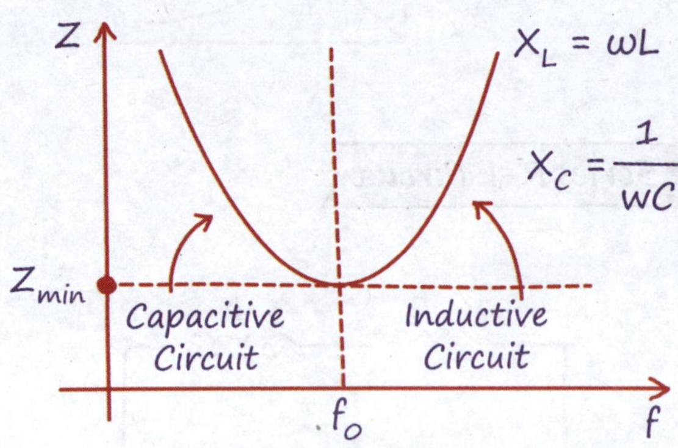

If Frequency increases from zero then impedence (Z) 1st decrease then increase

+ $f = 0\ Z = \infty$ $X_C = \infty\ X_L = 0$
+ $f = \infty\ Z = \infty$ $X_C = 0\ X_L = \infty$

Power drop :-

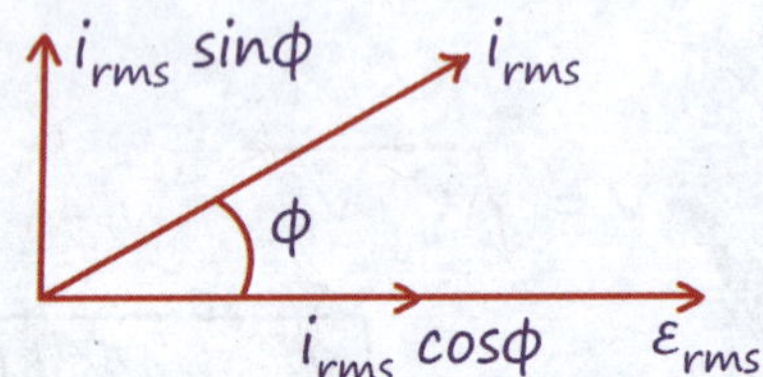

Vimp

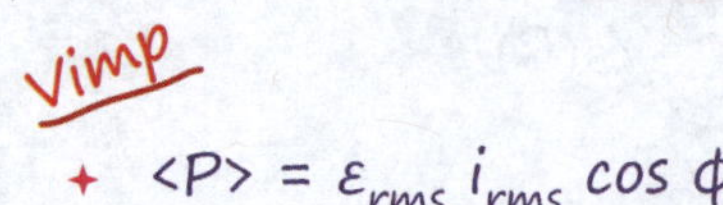

+ $\langle P \rangle = \varepsilon_{rms}\, i_{rms} \cos\phi$

Resonance in series LCR circuit :-

$X_L = X_C, \quad V_L = V_C$

$\omega L = \dfrac{1}{\omega C} \Rightarrow \omega = \dfrac{1}{\sqrt{LC}} \Rightarrow f = \dfrac{1}{2\pi\sqrt{LC}}$

$Z_{min} = R \Rightarrow V_{net} = V_R$

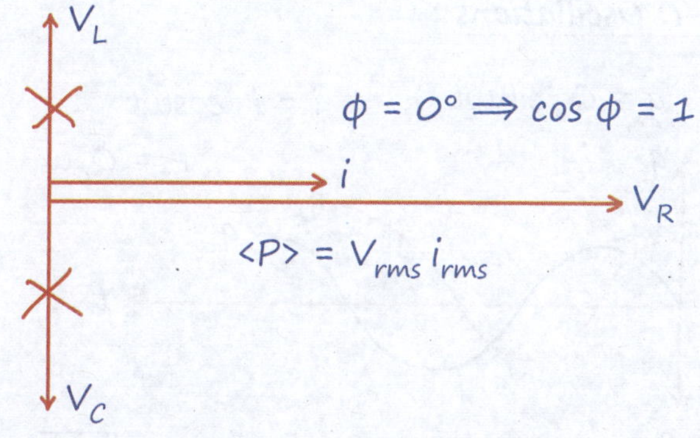

$\phi = 0° \Rightarrow \cos\phi = 1$

$\langle P \rangle = V_{rms}\, i_{rms}$

$i_{max} = \dfrac{\varepsilon}{R}$ $\qquad$ $P_{max} = \dfrac{\varepsilon_0^2}{2R}$

Vimp Graph :- MR*

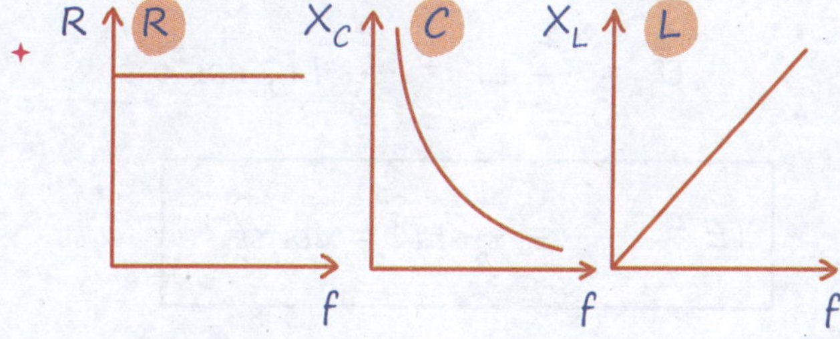

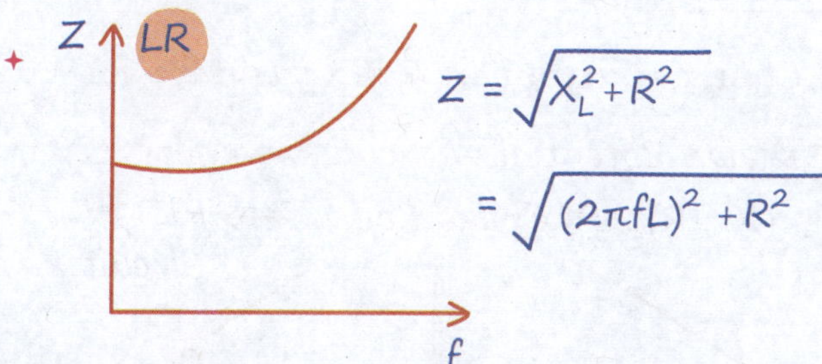

$$Z = \sqrt{X_L^2 + R^2}$$
$$= \sqrt{(2\pi f L)^2 + R^2}$$

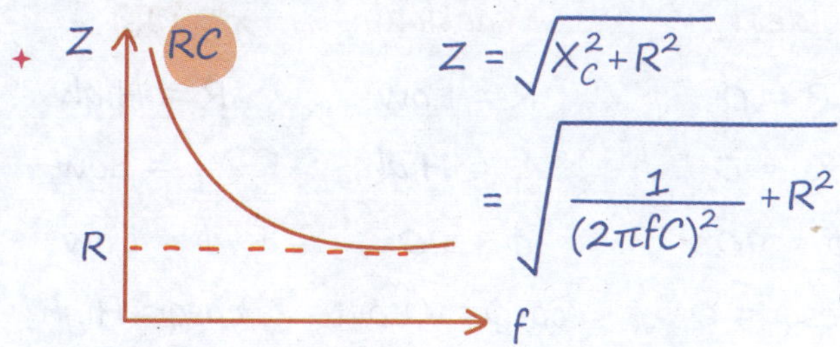

$$Z = \sqrt{X_C^2 + R^2}$$
$$= \sqrt{\frac{1}{(2\pi f C)^2} + R^2}$$

- L-C circuit

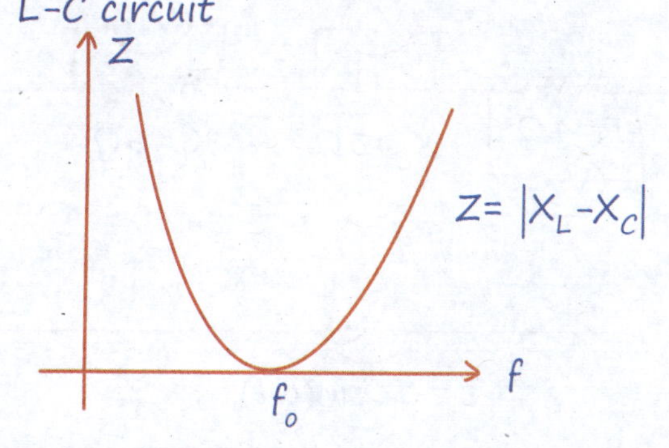

$$Z = |X_L - X_C|$$

Quality Factor :- (Q)

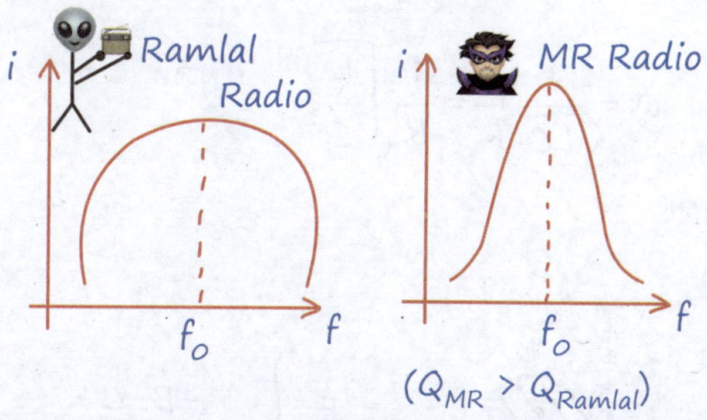

$(Q_{MR} > Q_{Ramlal})$

The sharper peak of MR Radio indicates a higher Q factor compared to Ramlal Radio. This means MR Radio has a better quality resonance.

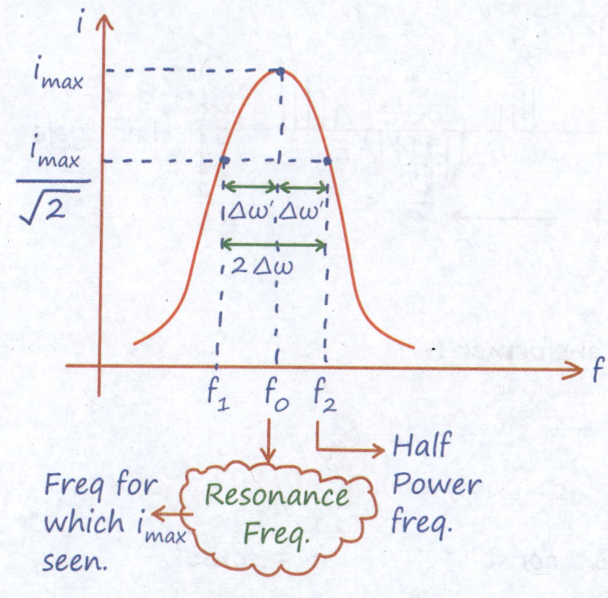

$$Q = \frac{\omega_0}{\Delta\omega} = \frac{\omega_0}{2\pi(f_2 - f_1)}$$

$$Q = \frac{\omega_0}{2\Delta\omega} = \frac{\omega_0}{2[2\pi(f_0 - f_1)}$$

$$2\Delta\omega = \frac{R}{L}$$

$$Q = \frac{\omega_0}{2\Delta\omega} = \frac{\omega_0 L}{R} = \frac{X_L}{R} = \frac{X_C}{R}$$

$$Q = \frac{\omega_0}{\Delta\omega} = \frac{\omega_0}{R/L} = \frac{\omega_0 L}{R} = \frac{X_L}{R}$$

$$Q = \frac{X_L}{R} = \frac{X_C}{R}$$

$$Q = \frac{1}{\sqrt{LC}}\frac{L}{R} = \frac{1}{R}\sqrt{\frac{L}{C}}$$

MR*

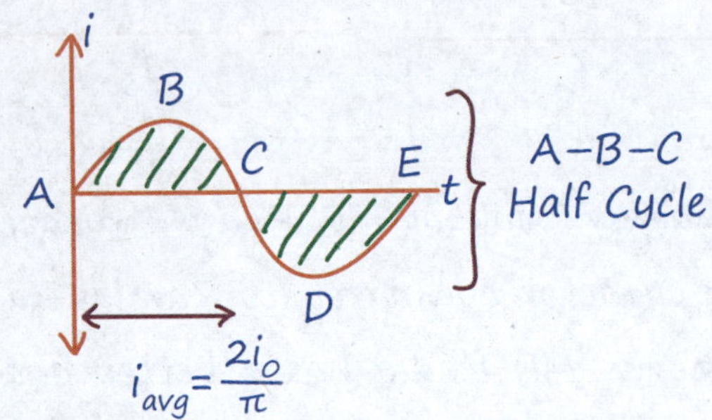

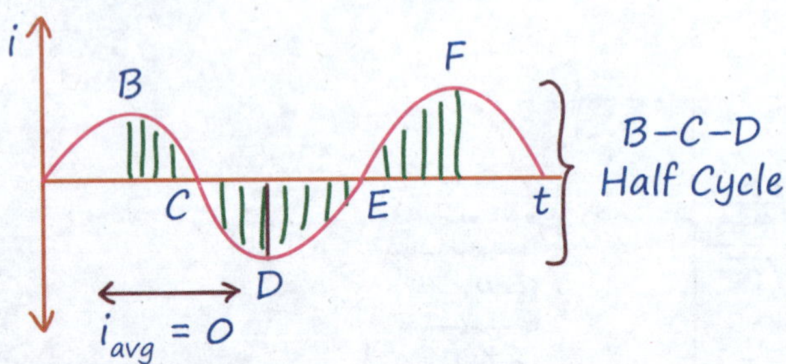

Transformer :-

- Voltage Regulator
- Based on principle of Mutual Inductance.

*$P = const^n$ $\quad$ $iV = const^n$

*$i \propto \frac{i}{V} \propto \frac{i}{N}$ $\quad$ *N = No. of turns.

* $$\frac{i_o}{i_{in}} = \frac{V_{in}}{V_{out}} = \frac{N_P}{N_S}$$

Step-UP	Step-DOWN
$V_o > V_{in}$	$V_o < V_{in}$
$N_S > N_P$	$N_S < N_P$
$i_S < i_P$	$i_S > i_P$

L-C Oscillations :-

$q = Q_o \sin\omega t$ $\quad$ $i = i_o \cos\omega t$

$i_o = Q_o\omega$

$$\frac{d^2q}{dt^2} + \frac{q}{LC} = 0 \Rightarrow \omega = \sqrt{\frac{1}{LC}}$$

$$U_E = \frac{q^2}{2C} = \frac{Q_o^2 \sin^2\omega t}{2C}$$

$$U_B = \frac{1}{2}Li^2 = \frac{1}{2}Li_o^2\cos^2\omega t$$

$$TE = \frac{Q_o^2}{2C} = \frac{1}{2}Li_o^2 = const^n$$

Choke Coil :-

- Control Current in A/C Circuit
- Divide Potential without Power loss.

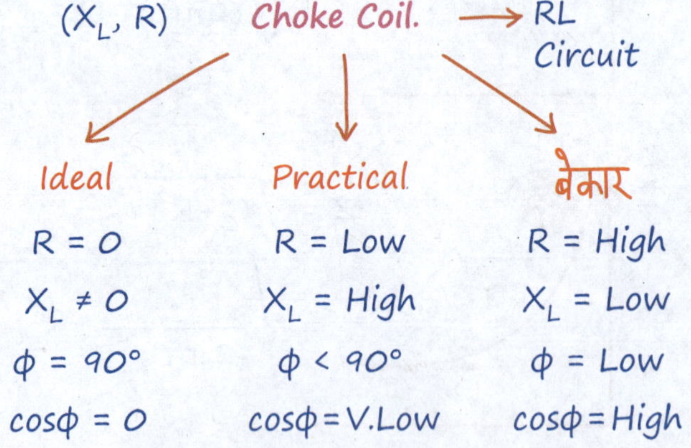

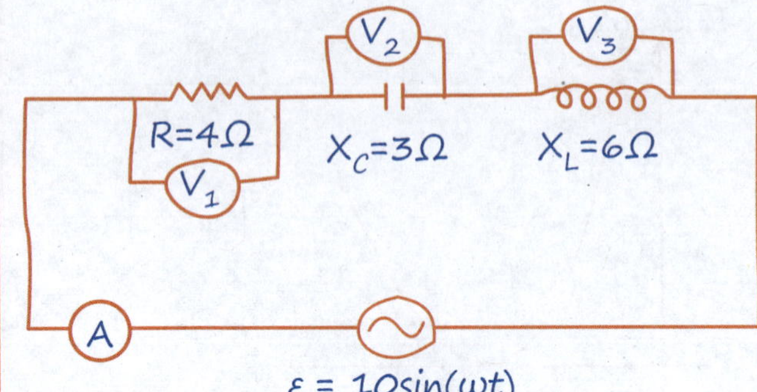

- $Z = \sqrt{R^2 + (X_L - X_C)^2} = \sqrt{4^2 + (6-3)^2}$
- $Z = 5\Omega$
- $i_0 = \frac{\varepsilon_0}{Z} = \frac{10}{5} = 2$ Amp
- i_{RMS} (Reading of Ammetre) $= \frac{i_0}{\sqrt{2}} = \sqrt{2}$
- Power factor $\cos\Phi = \frac{R}{Z} = \frac{4}{5}$

  $(\Phi = 37°)$
- $X_L > X_C \rightarrow$ Inductive circuit

  Voltage will leads by current by 37°
- $i = i_0 \sin(\omega t - \Phi)$

  $i = 2\sin(\omega t - 37°)$

Reading of $V_1 \Rightarrow \quad V_1 = i_{Rms} R = \sqrt{2}(4)$

Reading of $V_2 \Rightarrow \quad V_2 = i_{Rms} X_C = 3\sqrt{2}$

Reading of $V_3 \Rightarrow \quad V_3 = i_{Rms} X_L = 6\sqrt{2}$

RMS voltage across 'R' and 'C'

$$= \sqrt{(4\sqrt{2})^2 + (3\sqrt{2})^2} = 5\sqrt{2}$$

RMS voltage across 'C' and 'L'

$$= 6\sqrt{2} - 3\sqrt{2} = 3\sqrt{2}$$

RMS voltage across 'R' and 'L'

$$= \sqrt{(6\sqrt{2})^2 + (4\sqrt{2})^2} = \sqrt{104}$$

$\langle P \rangle = i_{rms} V_{rms} \cos\Phi$

$$= \sqrt{2} \times \frac{10}{\sqrt{2}} \times \frac{4}{5} = 8W$$

The Ultimate MR Star* Table

Circuit	Phase diff. Between i&V	Power factor cosϕ = R/Z	Impedance (Z)	Who leads!	Power loss
Pure resistive	0	1	R	Same Phase	$P = i_{rms}V_{rms}$
Pure Capacitive	$\pi/2$	Zero	$X_C = \frac{1}{\omega C}$	Current	Zero
Pure inductive	$\pi/2$	Zero	$Z_L = \omega L$	Voltage	Zero
RL	$0 < \phi < \frac{\pi}{2}$	b/n 0 to 1	$Z = \sqrt{R^2 + X_L^2}$	Voltage	$P = i_{rms}V_{rms}\frac{R}{Z}$
RC	$0 < \phi < \frac{\pi}{2}$	b/n 0 to 1	$Z = \sqrt{R^2 + X_C^2}$	Current	$P = i_{rms}V_{rms}\cos\phi$
LC	$\pi/2$	Zero	$Z = X_L - X_C$	Depends	Zero
Series LCR	$0 \le \phi < \frac{\pi}{2}$	1 or b/n 0 to 1	$Z = \sqrt{R^2 + (X_L - X_c)^2}$	Depends	$P = i_{rms}^2 R = \frac{V_{rms}^2}{R}$

ELECTRIC OSCILLATION
EQUIVALENT OF L.C. OSCILLATION :-

Mechanical system	Electrical system
Mass	Inductance
Displacement	Charge
Velocity	Current
Potential Energy	Electric Energy
Kinetic Energy	Magnetic Energy
$\frac{1}{K}$(K=Spring const.)	Capacitance

Q.1 In L.C. Oscillation Maximum Current is Io at t=o then find current in circuit when magnetic energy is half of electrical energy.

Sol. Total Energy Conserved

$$\frac{1}{2} L I_o^2 = Ue + Um$$

$$\frac{1}{2} L I_o^2 = 2Um + Um$$

$$\frac{1}{2} L I_o^2 = 3\,\frac{1}{2} LI^2$$

$$I = \frac{I_o}{\sqrt{3}}$$

Q.2 The number of turns in the primary and secondary coils of a step-down transformer are 200 and 50 respectively. If the power in the input is 100 watt at 1A then the output power and current will respectively be

Sol. Power in an transformer (ideal) does not change, the total electric power remains same.

Power = 100 Watt

for current, the proportionality is

$$\frac{N_1}{N_2} = \frac{V_1}{V_2} = \frac{I_2}{I_1}$$

$$\frac{200}{50} = \frac{I_2}{1}$$

$$I_2 = \frac{200}{50} = 4\text{Amp.}$$

MR*

'Aaj kuch karlo asa ki bhavishya me khud ko kosne ki nobat na aaye.'

29 Electromagnetic Wave

1 Behaviour of charges in Different conditions

Charge	E.F.	MF	EM Wave
Rest	✓	✗	✗
V = Const^n	✓	✓	✗
Accelerated	✓	✓	✓

2 Charging of Capacitor

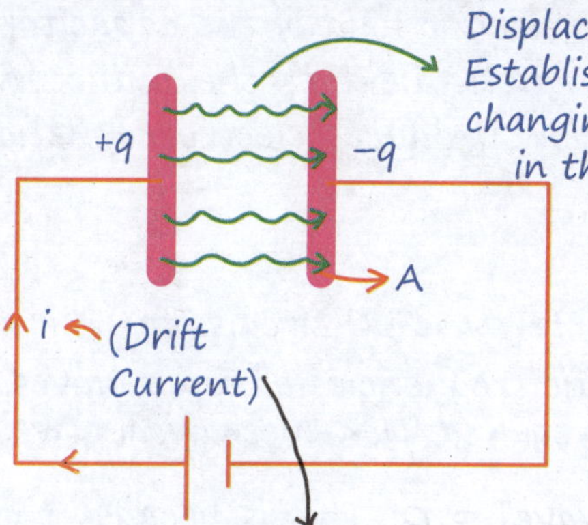

This is the conventional current flowing through the capacitor, causing the capacitor to charge.

$$E = \frac{\sigma}{\varepsilon_0} = \frac{q}{A\varepsilon_0}$$

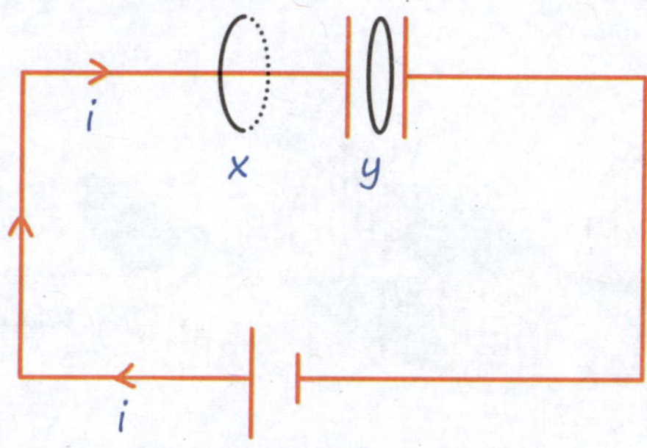

According to amperes law

for x: $\int \vec{B}_x \cdot d\vec{\ell} = \mu_0 i$

for y: $\int \vec{B}_y d\vec{\ell} = \mu_0 \not{i} 0$

hence $B_y = 0$

But maxwell found experimentally,

$B_y \neq 0$

and gives reason of this is **Displacement current** b/w the plate of capacitor.

Displacement Current :-

$Q = CV$ $\qquad$ $i_{Drift} = i_{Displacement}$

$$i = \frac{dQ}{dt} = C \cdot \frac{dV}{dt} = \frac{\varepsilon_0 A}{\ell}\frac{dV}{dt}$$

(ℓ = Distance between the plates)

$$i = \varepsilon_0 A \frac{dE}{dt} = \varepsilon_0 \frac{d\phi}{dt}$$

- Formed only due to changing E. Field.
- Not exist under steady current (ϕ = constant)
- Flows between c/s Area of Capacitor plate.

3 Maxwell Equation (4 equations)

Gauss law of Electrostat.

$$\phi = \oint \vec{E} \cdot \vec{ds} = \frac{q_{in}}{\varepsilon_0}$$

- Electric fields are generated by electric charges.

Gauss law of Magnetism.

$$\oint \vec{B} \cdot \vec{ds} = 0$$

- Magnetic fields have no beginning or end (no monopoles)

Induced E. Field :- Faraday's Law of induction

$$-\oint \vec{E}_{in} \cdot \vec{d\ell} = EMF = -\frac{d\phi}{dt}$$

$$\oint \vec{E}_{in} \cdot \vec{d\ell} = \frac{d\phi}{dt}$$

- A changing magnetic field creates an electric field..

Ampere's Maxwell Law :-

$$\oint \vec{B} \cdot \vec{d\ell} = \mu_0 [i_{Drift} + i_{Disp}]$$

$$= \mu_0 \left[i_{Drift} + \varepsilon_0 \frac{d\phi}{dt} \right]$$

- Electric currents and changing electric fields create magnetic fields.

4 Source of EMW

Accelerating Charge :-

When a charged particle accelerates, it emits electromagnetic waves. This is because the changing velocity of the charge causes disturbances in the electric and magnetic fields, which propagate as waves.

Example:

- An electron in an antenna oscillates back and forth, producing radio waves.

Transition of Electrons from n^{th} orbit :-

When an electron transitions from a higher energy orbit to a lower one in an atom, it releases energy in the form of electromagnetic radiation.

Example:

- The emission of light from hydrogen atoms when electrons move between energy levels.

Retardation of Electrons in High Atomic Weight Targets (X-rays) :-

When high-speed electrons are suddenly decelerated upon hitting a target of high atomic weight, X-rays are produced. This is known as Bremsstrahlung radiation.

Example:

- X-ray machines in medical imaging use this principle to produce X-rays by bombarding a metal target with electrons.

De-excitation of Nucleus in Radioactivity (γ-rays) :-

When the nucleus of a radioactive atom de-excites, it releases energy in the form of gamma rays. These are high-energy electromagnetic waves.

Example:

- Gamma rays emitted from radioactive materials used in cancer treatment (radiotherapy).

EM Wave in LC Oscillation :-

- LC Oscillation: In an LC circuit (inductor-capacitor circuit), the energy oscillates between the electric field of the capacitor and the magnetic field of the inductor. This oscillation produces electromagnetic waves.

Example:

Radio transmitters use LC circuits to generate specific frequencies of radio waves.

- No Medium required, Non-mechanical wave.
- Q_{net} [EM wave] = 0: This is because the wave consists of oscillating electric and magnetic fields, not actual moving charges.
- Transverse in nature
- E&B oscillate perpendicular to each other but in same phase.

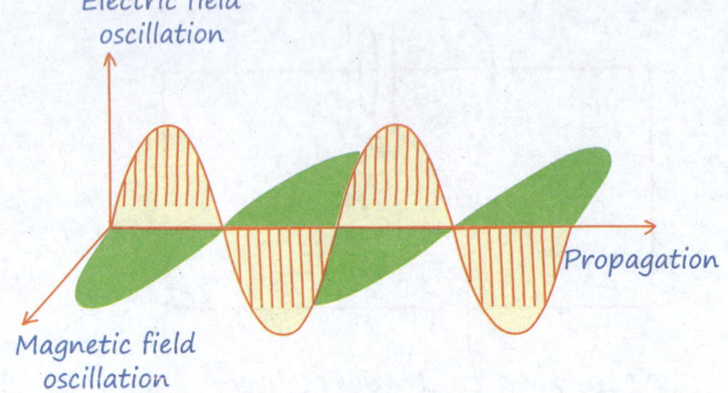

$E_y = E_0 \sin\omega t$

$B_z = B_0 \sin\omega t$

$E_0 = Q_0/A\varepsilon_0$

$E_0 = B_0 C$

E.F. & Mag. F are in Same phase

Direction of Wave :- $\vec{E} \times \vec{B}$

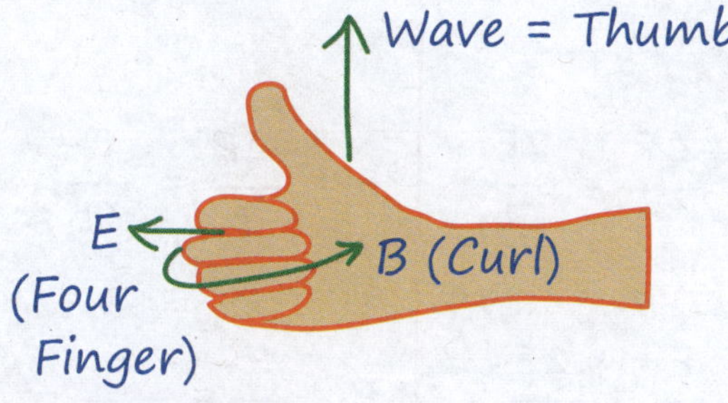

MR* for direction

Direction of wave = $\hat{E} \times \hat{B}$

Place Your four finger (with Palm) of right hand along electric field and Slap magnetic field. thumb will Indicate direction of wave.

5 Nature of EMW

Speed of EM-wave in Vaccum:

$$C = \frac{1}{\sqrt{\mu_0 \varepsilon_0}} = 3 \times 10^8 \text{ m/s}$$

Speed of EM-wave in a medium:

$$V = \frac{1}{\sqrt{\mu_m \varepsilon_m}} = \frac{C}{\sqrt{\mu_r \varepsilon_r}}$$

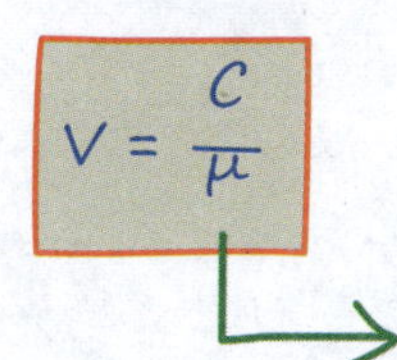

μ = Mag. Permeability

Refractive index

ε = Electric Permittivity.

+ Angular Frequency :-

$$\omega = \frac{2\pi}{T} = 2\pi f$$

+ Angular wave no :-

$$K = \frac{2\pi}{\lambda} = \frac{2\pi\nu}{C}$$

+ Speed of EMW :-

$$C = \lambda\nu = \frac{\lambda}{T} = \frac{\omega}{K}$$

If Medium is Changed !

λ = Changes

f = Same.

$C_1 = \lambda_1 \nu$

6 Energy Density

+ EM Wave (U) :-

$$(U) = \varepsilon_0 E^2 = \frac{B^2}{\mu_0}$$

$$U_{avg} = \langle U \rangle = \frac{\varepsilon_0 E_0^2}{2} = \frac{B_0^2}{2\mu_0}$$

+ Electrostatic (U_E) :-

$$U_E = \frac{1}{2}\varepsilon_0 E^2$$

$$\langle U_E \rangle_{avg} = \frac{1}{4}\varepsilon_0 E_0^2$$

+ Magnetic (U_B) :-

$$U_B = B^2/2\mu_0$$

$$\langle U_V \rangle_{avg} = B_0^2/4\mu_0$$

7 EM-Waves

The electromagnetic (EM) spectrum encompasses all types of electromagnetic radiation, from gamma rays to radio waves.
Gaddi SUV in My Range

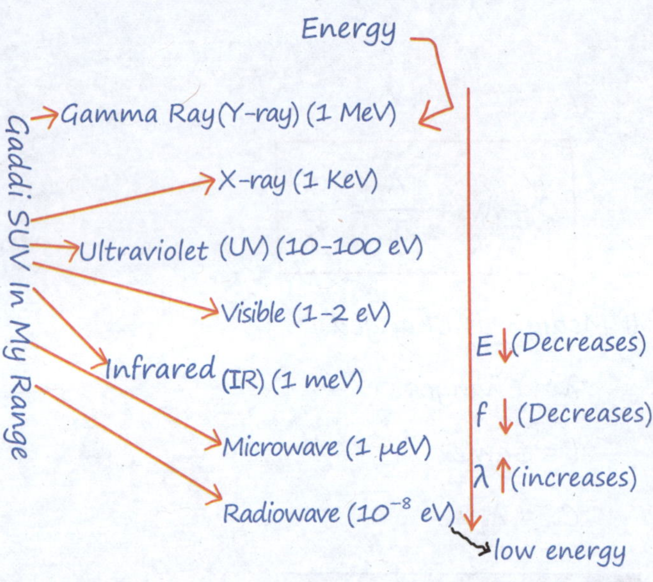

Intensity of EM Wave :-

The intensity of an electromagnetic wave represents the power per unit area carried by the wave.

$$\langle I\rangle = \frac{1}{2}\varepsilon_0 E_0^2 C = \frac{1}{2}\frac{B_0 E_0}{\mu_0}$$

8 Force & Radiation Pressure

Surface

Completely reflecting	Completely absorbing
Momentum :-	
$\lvert\Delta\vec{P}\rvert = 2P = \frac{2E}{C}$	$\lvert\Delta\vec{P}\rvert = P = \frac{E}{C}$
Force :-	
$F = \frac{\Delta P}{t} = \frac{2E}{Ct}$	$F = \frac{\Delta P}{t} = \frac{E}{Ct}$
Pressure :-	
$P = \frac{F}{A} = \frac{2E}{CAt}$	$P = \frac{F}{A} = \frac{E}{CAt}$
$P = \frac{2I}{C}$	$P = \frac{I}{C}$
I = Intensity	$\mu_r = \frac{\mu_m}{\mu_0}$
E = Energy	$\varepsilon_r = \frac{\varepsilon_m}{\varepsilon_0}$ = Dielec. Const^n

9 Poynting Vector

Direction :- $\parallel^{el}$ to wave & $\perp$ to electric field and the magnetic field.
Magnitude :- Intensity of EM Wave.

$$\vec{S} = \frac{\vec{E_0} \times \vec{B_0}}{2\mu_0}$$

Type of Radiation	Frequency Range (Hz)	Wavelength Range
gamma-rays	$10^{20} - 10^{24}$	$<10^{-12}$m
X-rays	$10^{17} - 10^{20}$	0.01nm – 10nm
ultraviolet	$10^{15} - 10^{17}$	400nm – 1pm
visible	$4 - 7.5 \times 10^{14}$	750 nm – 400nm
near-infrared	$1 \times 10^{14} - 4 \times 10^{14}$	2.5μm – 750nm
infrared	$10^{13} - 10^{14}$	25μm – 2.5μm
microwaves	$3 \times 10^{11} - 10^{13}$	1mm – 25μm
radio waves	$< 3 \times 10^{11}$	> 1 mm

	Radiowaves	Microwaves	IR	Visible	UV	X-Ray	Gamma-Rays
Production	Accl^n of e^- in Antenna	Klystron, Magneton & Gunn Diode	Heated Bodies	e^- Excitation	Transfer of e^- from E.S. to G.S.	X-ray Tube Innner Shell e^-	Radioactive Nucleus Nuclear Explosion
Detection	Receiver's Antenna.	Point Contact Diode	Bolometer, IR Photographic Plate, Photodiode	Human eye, Photographic Films, Photocells	Diodes & Photographic Films.	Gieger Counter Ionisat^n Chamber Photographic Plates	Gieger Counter Ionisat^n Chamber Photographic Plates
Uses	FM, AM, TV, Cellular Network	Radar Navigation, Measuring Speed of balls, Microwave Oven (3GHz, Reso. of H_2O).	Remotes, Hi Fi System	To See Objects	UV Fitter, Lasik Laser, Sanitization	Diagnostic tool, Radio-therapy.	Radio therapy High level Sanitization.

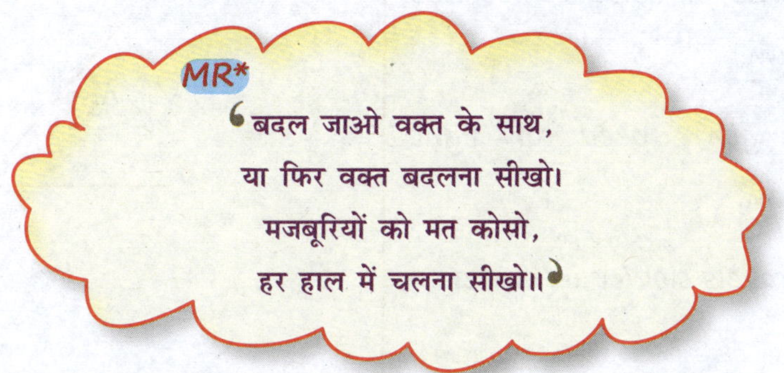

30 Ray Optics

1 Light

+ It is a form of electromagnetic energy which gives the Sensation of vision.
+ It is an essential part of the electromagnetic spectrum that is visible to the human eye.
+ Light itself is not visible; rather we see objects because they reflect or emit light.
+ Visible light has a wavelength range of approximately 380 nm to 700 nm, which corresponds to the portion of the electromagnetic spectrum that human eyes can detect."
+ [V I B G Y O R] : visible light Ka range

  The visible light spectrum includes colors in the following order: Violet (V), Indigo (I), Blue (B), Green (G), Yellow (Y), Orange (O), and Red (R)."
+ Speed of light does not depends on Speed of Source and speed of observer.
+ Frequency and sense of colour does not depends on medium.
+ Intensity, wavelength, speed of light depends on medium.

For example, light travels slower in water than in air."

Speed of Light in Vaccum

Speed of light in medium,

$$C = \frac{1}{\sqrt{\mu_0 \varepsilon_0}},\ V = \frac{1}{\sqrt{\mu_m \varepsilon_m}}$$

$$V = \frac{C}{\sqrt{\mu_r \varepsilon_r}} = \frac{C}{\mu}$$

$$\therefore \mu_1 V_1 = \mu_2 V_2$$

$$V = \lambda f,\ \lambda \propto \frac{1}{\mu} \propto V$$

MR*

Koi object jis light Ko emit Karega woh waisa dhikega.

Heat up

∴ Yellow balloon will burst.

2 Plane Mirror

Effect of Reflection & Refraction :-

Frequency of light

f = Medium independent, only depend on Source

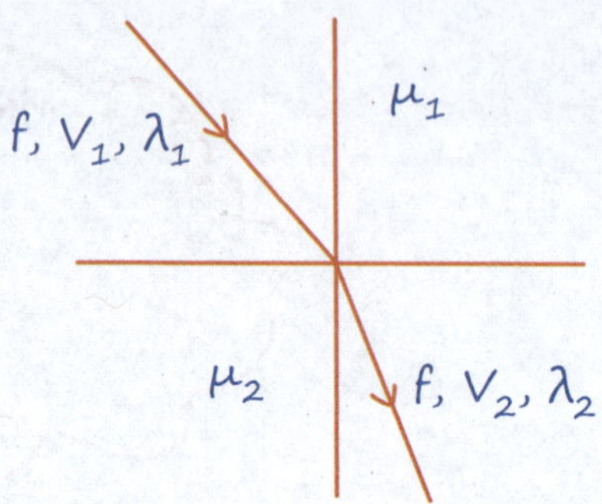

Reflection :-

- Laws of reflections :- $\angle i = \angle r$
  $\alpha = \beta$

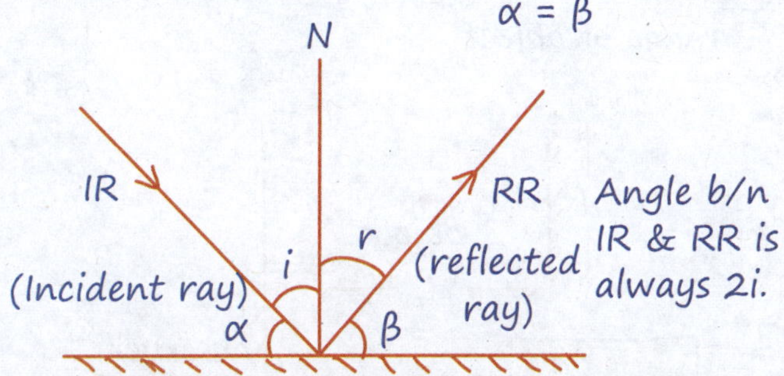

Angle b/n IR & RR is always 2i.

Deviation from Plane Mirror :-

(a) By a Mirror :-

$\delta = 180 - 2i$

Normal Incidence	Grazing Incidence
$\delta = 180°$	$\delta \approx 0$

(b) By two inclined Mirror :-

$\delta = 360 - 2\theta$

θ = Angle b/n Mirror.

Note :-

- If incident ray rotated "θ" then reflected ray rotates "θ" in Opposite Sense.
- Plane Mirror is rotated by "θ", reflected ray turns by "2θ" in same sense.
- If plane minor is rotated by θ about axis perpendicular to plane then No effect on reflected ray.

Image Formation by Plane Mirror :-

MR*

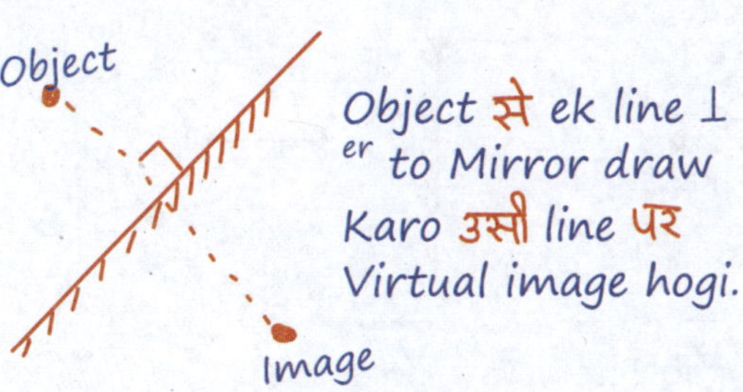

Object से ek line ⊥[er] to Mirror draw Karo उसी line पर Virtual image hogi.

Some Properties of Image formation by Plane mirror :-

- Focal length of Plane mirror is Infinite
- On covering part of mirror, No change in size of Image but brightness will decrease because less light is reflected."
- Real Object → Virtual Image: A plane mirror always forms a virtual image of a real object. The image appears to be behind the mirror and cannot be projected onto a screen.
- Virtual object → Real Image: A plane mirror can form a real image if the object is virtual. This occurs in certain optical setups, such as with converging lenses or other mirrors that create a virtual object.
- Object and Image always of same distance from mirror.
- Lateraly inverted image: The left and right sides of the image are reversed.
- Plane mirror can form Inverted Image of real object

Height of Mirror :-

(A) To see full height of object :-

$H_{mirror} = H_{object}/2.$

(B) To see full wall behind object ?

(Observer at Centre).

$$H_{mirror} = \frac{H_{wall}}{3}$$

(C) To see full wall behind object if man is not centre ?

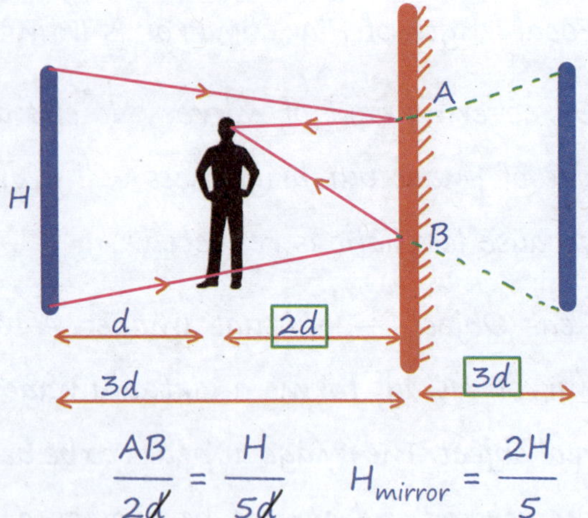

$$\frac{AB}{2d} = \frac{H}{5d} \qquad H_{mirror} = \frac{2H}{5}$$

Clock System :-

- The reflection inverts the left and right sides, effectively reversing the direction of the clock's hands." therefore the time shown in the image of a clock reflected in a plane mirror can be given as follow.
  - If time in clock → $(H_{hr} : M_{mint} : S_{sec})$ then mirrored time will be → $(11-H)_{hr} : (59-M)_{mint} : (60-S)_{sec}$
  - When only HR and MINT is given $= (11-H)_{hr} : (60-M)_{mint}$
  - If only HR is given mirror time $= (12-H)_{Hour}$

3 Velocity of Image

Object moving $\parallel^{el}$ to plane mirror :-

- Velocity of image w.r.t object, $\vec{V}_{IO} = 0$
- Velocity of object w.r.t. ground $\vec{V}_{OG}$ = Velocity of image w.r.t. ground $\vec{V}_{IG}$

- Observer is moving with constant speed u then time for which observer Can See image of object

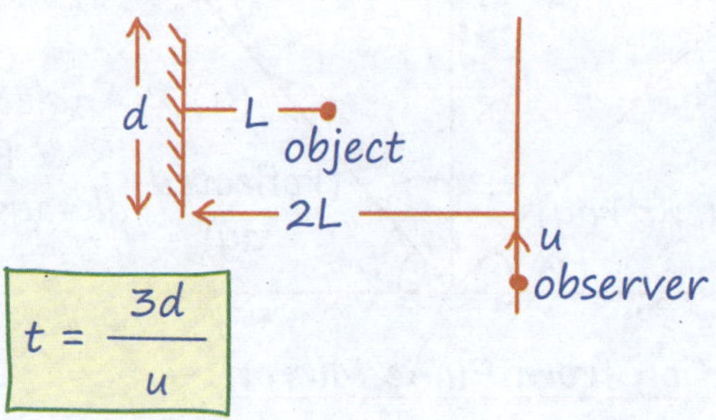

$$t = \frac{3d}{u}$$

Object is moving $\perp^{er}$ to plane mirror :-

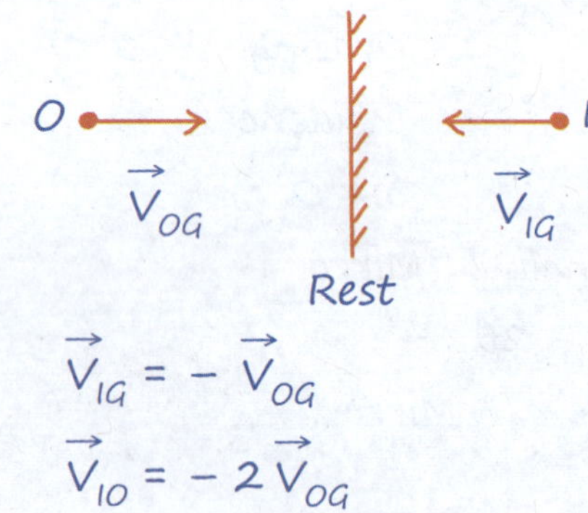

$$\vec{V}_{IG} = -\vec{V}_{OG}$$
$$\vec{V}_{IO} = -2\vec{V}_{OG}$$

Object is moving at an angle to mirror :-

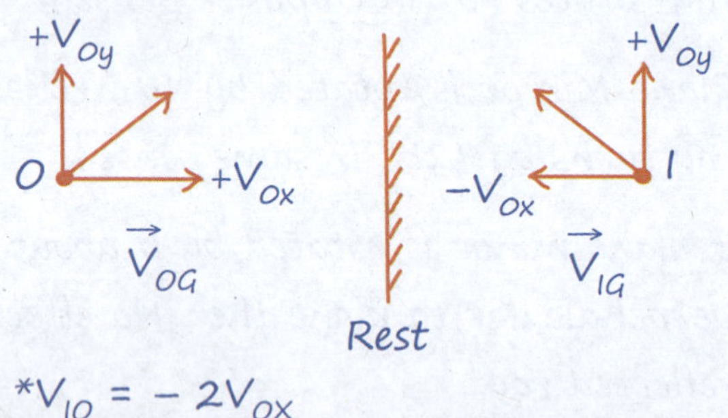

$$^*V_{IO} = -2V_{ox}$$

Object at rest & Mirror is moving with V_{MG}:-

V_{MG}

O rest

I

$$\vec{V}_{IO} = \vec{V}_{IG} = 2\vec{V}_{MG}$$

Image Formation by Two Plane Mirror :-

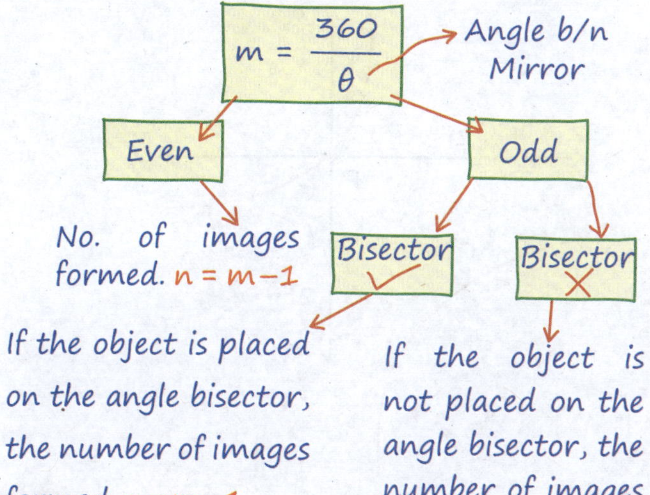

MR*

- $m = \frac{360°}{\theta}$ fraction then ush fraction Se choti integer lenge approximate Nahi lena,
- Ex M = 7.8 → m = 7 Image

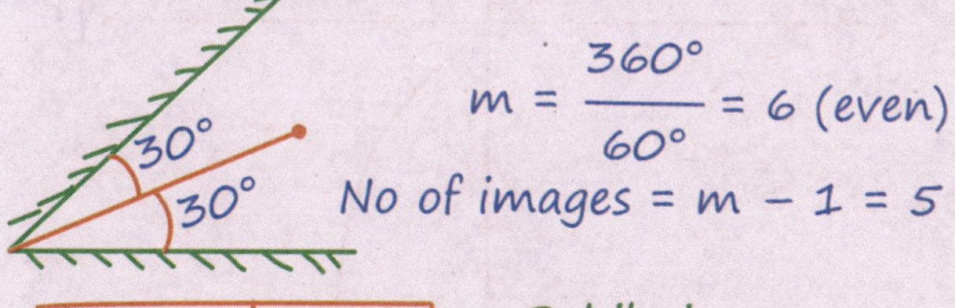

$m = \frac{360°}{60°} = 6$ (even)

No of images = m − 1 = 5

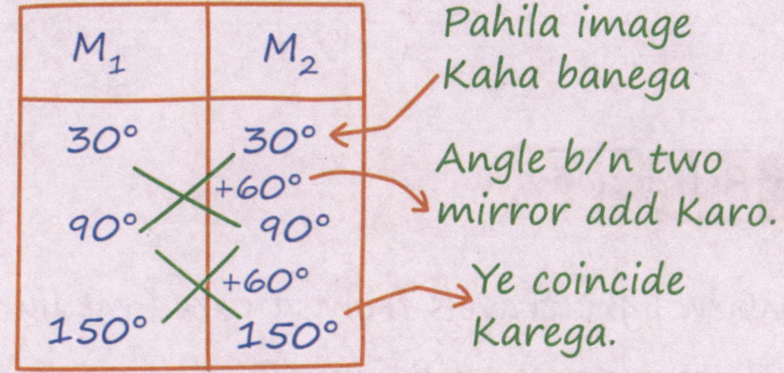

Pahila image Kaha banega

Angle b/n two mirror add Karo.

Ye coincide Karega.

Infinite Image When Two Plane Mirror is Parallel :-

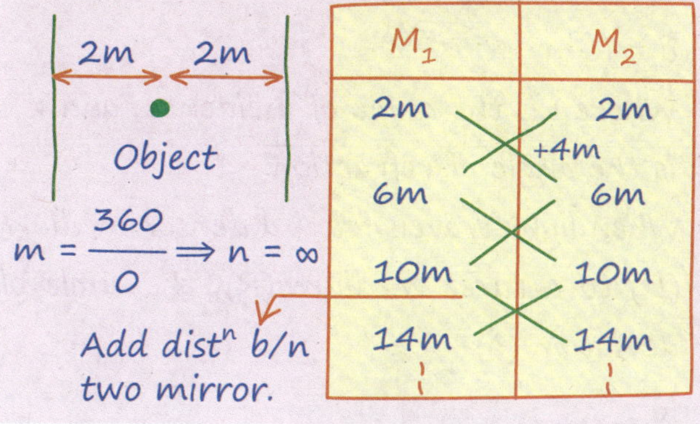

$m = \frac{360}{0} \Rightarrow n = \infty$

Add dist^n b/n two mirror.

4 Spherical Mirror

Property	Concave Mirror	Convex Mirror
Nature of mirror	Converging	Diverging
Focal length (f)	Negative (f = −ve)	Positive (f = +ve)
Object distance (U)	Negative (U = −ve) (Real object)	Negative (U = −ve) (Real object)
Image distance (V)	Negative (V = −ve) (Real Image)	Positive (V = +ve) (Virtual Image)

- Along Incident ray distance taken as +ve.

Mirror Equation :-

- $f = \frac{R}{2}$
- $\frac{1}{V} + \frac{1}{U} = \frac{1}{f}$

Transverse magnification

$$m_T = \frac{H_I}{H_O} = -\frac{V}{U} = \frac{f}{f-U} = \frac{f-V}{f}$$

$$m_L = \frac{-dV}{dU} = m_T^2 = \frac{V^2}{U^2}$$

* only valid for small object

Longitudinal Magnification

$$m = -\frac{V}{U}$$

m = +ve	m = −ve
> 1 = Erect & Magnified	> 1 = Inverted Magnified
< 1 = Erect & Diminished.	< 1 = Inverted & Diminished.

Joh Chahiye uska sign convention nai lete !

Image Formation by Concave Mirror :-

Object	Image
+ ∞	+ f[R, I, –ve]
+ b/n ∞ & C	+ b/n C & f [R, I, –ve]
+ A + C	+ At C[R, I, –1]
+ b/n C & f	+ b/n ∞ & C [R, I, –ve↑]
+ At f	+ At ∞. [R, I, –ve]
+ b/n f & Pole	+ Behind Mirrror [V, E, +ve]

Image Formation by Convex Mirror :-

Object	Image
+ at ∞	+ At focus. [V, E, m = +ve]
+ b/n ∞ & Pole	+ b/n Pole & Focus.

MR*

Tum mirror Ko kahi bhi rakho uska "f" change nai hoga lens Ka ho sakta hai.

Newton's Formula :-

$f = \sqrt{xy}$

x = Object dist^n from focus.

y = image dist^n from focus.

Velocity of Image in Case of Concave Mirror :-

1> Object is moving $\perp^{er}$ to Principle axis :-

- $V_I = m_T V_O$.

2> Object is moving $\parallel^{el}$ to Principle axis :-

- $V_I = m_L V_O = m_T^2 V_O$.

Graph for Concave Mirror :-

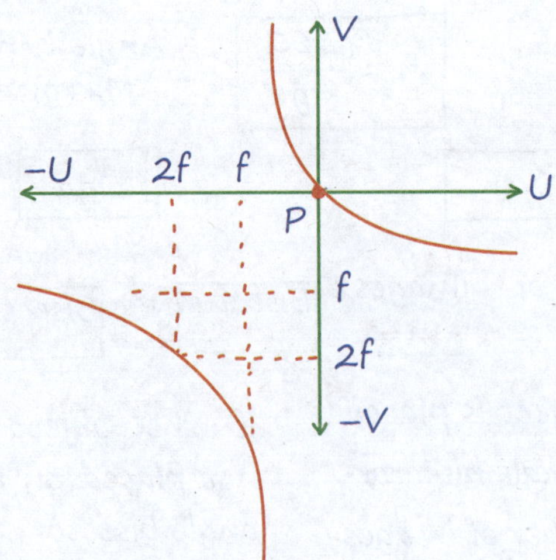

Graph for Convex Mirror :-

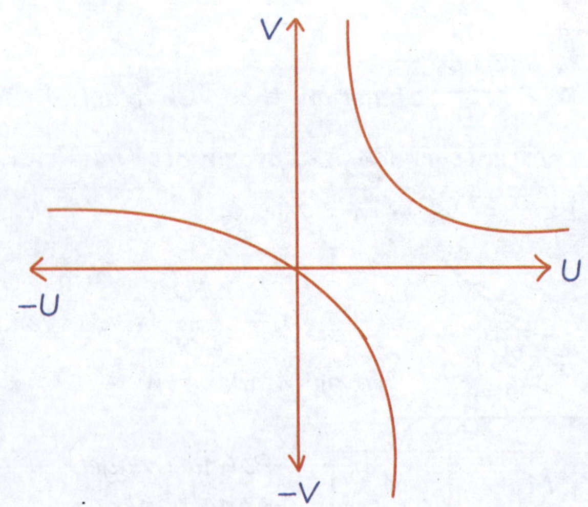

5 Refraction

+ When light travels from a rarer medium (R) to a denser medium (D), the angle of deviation (δ)

  $\delta = i - r$

  Where i is the angle of incidence, and r is the angle of refraction.

+ When light travels from a denser medium (D) to a rarer medium (R), the angle of deviation (δ)

  $\delta = r - i$

Snell's Law :-

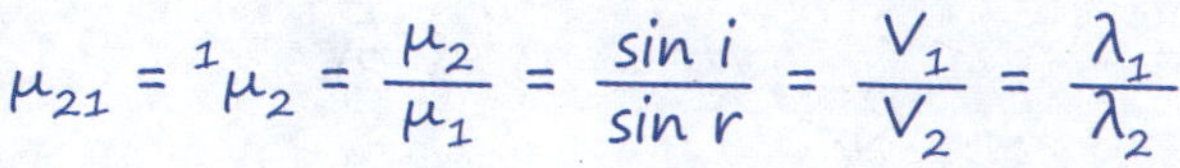

$$\mu_{21} = {}^{1}\mu_2 = \frac{\mu_2}{\mu_1} = \frac{\sin i}{\sin r} = \frac{V_1}{V_2} = \frac{\lambda_1}{\lambda_2}$$

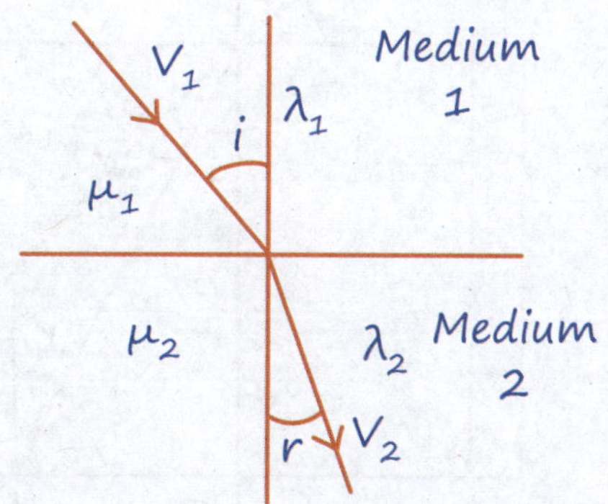

MR*

Snell's law can be directly applied b/n 1st & last medium irrespective of intermediate medium.

- "N" Number of waves in a "X" Distance if Wavelength is λ.

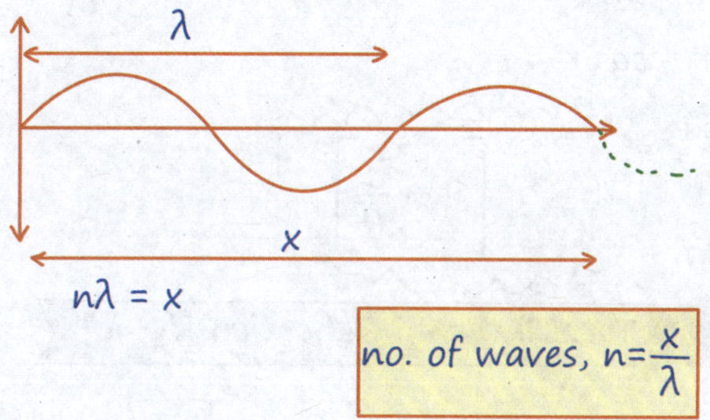

$n\lambda = x$

$$\text{no. of waves, } n = \frac{x}{\lambda}$$

Optical Path (d) :-

*d_{vacuum} & d_{medium} for same time.

$$d = \mu x$$

$x = d_{medium}$

$d = d_{vacuum}$

$$t = \frac{\mu x}{c}$$

6 Real & Apparent Depth

MR Special**

$$\frac{d_{app}}{\mu_{obr}} = \frac{d_{real}}{\mu_{obj}}$$

This means that apparent depth seen by an observer is influenced by the refractive indices of the media through which the light travels.

Apparent depth in multiple layers :-

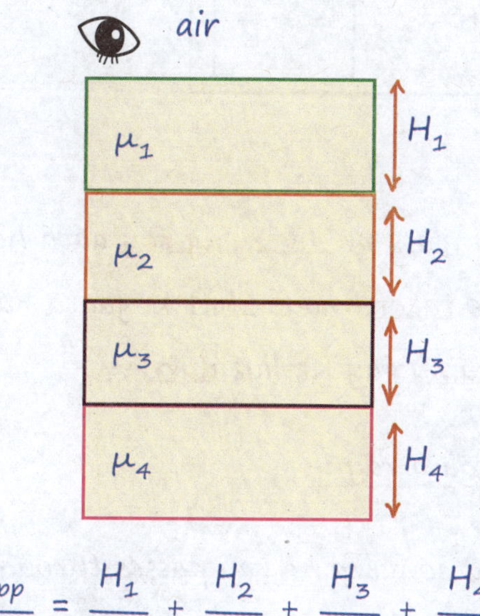

$$\frac{H_{app}}{1} = \frac{H_1}{\mu_1} + \frac{H_2}{\mu_2} + \frac{H_3}{\mu_3} + \frac{H_4}{\mu_4}$$

Diverging ray :-

- When a diverging ray passes through a medium with thickness t and refractive index μ, there is a lateral shift in the path of the ray.
- The real path is indicated by the solid line, and the apparent path is shown by the dotted line.

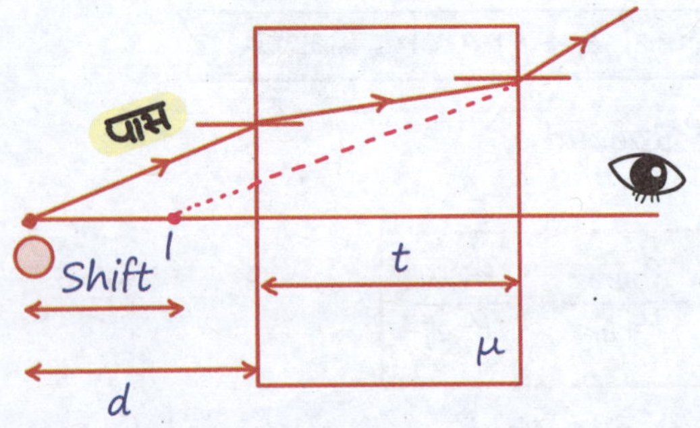

- $d_{real} = d + t$, $d_{app} = d + \frac{t}{\mu}$
- Shift = $d_{real} - d_{app}$

$$= (d+t) - \left(d + \frac{t}{\mu}\right)$$

$$\text{Shift} = t\left[1 - \frac{1}{\mu}\right]$$

MR*

Jab bhi glass pr diverging ray aata hai toh glass ke taraf Image Shift kr jaata hai. Aur Converging ray Ke liye durr.

Converging ray :-

When a converging ray passes through a medium with thickness t and refractive index μ, there is a lateral shift similar to the diverging ray.

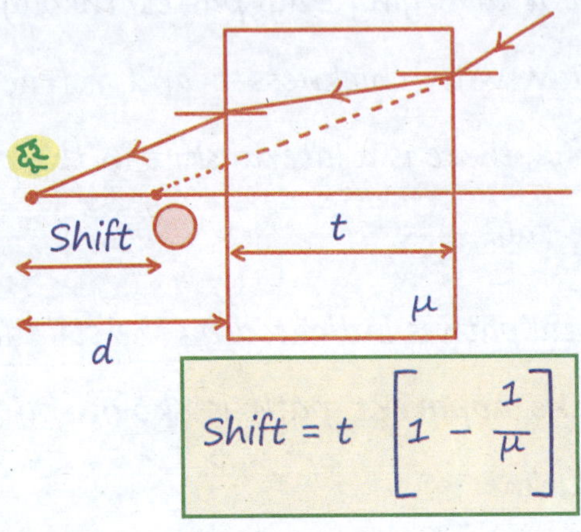

$$\text{Shift} = t\left[1 - \frac{1}{\mu}\right]$$

7 Lateral Shift

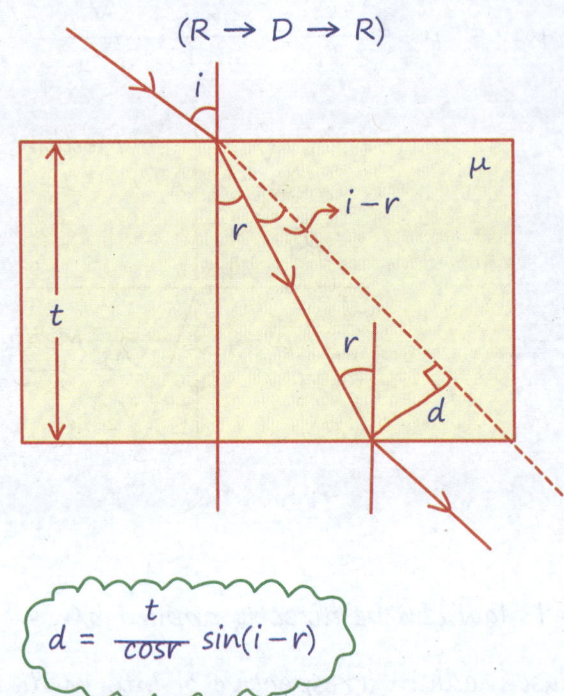

$$d = \frac{t}{\cos r}\sin(i - r)$$

- If $i \to$ small.

For small angles of incidence i, the sine and cosine functions can be approximated:

$\sin i \approx i$

$\cos r \approx i$

$$d = ti\left[1 - \frac{1}{\mu}\right]$$

8 Total Internal Reflection [i > ic]

- Total internal reflection occurs when a light ray traveling from a denser medium (D) to a rarer medium (R) strikes the boundary at an angle greater then critical angle ($i > i_c$).
- The light ray is completely reflected back into the denser medium, with no refraction occurring in the rarer medium.

- Condition for total internal reflection: light must travel from
  $D \rightarrow R$ Medium.
- Angle of incidence (i) must be greater than the critical angle (i_c)

Critical Angle :-

$$\sin i_c = \frac{\mu_{Kam}}{\mu_{Jyada}} = \frac{\mu_{rarer}}{\mu_{Denser}}$$

$$\sin i_c = \frac{\mu_2}{\mu_1} = \frac{V_1}{V_2} = \frac{\lambda_1}{\lambda_2}$$

R

D

i_c

Water – Air :- $i_c = 49°$

Glass – Air :- $i_c = 42°$

Radius of Visibility :-

The radius of visibility refers to the maximum distance at which an observer can see an object submerged in a medium (such as water) from above the surface.

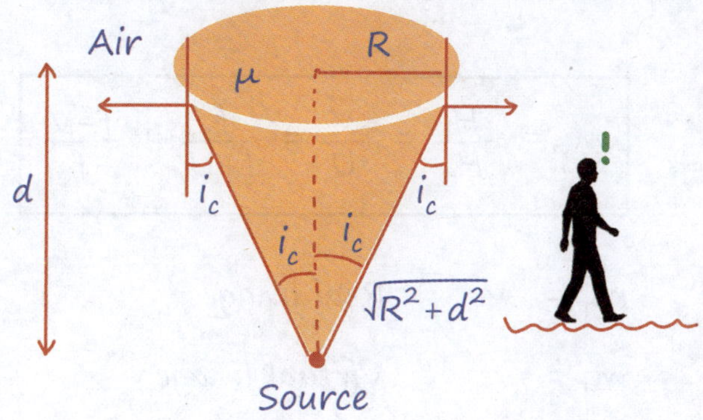

$$R = \frac{d}{\sqrt{\mu^2 - 1}}$$

- Water – Air

$$h = \frac{\sqrt{7}}{3} r$$

Velocity of Image in Refraction :-

1> Object is moving parallel to boundary :-

- $V_{IG} = V_{OG}$
- $V_{IO} = 0$

2> Object is moving $\perp^{er}$ to boundary :-

MR*

$$\frac{V_{app}}{\mu_{obr}} = \frac{V_{real}}{\mu_{obj}}$$

9 Refraction at Spherical Surfaces

MR* One stop solution

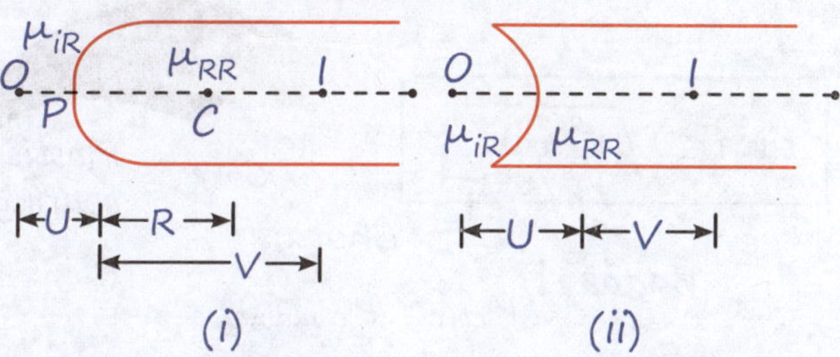

Spherical surface (i) convex (ii) concave

Refraction Formula:-

The refraction formula at a spherical surface separating two media with refractive indices μ_{IR} (incident ray medium) and μ_{RR} (refracted ray medium) is given by:

$$\frac{\mu_{RR}}{V} - \frac{\mu_{IR}}{U} = \frac{\mu_{RR} - \mu_{IR}}{R}$$

- Normal will pass through Center of Curvature.
- Nature of Surface will be decided by position of object.
- Concave :- R = –ve

  Convex :- R = +ve

Magnification :-

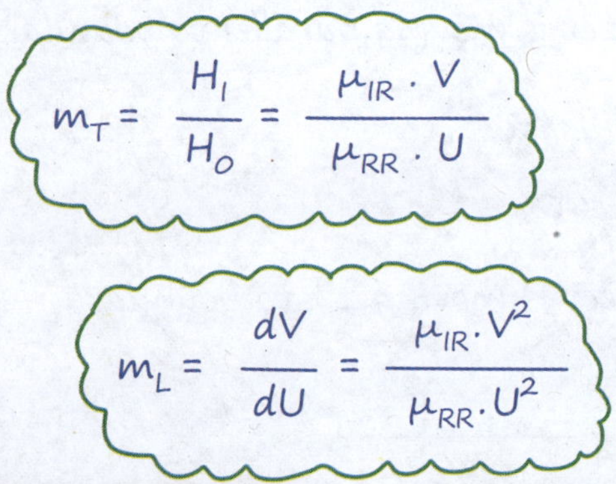

10 Optical Fiber

[Based on TIR]

The angle at which ray must be incident so as information gets transmitted :-

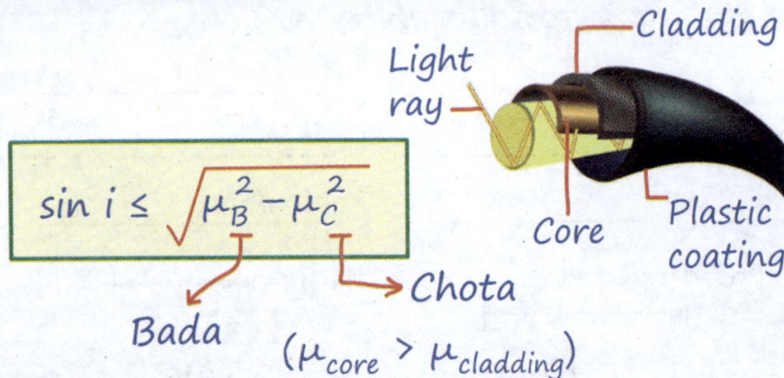

$(\mu_{core} > \mu_{cladding})$

11 Lens

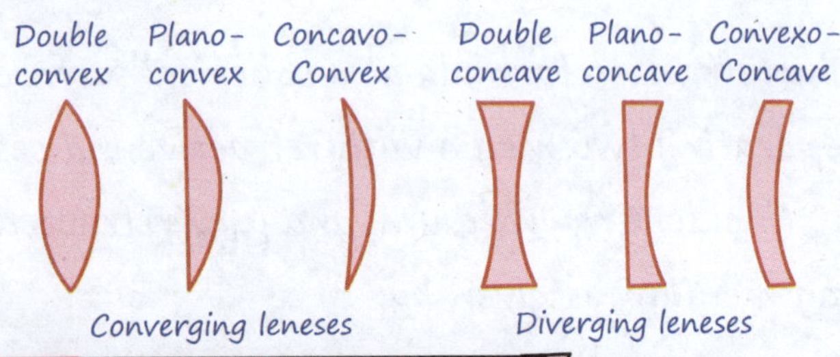

12 Lens-Maker Equation

$$\frac{1}{V} - \frac{1}{U} = \left(\frac{\mu_L}{\mu_M} - 1\right)\left[\frac{1}{R_1} - \frac{1}{R_2}\right]$$

Where:

μ_L = refractive index of the lens material.

μ_M = refractive index of the surrounding medium.

Lens Equation :-

$$\frac{1}{V} - \frac{1}{U} = \frac{1}{f}$$

$$\frac{1}{f} = \left(\frac{\mu_L}{\mu_M} - 1\right)\left[\frac{1}{R_1} - \frac{1}{R_2}\right]$$

Focal Length of lens :-

+ Focal length depends on medium.

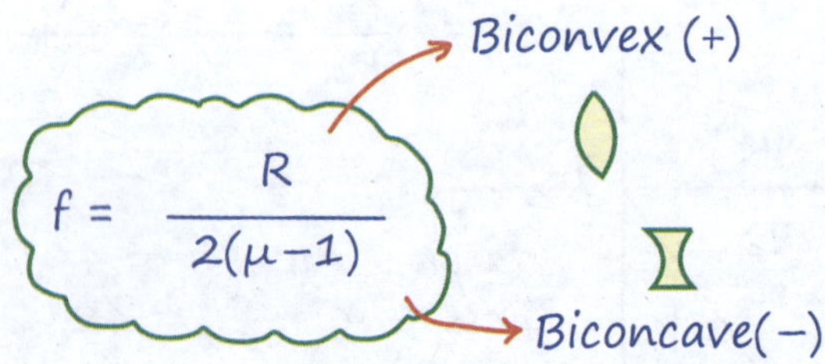

μ = refractive Index of lens w.r.t medium

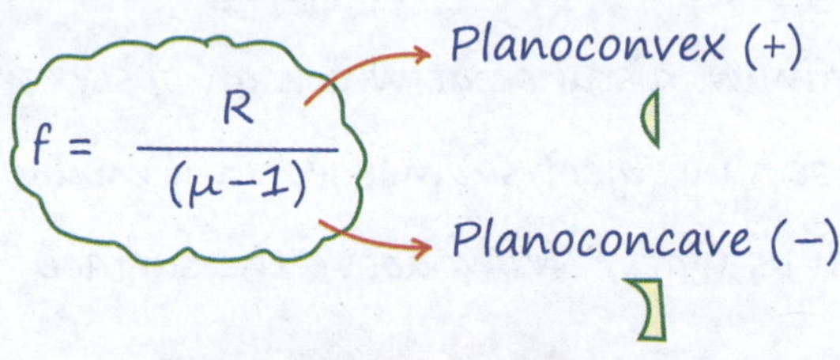

$$m_T = \frac{H_I}{H_O} = \frac{V}{U} = \frac{f}{f+U} = \frac{f-V}{f}$$

$m_T = -ve$ Real Image

$m_T = +ve$ Virtual Image

Small Object $\quad m_L = \frac{dV}{dU} = m_T^2$

Cutting of Lens :-

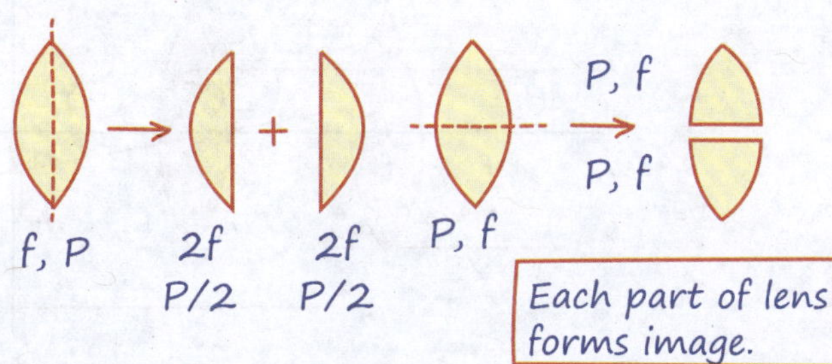

13 Ray-Diagram

(a) Biconvex lens :- $f = +ve$

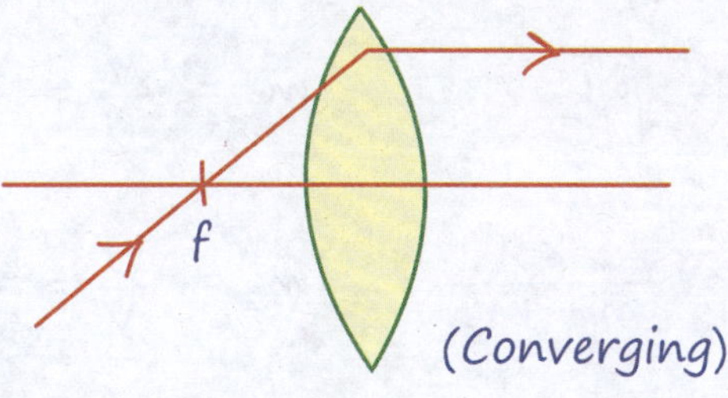

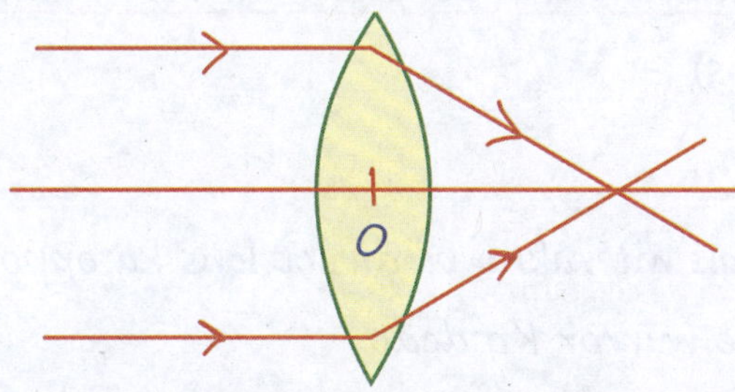

(b) Biconcave lens :- $f = -ve$

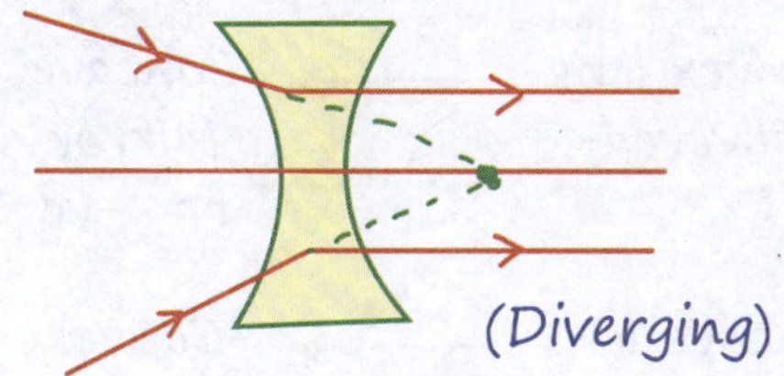

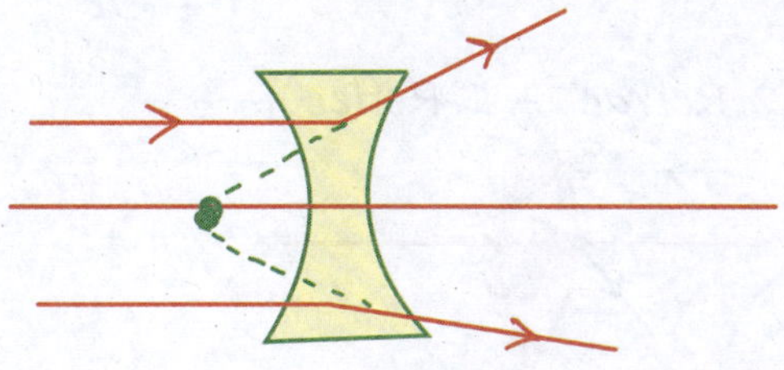

Nature of Lens Considering R.I. of Surrounding & Lens

Condition	Behaviour
$\mu_L = \mu_M$	Glass Plate: There is no refraction, and the lens does not converge or diverge light.
$\mu_L > \mu_M$	Same Nature: A convex lens converges light, and a concave lens diverges light.
$\mu_L < \mu_M$	Opposite Nature: A convex lens diverges light, and a concave lens converges light.

14 Image Formation by Equiconvex Lens

Object	Image
∞	f[R, I, –ve]
b/n ∞ & 2f	b/n 2f & f [R, I, –ve]
at 2f	at 2f [R, I, –1]
b/n 2f&f	b/n ∞ & 2f [R, I, –ve↑]
at f	at ∞ [R, I, –ve↑↑]
b/n f & Pole	on same side of lens [V, E, +ve]

Convex lens ⇅ Concave Mirror (Converging)

Can a Convex lens be have as diverging lens.

→ Yes, when $(\mu_m > \mu_L)$

15 Image Formation by Concave Lens

Object	image
at ∞	at focus. [V, E, m = +ve]
b/n ∞ & Pole	b/n Pole & Focus.

Concave Lens ⇌ Convex Mirror
(Diverging)

No. of Image Form :-

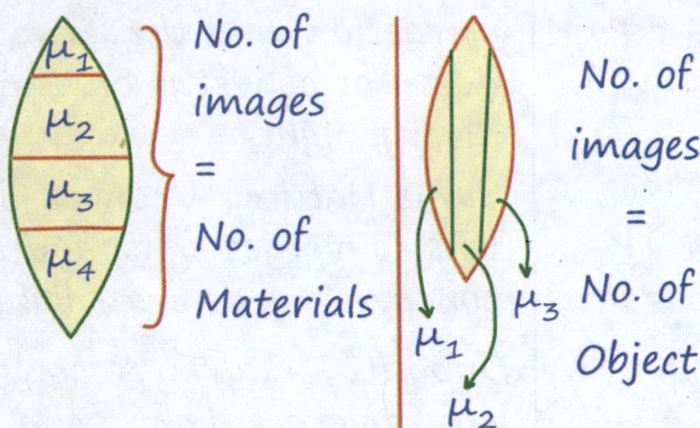

16 Combination of Lens

Power :-

Lens

$$P_L = \frac{1}{f_L}$$

Mirror

$$P_M = \frac{1}{f_M}$$

(a) When lens are in contact :-

$$P = P_1 + P_2$$

$$\frac{1}{f} = \frac{1}{f_1} + \frac{1}{f_2}$$

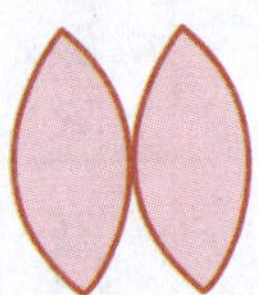

(b) When lens are separated by a distance "d":-

$$P = P_1 + P_2 - dP_1P_2$$

$$\frac{1}{f} = \frac{1}{f_1} + \frac{1}{f_2} - \frac{d}{f_1 f_2}$$

d

Take P with sign !

17 Displacement Method

Image Ka distⁿ Object Ko, Object Ka distⁿ Image Ko

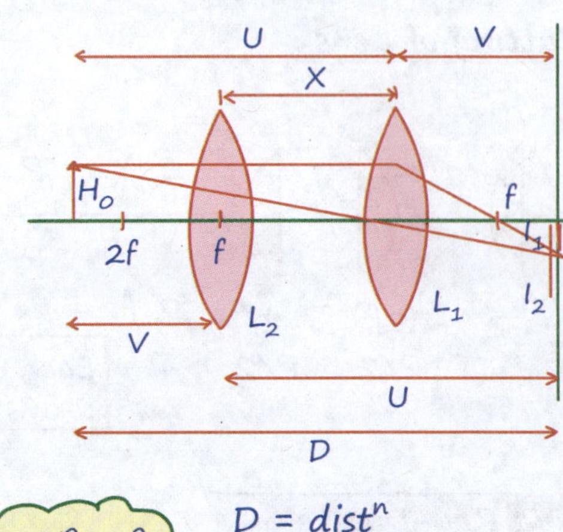

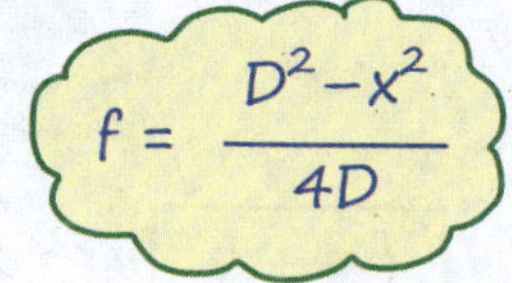

$$f = \frac{D^2 - x^2}{4D}$$

D = distⁿ b/n Object & Screen.

x = Distⁿ b/n two position of lens.

$$H_o = \sqrt{I_1 I_2}$$

$$m_1 = \frac{V}{U} = \frac{I_1}{H_o}$$

$$|f| = \frac{x}{m_1 - m_2}$$

$$m_2 = \frac{U}{V} = \frac{I_2}{H_o}$$

Combination of a Lens & a Mirror (Silvering of Lens) :-

MR*

Koi sign mt rakho bs direct lens ka opposite nature mirror Ko dedo.

$$\frac{1}{|f_{net}|} = \frac{2}{|f_L|} + \frac{1}{|f_M|}$$

Convex lens Silvered → Concave Mirror $f = -ve$

Concave lens Silvered → Convex Mirror $f = +ve$

[2 Refracⁿ + 1 Reflecⁿ]

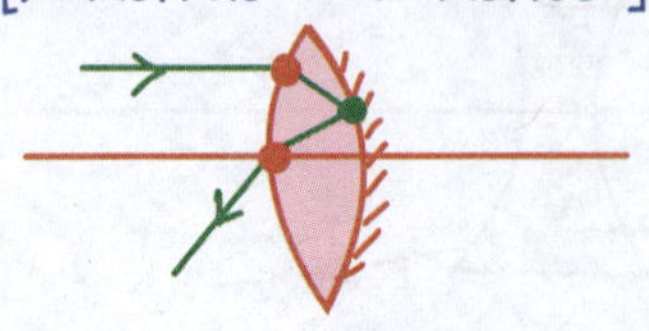

Case :-

(a) Convex Lens :-

$$f_{eq} = \frac{R}{2(2\mu-1)}$$ } Concave Mirror

(b) Plano-convex lens :-

$$f_{eq} = \frac{R}{2(\mu-1)}$$ } Concave Mirror

(c) Plano-convex lens :-

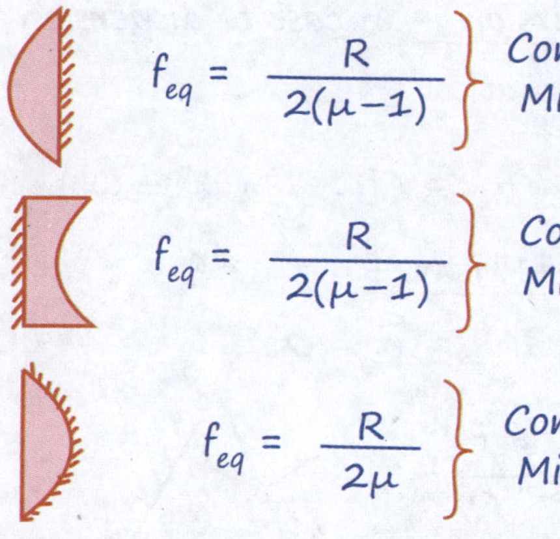

$$f_{eq} = \frac{R}{2(\mu-1)}$$ } Concave Mirror

$$f_{eq} = \frac{R}{2(\mu-1)}$$ } Convex Mirror

$$f_{eq} = \frac{R}{2\mu}$$ } Concave Mirror

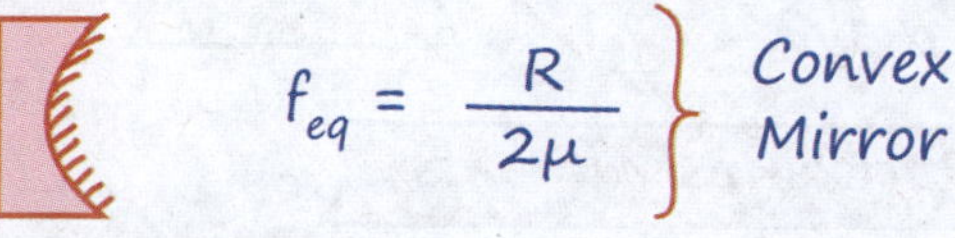

$$f_{eq} = \frac{R}{2\mu}$$ } Convex Mirror

18 Prism

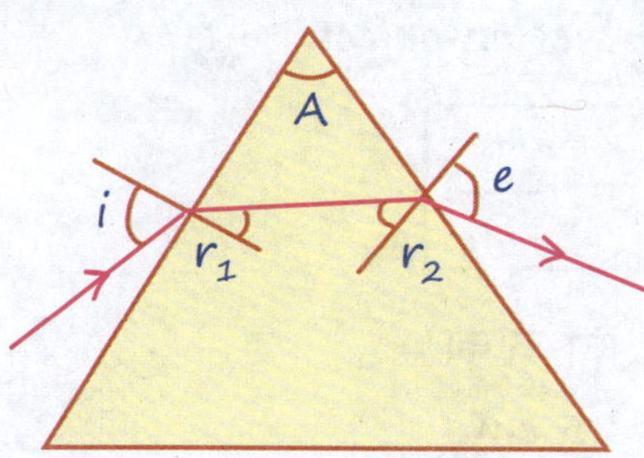

Angle of prism, $A = r_1 + r_2$ Total deviation $\delta_{Total} = i + e - A$

$$\mu = \frac{\sin i}{\sin r_1} \qquad \frac{1}{\mu} = \frac{\sin r_2}{\sin e}$$

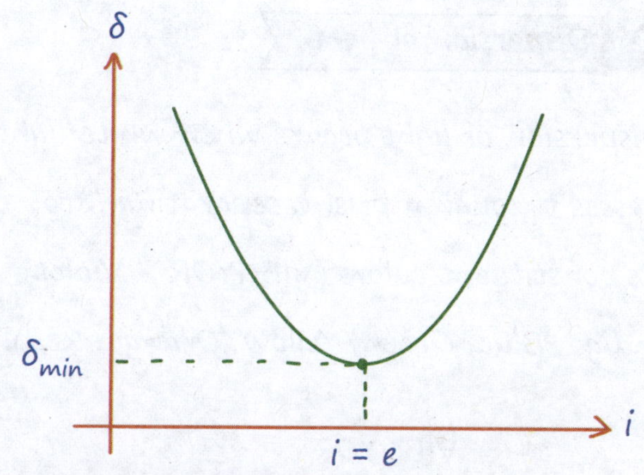

For Minimum deviation :-

$$r_1 = r_2 = \frac{A}{2} \qquad i = e$$

$$\delta_{min} = 2i - A, \quad \mu = \frac{\sin\left[\frac{\delta_m + A}{2}\right]}{\sin\left[\frac{A}{2}\right]}$$

For thin prism (A = small)

$$\delta_{min} = A(\mu - 1)$$

A = Refracting Angle.

✦ Half Angle Formula :-

$$\sin\theta = 2\sin\frac{\theta}{2} \cdot \cos\frac{\theta}{2}$$

Prism is placed in medium :-

μ_M μ_P

$$\delta = A\left[\frac{\mu_P}{\mu_M} - 1\right]$$

Condition for no deviation :-

A_1 μ_1 A_2 μ_2

$$\delta_1 + \delta_2 = 0$$

$$A_1(\mu_1 - 1) = -A_2(\mu_2 - 1)$$

19 Dispersion of light

Dispersion of light occurs when white light passes through a prism, separating into its constituent colors (VIBGYOR - Violet, Indigo, Blue, Green, Yellow, Orange, Red).

$\mu \propto \frac{1}{\lambda^2}$ $\xrightarrow{\text{VIBGYOR}}$ $\lambda\uparrow$ $\mu\downarrow$ $\delta\downarrow$

20 Angular dispersion (Q)

$$Q = \delta_V - \delta_R$$

$$Q = A\,[\mu_V - \mu_R]$$

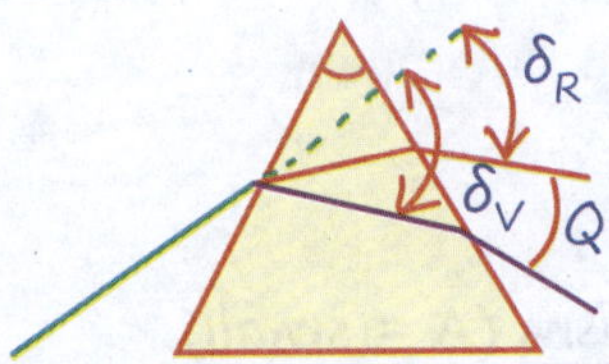

Mean deviation :-

$$\delta_{mean} = \frac{\delta_V + \delta_R}{2}$$

$$\delta_{mean} = A\left[\frac{\mu_R + \mu_V}{2} - 1\right]$$

$$\delta_{mean} = A\,[\mu_{mean} - 1]$$

$$\delta_{mean} = A\,[\mu_{yellow} - 1]$$

Dispersive Power :- (ω)

$$\omega = \frac{Q}{\delta_{mean}} = \frac{\delta_V - \delta_R}{\delta_{mean}}$$

$$\omega = \frac{\mu_V - \mu_R}{\mu_y - 1} = \frac{\delta_V - \delta_R}{\frac{\delta_V + \delta_R}{2}}$$

Dispersion without Deviation :-

$$\delta_y = -\delta_y'$$

$$A_1(\mu_1 - 1) = -A_2(\mu_2 - 1)$$

(emergent ray II^el incident ray)

Deviation without Dispersion :-

$$Q = -Q'$$

$$\delta_V - \delta_R = -[\delta_V' - \delta_R']$$

(White light bahar)

Total dispersion :- In case of dispersion without deviation

$$\theta = \theta_1 + \theta_2 = A[\mu - 1]\,(\omega_1 - \omega_2)$$

Silvering of Prism :-

$A = r_1$ $\quad$ $r_2 = 0$

$$\mu = \frac{\sin i}{\sin A}$$

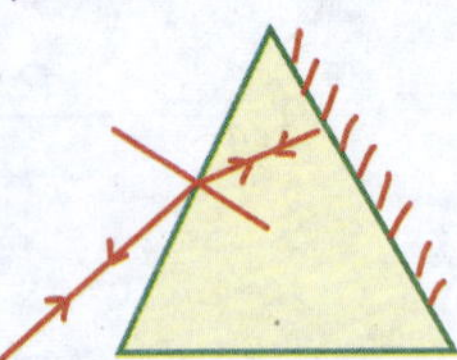

$\mu \propto 1/\lambda$

21 Optical Instruments

Visual Angle :-

Angle from horizontal with which a observer sees an object.

$$Q_o = \frac{H_o}{D}$$

Eye की औकात !

$D = 25\ cm$

Distance of Distinct Vision.

Myopia	Hypermetropia
Concave lens. Near Sightedness	Convex lens. Far-Sightedness

Simple Microscope

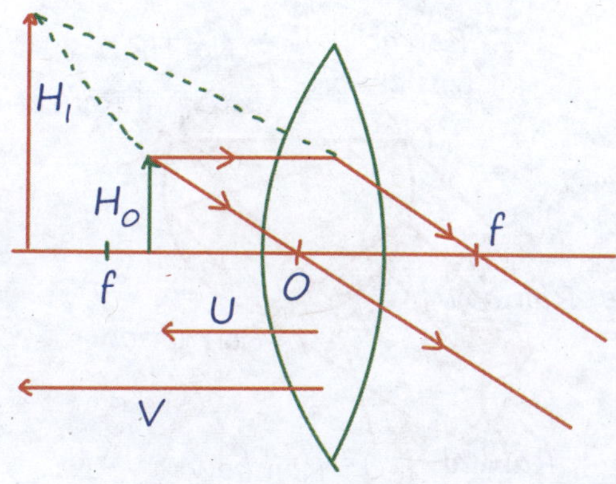

$$M = \frac{D}{U} = D\left[\frac{1}{f} + \frac{1}{V}\right]$$

I

- Strained eye position
- Near point
- $V = D$ (V: image distance; D: Distance of distinct vision)

$$M_{max} = 1 + \frac{D}{f}$$

II

- Relaxed eye position
- Far point
- $v = \infty$

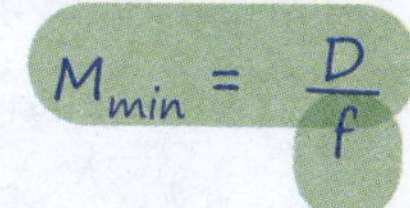

$$M_{min} = \frac{D}{f}$$

Compound Microscope :-

$f_o < f_e$

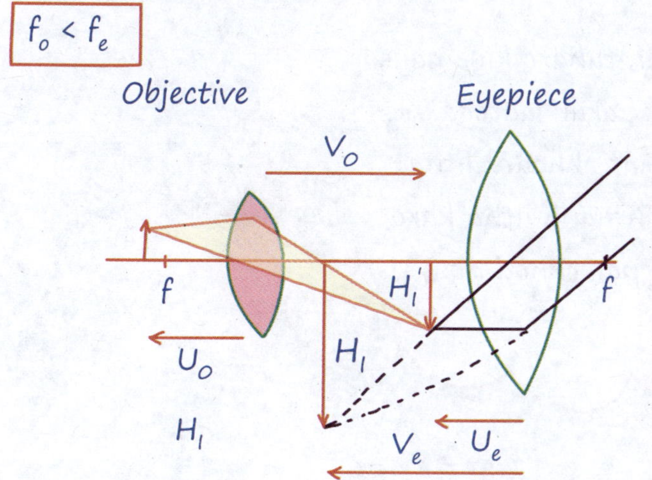

$$M = M_O M_e = \frac{V_O}{U_O}\,\frac{D}{U_e}$$

$$M = \frac{DV_O}{U_O}\left[\frac{1}{f_e} + \frac{1}{V_e}\right]$$

- Note :- $M = \frac{Q_I}{Q_O}$

I

- Strained eye position
- Near point
- $V = D$

$$M = \frac{V_O}{U_O}\left(1 + \frac{D}{f_e}\right)$$

II

- Relaxed eye position
- Far point
- $V = \infty$
- V_e = infinity

$u_e = f_e$

$$M = \frac{V_O D}{U_O f_e}$$

$$M = \frac{LD}{f_O f_e}$$

Length of Compound Microscope :-

$L = V_O + U_e \approx V_O$

In case of far point $L = V_O + f_e$

Astronomical Telescope

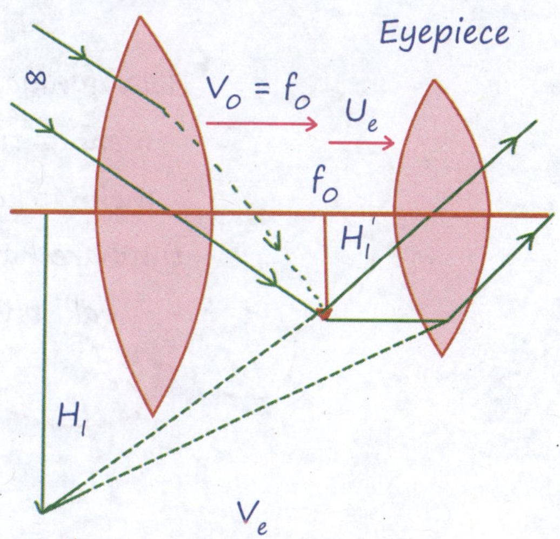

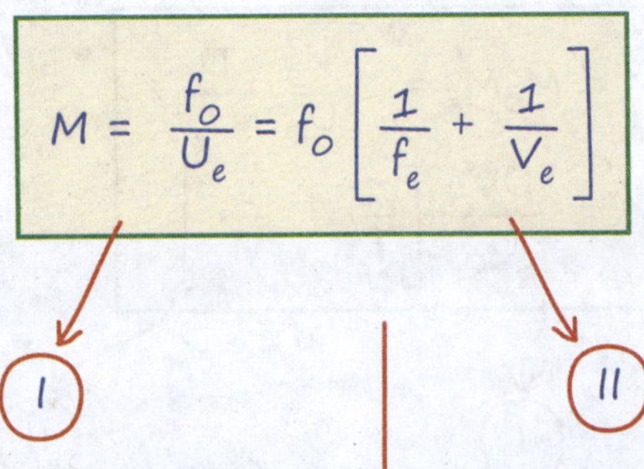

$$M = \frac{f_o}{U_e} = f_o\left[\frac{1}{f_e} + \frac{1}{V_e}\right]$$

I

- Strained eye position
- Near point
- $V = D$

$$M = f_o\left[\frac{1}{f_e} + \frac{1}{D}\right]$$

II

- Relaxed eye position.
- Far point
- $V = \infty$.

$$M = \frac{f_o}{f_e}$$

- **Length of Telescope :-**

$L = f_o + U_e$

$L = f_o + f_e$ → For image at ∞.

- **Galileo Telescope :-**

$L = f_o - f_e$

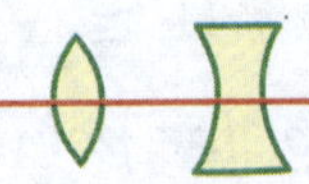

Rest Same.

22 Rainbow formation

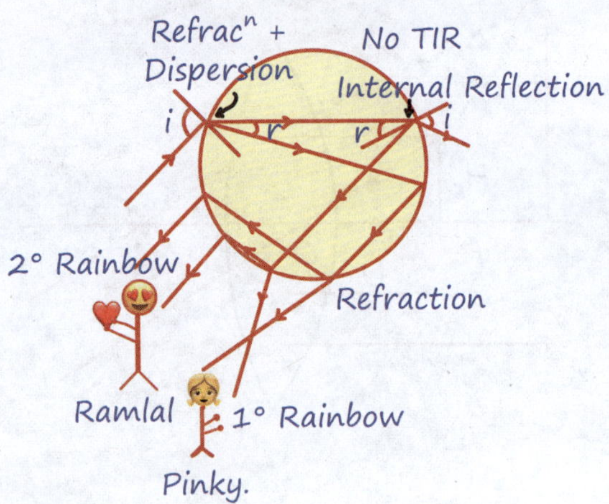

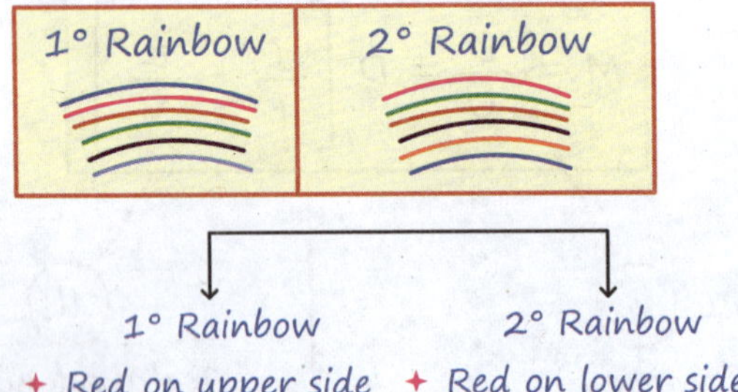

1° Rainbow	2° Rainbow
Red on upper side	Red on lower side
violet on lower	violet on upper

- **Scattering of Light :-**

E.g.:- Blue colour of Sky, Red Sky & sun on the time of sunrise or sunset.

Scatering $\propto \frac{1}{\lambda^4}$

MR*

"Beta duniya gol hai, tuhare kiye gaye mehnat tum tak zarur lautega.... mehnat karna mat chhodo...jitna tumhare hath main hai h utna karo, baki bhagwan par chhod do."

31 Wave Optics

(Mech + Longitudinal)

Huygen:- Wave

Einstein:- Particle } Light

Maxwell:- EM wave

(Non Mech + Transverse)

$$\frac{2\pi}{\phi} = \frac{\lambda}{\Delta x} = \frac{T}{\Delta t}$$

Equation of Wave for Light/ Sound / EM Wave / ac: -

$y = A\sin(\omega t + kx + \phi)$

Wave Sabhi Medium particle Ko SHM Deta hai!

1 Newton corpuscular theory

Explain:-

- Rectilinear propagation: Light travels in straight lines.
- Reflection: Bouncing of light from a surface.
- Refraction of Light: Bending of light when it passes from one medium to another.

Can't explain:-

- Interference: When two waves meet and combine.
- Diffraction: Bending of light around obstacles.
- Polarization: Orientation of light waves in particular directions.

2 Huygens wave theory

Explain:-

- Rectilinear propagation, Interference, Reflection, Refraction, Diffraction.

Can't explain:-

- Polarisation, PEE, Compton effect.

Types of waves:-

1. **Medium:-**

   Mechanical ✓ EMW ✗

2. **Propagation:-**

   Progressive ∞ Stationary Finite

3. **Vibration:-**

   Transverse ⊥ Longitudinal ||el

3 Wave front

A wavefront is a surface that connects all points vibrating in sync at the same phase of a wave. It helps visualize how waves propagate through different mediums.

Wavefront Type	Source of Light	Amplitude (A)	Intensity (I)
Spherical Wavefront	Point source	$A \propto \frac{1}{r}$	$I \propto (A)^2 \propto \frac{1}{r^2}$
Cylindrical Wavefront	Linear source	$A \propto \frac{1}{\sqrt{r}}$	$I \propto (A)^2 \propto \frac{1}{r}$
Plane Wavefront	Very distant source	$A \propto r^0$ (constant)	$I \propto (A)^2 \propto r^0$ (constant)

Intensity of wave:-

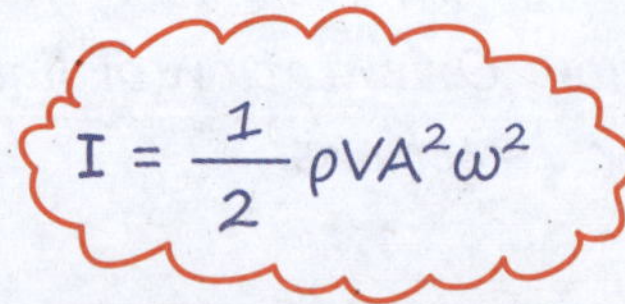

$$I = \frac{1}{2}\rho V A^2 \omega^2$$

4 Behaviour of plane wavefront on reflection & refraction

Plane Mirror:-

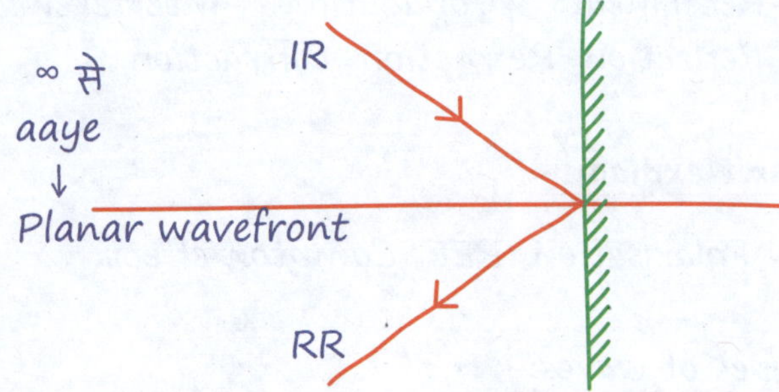

Convex Lens:-

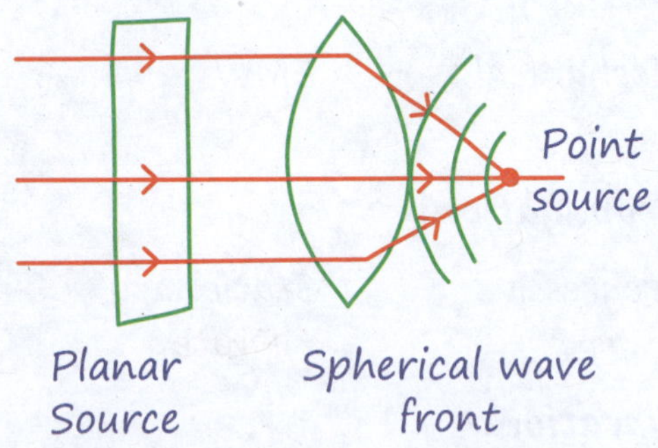

Concave lens:-

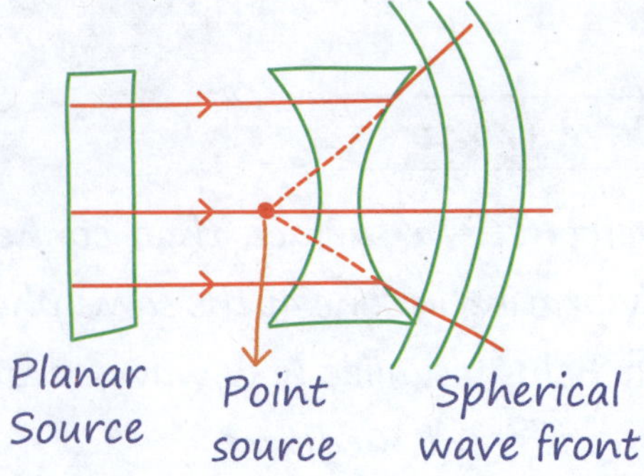

Prism:-

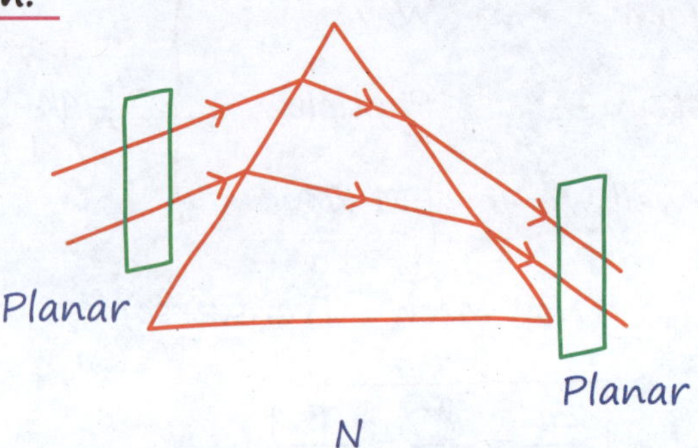

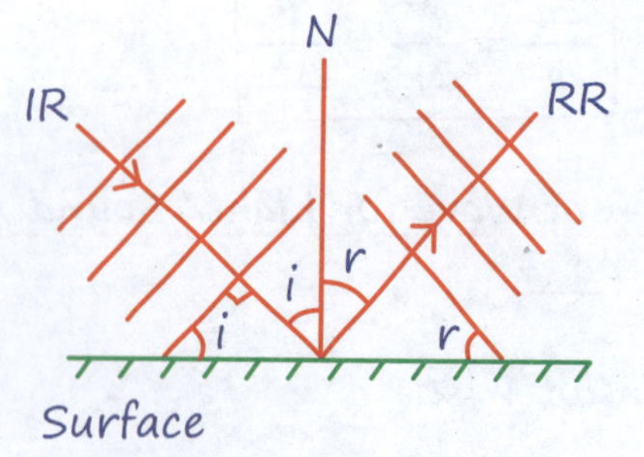

5 Sources

Phase difference

Time के साथ constant	Time के साथ Variable
Coherent	Incoherent
f_{req} = same	

Incoherent sources:-

$$\Delta\phi = [\omega_1 - \omega_2]t + [K_1x_1 - K_2x_2] + [\phi_1 - \phi_2]$$

Time Dependent Hai!

Coherent sources:-

$$\Delta\phi = K(x_1 - x_2] + \phi_1 - \phi_2$$

$$\Delta\phi = \frac{2\pi}{\lambda}.\Delta x. \quad \because \quad \phi_1 = \phi_2$$

Single wavelength | Monochromatic source

6 Interaction of light

- $\lambda << l_{object}$ = RAY :-

When the wavelength (λ) of light is much smaller than the size of the object it interacts with light can be treated as rays.

- $\lambda \approx l_{object}$ = WAVE

When the wavelength (λ) is comparable to the size of the object, light exhibits wave-like behavior.

- e−/p+/α-particle = PARTICLE

Light also behaves as particles (photons), particularly in interactions at the atomic or subatomic level, such as in the photoelectric effect or Compton scattering.
(Photon)

7 Interference

Interference is a phenomenon where two or more coherent light waves superimpose to form a resultant wave of greater, lower, or the same amplitude.

Requirements:

- For interference to occur, two coherent sources (sources with constant phase difference) are required.

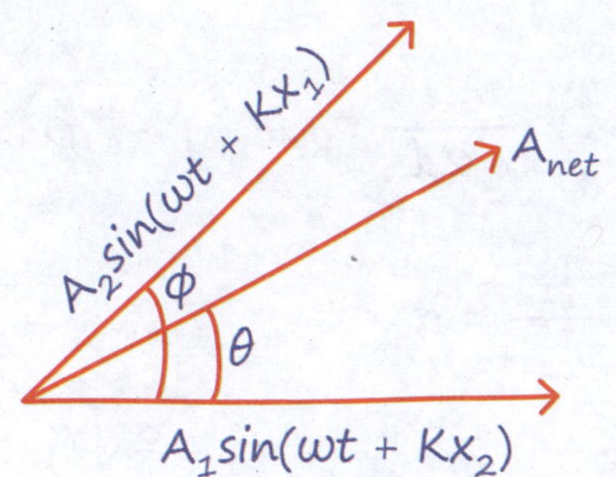

Resultant amplitude

$$A_{net} = \sqrt{A_1^2 + A_2^2 + 2A_1A_2\cos\phi}$$

Resultant intensity

$$I_{net} = I_1 + I_2 + 2\sqrt{I_1 I_2}\cos\phi$$

Phase Relation and angle (θ)

$$\tan\theta = \frac{A_2\sin\phi}{A_1 + A_2\cos\phi}$$

Fire concept MR*

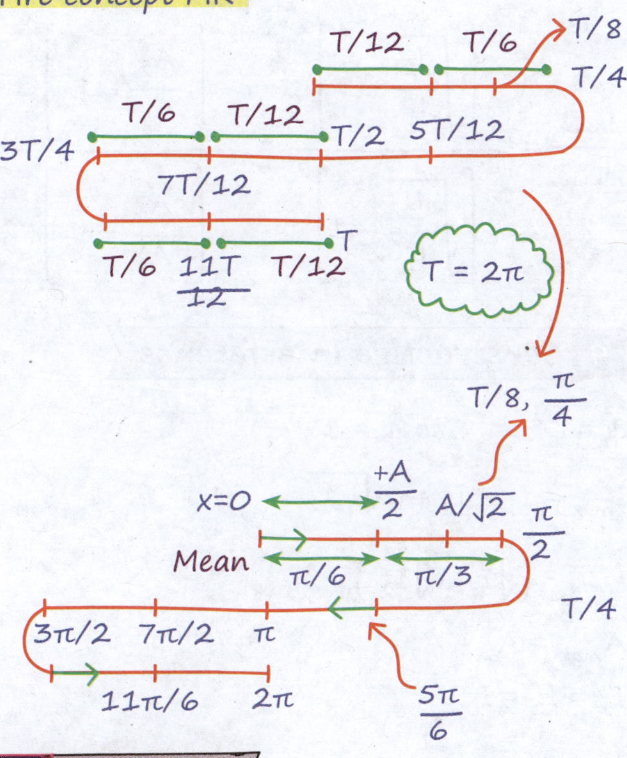

8 Intensity

$$I = \frac{1}{2}\rho V A^2 \omega^2$$

Where:

ρ = density of the medium

V = velocity of the wave

A = amplitude of the wave

ω = angular frequency of the wave

$I \propto d \propto A^2$ d = slit width

Average Intensity

$$I_{av} = \frac{I_{min} + I_{max}}{2}$$

Ratio of Maximum to minimum Intensity

$$\frac{I_{max}}{I_{min}} = \frac{\left(\sqrt{I_2}+\sqrt{I_1}\right)^2}{\left(\sqrt{I_2}-\sqrt{I_1}\right)^2} = \frac{\left(A_2 + A_1\right)^2}{\left(A_2 - A_1\right)^2}$$

$$\frac{I_{max}}{I_{min}} = \frac{\left[\sqrt{\frac{I_2}{I_1}}+1\right]^2}{\left[\sqrt{\frac{I_2}{I_1}}-1\right]^2} = \frac{\left[\frac{A_2}{A_1}+1\right]^2}{\left[\frac{A_2}{A_1}-1\right]^2}$$

9 Constructive Interference

$\phi = 0°$ $\cos\phi = 1$

$I_{max} = I_1 + I_2 + 2\sqrt{I_1 I_2}$

$I_{max} = \left(\sqrt{I_1}+\sqrt{I_2}\right)^2$

$A_{max} = A_1 + A_2$

$I_1 = I_2 = I_o$

$I_{max} = 4I_o$

$\phi = 2n(\pi) \rightarrow$ even

$\Delta x = n(\lambda) \rightarrow$ integral

$n = 0, 1, 2, 3, ...$

MR*

$\lambda = 2\pi$

- **YDSE:-**

$$Y = n\left(\frac{\lambda D}{d}\right)$$ Posit^n of nth Bright.

- **NOTE:-**

1> "n" source of same intensity I_o.
Find I_{Result}n:-
$I_{net} = nI_o$ (scalar Add^n)

2> "n" source of same intensity I_o then find I_{max}:- (Const. Inter.)
$*I_{max} = n^2 I_o = (n\sqrt{I_o})^2$

Fringe visibility:-

$$\text{Visibility Ratio} = \frac{I_{max} - I_{min}}{I_{max} + I_{min}}.$$

$$= \frac{\left(\sqrt{\frac{I_1}{I_2}}+1\right)^2 - \left(\sqrt{\frac{I_1}{I_2}}-1\right)^2}{\left(\sqrt{\frac{I_1}{I_2}}+1\right)^2 + \left(\sqrt{\frac{I_1}{I_2}}-1\right)^2}$$

10 Destructive interference

$\phi = 180°$ $\cos\phi = -1$

$I_{min} = I_1 + I_2 - 2\sqrt{I_1 I_2}$

$I_{min} = \left(\sqrt{I_1}-\sqrt{I_2}\right)^2$

$A_{min} = A_1 - A_2$

$I_1 = I_2 = I_o$

$I_{max} = 0, I_{min} = 0$

$\phi = (2n+1)\pi \rightarrow$ odd

$\Delta x = (2n+1)\frac{\lambda}{2} \rightarrow$ odd

$n = 0, 1, 2, 3, ...$

- **YDSE:-**

$$Y = (2n-1)\frac{\lambda D}{2d}$$ Posit^n of nth Dark.

IIT Neet 2016

$$\frac{I_1}{I_2} = \alpha$$

$$\text{Visibility Ratio} = \frac{2\sqrt{\alpha}}{1+\alpha}$$

☼ Two wave of "I_o" intensity I_{res}. If phase diff is ϕ:-

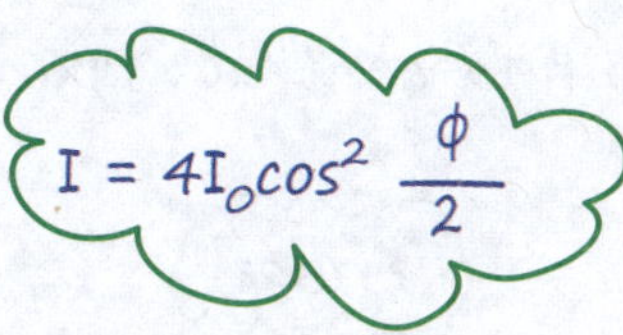

MR*

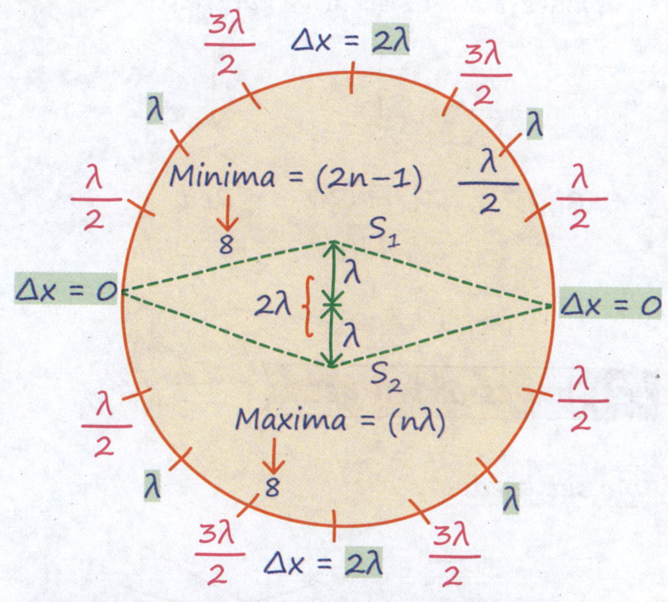

11 Young Double slit Experiment

Basic Setup :-

- S1 and S2 are two slits separated by a distance d.
- A screen is placed at a distance D from the slits.
- Light waves passing through the slits interfere and create a pattern of bright and dark fringes on the screen.

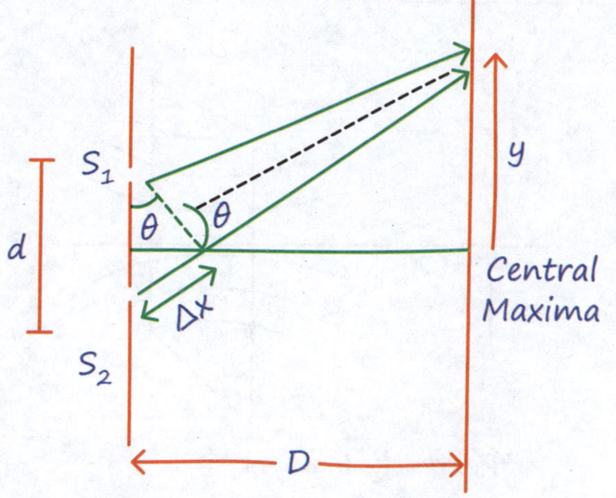

Path Difference (Δx) :-

$d \simeq \lambda$ $\quad \Delta x = d\sin\theta$

$= d\tan\theta$

$$\Delta x = \frac{Yd}{D}$$

Positions of Bright and Dark Fringes :-

- The position of the n-th bright fringe (nB) on the screen is given by:

$$y_n = \frac{n\lambda D}{d}$$

- The position of the n-th dark fringe is given by:

$$y_n = \frac{(n+\frac{1}{2})\lambda D}{d}$$

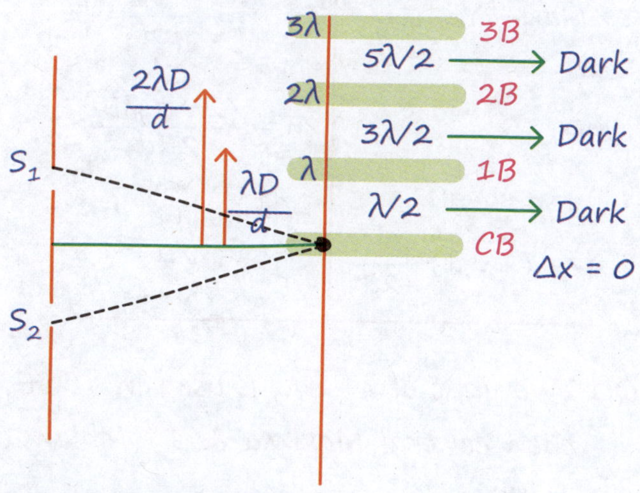

Fringe Width (β):-

$\beta = Yn^{th}$ Bright $- Y(n-1)^{th}$ Bright

$$\beta = \frac{\lambda D}{d}$$

Angular Fringe Width:-

$\theta = \frac{\beta}{D}$ $\quad \theta = \frac{\lambda}{d}$

MR*

☼ Angular kuch bhi puche aap "D" se divide kardena!

☼ $\mu_{air} > \mu_{vacuum}$

$\beta \propto \frac{1}{\mu}$ ∴ In vacuum, interference with large "β" seen!

YDSE in air	YDSE in liq.
1> $\beta = \frac{\lambda D}{d}$	1> $\beta' = \frac{\beta}{\mu}$
2> $\theta = \frac{\lambda}{d}$	2> $\theta' = \frac{\theta}{\mu}$
3> Max:- $Y_n = \frac{n\lambda D}{d}$	3> $Y_n' = \frac{n\lambda D}{\mu d}$
Min:- $Y_n = (2n-1)\frac{\lambda D}{2d}$	$Y_n' = \frac{(2n-1)\lambda D}{2\mu d}$
	f = same $\lambda' = \lambda/\mu$

Q.1 Two light of $\lambda_1 > \lambda_2$ is used in YDSE, then central Maxima & 1st Maxima will 2.

Sol. Central Maxima ⇒ At same position.

☼ 1st Maxima:- Not at same place.

↓

Jiska λ↑ Y↑ $\because Y = \frac{n\lambda D}{d}$

$\therefore \lambda_1$ = dur rahega!

White light:-

☼ sabse pahle maxima = violet ka ayega.

☼ sabse pahle dikhega = red.

☼ Used to find central maxima.

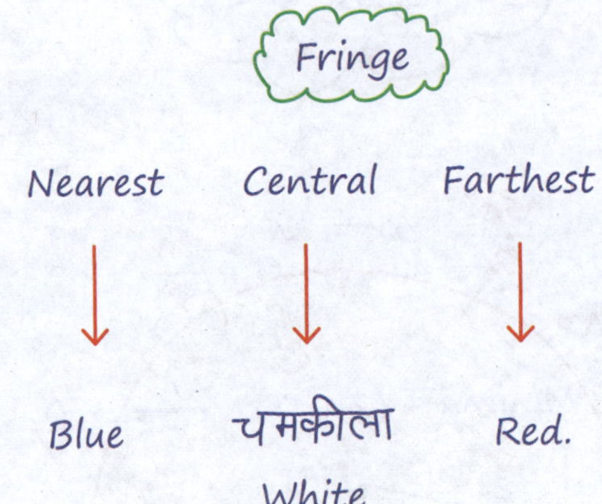

12 Shapes of fringes

Two slit used:-

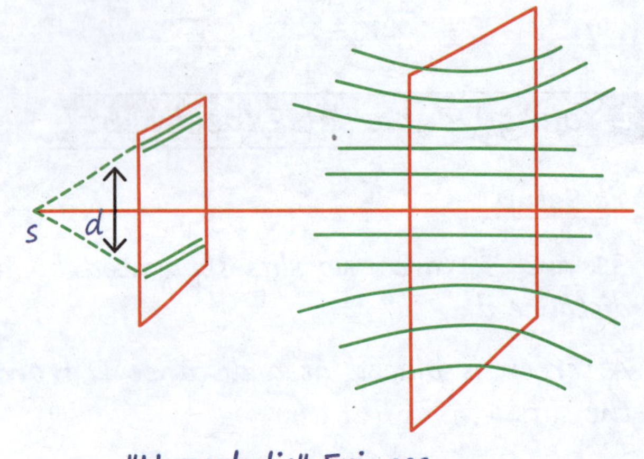

"Hyperbolic" Fringes

Two pin holed used:-

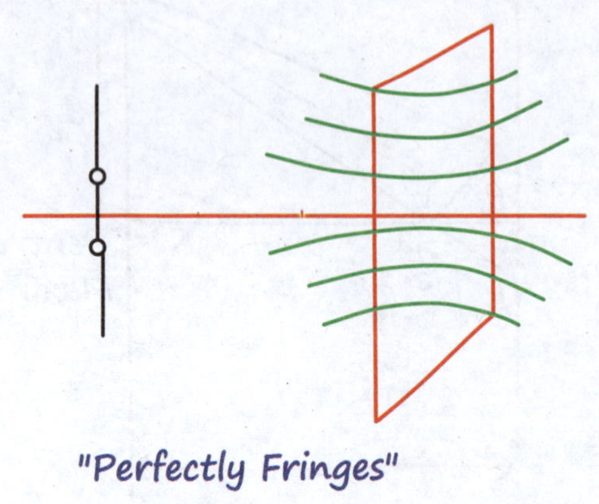

"Perfectly Fringes"

When two hole is along the line joining of source and screen:-

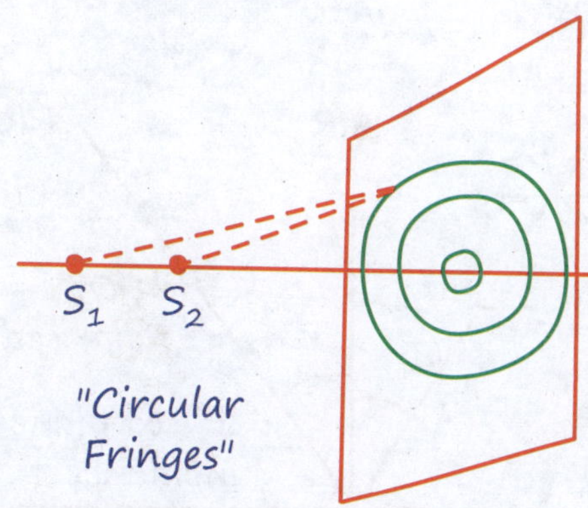

13 Optical path

air:- $\Delta\phi = \frac{2\pi}{\lambda} t$

Med^m:- $\lambda' = \frac{\lambda}{\mu}$ $\Delta\phi' = \frac{2\pi}{\lambda}(\mu t)$

YDSE when a slab of "μ" R.I. inserted in path of S_1:-

$d\sin\theta = t(\mu-1)$

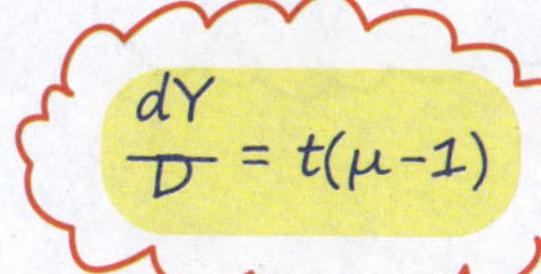

$\beta = \frac{\lambda D}{d}$

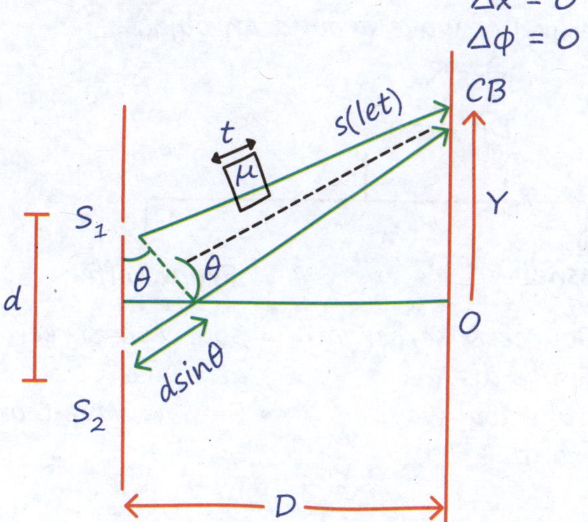

MR*

Jiske path mein slab jayega woh "μt" jyada chalega agar $\Delta x = 0$ krna hai toh usko niche lao yaneki kam karo toh S_2 ke path ko badao dono barabar hojayenge aur $\Delta x = 0$!

Position of CB:-

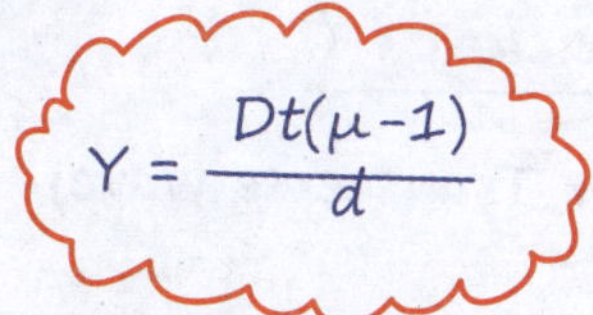

Kuch nai Normal YDSE ka

$\lambda = (\mu-1)t$ rakhdo!

Number of bright & dark fringes:-

Total No. of:-

(a) Bright Fringe = $\left[\frac{2d}{\lambda} + 1\right]$

(b) Dark Fringe = $2\left[\frac{d}{\lambda} + \frac{1}{2}\right]$

$\frac{d}{\lambda}$ = minimum integer } 2.5=2 2.9=2 3.6=3

Majduri se duri MR hai jaruri

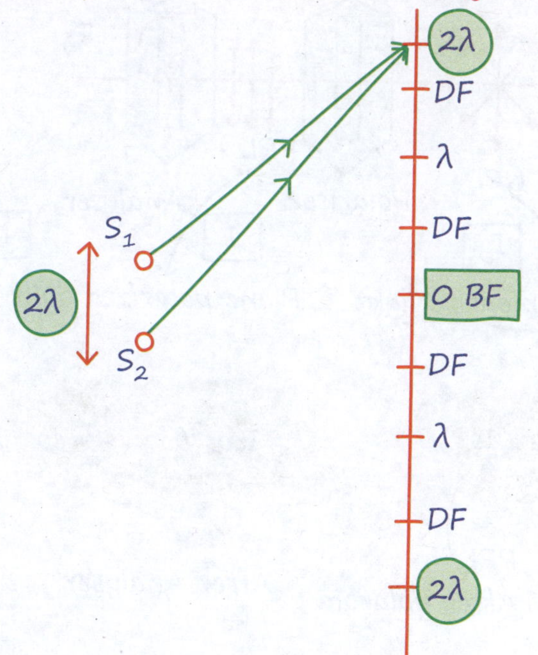

$$\text{No. of BF} = \frac{2(2\lambda)}{\lambda} + 1 = 5$$

$$\text{No. of DF} = 2\left[\frac{d}{\lambda} + \frac{1}{2}\right] \leftarrow \text{Integer}$$

$$= 2\,(2+0.5) = 2\,[2] = 4$$

14 Polarisation

(Only for Transverse wave)

E = cB → negligible compared to E.F

∴ Restricting of $\vec{E}$ = Polarisn.

Law of malus:-

I_o = Intensity of unpolarised light.

θ = Angle b/n Analyser & Polarizer.

I = Intensity of light transmitted through the polariser.

$*I = I_o \cos^2\theta.*$

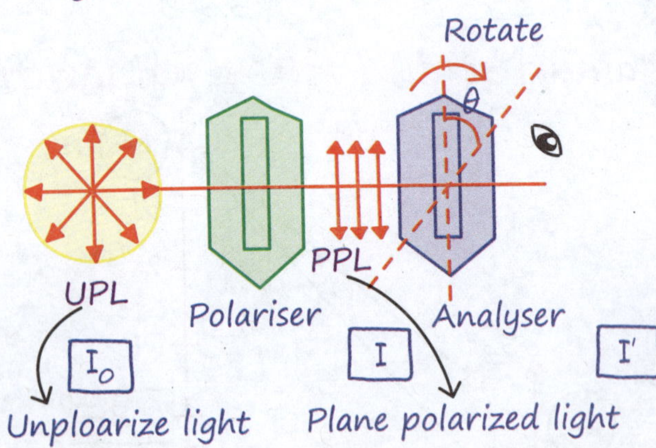

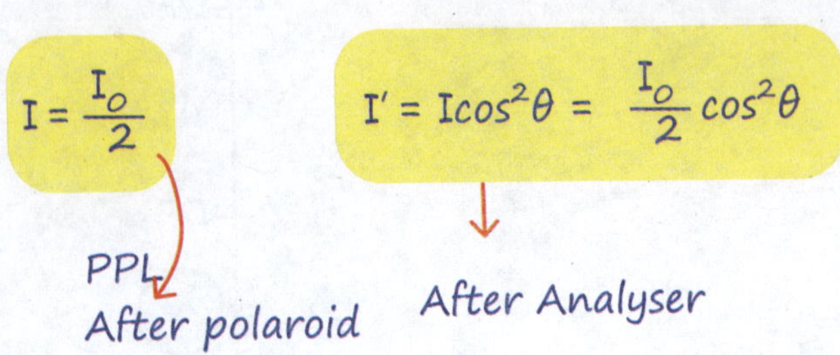

15 Brewster's Law (Polarisation by reflection)

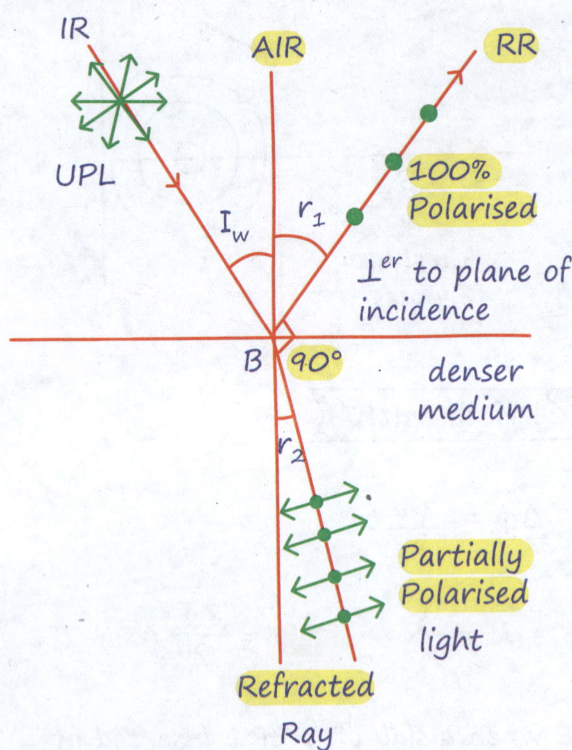

16 Diffraction

Bending of wave around an object.

जोरदार :- Diffracn $s \approx \lambda_{wave}$

Fresnel

- Source & screen at Finite distn
- Spherical wave front

Fraunhoffer

- Source & screen at ∞ distn
- Planar wave front

Fraunhoffer Diffraction:-

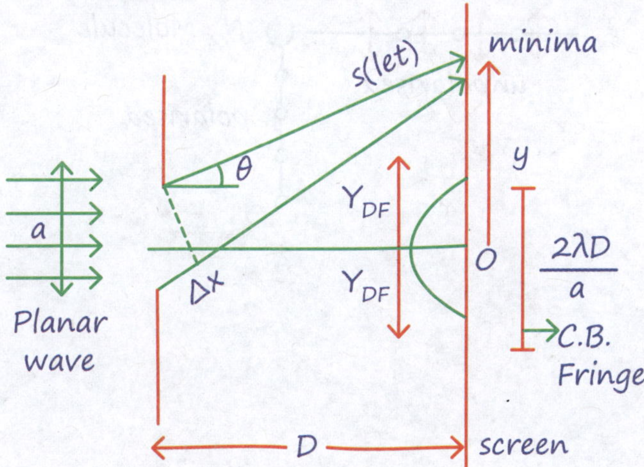

उलटा है YDSE का!

Neet

☼ Width of C.B. Fringe = $\frac{2\lambda D}{a}$

☼ Angular width of C.B.F. = $\frac{2\lambda}{a}$

Minima (D.I)	(C.I) Maxima
$a\sin\theta = n\lambda$	$a\sin\theta = (2n-1)\frac{\lambda}{2}$
$n = 1,2,3,4...$	$n = 2,3,4...$

$$\tan\theta = \frac{Y}{D}$$

☼ $a\sin\theta \approx a\tan\theta$

17 Validity of ray-optics

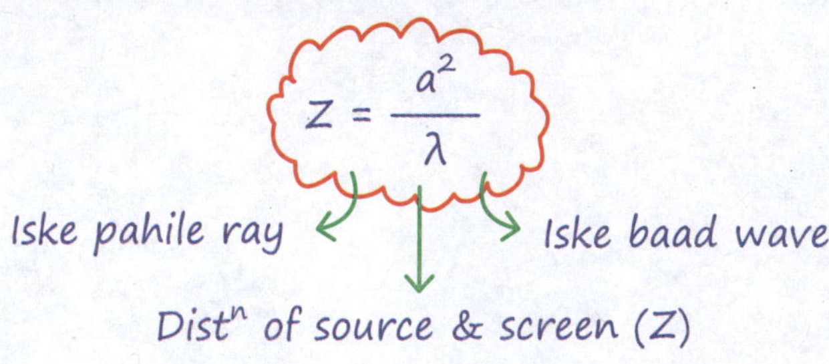

a = width of C.B. Fringe.

Important points:-

- 1 inch = 2.54 cm
- For max R.L. we use max λ.
- $\lambda_w = \frac{\lambda_{air}}{\mu}$
- * $d\theta_w = \frac{d\theta_a}{\mu}$ *
- If two coherent source of equal intensity are taken then fringe visibility is 100%. because $I_{min} = 0$.

18 Doppler's effect of light

$$r' = r\left[\frac{1 \pm \frac{V_r}{C}}{\sqrt{1-\left(\frac{V_r}{C}\right)^2}}\right] \Rightarrow \lambda' = \frac{\lambda}{1+\frac{V_r}{C}}$$

$$\frac{\Delta r}{r} = \frac{\Delta\lambda}{\lambda} = \pm\frac{V_r}{C}$$

19 Interference

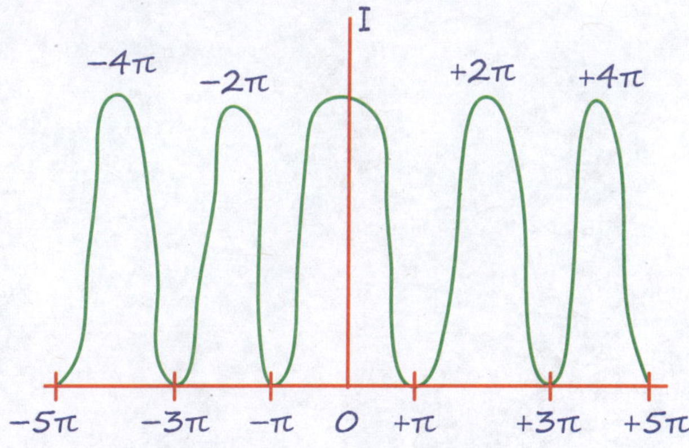

Interference due to thin filim.

- For reflected light:

Maxima – $2\mu t \cos r = (2n+1)\frac{\lambda}{2}$

Mininima – $2\mu t \cos r = n\lambda$

- For transmitted light:

Maxima – $2\mu t \cos r = n\lambda$

Minima – $2\mu t \cos r = (2n-1)\frac{\lambda}{2}$

20 Diffraction

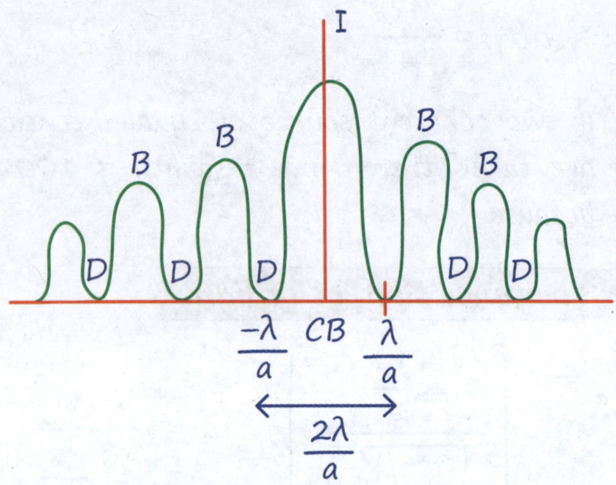

Polarisation by scattering

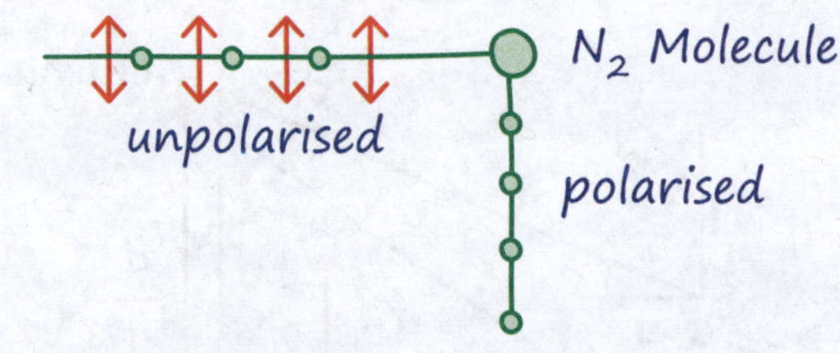

Dual Nature

1 Dual Nature of light:

Newton (Baba) → Light is a Particle (corpuscule)
→ Explain reflection / refraction

Huygen (Chacha) → Light is a mechanical wave
→ Ether medium Propsed by Huygen
→ Explain Diffraction / refraction.

Maxwell (Chacha) → Light is a Non-mechanical transverse wave.
→ No medium required.

de Broglie (Bhaiya) → Nature loves Symmetry
→ Light have dual Nature.

Davission & Geomer → e^- is a Wave exp^m verification

G.P. thomson → verification of electron as wave.

2 Photon

$$E = h\nu = \frac{hc}{\lambda} = \frac{2\times10^{-25}J}{\lambda} = \frac{12400\ eV}{\lambda(A^\circ)}$$

$$v_{photon} = \frac{1}{\sqrt{\mu_0\varepsilon_0}},$$ independent of frequency

$$(v_{photon})_{Med.} = \frac{c}{\sqrt{\mu_r\varepsilon_r}}$$

Energy of light beam:

$$E = nhf = \frac{nhc}{\lambda}$$ n= no. of Photons

$$1eV = 1.6\times10^{-19}\ joule$$

Properties of Photon:

- charge and rest mass =0
- Momentum of Photon $P = \frac{h}{\lambda}$
- Moving mass of photon $m = \frac{h}{\lambda C} = \frac{E}{c^2}$
- Photon does not deviate in electromagnetic field but deviate in gravitational field.

Power of light:

Total energy falling per second is called power of light beam.

$$P = \frac{E}{t} = \frac{nh\nu}{t}$$

No. of photons emitted per unit time $= \frac{n}{t}$

$$= \frac{P\lambda}{hc}$$

Intensity of light:

Total energy falling per unit area per-sec is called intensity.

$$I = \frac{nE}{At} = \frac{nh\nu}{At} = \frac{nhc}{\lambda At} = \frac{Power}{Area}$$

If source is same, frequency is same.

$I \propto$ no. of photons.

Q.1 If Intensity and frequency both becomes double then no. of Photon will be??

Ans. n = remains same

$I \propto nf$

Q.2 For a given source. Intensity becomes double then no. of Photon will?

Ans. becomes double $I \propto n$

Fractional change in frequency of photon when it travels "h" distance on earth surface:

$$\frac{\Delta\nu}{\nu} = \frac{gh}{c^2} \qquad h\nu_i + \frac{h\nu_i}{c^2}gh = h\nu_2 + 0$$

Radiation pressure:

✦ $P = \frac{I}{C}[2-\rho]$

*Complete Absorption : (ρ = 1)

$P = \frac{I}{C}$

$F = PA = \frac{IA}{C} = \frac{Power}{C}$

✦ $P = \frac{I}{C}[\sigma+1]$

*Complete Reflective : σ = 1

$P = \frac{2I}{C}$

$F = \frac{2IA}{C} = \frac{2Power}{C}$

Where,
I = Intensity of light Beam
c = Speed of light
P = Pressure
ρ = Absorption coefficient
σ = Reflection coefficient

3 Photoelectric effects:

✦ Conversion of light energy to electrical energy.

✦ Explained by Einstein

✦ Instantaneous process; time lag b/w falling of photon and emiting of electron is 10^{-9} sec.

✦ Efficiency = $\frac{\text{no of electron}}{\text{no of photon}} = 10^{-3}$ to 10^{-4}.

✦ One to one intraction one photon can eject only one electron.

✦ Work Function, ϕ = Material Properties $= h\nu_0 = \frac{hc}{\lambda_0}$

where ν_0 and λ_0 are thereshold frequency and wavelength.

✦ For Photoelectric effect :

$\nu_{light} \geq \nu_0$ [Jyada hogi]

$\lambda_{light} \leq \lambda_0$ [kam hogi]

$(KE)_{max} = eV_0$

V_0 = Stopping Potential

✦ $E_{photon} = \phi + (KE)_{max}$

$h\nu = h\nu_0 + eV_0$

$\frac{hc}{\lambda} = \frac{hc}{\lambda_0} + \frac{1}{2}MeV^2_{max}$

MR*

Intensity ∝ no. of Photon ∝ no of electron ∝ Photocurrent

Frequency ∝ energy of Photon ∝ energy of e^- (K.E) ∝ stopping potential

✦ Photocurrent depends on Intensity not on frequency.

✦ Stopping Potential depends on frequency not on Intensity

Graphs:

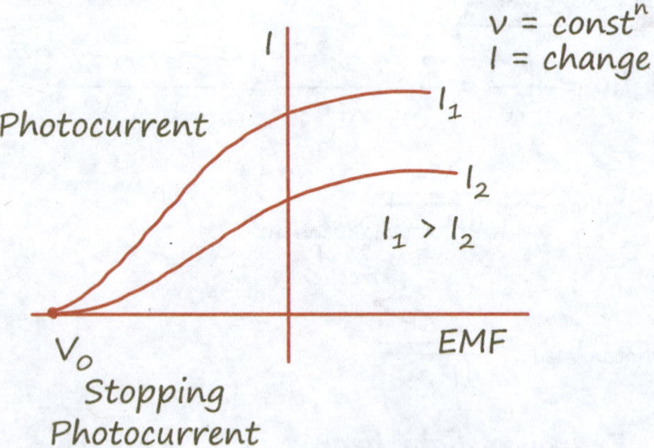

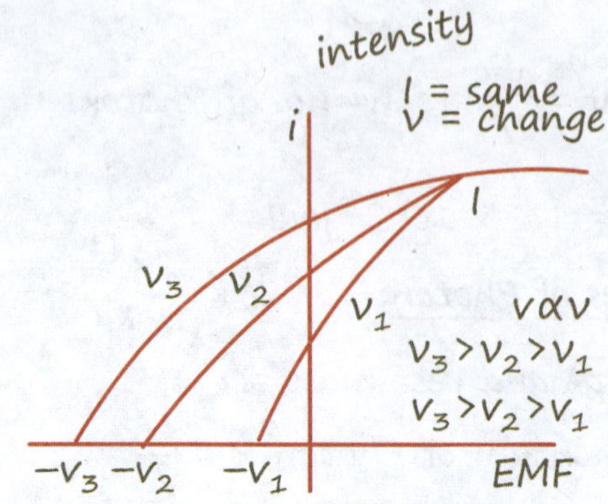

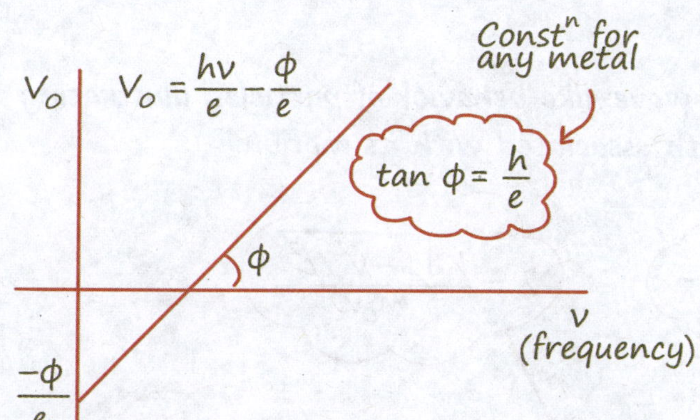

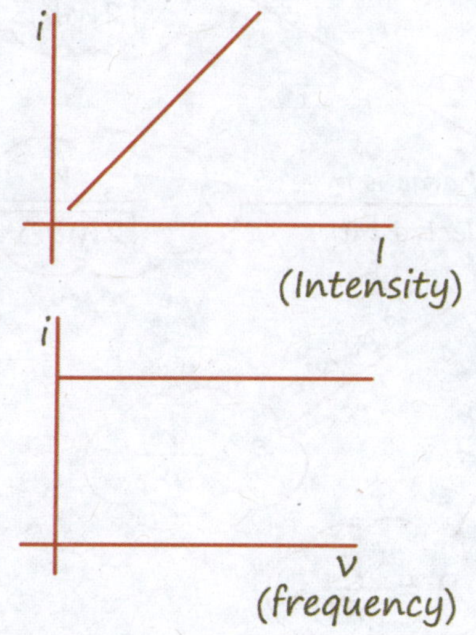

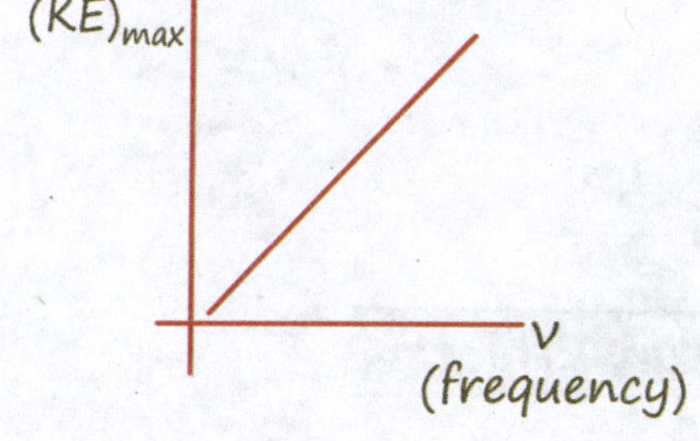

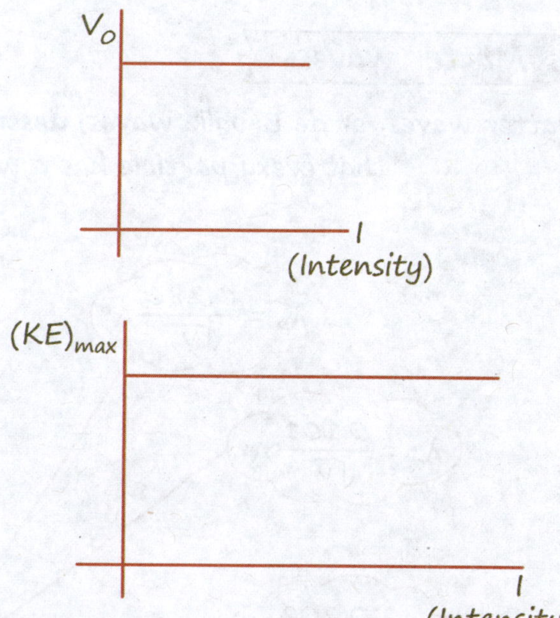

- If distance b/w Source and plate becomes double then stopping potential remains same but photocurrent becomes one fourth.
- If frequency becomes double then K.E of electrons becomes more than double.

Q.3 Saturation Current and stopping Potential are I_0 and V_0 when frequency of light is 1.5 times of threshold frequency. Now if frequency becomes half then saturation current and stopping potential will?

Sol. Both becomes zero, Photoelectric effect hoga hi Nahi

4 Matter waves:

Matter waves, or de Broglie waves, describe the wave-like behavior of particles. Illustrating that every particle has a wavelength associated with its motion.

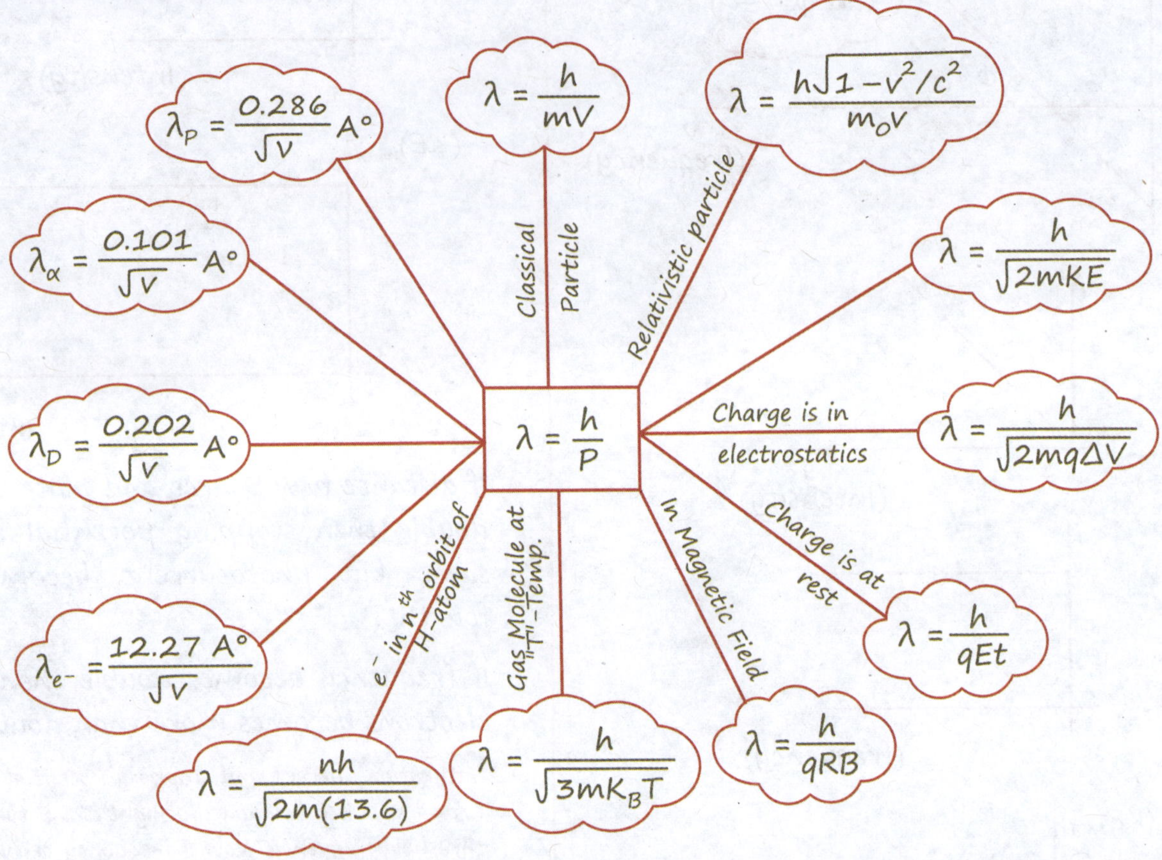

V. imp. λ v/s □ = Always Reactangular hyperbola : $K_B = 1.38 \times 10^{-28}$

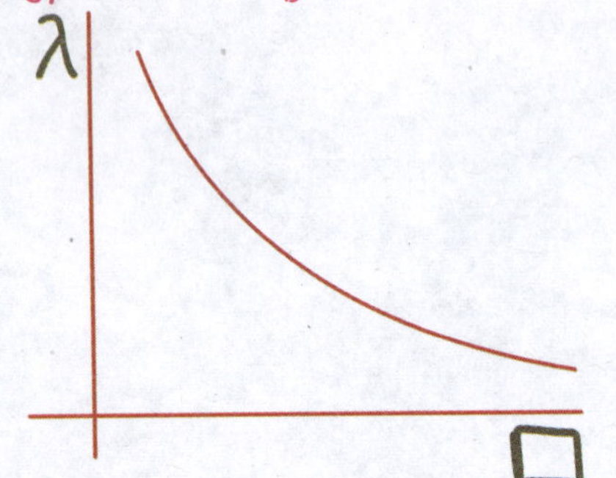

Wavelength of revolving e^- for Bohr's orbit:

$n\lambda = 2\pi r$

5 Compton Effect:

E = 2ev Visible Region

Photoelectric Effect:

E = 10eV X-ray Region

$\Delta\lambda = \lambda_2 - \lambda_1 = \frac{h}{m_e C}(1-\cos\phi)$, ϕ = scattering angle

$\frac{h}{m_e C} = 2.44$ pm

Pair Production: E = 1.02 MeV

6 X-ray:

(Neutral) $E = KeV$

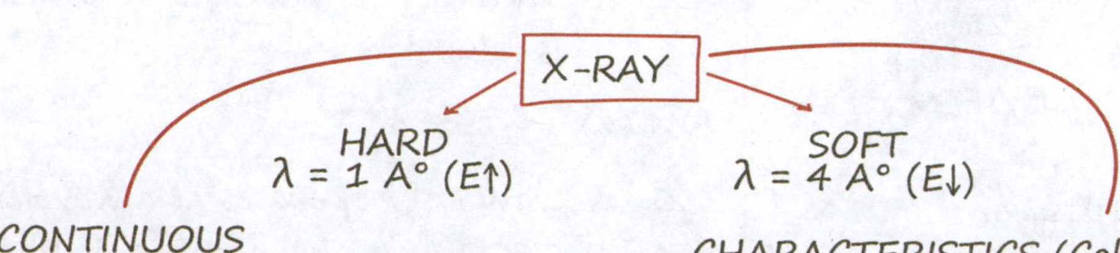

CONTINUOUS
(Retard^n of e^-)

CHARACTERISTICS (Collision b/n accelerated e^- and bonded e^-)

7 Inverse Phenomena of Photoectric effect

X-ray target atom → High atomic number, High melting Point, High Conductivity

Intensity of X-ray ∝ Tube Current	Frequency of X-ray ∝ Tube Voltage

X-RAY

CONTINUOUS
(Retard^n of e^-)

CHARACTERISTICS
(Collision b/n accelerated e^- and bonded e^-)

$$(KE)_{e^-} = \frac{hc}{\lambda_{cut\ off}}$$

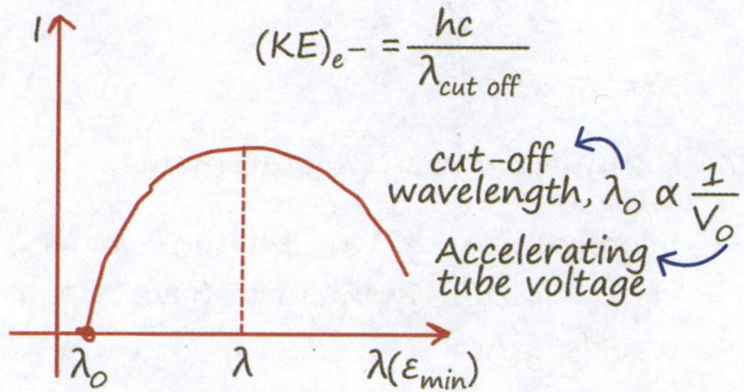

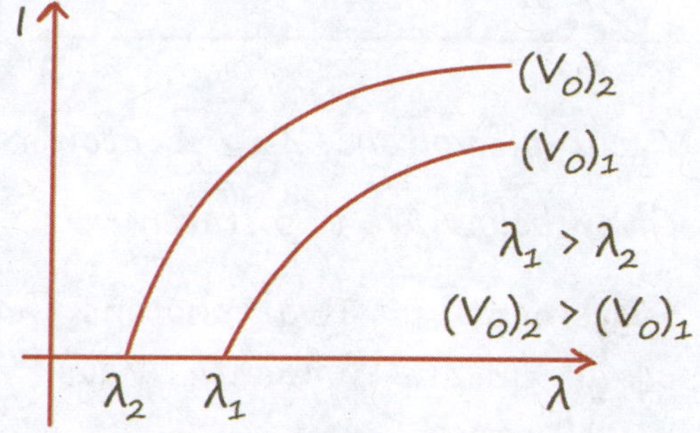

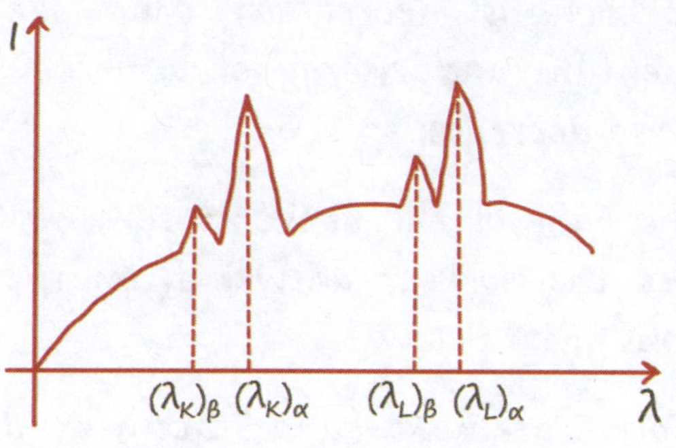

For K-Series:
$f = K^2(z-1)^2$

Q.4 Wavelength is λ for K_α line for atomic no $z = 43$ then find wavelength for K_α line for atomic no $z = 29$.

Sol. $\dfrac{\lambda}{\lambda^1} = \left(\dfrac{29-1}{43-1}\right)^2$, $\dfrac{\lambda}{\lambda^1} = \left(\dfrac{28}{42}\right)^2 = \left(\dfrac{2}{3}\right)^2$

$\dfrac{\lambda}{\lambda^1} = \dfrac{4}{9}$ $\quad \lambda^1 = \dfrac{9}{4}\lambda$

8 Heisenberg uncertainty principle:

$\Delta x . \Delta p \geq \frac{h}{4\pi}$ $\hbar = \frac{h}{2\pi}$

$\Delta L . \Delta Q \geq \frac{\hbar}{2}$ $\Delta t . \Delta E \geq \frac{\hbar}{2}$

Where,

Δx = uncertainity in position

Δp = uncertainity in momentum

ΔL = uncertainity in angular momentum

✦ Graph is given for two different atom A and of atomic no. Z_A and Z_B at cut-off voltage V_A and V_B

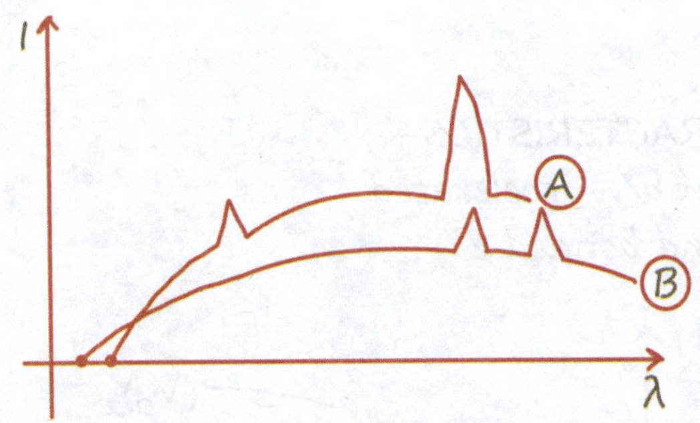

$V_A < V_B$ cut-off voltage, $Z_A > Z_B$ atomic no.

Q.5. Given below are two statements:

Statements-I: Two photons having equal linear momenta have equal wavelengths.

Statements-II: If the wavelength of photon is decreased, then the momentum and energy of a photon will also decrease.

In the light of the above statements, choose the correct answer from the options given below.

(a) Both Statement-I and Statement-II are fasle

(b) Statement-I is false but Statement-II is true

(c) Both Statement-I and Statement-II are true

(d) Statement-I is true but Statement-II is false.

Ans. (d)

Q.6. A 2mW laser operates at a wavelength of 1500 nm. The number of photons that will be emitted per second is

[Given Planck's constant $h = 6.6 \times 10^{-34}$ Js, speed of light $c = 3.0 \times 10^8$ m/s]

(a) 2×10^6 (b) 1.5×10^{16}

(c) 5×10^{15} (d) 1×10^{16}

Ans. (d) $P = 2mW = 2 \times 10^{-3}$ w $= \frac{nhc}{\lambda t}$

$$\frac{n}{t} = \frac{2\times10^{-3}\times\lambda}{hc}$$

$$= \frac{2\times10^{-3}\times500\times10^{-9}}{2\times10^{-25}}$$

$$= 5 \times 10^{15}$$

Q.7. Given below are two statements:

Statements-I: Two photons having equal linear momenta have equal wavelengths.

Statements-II: If the wavelength of photon is decreased, then the momentum and energy of a photon will also decrease.

In the light of the above statements, choose the correct answer from the options given below.

(a) Both Statement-I and Statement-II are false

(b) Statement-I is false but Statement-II is true

(c) Both statement-I and Statement-II are true

(d) Statement-I is true but statement-II is false.

Ans. (d) $P\uparrow = \frac{h}{\lambda\downarrow}$ & $E\uparrow = \frac{hc}{\lambda\downarrow}$, $\lambda = \frac{h}{P}$

Q.8. A 2 mW laser operates at a wavelength of 500 nm. The number of photons that will be emitted per second is

[Given Planck's constant $h = 6.6 \times 10^{-34}$ Js, speed of light $c = 3.0 \times 10^{8}$ m/s]

(a) 2×10^{6} (b) 1.5×10^{16}

(c) 5×10^{15} (d) 1×10^{16}

Ans. (c) $P = 2mW = 2 \times 10^{-3}\ w = \frac{nhc}{\lambda t}$

$$\frac{n}{t} = \frac{2 \times 10^{-3}\lambda}{hc}$$

$$= \frac{2 \times 10^{-3} \times 500 \times 10^{-9}}{2 \times 10^{-25}}$$

5×10^{15}

Q.9. From the photoelectric effect experiment, following observations are made. Identify which of these are correct

A. The stopping potential depends only on the work function of the metal.

B. The saturation current increases as the intensity of incident light increases.

C. The maximum kinetic energy of a photo electron depends on the intensity of the incident light.

D. Photoelectric effect can be explained using wave theory of light.

Choose the correct answer from the options given below:

(a) B, C only (b) A, C, D only

(c) B only (d) A, B, D only

Ans. (c) $E_p = \phi + eV_0$

Q.10. Given below are two statements:

Statements-I: Stopping potential in photoelectric effect does not depend on the power of the light source.

Statements-II: For a given metal, the maximum kinetic energy of the photoelectron depends on the wavelength of the incident light.

(a) Statement-I is incorrect but Statement-II is correct

(a) Both Statement-I and Statement-II are incorrect

(c) Statement-I is correct but Statement-II is incorrect

(d) Both Statement-I and Statement-II are correct

Sol. (d)

$$P = \frac{n\,hf}{t}$$

$$I = \frac{n\,hf}{At}$$

Stopping potential,

$$V_s = \frac{\frac{hc}{\lambda} - \phi}{e}$$

$$= \frac{(K.E)_{max} - \phi}{e}$$

Q.11. In photo electric effect

A. The photocurrent is proportional to the intensity of the incident radiation

B. Maximum Kinetic energy with which photoelectrons are emitted depends on the intensity of incident light.

C. Max. K.E with which photoelectrons are emitted depends on the frequency of incident light.

D. The emission of photoelectrons require a minimum threshold intensity of incident

E. Max. K.E of the photoelectrons is independent of the frequency of the incident light Choose the correct answer from the options given below:

(a) A and C only (b) A and E only

(c) B and C only (d) A and B only

Ans. (a)

Q.12. The threshold frequency of metal is f_o When the light of frequency $2f_o$ is incident on the metal plate, the maximum velocity of photoelectron is v_1 When the frequency of incident radiation is increased to $5f_o$ The maximum velocity of photoelectrons emitted is v_2 The ratio of v_1 to v_2 is:

(a) $\frac{v_1}{v_2} = \frac{1}{2}$ (b) $\frac{v_1}{v_2} = \frac{1}{8}$

(c) $\frac{v_1}{v_2} = \frac{1}{16}$ (d) $\frac{v_1}{v_2} = \frac{1}{4}$

Ans. $2hf_o = hf_o + \frac{1}{2} mV_1^2$

$5hf_o = hf_o + \frac{1}{2} mV_2^2$

$\frac{hf_o}{4hf_o} = \left(\frac{V_1}{V_2}\right)^2$

$\sqrt{\frac{1}{4}} = \frac{V_1}{V_2}$

Q.13. The stopping potential in the context of photoelectric effect depends on the following property of incident electromagnetic radiation

(a) Frequency

(b) Amplitude

(c) Intensity

(d) Phase

Ans. (a)

Q.14. Two identical photocathodes receive the light of frequencies f_1 and f_2 respectively. If the velocities of the photo-electrons coming out are v_1 and v_2 respectively, then

(a) $v_1^2 - v_2^2 = \frac{2h}{m}[f_1 - f_2]$

(b) $v_1^2 + v_2^2 = \frac{2h}{m}[f_1 + f_2]$

(c) $v_1 + v_2 = \left[\frac{2h}{m}(f_1 + f_2)\right]^{1/2}$

(d) $v_1 - v_2 = \left[\frac{2h}{m}(f_1 - f_2)\right]^{1/2}$

Ans. (d)

Q.15. When radiation of wavelength λ is incident on a metallic surface, the stopping potential of ejected photoelectrons is 4.8 V. If the same surface is illuminated by radiation of double the previous wavelength, then the stopping potential becomes 1.6 V. The threshold wavelength of the metal is

(a) 6λ (b) 8λ

(c) 2λ (d) 4λ

Ans. (c) $$\frac{\frac{hc}{\lambda}-\frac{hc}{\lambda_0}}{\frac{hc}{2\lambda}-\frac{hc}{\lambda_0}}=\frac{48ev}{16ev}=3$$

$$hc\left(\frac{1}{\lambda}-\frac{1}{\lambda_0}\right)=hc\left(\frac{3}{2\lambda}-\frac{3}{\lambda_0}\right)$$

$$\frac{3}{2\lambda}-\frac{1}{\lambda}=\frac{3}{\lambda_0}-\frac{1}{\lambda_0}$$

$$\frac{1}{\lambda}=\frac{2}{\lambda_0}\ \lambda_0=2\lambda$$

Q.16. Electron beam used in an electron microscope, accelerated by a voltage of 20 kV has a de-Broglie wavelength of λ_0. If the voltage is increased to 40 kV then the de-Broglie wavelength associated with the electron beam would be:

(a) $3\lambda_0$ (b) $9\lambda_0$

(c) $\lambda_0/2$ (d) $\lambda_0/\sqrt{2}$

Ans. (d) $\lambda \propto \frac{1}{\sqrt{2}}$

Q.17. An α-particle, a proton and an electron have the same kinetic energy. Which one of the following is correct in case of their De-Broglie wavelength

(a) $\lambda_\alpha > \lambda_p > \lambda_e$ (b) $\lambda_\alpha < \lambda_p < \lambda_e$

(c) $\lambda_\alpha = \lambda_p = \lambda_e$ (d) $\lambda_\alpha > \lambda_p < \lambda_e$

Ans. (b) $\lambda = \frac{h}{\sqrt{2m\,K.E}}$

Q.18. An electron moving with speed v and a photon moving with speed c have same de-Broglie wavelength. The ratio of kinetic energy of electron to that of photon is

(a) $\frac{v}{2c}$ (b) $\frac{2c}{v}$

(c) $\frac{v}{3c}$ (d) $\frac{3c}{v}$

Ans. (a) $e^- \rightarrow v$

$$\lambda_e=\frac{h}{m_eV}$$

$$E_P=\frac{hc}{\lambda_P}$$

$$\lambda_P=\frac{hc}{\lambda_P}=\frac{h}{m_eV}$$

$$E_p = meVc$$

$$\frac{K.E_e}{E_P}=\frac{\frac{1}{2}m_eV^2}{m_eV\,C}=\frac{V}{2c}$$

Q.19. A particle is traveling 4 times as fast as an electron. Assuming the ratio of de-Broglie wavelength of a particle to that of electron is 2 : 1, the mass of the particle is:

(a) 1/16 times the mass of e^-

(b) 8 times the mass of e^-

(c) 1/8 times the mass of e^-

(d) 16 times the mass of e

Ans. (c) $V_P = 4V_e$

$$\frac{\lambda_P}{\lambda_e}=\frac{2}{1}$$

$$\lambda_e=\frac{h}{m_eV_e},\ \lambda_P=\frac{h}{m_PV_P}$$

$$\frac{\lambda_P}{\lambda_e}=\frac{m_eV_e}{m_PV_P} \Rightarrow \frac{2}{1}=\frac{m_e}{m_P}\ \frac{1}{4}$$

$$\Rightarrow \frac{m_e}{m_P}=8$$

Q.20. An electron (of mass m) and a photon have the same energy E in the range of a few eV. The ratio of the de-Broglie wavelength associated with the electron and the wavelength of the photon is (c = speed of light in vacuum)

(a) $\left(\frac{E}{2m}\right)^{1/2}$ (b) $\frac{1}{c}\left(\frac{E}{2m}\right)^{1/2}$

(c) $c(2mE)^{1/2}$ (d) $\frac{1}{c}\left(\frac{2E}{m}\right)^{1/2}$

Ans. (b) $\frac{\lambda_e}{\lambda_p} = \frac{\frac{h}{\sqrt{2mE}}}{\frac{hc}{E}} = \frac{E}{c\sqrt{2Em}} = \frac{1}{c}\sqrt{\frac{E}{2m}}$

Q.21. An electron, a doubly ionized helium ion (He^{++}) and a proton are having the same kinetic energy. The relation between their respective de-Broglie wavelengths λ_e, $\lambda_{He^{++}}$ and λ_p, is

(a) $\lambda_e > \lambda_{He^{++}} > \lambda_p$ (b) $\lambda_e < \lambda_p < \lambda_{He^{++}}$

(c) $\lambda_e > \lambda_p > \lambda_{He^{++}}$ (d) $\lambda_e < \lambda_{He^{++}} = \lambda_p$

Ans. (c) $\lambda \propto \frac{1}{\sqrt{m}}$

Since,

$m_{He^{++}} > m_p > m_e$

Therefore (c) is correct

Q.22. Two particle move at right angle to each other. Their de-Broglie wavelengths are λ_1 and λ_2 respectively. The particles suffer perfectly inelastic collision. The de-Broglie wavelength λ, of the final particle, is given by

(a) $\lambda = \frac{\lambda_1 + \lambda_2}{2}$

(b) $\frac{2}{\lambda} = \frac{2}{\lambda_1} + \frac{1}{\lambda_2}$

(c) $\lambda = \sqrt{\lambda_1 \lambda_2}$

(d) $\frac{1}{\lambda^2} = \frac{2}{\lambda_1^2} + \frac{1}{\lambda_2^2}$

Ans. (d)

P_1

P_2

$P_1\hat{i} + P_2\hat{j} = P, \sqrt{P_1^2 + P_1^2} = P$

$\frac{h^2}{\lambda_1^2} + \frac{h^2}{\lambda_2^2} = \frac{h^2}{\lambda_2}$

$\frac{1}{\lambda^2} = \frac{1}{\lambda_1^2} + \frac{1}{\lambda_2^2}$

MR*

"Kismat sath de ya na de lekin kabiliyat jarur sath deti hai."

J.J Thomson recived the nobel prize for discovery of electron and his son G.P Thomson get nobel prize for experimental varification of wave nature of electron, whole thomson family in the business of electron

33 Atoms

1 Postulates of Thomson's atomic model

Postulate 1: An atom consists of a positively charged sphere with electrons embedded in it

Postulate 2: An atom as a whole is electrically neutral because the negative and positive charges are equal in magnitude

Thomson atomic model is compared to watermelon. Where he considered:

- Watermelon seeds as negatively charged particles
- The red part of the watermelon as positively charged

Limitations of Thomson's atomic model :-

- It failed to explain the stability of an atom because his model of atom failed to explain how a positive charge holds the negatively charged electrons in an atom. Therefore, This theory also failed to account for the position of the nucleus in an atom
- Thomson's model failed to explain the scattering of alpha particles by thin metal foils
- No experimental evidence in its support

2 Rutherford Alpha Particle Scattering Exp

Rutherford and his team took a thin gold foil having a thickness of 2.1×10^{-7} m and placed it in the centre of a rotatable'. detector made of zinc sulfide and a microscope. Then, they directed a beam of 5.5MeV alpha particles emitted from a radioactive source at the foil. Lead bricks collimated these alpha particles as they passed through them.

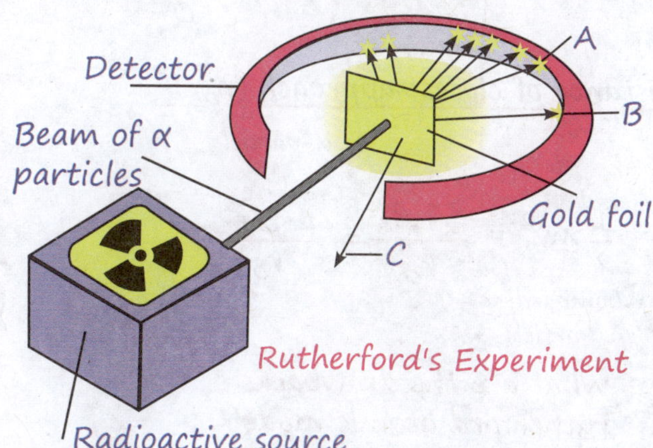

Rutherford's Experiment

After hitting the foil, the scattering of these alpha particles could be studied by the brief flashes on the screen. They expected to learn more about the structure of the atom from the results of this experiment:

- Most of the positively charged alpha particles went undeflected through the foil. This shows that most of the space in an atom is empty.
- Few positively charged alpha particles deflected through small and large angles. This shows that there is presence of positive center in the atom. This positive center is known as nucleus.
- Very few positively charged alpha particles bounced back. This is because the nucleus is very dense and does not allow the alpha particles to pass through it.
- The volume occupied by the nucleus is negligible compared to the total volume of the atom. This shows that radius of atom is much higher than that of the nucleus.

No. of Scattered α-Particle:

$$N \propto \frac{1}{\sin^4(\theta/2)}$$

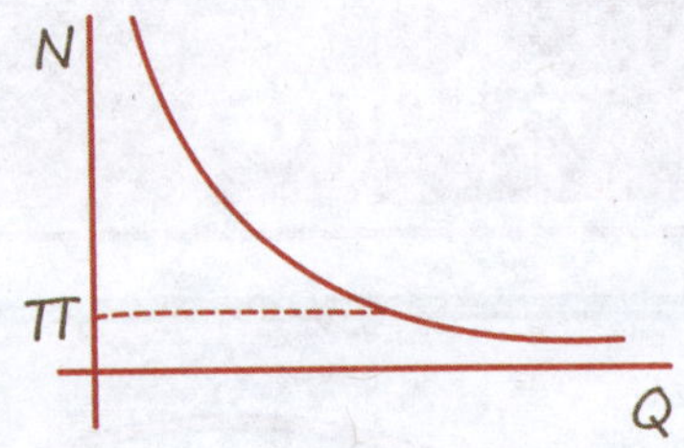

Distance of closest approach (r_o) :-

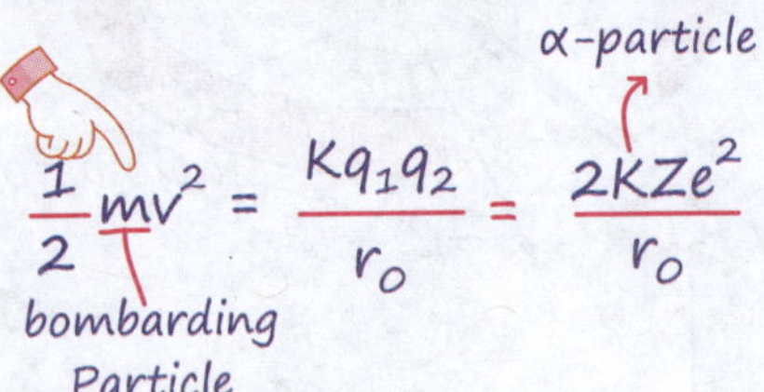

$$\frac{1}{2}mv^2 = \frac{Kq_1q_2}{r_o} = \frac{2KZe^2}{r_o}$$

Q.1 What are the drawbacks of the rutherford atomic model?

Ans. Rutherford's atomic model failed to explain the stability of electrons in a circular path. He stated that electrons revolve around the nucleus in a circular path, but particles in motion would undergo acceleration and cause energy radiation. Eventually, electrons should lose energy and fall into the nucleus. But it never happens.

3 Bohr's Atomic Model

- Bohr proposed that in an atom electron revolves around the nucleus in a definite circular path called orbits or shells.
- These orbits are called 'stationary orbits' and each orbit or shell possesses fixed energy.
- The energy levels are represented by an integer called quantum numbers where these integers indicate principal quantum numbers i.e. orbits that can be assigned as K, L, M, N, corresponding integer no. n = 1, 2, 3, 4
- The change in energy occur when the electron jumps from one orbit to another orbit.'
- The electron gains energy when it jumps from a lower energy level to a higher energy level and the electron loses its energy as it jumps from the higher energy to a lower energy level.
- The electron's angular momentum is quantized in its orbitals. So, electrons can move only those permissible orbits that should be an integral multiple of $\frac{h}{2\pi}$, where h is planks constant

(i.e., $L = \frac{nh}{2\pi}$)

Postulate 1 : $\frac{mv^2}{r} = \frac{KZe^2}{r^2}$

Postulate 2 : $mvr = \frac{nh}{2\pi}$

Radius of Nth Orbit :

$$r = \frac{\varepsilon_0 h^2}{\pi m e^2 Z} n^2 = 0.53\frac{n^2}{Z}$$

Velocity of Nth Orbit :

$$V_n = \frac{nh}{2\pi m r_n} = \frac{e^2}{2\varepsilon_0 h}\frac{Z}{n}$$

$$V_n = 2.18 \times 10^6 \frac{Z}{n} \text{ m/s}$$

Energies of Nth orbit :

$$KE = \frac{1}{2}mV_n^2 = \frac{me^4Z^2}{8\varepsilon_0^2h^2n^2}$$

Potential energy: Kinetic energy : Total energy $= -1 : \frac{1}{2} : -\frac{1}{2}$

$= -2 : 1 : -1$

- T.E. $= E_n = -13.6\frac{Z^2}{n^2}$ eV
- $V_1 = \frac{E_n}{e}$ = Volt

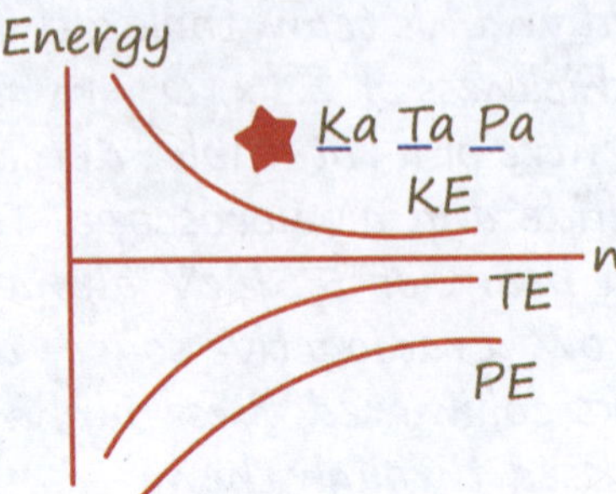

Special Relations :-

- **Time Period :**

$$T = \frac{2\pi r}{V} \qquad T \propto \frac{n^3}{z^2}$$

- **Acceleration :**

$$a_c = \frac{v^2}{r} \qquad a_c \propto \frac{z^3}{n^4}$$

- **Angular Velocity :**

$\omega = 2\pi/T$

$\omega \propto z2/n3$

- **Angular freq^n :**

$$w = 2\pi f \qquad f \propto \frac{z^2}{n^3}$$

- **Current :**

$$i = \frac{q}{t} \qquad i \propto \frac{z^2}{n^3}$$

- **Magnetic dipole Moment :**

$$M = i.A = \frac{eV}{2\pi r} \cdot \pi r2 = \frac{eVr}{2}$$

$M \propto n'$

Relation B/N Magnetic Moment & Angular Momentum :-

Magnetic Field $B \propto \frac{1}{n^5}$

$$M = \frac{eVr}{2} = \frac{eL}{2m} \; (L = mvr)$$

$$\vec{M} = \frac{-e}{2m}\vec{L}$$

Radiation Energy :-

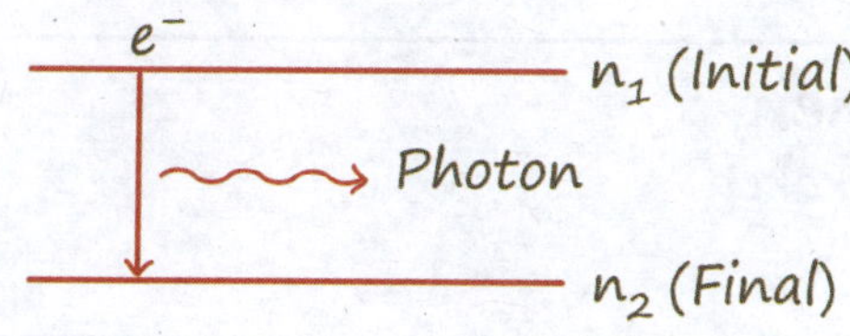

$R = 1/hc$

$$\bar{\nu} = \frac{1}{\lambda} = \frac{13.6\, z^2}{hc}\left[\frac{1}{n_2^2} - \frac{1}{n_1^2}\right]$$

* $R = \frac{me^4}{8\varepsilon_0^2 h^3 c} = 1.097 \times 10^7/m$

Energy Level Diagram :-

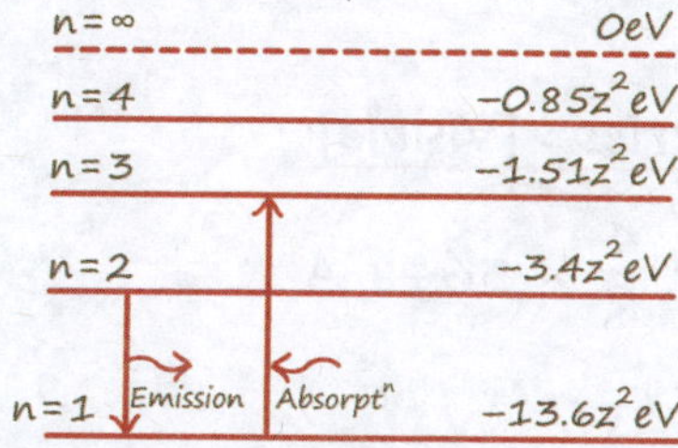

* No. of Spectral Line :

$$N = \frac{(n_2 - n_1)(n_2 - n_1 + 1)}{2} \quad n_2 \to n_1$$

$$N = \frac{n(n-1)}{2} \quad n_2 \to \text{G.S.}$$

4 Recoiling of an atom

When photon of momentum P emitted from stationary atom then, recoil momentum and energy of atom can be given as follow:

+ Recoil Momentum

$$P = \frac{h}{\lambda} = Rz^2h\left[\frac{1}{n_1^2} - \frac{1}{n_2^2}\right]$$ (Momentum conserves here)

+ Recoil Energy :

$$E = \frac{P^2}{2m} = \frac{h^2}{2m\lambda^2}$$

m : mass of Recoil atom.

5 H-Spectrum

For lyman series (K-series)

λ_{min} = E_{max} ($n_2 = \infty \to n_1 = 1$) [Limit Line: last line of K-series]

$\lambda_{max} = E_{min}$ ($n_2 = 2 \to n_1 = 1$) [1st Line of K-series]

Lyman Series : [UV]

$$\lambda_{min} = \boxed{\frac{1}{R} = 912\ A^\circ}$$

$\lambda_{max} = \frac{4}{3R} = 1216\ A^\circ$

Balmer Series : [Visible]

$\lambda_{min} = \frac{4}{R} = 3648\ A^\circ$

$\lambda_{max} = \frac{36}{5R} = 6565\ A^\circ$

Paschen Series : [IR]

$\lambda_{min} = \frac{9}{R} = 8208\ A^\circ$

$\lambda_{max} = \frac{144}{7R} = 18761.1\ A^\circ$

Brackelt Series : [IR]

$\lambda_{min} = \frac{16}{R} = 14592\ A^\circ$

$\lambda_{max} = \frac{400}{9R} = 40533\ A^\circ$

Bohr's Quantum Condition from Debroglie Hypothesis :

$2\pi r = n\lambda = \frac{nh}{mv}$

$L = \frac{nh}{2\pi}$

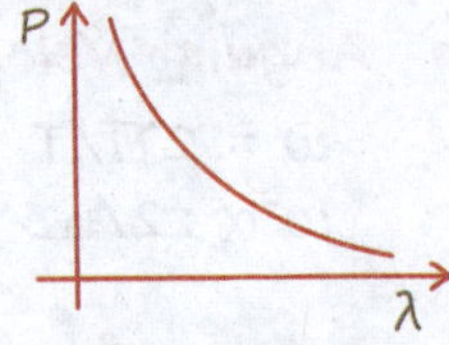

- When H-atom is raised from the ground state to 3rd excited state then → potential energy increases and K.E. decreases.

Spectral Series of Atom

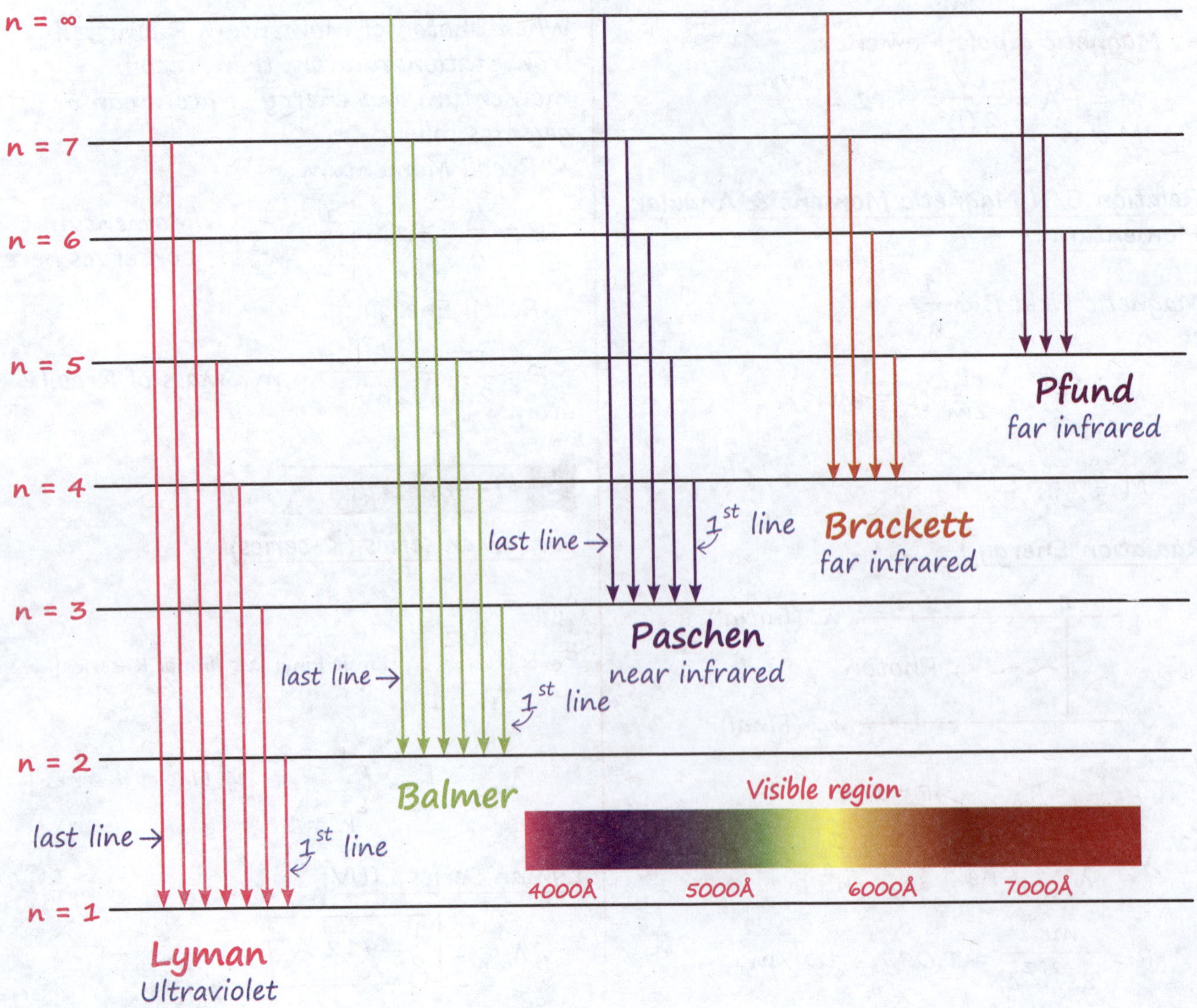

n = ∞

n = 7

n = 6

n = 5

n = 4

n = 3

n = 2

n = 1

Pfund
far infrared

last line →
1st line
Brackett
far infrared

Paschen
near infrared

last line →
1st line

Balmer

Visible region
4000Å 5000Å 6000Å 7000Å

last line →
1st line

Lyman
Ultraviolet

Q.2 The radius of electron's second stationary orbit in Bohr's atom is R. The radius of 3rd orbit will be

(a) R/3 (b) 2.25 R

(c) 3R (d) 9R

Sol. (b) (n = 2, R), (n = 3, R1)

Since, $R \propto n^2$, therefore

$$\frac{R \propto (2)^2}{R^1 \propto (3)^2} \Rightarrow \frac{R}{R^1} = \frac{4}{9}$$

$$R^1 = \frac{9}{4}R = 2.25R$$

Q.3 A small particle of mass m moves in such a way that its potential energy $U = \frac{1}{2}m\omega^2 r^2$ where ω is constant and r is the distance of the particle from origin. Assuming Bohr's quantization of momentum and circular orbit the radius of nth orbit will be proportional to:

(a) $\sqrt{n}$ (b) n

(c) n^2 (d) 1/n

Sol. (a) $U = \frac{1}{2}m\omega^2 r^2$

$$F = -\frac{dU}{dr} = -m\omega^2 r$$

$$\frac{mv^2}{r} = -m\omega^2 r$$

$$v^2 \propto r^2$$

$$v \propto r$$

$$mvr = \frac{nh}{2\pi}$$

$$mr \times r = \frac{nh}{2\pi}$$

$$r^2 \propto n$$

$$r \propto \sqrt{n}$$

Q.4 The radius of 2nd orbit of He+ of Bohr's model is r_1 and that of fourth orbit of Be^{3+} is represented as r_2. Now the ratio r_2/r_1 is x : 1. The value of x is ______.

Sol. $r \propto \frac{n^2}{2}$, $\frac{r_2}{r_1} = \frac{4^2}{4} \times \frac{2}{2^2}$

$$\frac{r_2}{r_1} = \frac{2}{1} = \frac{x}{1} \Rightarrow x = 2$$

Q.5 The ratio for the speed of the electron in the 3rd orbit of He^+ to the speed of the electron in the 3rd orbit of hydrogen atom will be:

(a) 1 : 1 (b) 1 : 2

(c) 4 : 1 (d) 2 : 1

Sol. (d) $v \propto \frac{z}{n} \Rightarrow v \propto z$

$$\frac{V_{He+}}{H_{H+}} = \frac{2}{1}$$

Q.6 A hydrogen atom in its ground state absorbs 10.2 eV of energy. The angular momentum of electron of the hydrogen atom will increase by the value of: (Given, Planck's constant $= 6.6 \times 10^{-34}$ Js).

(a) 2.10×10^{-34} Js

(b) 1.05×10^{-34} Js

(c) 3.15×10^{-34} Js

(d) 4.2×10^{-34} Js

Sol. (b) Change in angular momentum

$$\Delta L = \frac{2h}{2\pi} - \frac{h}{2\pi} = \frac{h}{2\pi}$$

$$= \frac{6.6 \times 10^{-34}}{2 \times 3.14} = 1.05 \times 10^{-34} \text{ J.s}$$

Q.7 The momentum of an electron revolving in nth orbit is given by: (Symbols have their usual meanings)

(a) $\frac{nh}{2\pi r}$ (b) $\frac{nh}{2r}$

(c) $\frac{nh}{2\pi}$ (d) $\frac{2\pi r}{nh}$

Sol. (a) $L = mvr = \frac{nh}{2\pi}$

$$P = \frac{nh}{2\pi r}$$

Q.8 In Bohr's atomic model of hydrogen, let K, P and E are the kinetic energy, potential energy and total energy of the electron respectively. Choose the correct option when the electron undergoes transitions to a higher level:

(a) All K, P and E increase

(b) K decreases, P and E increase.

(c) P decreases, K and E increase.

(d) K increases, P and E decrease.

Sol. (b)

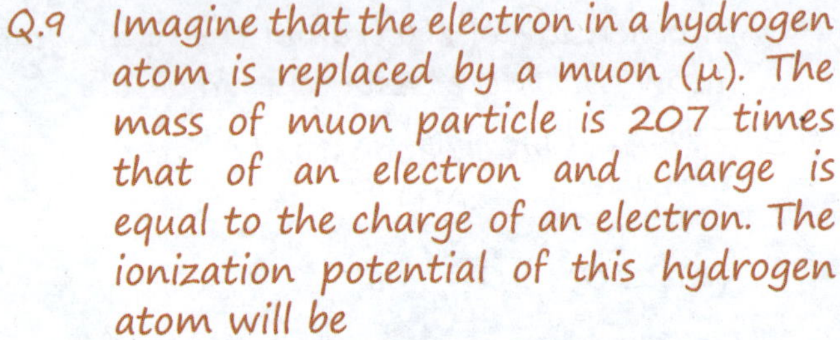

Q.9 Imagine that the electron in a hydrogen atom is replaced by a muon (μ). The mass of muon particle is 207 times that of an electron and charge is equal to the charge of an electron. The ionization potential of this hydrogen atom will be

(a) 13.6 eV (b) 331.2 eV

(c) 2815.2 eV (d) 27.2 eV

Sol. (c) I.P = T.E = $|K.E|$

$$= \frac{me^4 z^2}{8\varepsilon_0^2 h^2 n^2} = 13.6 \text{ eV} \times 207$$

Q.10 Which level of the single ionized carbon has the same energy as the ground state energy of hydrogen atom?

(a) 6 (b) 8

(c) 4 (d) 1

Sol. (a) For Hydrogen atom,

$$E = -13.6\frac{Z^2}{n^2} = -13.6\frac{(1)^2}{(1)^2} = -13.6$$

For Carbon atom, z = 6

$$E = -13.6\frac{z^2}{n^2}$$

$$= -13.6 = -13.6\frac{z^2}{n^2} \Rightarrow n = Z = 6$$

Q.11 The time period of revolution of electron in its ground state orbit in a hydrogen atom is 1.6×10^{-16} s. The frequency of revolution of the electron in its first excited state (in s^{-1}) is:

(a) 6.2×10^{15} (b) 5.6×10^{12}

(c) 7.8×10^{14} (d) 1.6×10^{14}

Sol. (c) $T = T_0 = 1.6 \times 10^{-16} = \frac{n^3}{2^2}$

$$f = f_0\frac{z^2}{n^3} = \frac{2}{3} \times 10^{16} \times \frac{1}{(2)^3}$$

$$= \frac{2 \times 10^{16}}{3 \times 8}$$

$$= \left(\frac{200}{24}\right) \times 10^{14} \approx 7.8 \times 10^{14} \text{ s}^{-1}$$

Q.12 A hydrogen atom, initially in the ground state is excited by absorbing a photon of wavelength 980 Å. The radius of the atom in the excited state, in terms of Bohr radius a_0, will be (h_c = 12500 eV – Å)

(a) $25a_0$ (b) $9a_0$

(c) $16a_0$ (d) $4a_0$

Sol. (c) $E_n - E_1 = \frac{12500}{980}$

$$-13.6\left[\frac{1}{n^2} - \frac{1}{1^2}\right] = 12.75$$

$n = 4$

New radius, $r = n^2 a_0 = 16a_0$

Q.13 A He+ ion is in its first excited state. Its ionization energy is:

(a) 6.04 eV (b) 13.60 eV

(c) 54.40 eV (d) 48.36 eV

Sol. (d) $|T.E| = \left(\frac{z^2}{n^2}\right) = 13.6 \text{ eV}$

Q.14 Speed of an electron in Bohr's 7th orbit for Hydrogen atom is 3.6×10^6 m/s. The corresponding speed of the electron in 3rd orbit, in m/s is:]

(a) (1.8×10^6)

(b) (7.5×10^6)

(c) (3.6×10^6)

(d) (8.4×10^6)

Sol. (d) $V \propto \frac{1}{n}$

$$\frac{3.6 \times 10^6}{V_0} = \frac{3}{7} \Rightarrow V_0 = 8.4 \times 10^6$$

Q.15 If λ_1 and λ_2 are the wavelength of the third member of Lyman and first member of the Paschen series respectively, then the value of $\lambda_1 : \lambda_2$ is:

(a) 1 : 3

(b) 7 : 108

(c) 1 : 9

(d) 7 : 135

Sol. (d)

$n_1 = 4$, $n_2 = 1$	$n_1 = 4$, $n_2 = 3$
$\frac{1}{\lambda_2} \quad R\left(1 - \frac{1}{16}\right)$	$\frac{1}{\lambda_2} = R\left(\frac{1}{9} - \frac{1}{16}\right)$
$\lambda_1 = \frac{16}{15R}$	$\lambda_2 = \frac{144}{7R}$

$$\lambda_1 : \lambda_2 = 7 : 135$$

Q.16 The electron in a hydrogen atom first jumps from the third excited state to the second excited state and subsequently to the first excited state. The ratio of the respective wavelengths, λ_1/λ_2, of the photons emitted in this process is

(a) 20/7 (b) 7/5

(c) 9/7 (d) 27/5

Sol. (a) Since, $\frac{1}{\lambda} \propto \left(\frac{1}{n_1^2} \quad \frac{1}{n_2^2}\right)$

Therefore,

$$\frac{\lambda_1}{\lambda_2} = \frac{\left(\frac{1}{2^2} - \frac{1}{3^2}\right)}{\left(\frac{1}{3^2} - \frac{1}{4^2}\right)} = \frac{20}{7}$$

MR*

"Kaam karo aisa ki pehchan ban jaye...chalo to aisa ki nishaan ban jaye..., are zindagi to har koi kaat leta h yaha...., agar dam hai to jiyo aise ki misaal ban jaye."

34 Nuclei

1 Nucleus

The nucleus is the tiny center of an atom where almost all the positive charge and mass are located.

Composition:

- The nucleus is made up of protons and neutrons. Together, these particles are known as nucleons.
- **Protons:** These carry a positive charge.
- **Neutrons:** These are neutral, meaning they have no charge.

Mass of Proton :-

$$m_p = 1.67 \times 10^{-27}\ Kg = 1.0073\ amu$$

Mass of Neutron :-

$$m_n = 1.67 \times 10^{-27}\ Kg = 1.0087\ amu$$

Mass of an electron :-

$$m_e = 9.1 \times 10^{-31}\ Kg$$

Mass of comparison :-

$$m_\alpha > m_n > m_p > m_e$$

2 Important points related to nucleus

Atomic Number (Z)

- This is the number of protons in the nucleus of an atom. It defines the identity of the element.

Mass Number (A)

- This is the total number of protons and neutrons in the nucleus of an atom. It represents the mass of the nucleus.

Rest Mass Energy :

Energy stored in the body due to their Mass

$$E = mc^2$$

Q.1 Find rest mass energy of object having mass 0.5 kg.

Sol. $E = mc^2$

$$= 0.5\ (3 \times 10^8)^2 = 4.5 \times 10^{16}\ J$$

Atomic Mass Unit :

$$1\ amu = \frac{931.5\ MeV}{C^2} = \frac{m_6 C^{12}}{12}$$

Nuclear Size :

$$R = R_0 A^{1/3} \qquad R_0 = 1.2 f_m\ (1\ f_m = 10^{-15} m)$$

Nuclear Density :

$$\rho = 10^{17}\ Kg/m^3$$ Constant because it does not depend on the mass number.

Volume $\propto$ A

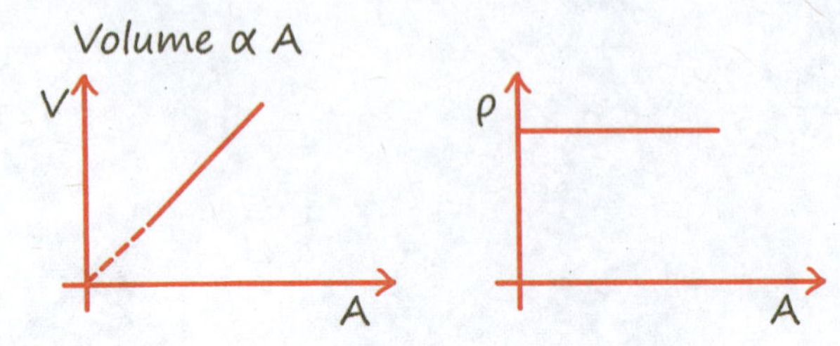

3 Nuclear Binding Energy (Mass Defect)

Mass defect = (Mass of all the nucleons of nucleus) − (Mass of nucleus)

$$\Delta m = [ZM_p + (A-Z)M_n - M_{nucleus}]$$

$$BE = \Delta mc^2$$

Stability $\alpha \; \frac{B.E.}{A}$ (Binding energy per mass number)

Q.2 Which is more stable $_4X^6$ and $_6Y^{12}$ having B.E. 24eV and 36eV respectively.

Sol. $\left(\frac{B.E.}{A}\right)_X = \frac{24}{6} = 4ev$

$\left(\frac{B.E.}{A}\right)_Y = \frac{36}{12} = 3ev$ (X is more stable than Y)

Binding Energy Curve:-

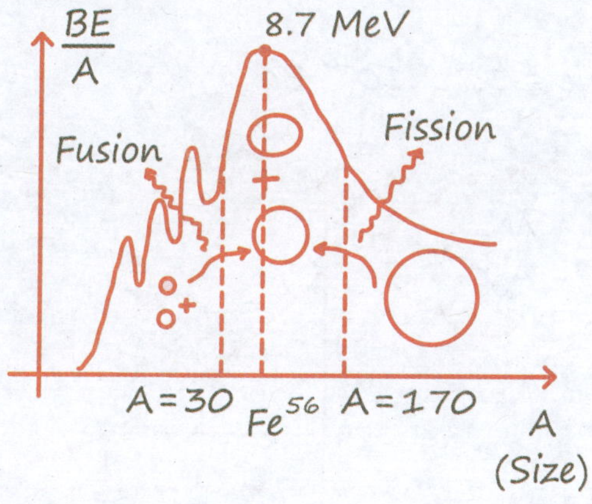

- As mass number increases then Ist stability increase then decreases.
- Nucleons of lower mass number fuse for stability and release energy.
- Nucleons of higher mass number break (fission) for stability.

4 Nuclear Force

→ Short range, Non-central, Nonconservative

- Does not follow inverse square law
- Charge independent
- Saturation dependent,
- $F_{nuclear} = 10^{39} F_{gravity}$
- Weak nuclear force → shortest range (10^{-16} m), repulsive, Mediated by Boson
- Strong nuclear force → Range 10^{-15} m, attractive, mediated by Meson

  Always attractive.

  $F_{NN} = F_{PP} = F_{NP}$

  Range = 1 Fm

  $F_{Nuclear} = 100\ E_{electrostatics}$!

5 Q-Value

$Q = [BE_P - BE_R]$

Q = +ve (Exo) Q = −ve(Endo)

$BE_P \uparrow$ = Energy Release

$Q = [M_{Reactant} - M_{Product}]\,c^2$

Nuclear reaction coresponding given graph

- A ⟶ B+D

  Energy released
- C ⟶ D+B

  Energy released
- 2B ⟶ C

  Energy absorbed

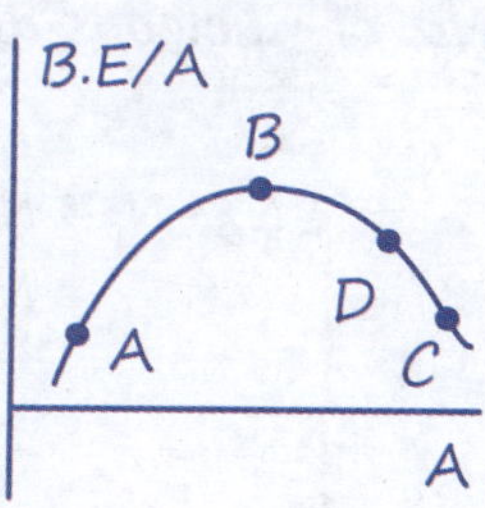

Q-Value of the reaction

A+B ⟶ C+D+Q

$Q = [(m_A+m_B) - (m_C+m_D)] \times c^2$

Q = [B.E(C) + B.E(D)] − [(B.E(A) + B.E. (B)]

- If $[m_A+m_B] > (m_C+m_D)$
- If $[B.E.(C+D)] > B.E.(A+B)$

→ energy released

- If $m_A+m_B < m_C+m_D$
- If $B.E.(C+D) < B.E.(A+B)$

→ energy absorbed

6 Law of Radioactive Decay

Number of nucleon becomes less and less but always some left.

Rate of decay ∝ remaining no. of nucleons

$$\frac{-dN}{dt} = \lambda N$$

λ = decay const^n → Material properties

$$N_t = N_0 e^{-\lambda t}$$

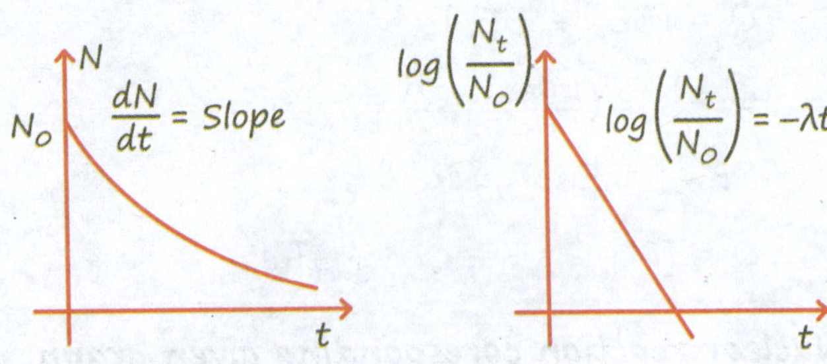

No. of nucleons decayed :

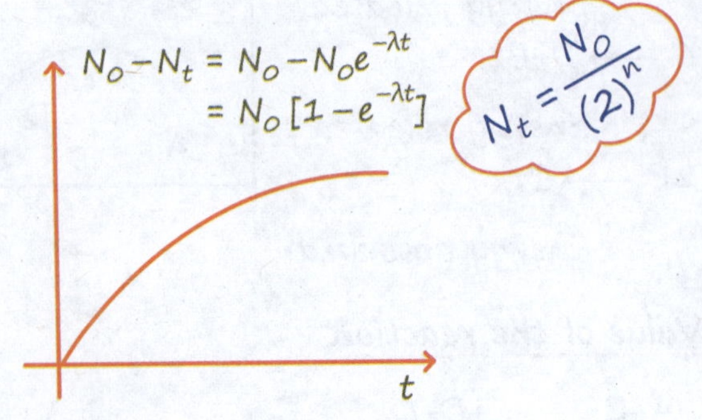

$$N_0 - N_t = N_0 - N_0 e^{-\lambda t}$$
$$= N_0[1 - e^{-\lambda t}]$$

$$N_t = \frac{N_0}{(2)^n}$$

Mean Life :-

$\tau = \frac{1}{\lambda}$ No. of nucleons becomes 37% of initial.

Half-Life :-

No. of Nucleons $\longrightarrow \frac{1}{2} N_0$;

$$t_{1/2} = \frac{\ln 2}{\lambda} = \frac{0.693}{\lambda}$$

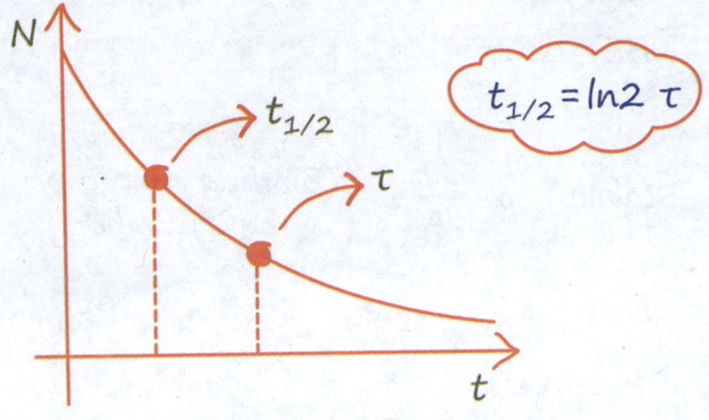

$$t_{1/2} = \ln 2\ \tau$$

7 Activity

(Rate of disintegrat^n of a sample).

1Bq = 1 decay/s

1 Rutherford = 10^6 dps

1 Curie = 3.7×10^{10} Bq

$$A_t = \lambda N_t$$
$$A_t = A_0 e^{-\lambda t}$$

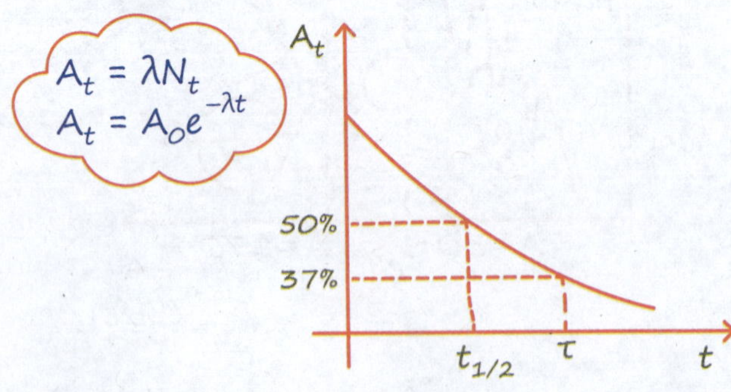

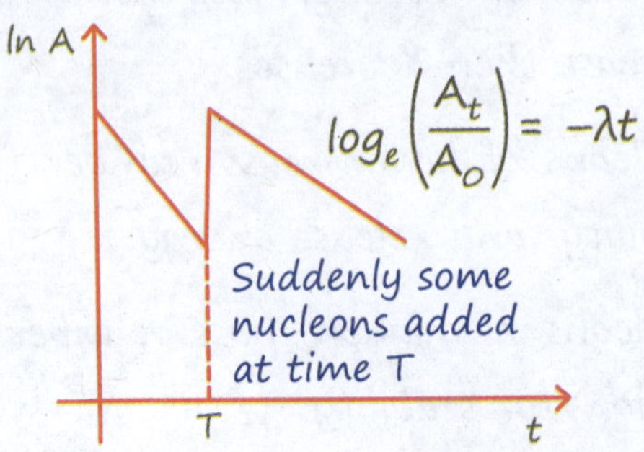

$$\log_e\left(\frac{A_t}{A_0}\right) = -\lambda t$$

Time	No. of nuclei at "t" N_t (Remaining)	No. of dis-integrated Nuclei $(N_o - N_t)$	$\frac{N_t}{N_o - N_t}$ $\left[\frac{\text{Remain}}{\text{decay}}\right]$	$\frac{N_o}{N_t}$ $\left[\frac{\text{initial}}{\text{remain}}\right]$	$\frac{N_o}{N_o - N_t}$ $\left[\frac{\text{initial}}{\text{decay}}\right]$
$t = 0$	N_o	0	–	–	–
$t = 1_{t/2}$	$N_o/2$	$N_o/2$	1 : 1	2 : 1	2 : 1
$t = 2_{t/2}$	$N_o/4$	$3N_o/4$	1 : 3	4 : 1	4 : 3
$t = 3_{t/2}$	$N_o/8$	$7N_o/8$	1 : 7	8 : 1	8 : 7
$t = 4_{t/2}$	$N_o/16$	$15N_o/16$	1 : 15	16 : 1	16 : 15

$= N_o - \frac{N_o}{2^n}$ $(1 : 2^n - 1)$ $(2^n : 1)$

$nt_{1/2} - \frac{N_o}{2^n} = \frac{N_o}{2^{t/T}}$

α-decay :

- Recoil velocity of daughter Nuclei :

  Q = Released energy in α-decay

  A = Mass no. of original nuclei

$$\vec{V}_D = \frac{-4\vec{V}_\alpha}{(A-4)}$$

- $KE_\alpha = \left[\frac{A-4}{A}\right]Q$
- $KE_D = \left[\frac{4Q}{A}\right]$

β-decay :

- Q-value in terms of Atomic Mass :
- $\beta^- : Q = [M_R - M_P]c^2$ } same!
- $\beta^+ : Q = [M_R - M_P - 2M_e]\,c^2$

Parallel Disintegration :

In parallel disintegration

$\lambda_{eq} = \lambda_1 + \lambda_2$

$(ty_2) = \frac{t_1 t_2}{t_1 + t_2}$

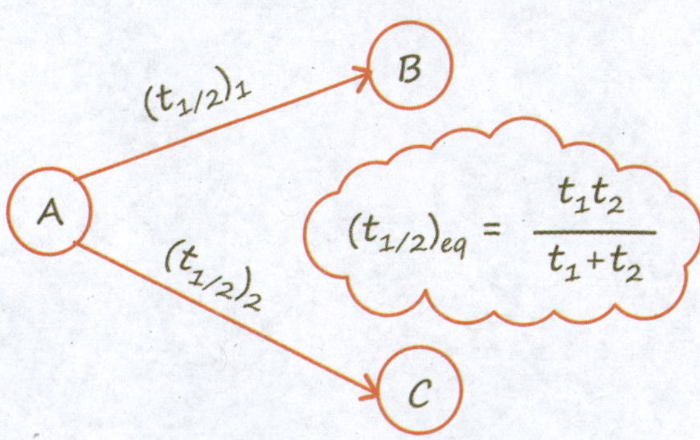

–: Nuclear Reactor :–

n = 2.5 nucleons/fission

$$K = \frac{\text{Rate of Prod}^n \text{ of nucleon}}{\text{Rate of loss of neutron}}$$

$K = 1$	Critical
$K < 1$	Controlled
$K > 1$	Uncontrolled

–: Power of Reactor :–

$$P = \frac{nE}{t} = \frac{nMc^2}{t}$$

- Neutron v/s Proton graph :

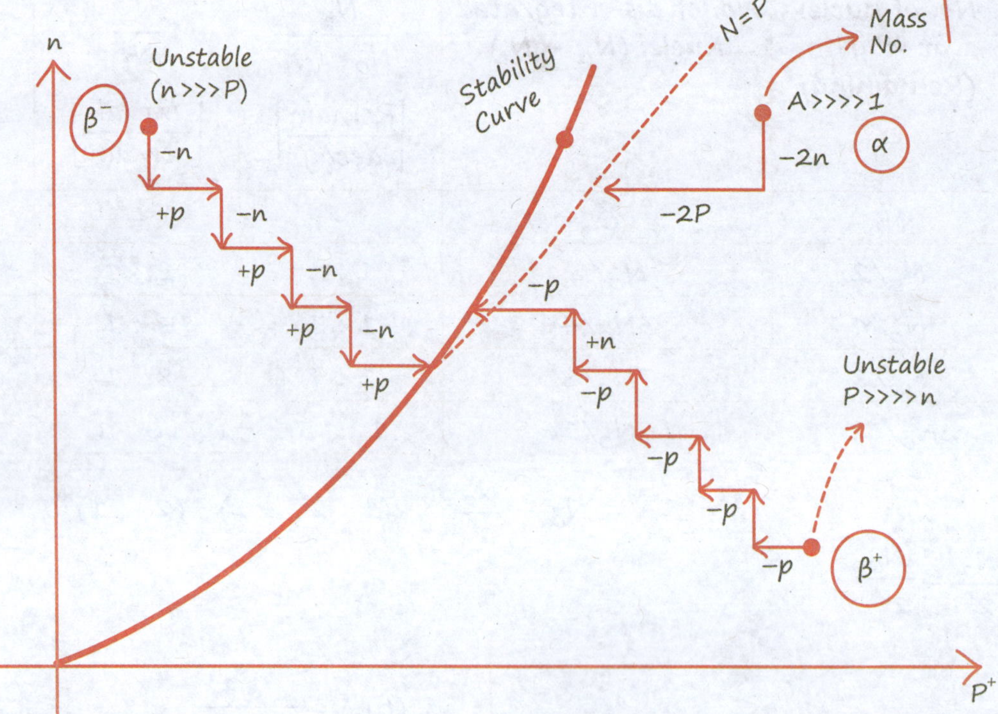

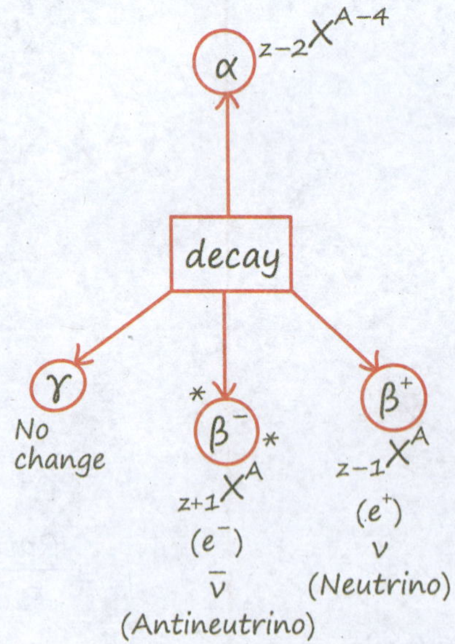

Condition of fusion

1. High temperature = 10^7K
2. High Pressure

Thermal Power Plant

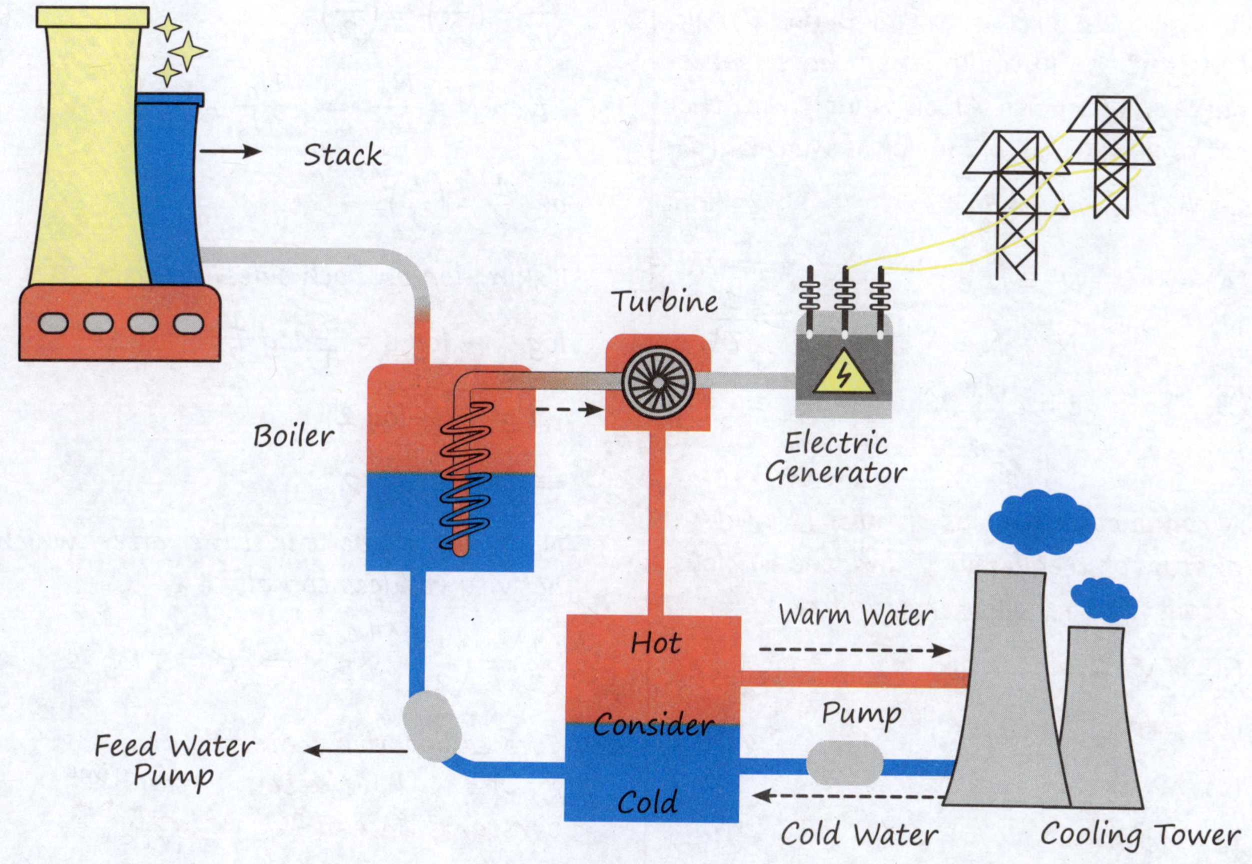

–: Nuclear Bomb :–

H-bomb	Atom Bomb
(Fusion)	(Fission)
→E = 26.7 MeV	E = 200 MeV
Uncontrolled	Controlled!
Chain rx^n	

IMP PYQ

Q.1 Radioactive material A has decay constant 8λ and material B has decay constant λ. Initially, they have same number of nuclei. After what time, the ratio of number of nuclei of material B to that A will be $\frac{1}{e}$?

(A) → N_o $\quad N^{+} = N_O e^{-8\lambda t}$

(B) → No $\quad N^{+} = N_O e^{-\lambda t}$

$e = e^{-8\lambda t + \lambda t}$

$e = e^{-7\lambda t}$

$1 = -7\lambda t$

$t = \frac{1}{7\lambda}$

Q.2 A radioactive nucleus of mass M emits a photon of frequency ν and the nucleus recoils. The recoil energy will be

(a) $h^2\nu^2/2Mc^2$

(b) zero

(c) $h\nu$

(d) $Mc^2 - h\nu$

Ans. (a)

Momentum of a photon

$$p = \frac{h\nu}{c}$$

Hence, recoil energy, $E = \frac{p^2}{2M}$

$$\therefore \quad E = \frac{\left(\frac{h\nu}{c}\right)^2}{2M} \quad \text{or} \quad E = \frac{h^2\nu^2}{2Mc^2}$$

Q.3 The activity of a radioactive sample is measured as N_0 counts per minute at $t = 0$ and N_0/e counts per minute at $t = 5$ min. The time (in minute) at which the activity reduces to half its value is

(a) $\log_e 2/5$ (b) $\frac{5}{\log_e 2}$

(c) $5 \log_{10} 2$ (d) $5 \log_e 2$

Ans. (d)

Fraction remains after n half lives

$$\frac{N}{N_0} = \left(\frac{1}{2}\right)^n = \left(\frac{1}{2}\right)^{t/T}$$

Given $N = \frac{N_0}{e} \Rightarrow \frac{N_0}{eN_0} = \left(\frac{1}{2}\right)^{5/T}$

or $\frac{1}{e} = \left(\frac{1}{2}\right)^{5/T}$

Taking log on both sides, we get

$$\log 1 - \log e = \frac{5}{T} \log \frac{1}{2};$$

$$-1 = \frac{5}{T}(-\log 2)$$

$$\Rightarrow T = 5 \log_e 2$$

Now, let t' be the time after which activity reduces to half

$$\left(\frac{1}{2}\right) = \left(\frac{1}{2}\right)^{t'/5\log_e 2} \Rightarrow t' = 5 \log_e 2$$

Q.4 Two radioactive materials X_1 and X_2 have decay constants 5λ and λ respectively. If initially they have the same number of nuclei, then the ratio of the number of nuclei of X_1 to that of X_2 will be $\frac{1}{e}$ after a time

Save Time — Trick: $t = \frac{\text{Power of}}{\Delta\lambda}$

(a) λ (b) $\frac{1}{2}\lambda$

✓(c) $\frac{1}{4\lambda}$ (d) $\frac{e}{\lambda}$

Q.5 In compound $X(n, \alpha) \rightarrow {}_3Li^7$, the element X is

(a) ${}_2He^4$ (b) ${}_5B^{10}$

(c) ${}_5B^9$ (d) ${}_4Be^{11}$

Ans. (b)

The given nuclear reaction can be written as

$${}_ZX^A + {}_0n^1 \longrightarrow {}_3Li^7 + {}_2He^4$$

Conservation of mass number gives,

$A + 1 = 7 + 4 \Rightarrow A = 10$

Conservation of charge number/Atomic No. gives,

$Z + 0 = 2 + 3 \Rightarrow Z = 5$

Hence, Z = 5, A = 10 corresponds to boron ($_5B^{10}$).

Q.6 The activity of a radioactive sample is measured as 9750 counts/min at t = 0 and as 975 counts/min at t = 5 min. The decay constant is approximately

(a) 0.922/min (b) 0.691.min

(c) 0.461/min (d) 0.230/min

Ans. (c)

According to law of radioactivity

$$\frac{N}{N_0} = e^{-\lambda t} \quad ...(i)$$

$$\Rightarrow \frac{N_0}{N} = e^{\lambda t}$$

[N = final concentration
N_0 = initial concentration
λ = decay constant]

Taking logarithm on both sides of Eq. (i), we have

$$\log_e\left(\frac{N_0}{N}\right) = \log_e(e^{\lambda t}) = \lambda t \log_e e = \lambda t$$

As we know that,

$\log_e x = 2.3026 \log_{10} x$

Making substitution, we get

$$\lambda = \frac{2.3026 \log_{10}\left(\frac{9750}{975}\right)}{5}$$

[$\because N_0$ = 9750 counts/min and N = 975 counts/min]

$$= \frac{2.3026}{5}\log_{10} 10 = \frac{2.3026}{5}\text{ min}^{-1}$$

$$= 0.461 \text{ min}^{-1}$$

Q.7 Nuclear fission can be explained by

(a) proton-proton cycle

(b) liquid drop model of nucleus

(c) independent of nuclear particle model

(d) nuclear shell model

Ans. (b)

Q.8 Which of the following is used as a moderator in nuclear reactors?

(a) Plutonium (b) Cadmium

(c) Heavy water (d) Uranium

Ans. (c)

Q.9 Energy released in the fission of a single $_{92}U^{235}$ nucleus is 200 MeV. The fission rate of a $_{92}U^{235}$ filled reactor operating at a power level of 5 W is

(a) $1.56 \times 10^{-10}\ s^{-1}$

(b) $1.56 \times 10^{11}\ s^{-1}$

(c) $1.56 \times 10^{-16}\ s^{-1}$

(d) $1.56 \times 10^{-17}\ s^{-1}$

Ans. (b)

$$P = nE/t \left(\frac{n}{t} = P/E\right)$$

$$\text{Fission rate} = \frac{\text{total nuclear power}}{\text{energy produced/fission}}$$

Here, total nuclear power = 5W

Energy released per fission = 200 MeV

$$\therefore \text{ Fission rate} = \frac{5}{200 \text{ MeV}}$$

$$= \frac{5}{200 \times 1.6 \times 10^{-13}} \quad [\because 1 \text{ MeV J}]$$

$$= 1.56 \times 10^{11}\ s^{-1}$$

Q.10 Read the following statements:

A. Volume of the nucleus is directly proportional to the mass number.

B. Volume of the nucleus is independent of mass number.

C. Density of the nucleus is directly proportional to the mass number.

D. Density of nucleus is directly proportional to the cube root of the mass number.

E. Density of the nucleus is independent of the mass number.

Choose the correct option from the following options.

(a) (A) and (D) only.

(b) (A) and (E) only.

(c) (B) and (E) only.

(d) (A) and (C) only

Ans. (b) Volume, $V \propto R^3 \propto A$

Q.11 The ratio of the density of oxygen nucleus ($^{16}_{8}O$) and helium nucleus ($^{4}_{2}He$) is

(a) 4:1 (b) 8:1

(c) 1:1 (d) 2:1

Ans. (c) Density is independent of mass number (A), Therefore ratio would be 1 : 1

Q.12 A nucleus disintegrates into two smaller parts, which have their velocities in the ratio 3 : 2. The ratio of their nuclear sizes will be $\left(\frac{x}{3}\right)^{1/3}$. The value of 'x' is ________.

Ans. $\frac{V_1}{V_2} = \frac{3}{2}$

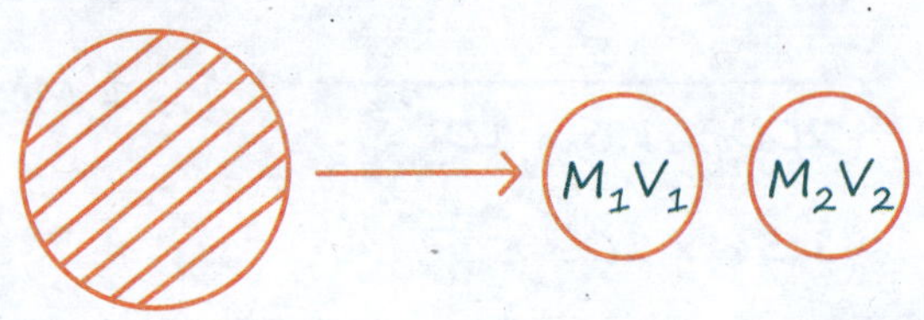

$M_1V_1 = M_2V_2$

$\frac{m_1}{m_2} = \frac{V_1}{V_2} = \frac{2}{3}$

From the formula of Nuclear mass density, $\frac{m}{R^3}$ = constant

Therefore,

$$\frac{R_1}{R_2} = \left(\frac{2}{3}\right)^{1/3} = \left(\frac{x}{3}\right)^{1/3}$$

$\boxed{x = 2}$

Q.13 Given below are two statements: one is labelled as Assertion A and the other is labelled as Reason R

Assertion A: The binding energy per nucleon is practically independent of the atomic number for nuclei of mass number in the range 30 to 170.

Reason R: Nuclear force is short ranged.

In the light of the above statements, choose the correct answer from the options given below

(a) Both A and R are true but R is not the correct explanation of A

(b) A is true but R is false

(c) A is false but R is true

(d) Both A and R are true and R is the correct explanation of A

Ans. (d)

Q.14 If the Net force between two protons, two neutrons and between a proton and a neutron is denoted by F_{pp}, F_{nn} and F_{pn}, respectively, then:

(a) $F_{pp} \approx F_{nn} \approx F_{pn}$

(b) $F_{pp} \neq F_{nn}$ and $F_{pn} = F_{nn}$

(c) $F_{pp} = F_{nn} = F_{pn}$

(d) $F_{pp} \neq F_{nn} \neq F_{pn}$

Ans. (b)

Q.15 Energy released in the fission of a single $_{92}U^{235}$ nucleus is 200 MeV. The fission rate of a $_{92}U^{235}$ filled reactor operating at a power level of 5 W is

(a) $1.56 \times 10^{-10}\ s^{-1}$

(b) $1.56 \times 10^{11}\ s^{-1}$

(c) $1.56 \times 10^{-16}\ s^{-1}$

(d) $1.56 \times 10^{-17}\ s^{-1}$

Ans. (b) $P = \left(\frac{n}{t}\right) E$

$$\frac{n}{t} = \frac{P}{E} = \frac{5}{200 \times 10^{6} \times 1.6 \times 10^{-13}}$$

$= 1.56 \times 10^{11}\ s^{-1}$

Q.16 The power obtained in a reactor using U^{235} disintegration is 1000 kW. The mass decay of U^{235} per hour is:

(a) 1 microgram (b) 10microgram

(c) 20 microgram (d) 40microgram

Ans. (d) $P = 10^{3} \times 10^{3} = \frac{mc^{2}}{t}$

$$m = \frac{10^{6} \times t}{c^{2}}$$

$$= \frac{10^{6} \times 60 \times 60}{(3 \times 10^{8})^{2}}$$

$m = 4 \times 10^{-8}$ kg

$m = 4 \times 10^{-2}$ kg $\times 10^{-6}$

$= 40 \times 10^{-3}\ v_o\ 10^{-6} = 40\ \mu g$

Q.17 If in nuclear reactor using U^{235} as fuel, the power output is 4.8 MW, the number of fissions per second is:

(Energy released per fission of U^{235} = 200 MeV watts, 1 eV = 1.6×10^{-19} J)

(a) 1.5×10^{17} (b) 3×10^{19}

(c) 1.5×10^{25} (d) 3×10^{25}

Ans. (a) $4.8 \times 10^{6} = \left(\frac{n}{t}\right) 200 \times 1.6 \times 10^{-13}$.

Q.18 Fusion reaction takes place at a higher temperature because:

(a) atoms get ionized at high temperatures.

(b) Kinetic energy is high enough to overcome the Coulomb repulsion between nuclei.

(c) molecules break up at a high temperature.

(d) nuclei break up at a high temperature.

Ans. (b)

Q.19 For the nuclear reaction:

$_{92}U^{235} + {}_{56}Ba^{144} + \ldots + 3\,{}_{0}n^{1}$

The blank space can be filled by:

(a) $_{26}Kr^{89}$ (b) $_{36}Kr^{89}$

(a) $_{26}Sr^{90}$ (a) $_{38}Sr^{89}$

Ans. (a) 236 92

−147 −56

+ 89 26

Q.20 How many alpha and beta particles are emitted when Uranium $_{92}U^{238}$ decays to lead $_{82}Pb^{206}$?

(a) 3 alpha particles and 5 beta particles

(b) 6 alpha particles and 4 beta particles

(c) 4 alpha particles and 5 beta particles

(d) 8 alpha particles and 6 beta particles

Ans. ${}_{92}U^{238} \rightarrow {}_{82}Pb^{206}$

238	No. of alpha-particle
-206	emitted = $\frac{32}{4}$ = 8
32	No. of Beta-particle
No	emitted = 82 − 76 = 6

Q.21 In the following nuclear reaction

$D \xrightarrow{\alpha} D_1 \xrightarrow{\beta^-} D_2 \xrightarrow{\alpha} D_3 \xrightarrow{\gamma} D_4$

Mass number of D is 182 and atomic number is 74. Mass number and atomic number of D_4 respectively will be

(a) 174 and 71 (b) 174 and 69

(c) 172 and 69 (d) 172 and 71

Ans. (a) ${}^{182}_{74}D \xrightarrow{\alpha} {}^{178}_{72}D_1 \xrightarrow{\beta^-} {}^{178}_{73}D_2 \xrightarrow{\alpha} {}^{174}_{71}D_3 \xrightarrow{\gamma} {}^{174}_{71}D_4$

Q.22 Consider the following radioactive decay process

${}^{218}_{84}A \xrightarrow{\alpha} A_1 \xrightarrow{\beta^-} A_2 \xrightarrow{\gamma} A_3 \xrightarrow{\alpha} A_4 \xrightarrow{\beta^+} A_5 \xrightarrow{\gamma} A_6$

The mass number and the atomic number A_6 are given by:

(a) 210 and 82 (b) 210 and 84

(c) 210 and 80 (d) 211 and 80

Ans. (c)

Q.23 The Q-value of a nuclear reaction and kinetic energy of the projectile particle, K_p are related as:

(a) $Q = K_p$ (b) $(K_p + Q) < 0$

(c) $Q < K_p$ (d) $(K_p + Q) > 0$

Ans. (a) K + Q = 0

$$Q = K.E_1 + K.E_2$$

Q.24 Two lighter nuclei combine to form a comparatively heavier nucleus by the relation given below:

${}^2_1X + {}^2_1X = {}^4_2Y$

The binding energies per nucleon for ${}^2_1X$ and ${}^4_2Y$ are 1.1 MeV and 7.6 MeV respectively. The energy released in this process is ______ MeV.

Ans. Q = 4(7.6)MeV − 4(1.1)MeV

= (30 . 4 − 4 . 4) = 26 Mev

Q.25 The mass of proton, neutron and helium nucleus are respectively 1.0073 u, 1.0087 u and 4.0015 u. The binding energy of helium nucleus is:

(a) 14.2 MeV (b) 28.4 MeV

(c) 56.8 MeVX (d) 7.1 MeV

Ans. BE = [$2m_p + 2m_p - m({}^4_2He) \times c^2$

= (2 × 1.0073 + 2 × 1.0087 − 4.0015)

= 2.0146 4.0320

2.0174 4.0015

0.0305

B.E = 0.0205 ×931.5 MeV

0.305 × 93.15

= 28.4 MeV

MR*

"Parinde ruk mat, tujhme jaan baki hai, manjil dur hai abhi, bahut udan baki hai"

35 Semiconductor

1 Types of Material

Conductor :

- Materials that allow electricity to flow easily through them.
- CB & VB Overlap: The energy levels of the electrons (called the Conduction Band and the Valence Band) overlap, so electrons can move freely.
- Conductivity, $\sigma = 10^2$ to 10^8 Sm^{-1} (Siemens per meter)

Semi-Conductor :

Materials that can conduct electricity under certain conditions but not as well as conductors.

Examples: Silicon (Si) and Germanium (Ge).

- Energy gap, $E_g < 3eV$ Si & Ge
- $\sigma = 10^5$ to 10^{-6} Sm^{-1}
- $E_g = (E_{CB})_{min} - (E_{VB})_{max}$
- At 0 K current can flow through conductor but can not flow through semi-conductor.

Insulators :

Materials that do not allow electricity to flow through them easily.

Examples: Rubber, glass.

- Energy gap, $E_g \geq 3eV$
- $\sigma = 10^{-11}$ to 10^{-19} Sm^{-1}

Intrinsic Semi-Conduct :

Pure semiconductors without any added impurities.

Examples: Pure Silicon (Si) and Germanium (Ge).

[Pure] ⟶ Si & Ge

At T = 0 K insulator.

Band diagram

$E_g = (E_c)_{min} - (E_v)_{max}$

Conduction Band
all electrons are free & current flows due to these electrons

$(E_c)_{min}$

$(E_v)_{max}$

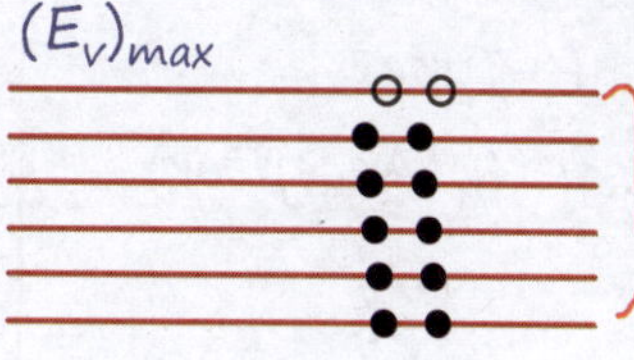

Valance Band
all electrons are fixed, current does not flows due to these electrons

Electric Current in intrinsic Semi-conductor

$n_e = n_h = n$ = no. of intrinsic charge density

$\vec{J} = \sigma \vec{E}$

$J = en(\mu_e + \mu_h)E$; $i = i_e + i_h$

$\sigma = en(\mu_e + \mu_h)$

$i_e > i_h$ $\quad \mu_e > \mu_h$

$\mu = \frac{e\tau}{m}$ $\quad (M_{eff})_{hole} > (M_{eff})_{e-}$

* Condition for intrinsic Semi-Conductor :

$n_i = n_e = n_h$; $\sigma = en_i(\mu_e + \mu_h)$

Extrinsic Semi-Conductor :

intrinsic S.C. + impurities = E.S.C

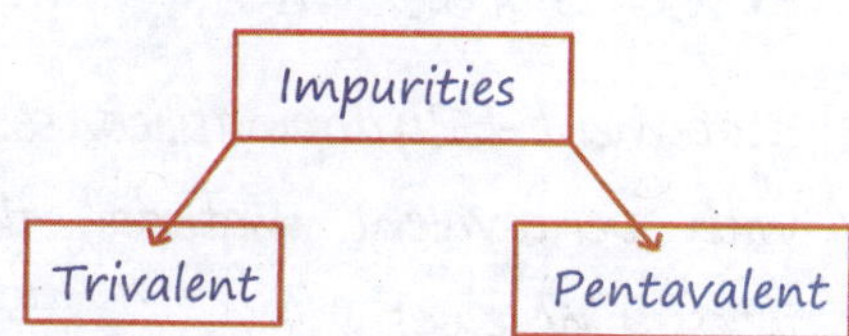

P-type (+)	N-type (−)
$n_h >>> n_e$	$n_e >>> n_h$
"B-Family"	"N-Family"
Neutral	Neutral
$n_h \geq N_A$	$n_{e^-} \geq N_D$
N_A = no. of acceptor ion density	N_D = no. of donor ion density
Fixed −ve ion	Fixed +ve ion
Acceptor $1e^-$ accept udhar lega Ge/Si se phir bond bnayga	Donor $1e^-$ donate kar ke Ge/Si se bond bna lega

2 Law of Mass Action

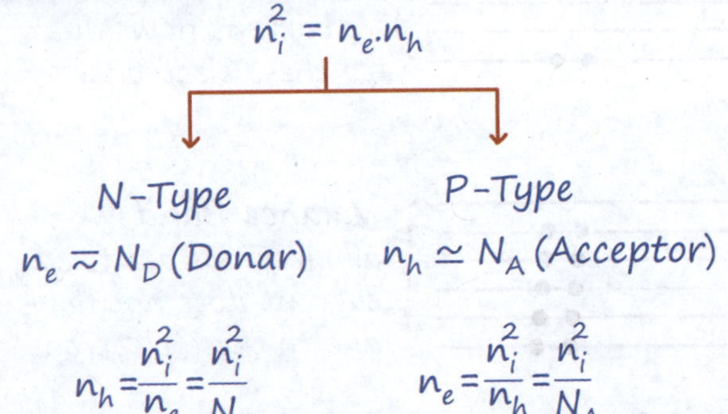

Q.1 In a semiconductor, the number density of intrinsic charge carriers at 27°C is 1.5×10^{16} m^{-3}. If the semiconductor is doped with impurity atom, the hole density increases to 4.5×10^{22} m^{-3}. The electron density in the doped semiconductor is

Ans. $n_e n_h = n_i^2$

$$\Rightarrow n_e = \frac{n_i^2}{n_h} = \frac{(1.5 \times 10^{16})^2}{4.5 \times 10^{22}}$$

$$\Rightarrow n_e = \frac{1.5 \times 1.5 \times 10^{32}}{4.5 \times 10^{22}}$$

$$\Rightarrow n_e = 5 \times 10^9 \ m^{-3}$$

Q.2 Statement-I: By doping silicon semiconductor with pentavalent material, the electrons density increases.

Statement-II: The n-type semiconductor has net negative charge. In the light of the above statements, choose the most appropriate answer from the options given below:

(a) Statement-I is false but statement-II is true

(b) Statement-I is true but statement-II is false

(c) Both statement-I and statement-II are true

(d) Both statement-I and statement-II is false

Ans. (b)

(I) Doping Silicon with pentavalent impurity result in an extra electron, therefore electron density increases.

(II) The net charge on either type of semiconductor is zero.

Q.3 The effect of increase in temperature on the number of electrons in conduction band (n_e) and resistance of a semiconductor will be as:

(a) Both n_e and resistance decrease

(b) Both n_e and resistance increase

(c) n_e increases, resistance decrease

(d) n_e decreases, resistance increases

Ans. (c) When the temperature rises, electrons in the valence band move to the conduction band, increasing the number of electrons there. As a result, the resistivity of a semiconductor decreases, leading to lower resistance as the temperature goes up.

3 P-N Junction Diode

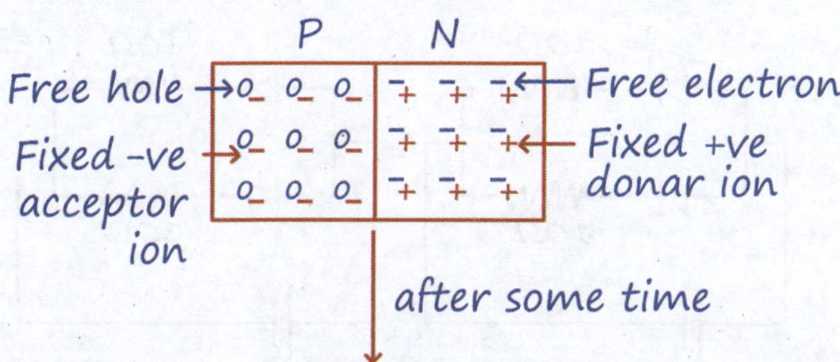

Diffusion current from P to N due to

{some electron diffuse from N to P side & some hole diffuse from P to N side}

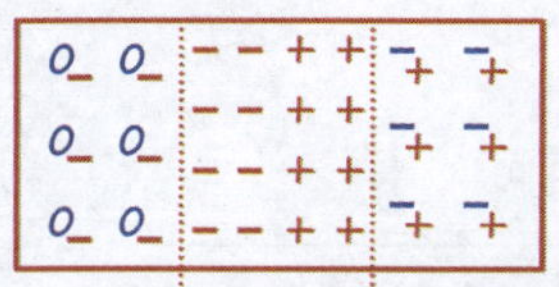

Depletion region
- no free carriers
- Contains fixed ions creating an electric field and a potential barrier.

Potential barrier → Formed due to immobile fixed acceptor and donor ion

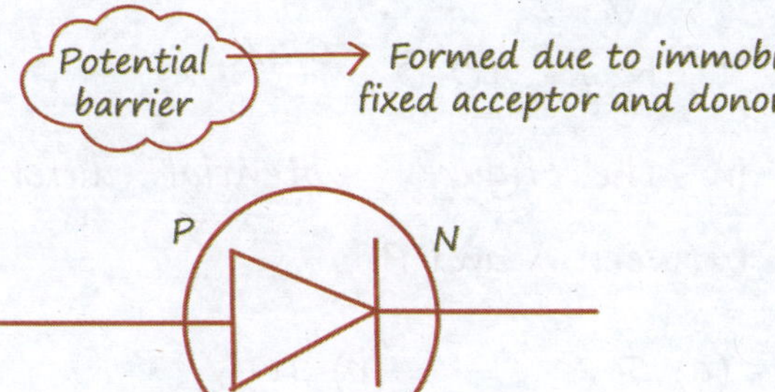

- Electric Field formed from N to P side within Depletion region :
- Width of depletion $\propto \frac{1}{\text{doping level}} \propto \text{Temp}^r$
- Drift current now from N to P due to electric field
- At equilibrium $i_{Drift} = i_{Diffusion}$

Graphs :

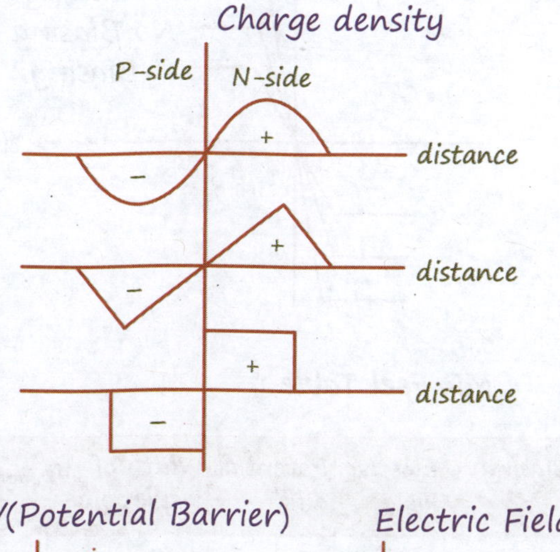

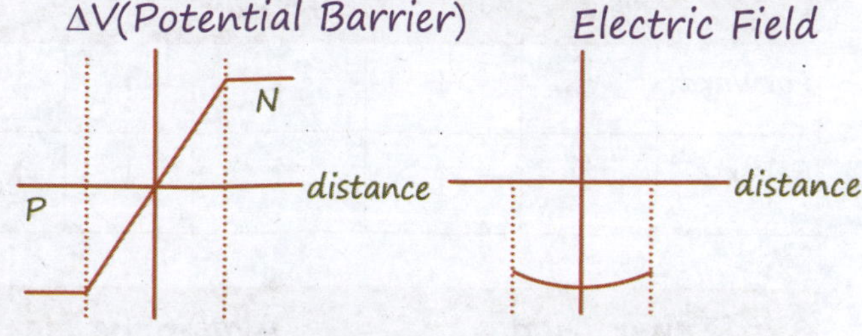

Knee Voltage : (V_B)

– V_{min} to flow e^- & hole pairs

– "V" above wiz. current rises rapidly.

Si = 0.7 V = 0.7 eV

Ge = 0.34 V = 0.34 eV

Biasing :

Forward	Reverse
$i = ✓$ P ⟶ N High Low i, Ge, Si, V 0.34V 0.7V	$i = 0$ P ⟶ N Low High i F.C. (mA) Breakdown Voltage ΔI ΔV 0.7V V V_B = Knee Voltage R.C. (μA) Dynamic Resistance $R_B = \frac{\Delta V}{\Delta I}$ Depl^n layer width "0.5μm to 1μm"

- 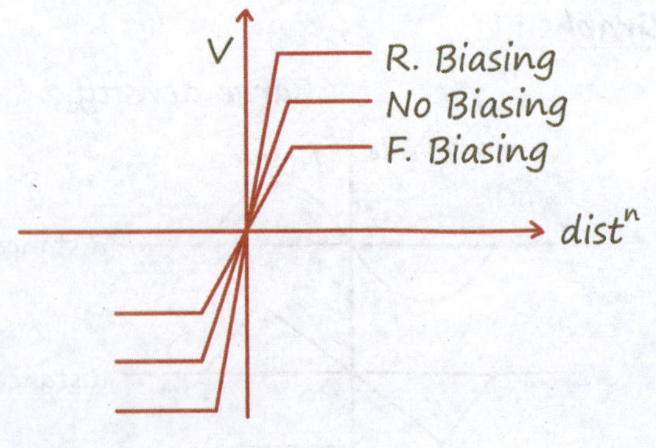

- **MR Feel Table :**

Biasing	Electric Field	Potential diff^n	Width of Depletion	$i_{Diffusion}$	i_{Drift}
Forward	↓	↓	↓	↑	↓
Reverse	↑	↑	↑	↓	↑

Zener Diode	Avalanche Breakdown
In high doped semi-C	In low doped semi-C
In reverse bias as V↑ the e⁻s hole become free due to breaking of co-valent bond	In revese bias at vary high voltage
Reversible	**Not reversible**

Diode	Biasing
−2V (diode, to ground)	R.B.
+2V (diode, to ground)	R.B.
−5V (diode) −7V	F.B.
(ground, diode) −2V	F.B.
5V (diode) 7V	R.B.

Q.4 Find the current through the battery in each of the circuits shown in figure.

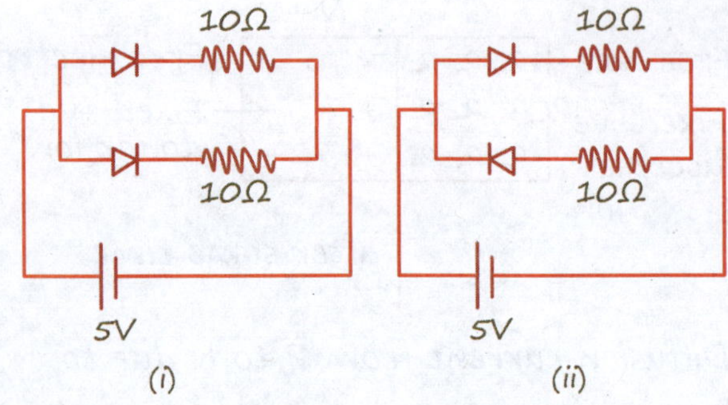

Ans. In fig. (i) Both diodes are forward biased. Thus the net diode resistance is 0.

$$i = \frac{5}{(10 \times 10)/10 \times 10} = \frac{5}{5} = 1A$$

In fig (ii) One diode is forward biased and other is reverse biased. Current passes through the forward biased diode only.

$$i = \frac{V}{R_{net}} = \frac{5}{10+0} = 0.5A$$

Q.5 In the figure, potential difference between A and B is

(a) 5 V (b) 10 V

(c) zero (d) 15 V

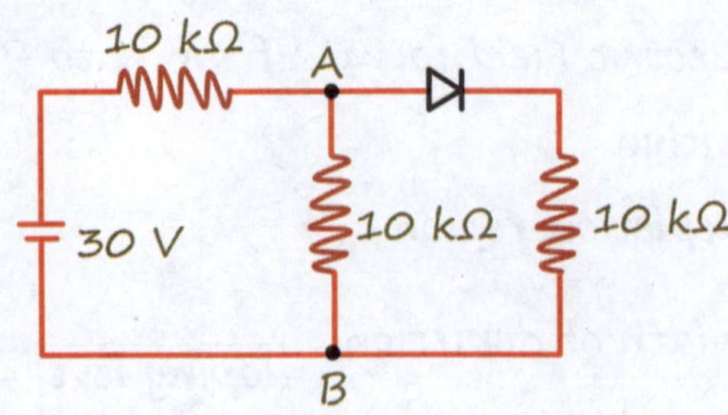

Ans. IHere, diode is in forward biased, therefore, it behaves like a conductive wire, potential difference between A and B:

$$V_A - V_B = V \times \frac{R_{AB}}{R_{total}}$$

$$= 30 \times \frac{\frac{10 \times 10}{10 + 10}}{\left(\frac{10 \times 10}{10 + 10} + 10\right)} = 10V$$

4 Rectifier

Half-wave rectifier : (1 Diode use)

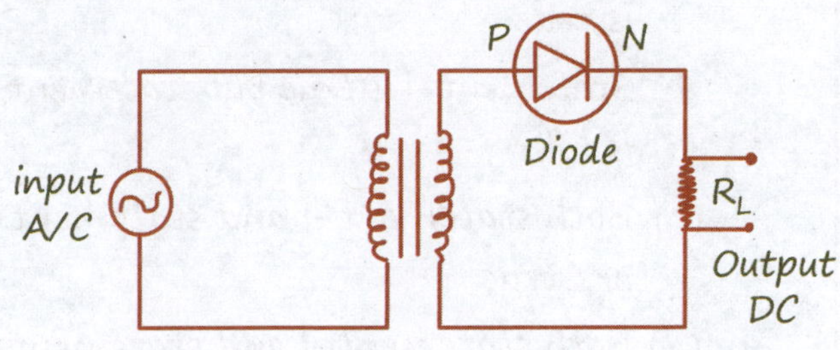

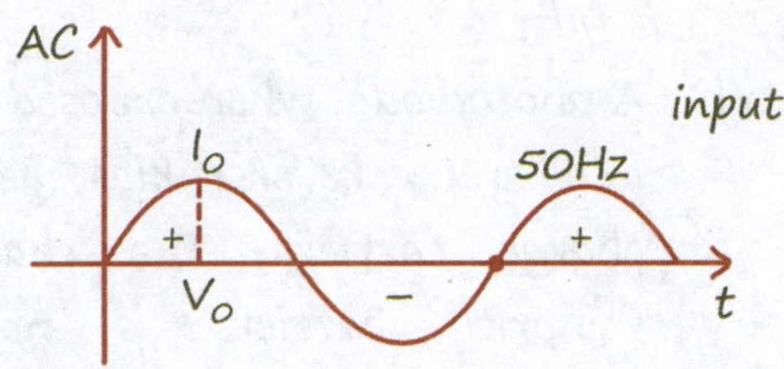

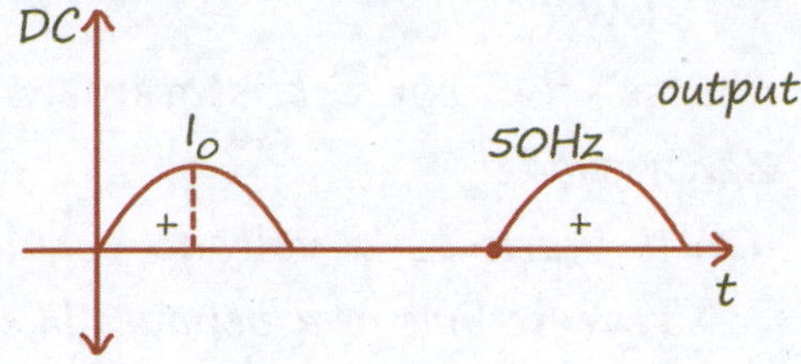

$$i_{DC} = \frac{i_0}{\pi} \qquad V_{DC} = \frac{V_0}{\pi}$$

$$i_{rms} = \frac{i_0}{2} \qquad V_{rms} = \frac{V_0}{2}$$

$$\eta = \frac{P_0}{P_i} \times 100 \qquad P = iV = \frac{V^2}{R}$$

Full wave rectifier :

- **Center tapped rectifier : (2-Diode use)**

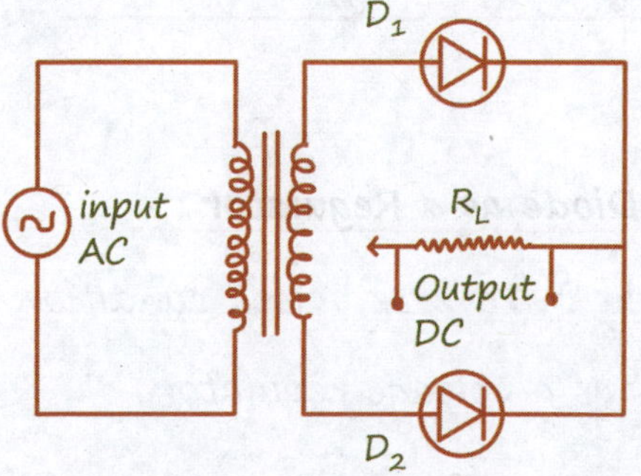

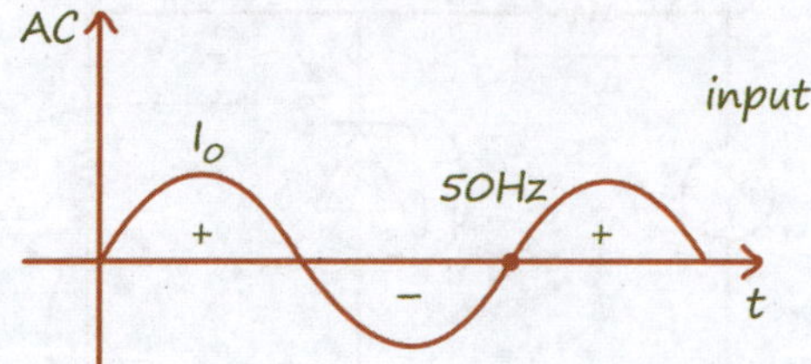

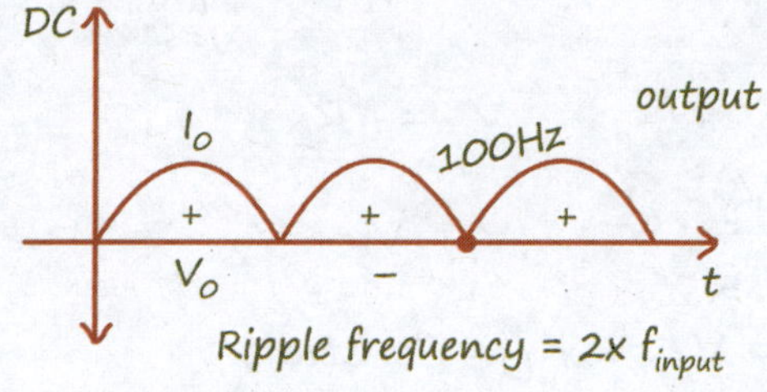

Ripple frequency = 2x f_{input}

$$\eta = P_o/P_i \times 100$$

$$i_{avg} = \frac{2i_0}{\pi} \qquad V_{avg} = \frac{2V_0}{\pi}$$

$$i_{rms} = \frac{i_0}{\sqrt{2}} \qquad V_{rms} = \frac{V_0}{\sqrt{2}}$$

- When capacitor is connected parallel with load resistance then output voltage remains constant.

Bridge rectifier : (4-Diode use)

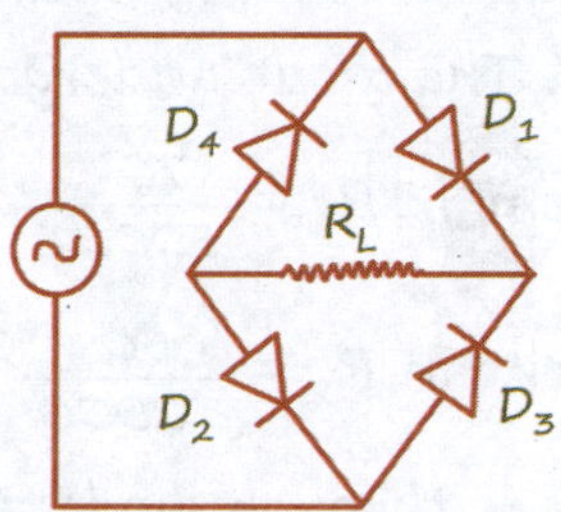

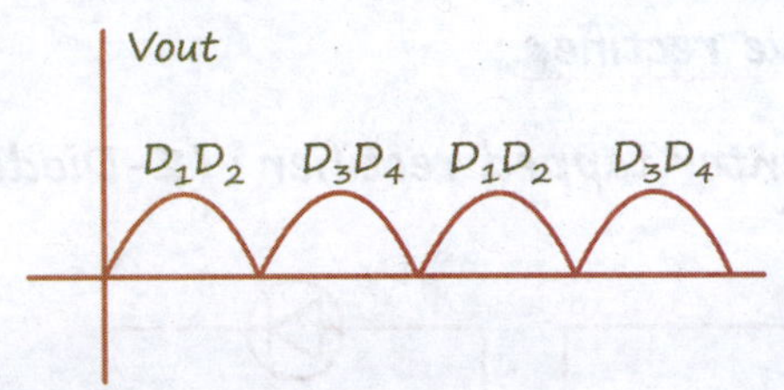

Zener Diode as a Regulator :

- Works in reverse baised condition
- Acts as a voltage regulator.

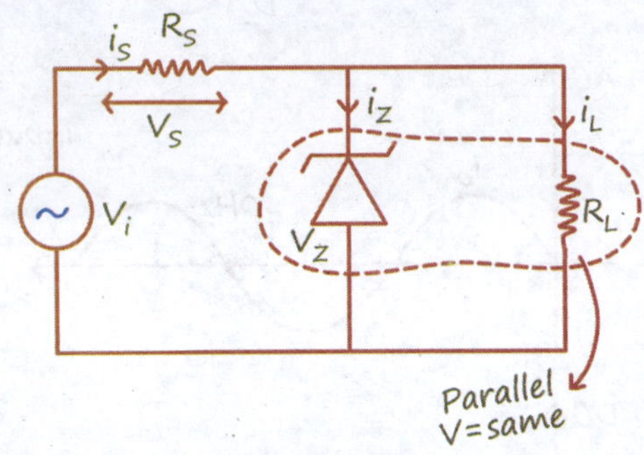

$i_S = i_Z + i_L$ $\quad V_Z = i_L R_L$

$i_S = \frac{V_S}{R_S}$ $\quad V_i = V_S + V_Z$

$V_{out} = V_Z = i_L R_L =$ Constant

When diode is working.

Q.6 For the circuit shown below, the current through the zener diode is:

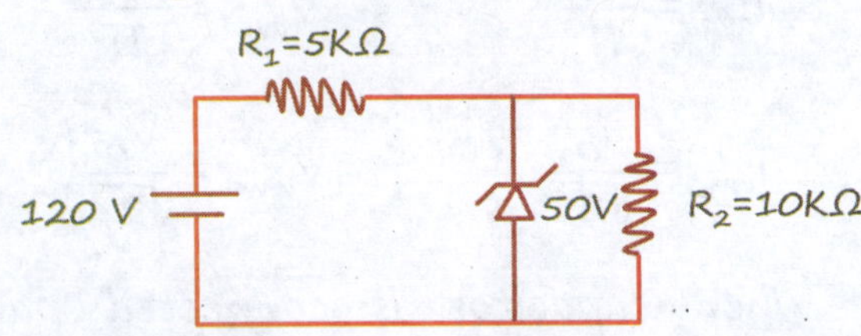

Ans. Assuming Z.D. does not undergo breakdown, current in circuit $= \frac{120}{15000}$ = 8mA

Voltage drop across diode = 80V > 50 V. The diode undergo breakdown.

Current in $R_1 = \frac{70}{5000} = 14$ mA

Current in $R_2 = \frac{50}{10000} = 5$ mA

Current through diode = 9 mA

Q.7 Statement-I: To get a steady dc output from the pulsating voltage received from a full wave rectifier we can connect a capacitor across the output parallel to the load RL.

Statement-II: To get a steady dc output from the pulsating voltage received from a full wave rectifier we can connect an inductor in series with RL. In the light of the above statements, choose the most appropriate answer from the options given below:

(a) Statement-I is false but statement-II is true

(b) Statement-I is true but statement-II is false

(c) Both statement-I and statement-II are true

(d) Both statement-I and statement-II is false

Ans. (b) A photodiode, when exposed to light and under reverse bias, generates charge carriers. The change in minority carriers is noticeable because the starting current is very low.

Q.8 Choose the correct statement about Zener diode:

(a) It work as a voltage regulator in reverse bias and behaves like simple p-n junction diode in forward bias.

(b) It works as a voltage regulator in both forward and reverse bias.

(c) It works a voltage regulator only in forward bias.

(d) It works as a voltage regulator in forward bias and behaves like simple p-n junction diode in reverse bias.

Ans. (a)

5 Application of Diodes

Photodiode	Solar Cell	LED
$h\nu > E_g$	$h\nu > E_g$	$h\nu \le E_g$
Reverse Biased	Conneted in No Biasing	Forward Biasing
i_1, i_2, i_3, i_4, i, V	i, V, V_{OC} open circuit voltage, i_{SC} Short circuit current	i, V
$i_4 > i_3 > i_2 > i_1$ Intensity ↑ Photo Current ↑ • Act as Light Sensor		Ga – As = Infrared Ga – As – P → Red, Yellow Ga, P → Red, Green

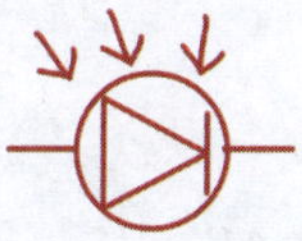

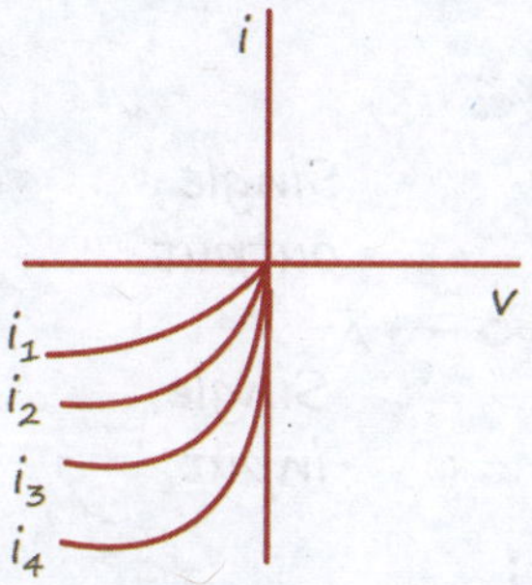

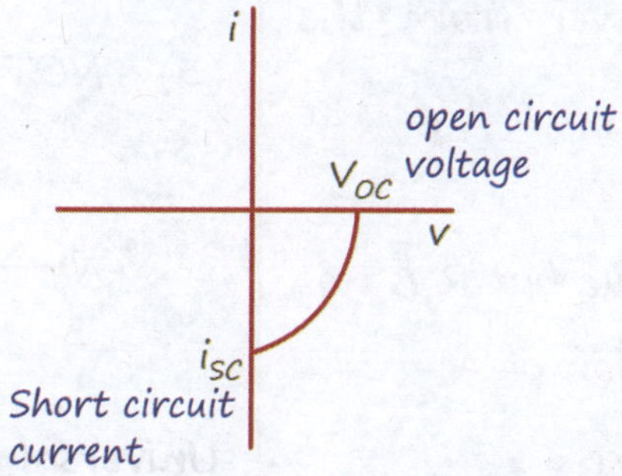

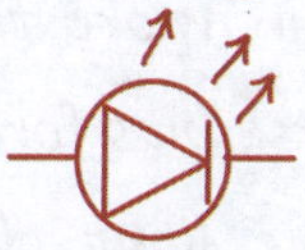

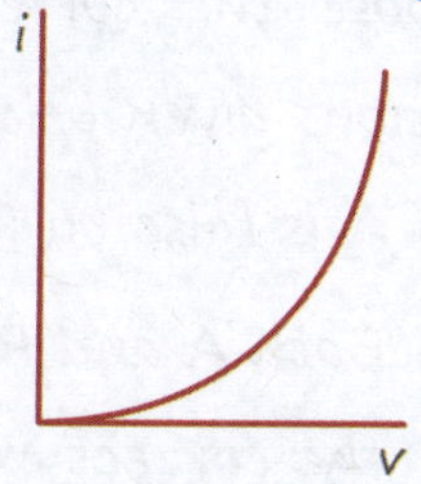

Q.9 Which of the following statement is not correct in the case of light emitting diodes?

A. It is a heavily doped p-n junction.

B. It emits light only when it is forward biased.

C. It emits light only when it is reverse biased.

D. The energy of the light emitted is equal to or slightly less than the energy gap of the semiconductor used.

Choose the correct answer from the options given below:

(a) C and D (b) A only

(c) C only (d) B only

Ans. (c)

Q.10 The photodiode is used to detect the optical signals. These diodes are preferably operated in reverse biased mode because

(a) Fractional change in majority carriers produce higher forward bias current

(b) Fraction change in majority carriers produce higher reverse bias current

(c) Fractional change in majority carriers produce higher forward bias current

(d) Fractional change in minority carriers produce higher reverse bias current

Ans. (d)

Q.11 Assertion(A): Photodiodes are preferably operated in reverse bias condition for light intensity measurement.

Reason (R): The current in the forward bias in more than the current in the reverse bias for a p-n junction diode.

In the light of the above statement, choose the correct answer from the options given below:

(a) A is false bu R is true.

(b) Both A and R are true but R is not the correct explanation of A

(c) A is true but R is false

(d) Both A and R are true and R is the correct explanation of A

Ans. (b)

6 Logic Gates

Q. 12 MR and Ramlal ne bank me joint account open kiya, dono ko different atm password mila. Atm me dono ka password match hone ke bad paisa milega then atm me Koin sa gate use huaa hai..?

Ans. AND gate

Q. 13 MR audi car se ja raha hai, Ramlal apna truck le ke nikla MR ko takkar marne, Jo aage ya piche khi se takkar mar skta hai, air bag open karne ke liya car me koin sa gate use hoga ....?

Ans. OR gate

Fundamental Gates :

(1) OR-Gate :

A, B → $A + B$

(2) AND-Gate :

A, B → AB

(3) NOT-Gate :

Single output

A → $\overline{A}$

Single input

Universal Gate :

(1) NAND-Gate :

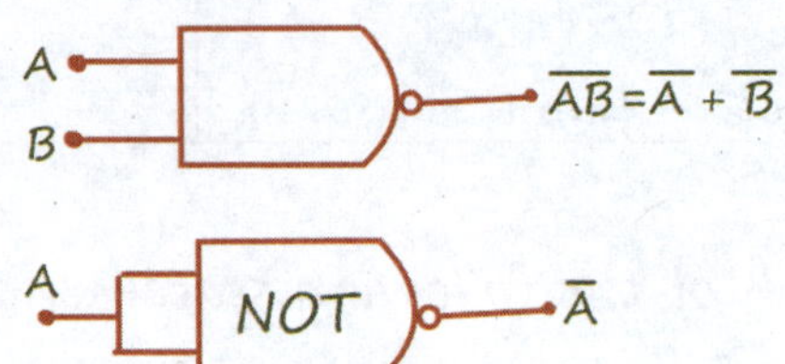

(2) NOR-Gate :

A, B → $\overline{A+B} = \overline{A} \cdot \overline{B}$

A → NOT → $\overline{A}$

Exclusive Gates :

(1) XNOR :

A, B → $AB + \overline{A}\overline{B}$ = Y(output)

(2) XOR :

A, B → $\overline{A}B + A\overline{B}$ = Y(output)

Truth table for all gate :-

Input		Output					
A	B	OR Gate	NOR Gate	AND Gate	NAND Gate	X-OR Gate	X-NOR Gate
0	0	0	1	0	1	0	1
1	0	1	0	0	1	1	0
0	1	1	0	0	1	1	0
1	1	1	0	1	0	0	1

Time Scale for different gate :-

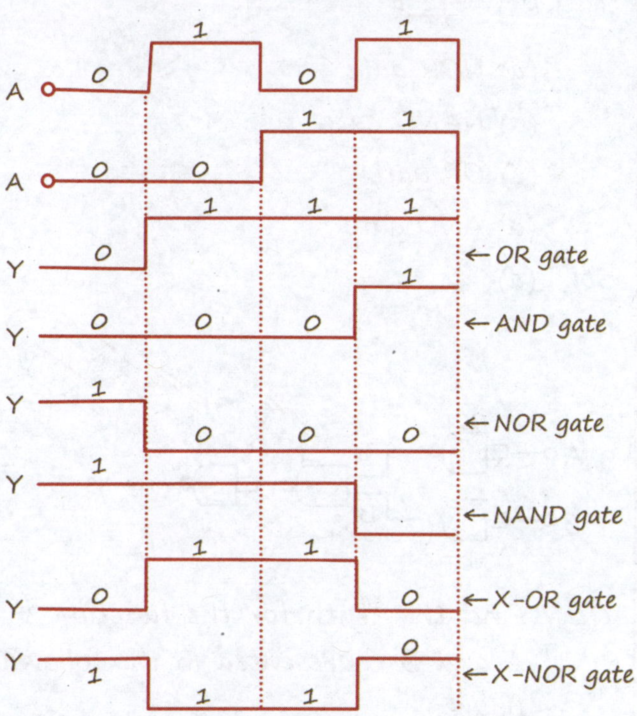

7 Basic Boolean Exp

0 + A = A	0.A = 0	0.1 = 0
1 + A = 1	A.A = A	1.1 = 1
A + A = A	1.A = A	$\overline{1} = 0$
$A + \overline{A} = 1$	$A.\overline{A} = 0$	$\overline{\overline{1}} = 1$
1 + 0 = 1		$\overline{0} = 1$
1 + 1 = 1	$\overline{\overline{A}} = A$	$\overline{\overline{0}} = 0$

De Morgan Principle :

$$\overline{A + B} = \overline{A} \cdot \overline{B}$$

$$\overline{A \cdot B} = \overline{A} + \overline{B}$$

Special Case :

$$\overline{\overline{A} + \overline{B}} = A \cdot B$$

$$\overline{\overline{A}\,\overline{B}} = A + B$$

Electrical equivalent circuit :

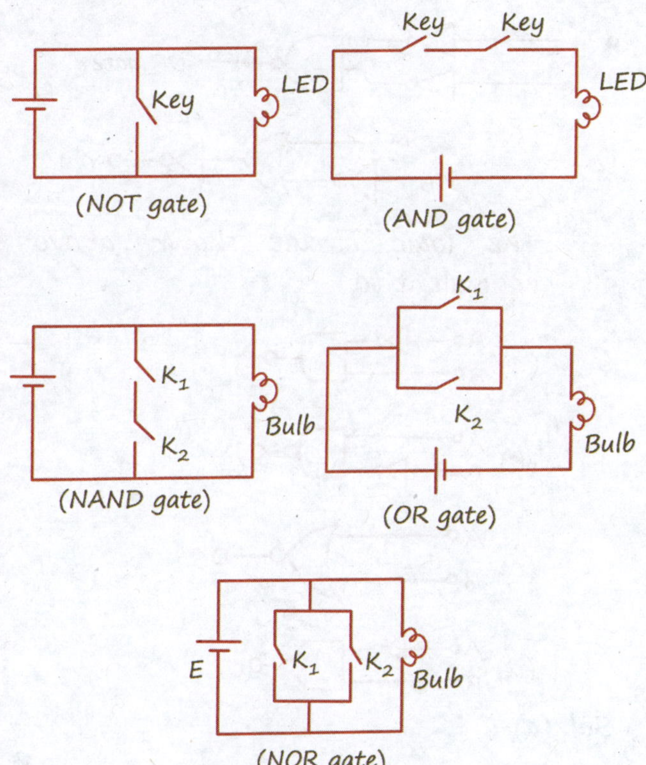

Formation of Different gates using NAND gate:

Gate	NOT	AND	OR	NOR
No. of NAND gate required	1	2	3	4

Formation of Different gates using NOR gate:

Gate	NOT	AND	OR	NAND
No. of NOR gate required	1	3	2	4

- 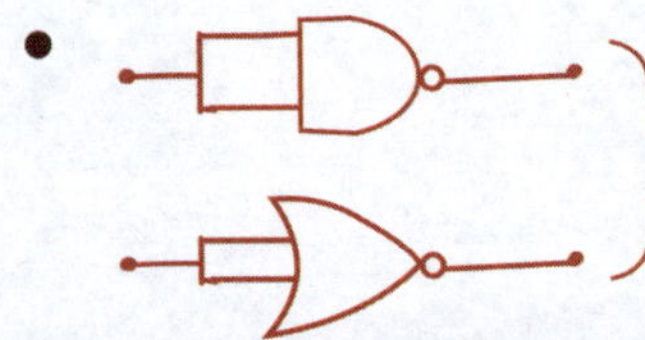

Single input NAND and Single input NOR gate will behave as NOT gate

- 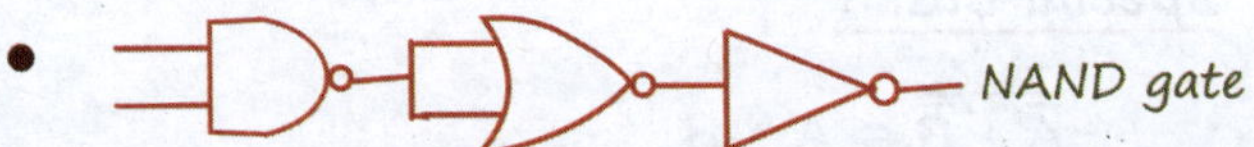

- AND gate

- OR gate

Q.14 The logic circuit shown above is equivalent to:

(a)

(b)

(c)

(d)

Sol. (a)

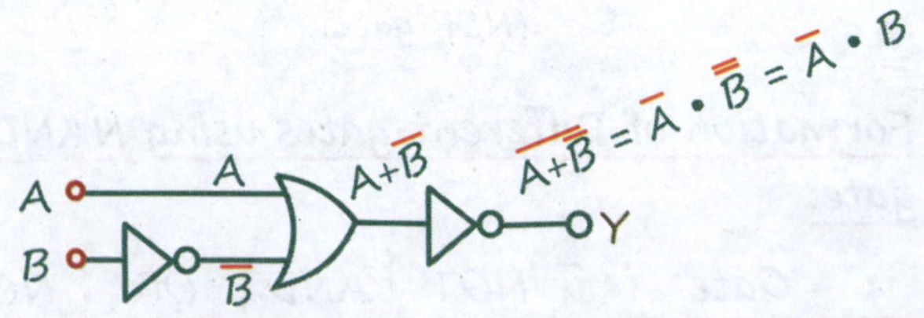

Checking with option (a)

$\bar{A}\cdot B$

Q.15 Identify the logic operation carried out by the given circuit:

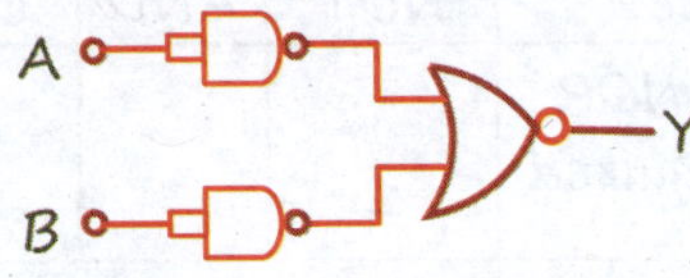

(a) NAND (b) NOR

(c) AND (d) OR

Sol. (c)

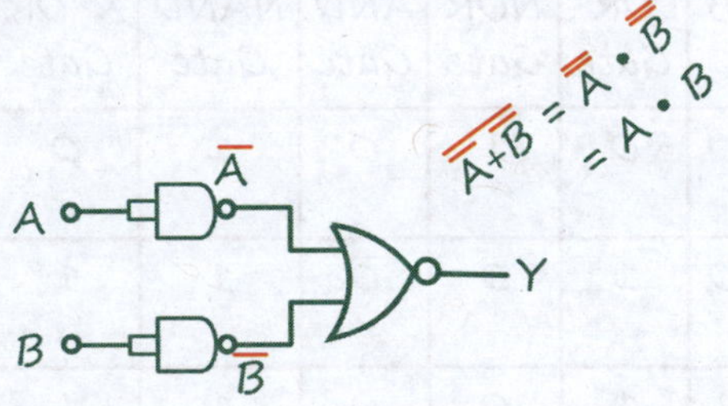

Q.16 The following logic gate is equivalent to:

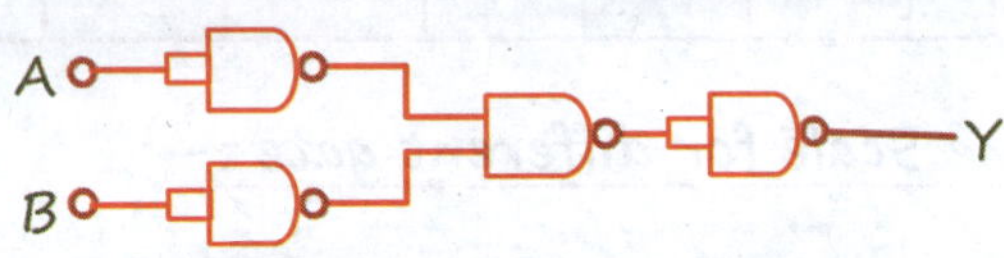

(a) NOR gate

(b) NAND gate

(c) OR gate

(d) AND gate

Sol. (a)

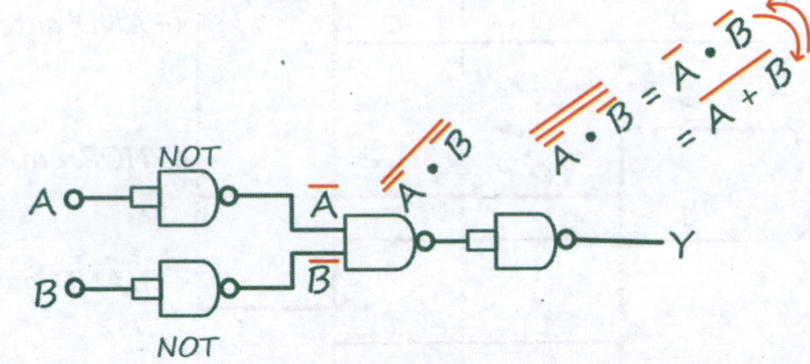

Q.17 Find the truth for the function Y of A and B represented in the following figure:

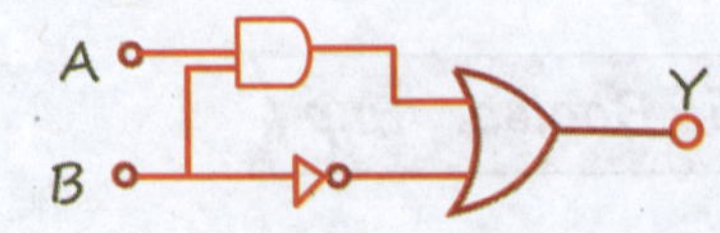

(a)

A	B	Y
0	0	0
0	1	1
1	0	0
1	1	0

(b)

A	B	Y
0	0	1
0	1	0
1	0	1
1	1	1

(c)

A	B	Y
0	0	0
0	1	0
1	0	0
1	1	1

(d)

A	B	Y
0	0	0
0	1	1
1	0	1
1	1	1

Sol. (b)

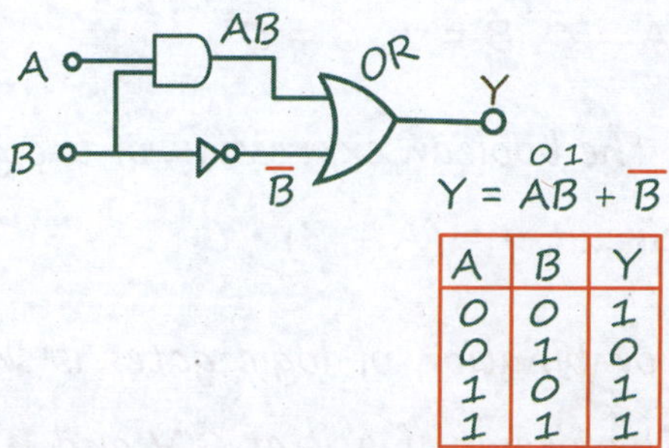

A	B	Y
0	0	1
0	1	0
1	0	1
1	1	1

Q.18 The logic operations performed by the given digital circuit is equivalent to:

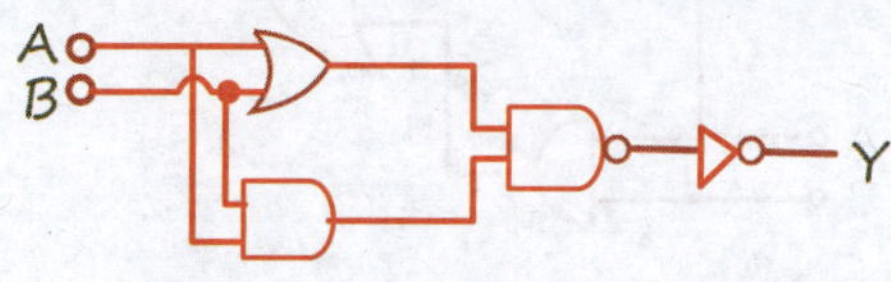

(a) AND (b) NOR

(c) OR (d) NAND

Sol. (a)

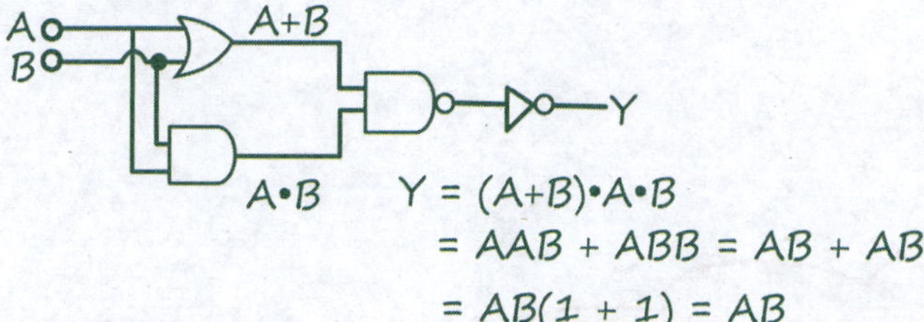

$$Y = (A+B)\cdot A\cdot B$$
$$= AAB + ABB = AB + AB$$
$$= AB(1 + 1) = AB$$

Q.19 A logic gate circuit has two inputs A and B and output Y. The voltage waveforms of A, B and Y are shown below. The logic gate circuit is:

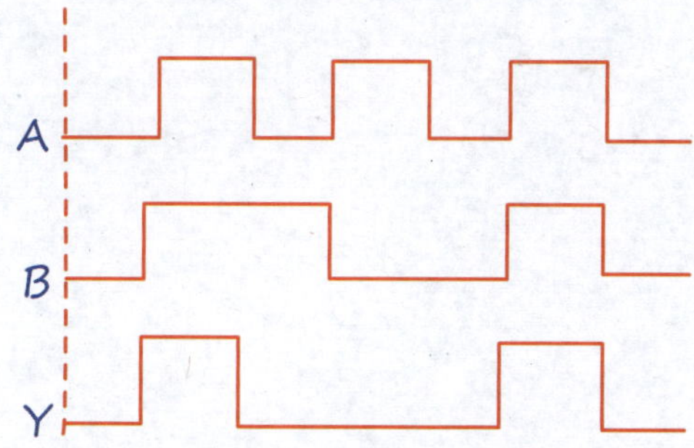

(a) AND gate (b) OR gate

(c) NOR gate (d) NANA gate

Sol. (a)

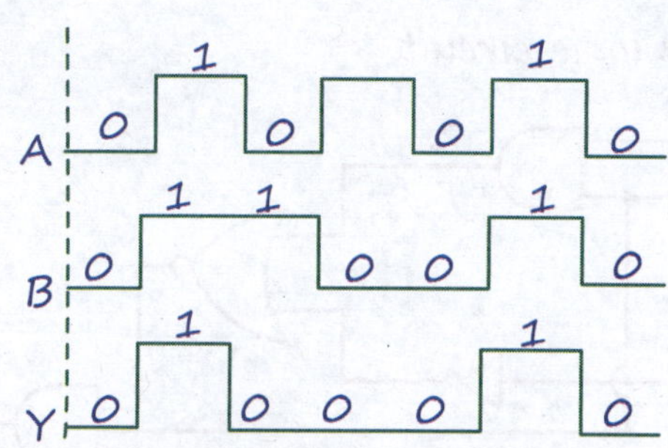

Q.20 Name the logic gate equivalent to the diagram attached.

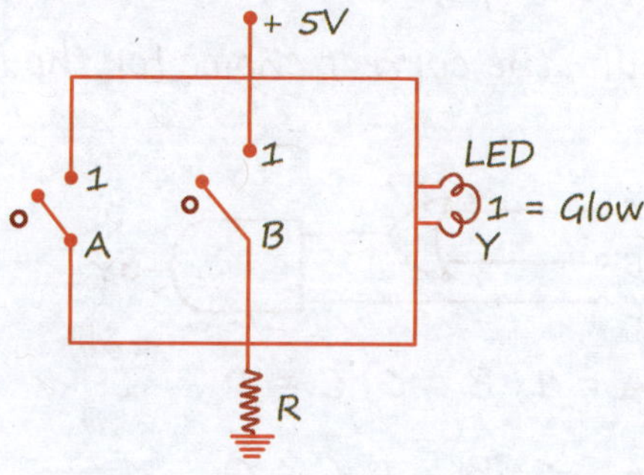

(a) OR

(b) NOR

(c) NAND

(d) AND

Sol. (b)

Q.21 The logic gate equivalent to the given circuit diagram is:

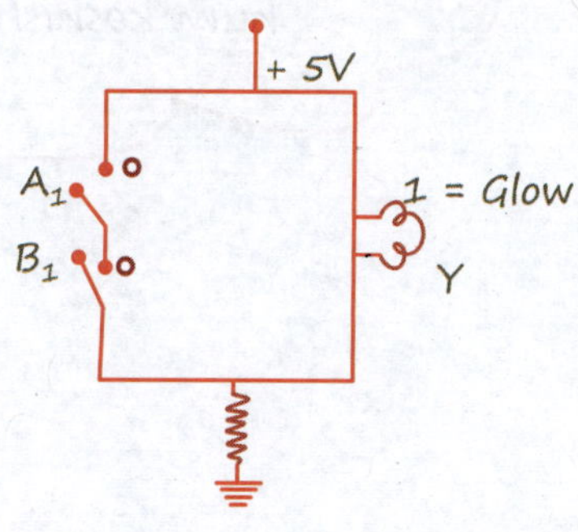

(a) OR

(b) NAND

(c) NOr

(d) AND

Sol. (b)

Q.22 Find the output boolean function for the logic circuit.

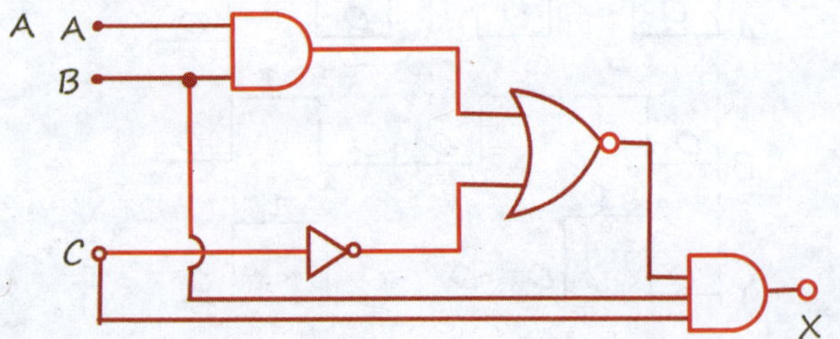

Sol. $X = \overline{A}BC$

Q.23 To get output Y = 1 for the following circuit, the correct choice for the input is:

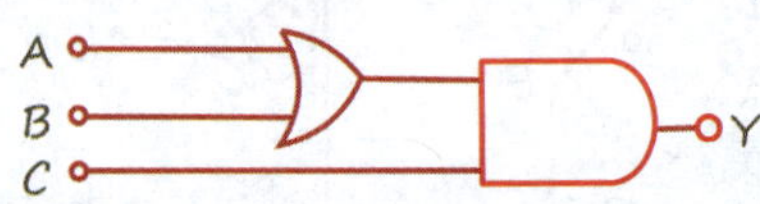

(a) A = 1, B = 0, C = 0

(b) A = 1, B = 1, C = 0

(c) A = 1, B = 0, C = 1

(d) A = 0, B = 1, C = 0

Ans. (c) The boolean expression of the given logic circuit $y = (A + B) \cdot C$.

Q.24 A combination of logic gates is shown in the circuit. If A is at 0 V and B is at 5V, then the potential of R is:

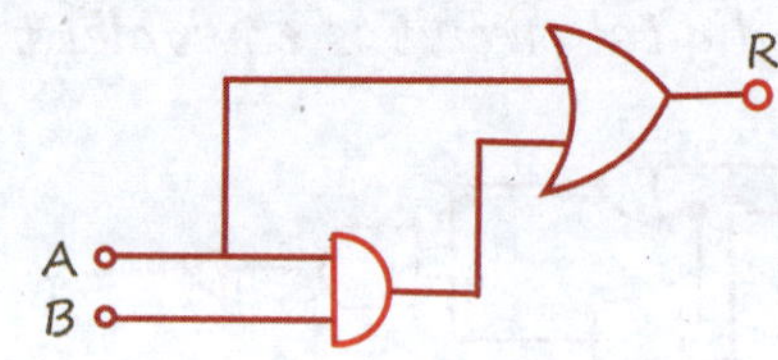

(a) 0 V (b) 5 V

(c) 10 V (d) Any of these

Ans. (a)

MR*

"Manjil Mile na mile yah to kismat ki bat h hum koshish hi na kare ye to galt bat hai."